Engineering Mathematics Pocket Book

Fourth Edition

John Bird

Engineering Mathematics Pocket Book

Fourth edition

John Bird BSc(Hons), CEng, CSci, CMath, FIMA, FIET, MIEE, FIIE, FCollT

LONDON AND NEW YORK

First published by Newnes
This edition published 2011 by Routledge
2 Park Square, Milton Park, Abingdon, Oxon OX14 4RN

Simultaneously published in the USA and Canada by Taylor & Francis Group,
711 Third Avenue, New York, NY 10017, USA

Routledge is an imprint of the Taylor & Francis Group, an informa business

First published as the *Newnes Mathematics for Engineers Pocket Book* 1983
Reprinted 1988, 1990 (twice), 1991, 1992, 1993
Second edition 1997
Third edition as the *Newnes Engineering Mathematics Pocket Book* 2001
Fourth edition as the *Engineering Mathematics Pocket Book* 2008

British Library Cataloguing in Publication Data
A catalogue record for this book is available from the British Library

Library of Congress Cataloguing in Publication Data
A catalogue record for this book is available from the Library of Congress

ISBN: 978-0-7506-8153-7

Contents

Preface

Engineering Mathematics Pocket Book 4th Edition is intended to provide students, technicians, scientists and engineers with a readily available reference to the essential engineering mathematics formulae, definitions, tables and general information needed during their studies and/or work situation – a handy book to have on the bookshelf to delve into as the need arises.

In this 4th edition, the text has been re-designed to make information easier to access. Essential theory, formulae, definitions, laws and procedures are stated clearly at the beginning of each section, and then it is demonstrated how to use such information in practice.

The text is divided, for convenience of reference, into sixteen main chapters embracing engineering conversions, constants and symbols, some algebra topics, some number topics, areas and volumes, geometry and trigonometry, graphs, vectors, complex numbers, matrices and determinants, Boolean algebra and logic circuits, differential and integral calculus and their applications, differential equations, statistics and probability, Laplace transforms and Fourier series. To aid understanding, over 500 application examples have been included, together with over 300 line diagrams.

The text assumes little previous knowledge and is suitable for a wide range of courses of study. It will be particularly useful for students studying mathematics within National and Higher National Technician Certificates and Diplomas, GCSE and A levels, for Engineering Degree courses, and as a reference for those in the engineering industry.

John Bird
Royal Naval School of Marine Engineering, HMS Sultan, formerly University of Portsmouth and Highbury College, Portsmouth

1 Engineering Conversions, Constants and Symbols

1.1 General conversions

Length (metric)

1 kilometre (km) = 1000 metres (m)
1 metre (m) = 100 centimetres (cm)
1 metre (m) = 1000 millimetres (mm)
1 cm = 10^{-2} m
1 mm = 10^{-3} m
1 micron (μ) = 10^{-6} m
1 angstrom (A) = 10^{-10} m

Length (imperial)

1 inch (in) = 2.540 cm or 1 cm = 0.3937 in
1 foot (ft) = 30.48 cm
1 mile (mi) = 1.609 km or 1 km = 0.6214 mi
1 cm = 0.3937 in
1 m = 39.37 in = 3.2808 ft = 1.0936 yd
1 km = 0.6214 mile
1 nautical mile = 1.15 mile

Area (metric)

1 m^2 = 10^6 mm^2
1 mm^2 = 10^{-6} m^2
1 m^2 = 10^4 cm^2
1 cm^2 = 10^{-4} m^2
1 hectare (ha) = 10^4 m^2

Area (imperial)

1 m^2 = 10.764 ft^2 = 1.1960 yd^2
1 ft^2 = 929 cm^2
1 $mile^2$ = 640 acres
1 acre = 43560 ft^2 = 4840 yd^2
1 ha = 2.4711 acre = 11960 yd^2 = 107639 ft^2

Volume

1 litre (l) = 1000 cm^3
1 litre = 1.057 quart (qt) = 1.7598 pint (pt) = 0.21997 gal

1 m^3 = 1000 l

	1 British gallon = 4 qt = 4.545 l = 1.201 US gallon
	1 US gallon = 3.785 l
Mass	1 kilogram (kg) = 1000 g = 2.2046 pounds (lb)
	1 lb = 16 oz = 453.6 g
	1 tonne (t) = 1000 kg = 0.9842 ton
Speed	1 km/h = 0.2778 m/s = 0.6214 m.p.h.
	1 m.p.h. = 1.609 km/h = 0.4470 m/s
	1 rad/s = 9.5493 rev/min
	1 knot = 1 nautical mile per hour = 1.852 km/h = 1.15 m.p.h.
	1 km/h = 0.540 knots
	1 m.p.h. = 0.870 knots
Angular measure	1 rad = 57.296°

1.2 Greek alphabet

Letter Name	Upper Case	Lower Case
Alpha	A	α
Beta	B	β
Gamma	Γ	γ
Delta	Δ	δ
Epsilon	E	ε
Zeta	Z	ζ
Eta	H	η
Theta	θ	θ
Iota	I	ι
Kappa	K	κ
Lambda	Λ	λ
Mu	M	μ
Nu	N	ν
Xi	Ξ	ξ

Omicron	O	o
Pi	Π	π
Rho	P	ρ
Sigma	Σ	σ
Tau	T	τ
Upsilon	Y	υ
Phi	Φ	ϕ
Chi	X	χ
Psi	Ψ	ψ
Omega	Ω	ω

1.3 Basic SI units, derived units and common prefixes

Basic SI units

Quantity	Unit
Length	metre, m
Mass	kilogram, kg
Time	second, s
Electric current	ampere, A
Thermodynamic temperature	kelvin, K
Luminous intensity	candela, cd
Amount of substance	mole, mol

SI supplementary units

Plane angle	radian, rad
Solid angle	steradian, sr

Derived units

Quantity	Unit
Electric capacitance	farad, F
Electric charge	coulomb, C
Electric conductance	siemens, S
Electric potential difference	volts, V
Electrical resistance	ohm, Ω
Energy	joule, J
Force	Newton, N
Frequency	hertz, Hz
Illuminance	lux, lx
Inductance	henry, H
Luminous flux	lumen, lm
Magnetic flux	weber, Wb
Magnetic flux density	tesla, T
Power	watt, W
Pressure	pascal, Pa

Some other derived units not having special names

Quantity	Unit
Acceleration	metre per second squared, m/s^2
Angular velocity	radian per second, rad/s
Area	square metre, m^2
Current density	ampere per metre squared, A/m^2
Density	kilogram per cubic metre, kg/m^3
Dynamic viscosity	pascal second, Pa s
Electric charge density	coulomb per cubic metre, C/m^3
Electric field strength	volt per metre, V/m
Energy density	joule per cubic metre, J/m^3
Heat capacity	joule per Kelvin, J/K
Heat flux density	watt per square metre, W/m^3
Kinematic viscosity	square metre per second, m^2/s
Luminance	candela per square metre, cd/m^2

Magnetic field strength	ampere per metre, A/m
Moment of force	newton metre, Nm
Permeability	henry per metre, H/m
Permittivity	farad per metre, F/m
Specific volume	cubic metre per kilogram, m^3/kg
Surface tension	newton per metre, N/m
Thermal conductivity	watt per metre Kelvin, W/(mK)
Velocity	metre per second, m/s^2
Volume	cubic metre, m^3

Common prefixes

Prefix	Name	Meaning
Y	yotta	multiply by 10^{24}
Z	zeta	multiply by 10^{21}
E	exa	multiply by 10^{18}
P	peta	multiply by 10^{15}
T	tera	multiply by 10^{12}
G	giga	multiply by 10^{9}
M	mega	multiply by 10^{6}
k	kilo	multiply by 10^{3}
m	milli	multiply by 10^{-3}
μ	micro	multiply by 10^{-6}
n	nano	multiply by 10^{-9}
p	pico	multiply by 10^{-12}
f	femto	multiply by 10^{-15}
a	atto	multiply by 10^{-18}
z	zepto	multiply by 10^{-21}
y	yocto	multiply by 10^{-24}

1.4 Some physical and mathematical constants

Below are listed some physical and mathematical constants, each stated correct to 4 decimal places, where appropriate.

Quantity	Symbol	Value
Speed of light in a vacuum	c	2.9979×10^{8} m/s
Permeability of free space	μ_0	$4\pi \times 10^{-7}$ H/m
Permittivity of free space	ε_0	8.8542×10^{-12} F/m
Elementary charge	e	1.6022×10^{-19} C
Planck constant	h	6.6261×10^{-34} J s
	$\hbar = \dfrac{h}{2\pi}$	1.0546×10^{-34} J s
Fine structure constant	$\alpha = \dfrac{e^2}{4\pi\varepsilon_0 \hbar c}$	7.2974×10^{-3}
Coulomb force constant	k_e	8.9875×10^{9} Nm^2/C^2
Gravitational constant	G	6.6726×10^{-11} m^3/kg s^2
Atomic mass unit	u	1.6605×10^{-27} kg
Rest mass of electron	m_e	9.1094×10^{-31} kg
Rest mass of proton	m_p	1.6726×10^{-27} kg
Rest mass of neutron	m_n	1.6749×10^{-27} kg
Bohr radius	a_0	5.2918×10^{-11} m
Compton wavelength of electron	λ_C	2.4263×10^{-12} m
Avogadro constant	N_A	6.0221×10^{23}/mol
Boltzmann constant	k	1.3807×10^{-23} J/K
Stefan-Boltzmann constant	σ	5.6705×10^{-8} W/m^{2}K^4
Bohr constant	μ_B	9.2740×10^{-24} J/T
Nuclear magnetron	μ_N	5.0506×10^{-27} J/T
Triple point temperature	T_t	273.16 K
Molar gas constant	R	8.3145 J/K mol
Micron	μm	10^{-6} m
Characteristic impedance of vacuum	Z_o	376.7303 Ω

Astronomical constants

Mass of earth	m_E	5.976×10^{24} kg
Radius of earth	R_E	6.378×10^{6} m
Gravity of earth's surface	g	9.8067 m/s^2
Mass of sun	$M_\odot$	1.989×10^{30} kg
Radius of sun	$R_\odot$	6.9599×10^{8} m
Solar effective temperature	T_e	5800 K
Luminosity of sun	$L_\odot$	3.826×10^{26} W
Astronomical unit	AU	1.496×10^{11} m
Parsec	pc	3.086×10^{16} m
Jansky	Jy	10^{-26} W/m^{2}HZ
Tropical year		3.1557×10^{7} s
Standard atmosphere	atm	101325 Pa

Mathematical constants

Pi (Archimedes' constant)	π	3.1416
Exponential constant	e	2.7183
Apery's constant	ς (3)	1.2021
Catalan's constant	G	0.9160
Euler's constant	γ	0.5772
Feigenbaum's constant	α	2.5029
Feigenbaum's constant	δ	4.6692
Gibb's constant	G	1.8519
Golden mean	ϕ	1.6180
Khintchine's constant	K	2.6855

1.5 Recommended mathematical symbols

equal to	$=$
not equal to	$\neq$
identically equal to	$\equiv$
corresponds to	$\triangleq$
approximately equal to	$\approx$
approaches	$\rightarrow$
proportional to	$\propto$

infinity	∞
smaller than	$<$
larger than	$>$
smaller than or equal to	$\leq$
larger than or equal to	$\geq$
much smaller than	$\ll$
much larger than	$\gg$
plus	$+$
minus	$-$
plus or minus	$\pm$
minus or plus	$\mp$
a multiplied by b	ab or $a \times b$ or $a \cdot b$
a divided by b	$\frac{a}{b}$ or a/b or ab^{-1}
magnitude of a	$\lvert a \rvert$
a raised to power n	a^n
square root of a	$\sqrt{a}$ or $a^{\frac{1}{2}}$
n'th root of a	$\sqrt[n]{a}$ or $a^{\frac{1}{n}}$ or $a^{1/n}$
mean value of a	$\bar{a}$
factorial of a	$a!$
sum	Σ
function of x	$f(x)$
limit to which f(x) tends as x approaches a	$\lim_{x \to a} f(x)$
finite increment of x	Δx
variation of x	δx
differential coefficient of f(x) with respect to x	$\frac{df}{dx}$ or df/dy or $f'(x)$
differential coefficient of order n of f(x)	$\frac{d^n f}{dx^n}$ or $d^n f/dx^2$ or $f^n(x)$

partial differential coefficient of f(x, y, ...) w.r.t. x when y, ... are held constant	$\frac{\partial f(x,y,...)}{\partial x}$ or $\left(\frac{\partial f}{\partial x}\right)_y$ or f_x
total differential of f	df
indefinite integral of f(x) with respect to x	$\int f(x)dx$
definite integral of f(x) from x = a to x = b	$\int_a^b f(x)dx$
logarithm to the base a of x	$\log_a x$
common logarithm of x	lg x or $\log_{10} x$
exponential of x	e^x or exp x
natural logarithm of x	ln x or $\log_e x$
sine of x	sin x
cosine of x	cos x
tangent of x	tan x
secant of x	sec x
cosecant of x	cosec x
cotangent of x	cot x
inverse sine of x	$\sin^{-1} x$ or arcsin x
inverse cosine of x	$\cos^{-1} x$ or arccos x
inverse tangent of x	$\tan^{-1} x$ or arctan x
inverse secant of x	$\sec^{-1} x$ or arcsec x
inverse cosecant of x	$\mathrm{cosec}^{-1} x$ or arccosec x
inverse cotangent of x	$\cot^{-1} x$ or arccot x
hyperbolic sine of x	sinh x
hyperbolic cosine of x	cosh x
hyperbolic tangent of x	tanh x
hyperbolic secant of x	sech x
hyperbolic cosecant of x	cosech x
hyperbolic cotangent of x	coth x
inverse hyperbolic sine of x	$\sinh^{-1} x$ or arsinh x
inverse hyperbolic cosine of x	$\cosh^{-1} x$ or arcosh x
inverse hyperbolic tangent of x	$\tanh^{-1} x$ or artanh x
inverse hyperbolic secant of x	$\mathrm{sech}^{-1} x$ or arsech x
inverse hyperbolic cosecant of x	$\mathrm{cosech}^{-1} x$ or arcosech x
inverse hyperbolic cotangent of x	$\coth^{-1} x$ or arcoth x

complex operator	i, j
modulus of z	|z|
argument of z	arg z
complex conjugate of z	z*
transpose of matrix A	A^T
determinant of matrix A	|A|
vector	**A** or $\vec{A}$
magnitude of vector **A**	|**A**|
scalar product of vectors **A** and **B**	**A • B**
vector product of vectors **A** and **B**	**A × B**

1.6 Symbols for physical quantities

(a) Space and time

angle (plane angle)	$\alpha, \beta, \gamma, \theta, \phi$, etc.
solid angle	Ω, ω
length	l
breadth	b
height	h
thickness	d, δ
radius	r
diameter	d
distance along path	s, L
rectangular co-ordinates	x, y, z
cylindrical co-ordinates	r, ϕ, z
spherical co-ordinates	r,θ, ϕ
area	A
volume	V
time	t
angular speed, $\frac{d\theta}{dt}$	ω
angular acceleration, $\frac{d\omega}{dt}$	α
speed, $\frac{ds}{dt}$	u, v, w

acceleration, $\frac{du}{dt}$	a
acceleration of free fall	g
speed of light in a vacuum	c
Mach number	Ma

(b) Periodic and related phenomena

period	T
frequency	f
rotational frequency	n
circular frequency	ω
wavelength	λ
damping coefficient	δ
attenuation coefficient	α
phase coefficient	β
propagation coefficient	γ

(c) Mechanics

mass	m
density	ρ
relative density	d
specific volume	v
momentum	p
moment of inertia	I, J
second moment of area	I_a
second polar moment of area	I_p
force	F
bending moment	M
torque; moment of couple	T
pressure	p, P
normal stress	σ
shear stress	τ
linear strain	ε, e
shear strain	γ
volume strain	θ
Young's modulus	E
shear modulus	G
bulk modulus	K

Poisson ratio	μ, ν
compressibility	κ
section modulus	Z, W
coefficient of friction	μ
viscosity	η
fluidity	ϕ
kinematic viscosity	ν
diffusion coefficient	D
surface tension	γ, σ
angle of contact	θ
work	W
energy	E, W
potential energy	E_p, V, Φ
kinetic energy	E_k, T, K
power	P
gravitational constant	G
Reynold's number	Re

(d) Thermodynamics

thermodynamic temperature	T, Θ
common temperature	t, θ
linear expansivity	α, λ
cubic expansivity	α, γ
heat; quantity of heat	Q, q
work; quantity of work	W, w
heat flow rate	Φ, q
thermal conductivity	λ, k
heat capacity	C
specific heat capacity	c
entropy	S
internal energy	U, E
enthalpy	H
Helmholtz function	A, F
Planck function	Y
specific entropy	s
specific internal energy	u, e
specific enthalpy	h
specific Helmholz function	a, f

(e) Electricity and magnetism

Electric charge; quantity of electricity	Q
electric current	I
charge density	ρ
surface charge density	σ
electric field strength	E
electric potential	V, ϕ
electric potential difference	U, V
electromotive force	E
electric displacement	D
electric flux	ψ
capacitance	C
permittivity	ε
permittivity of a vacuum	ε_0
relative permittivity	ε_r
electric current density	J, j
magnetic field strength	H
magnetomotive force	F_m
magnetic flux	Φ
magnetic flux density	B
self inductance	L
mutual inductance	M
coupling coefficient	k
leakage coefficient	σ
permeability	μ
permeability of a vacuum	μ_0
relative permeability	μ_r
magnetic moment	m
resistance	R
resistivity	ρ
conductivity	γ, σ
reluctance	R_m, S
permeance	Λ
number of turns	N
number of phases	m
number of pairs of poles	p
loss angle	δ
phase displacement	ϕ
impedance	Z
reactance	X

resistance	R
quality factor	Q
admittance	Y
susceptance	B
conductance	G
power, active	P
power, reactive	Q
power, apparent	S

(f) Light and related electromagnetic radiations

radiant energy	Q, Q_e
radiant flux, radiant power	Φ, Φ_e, P
radiant intensity	I, I_e
radiance	L, L_e
radiant exitance	M, M_e
irradiance	E, E_e
emissivity	e
quantity of light	Q, Q_v
luminous flux	Φ, Φ_v
luminous intensity	I, I_v
luminance	L, L_v
luminous exitance	M, M_v
illuminance	E, E_v
light exposure	H
luminous efficacy	K
absorption factor, absorptance	α
reflexion factor, reflectance	ρ
transmission factor, transmittance	τ
linear extinction coefficient	μ
linear absorption coefficient	a
refractive index	n
refraction	R
angle of optical rotation	α

(g) Acoustics

speed of sound	c
speed of longitudinal waves	c_l
speed of transverse waves	c_t

group speed	c_g
sound energy flux	P
sound intensity	I, J
reflexion coefficient	ρ
acoustic absorption coefficient	α, α_a
transmission coefficient	τ
dissipation coefficient	δ
loudness level	L_N

(h) Physical chemistry

atomic weight	A_r
molecular weight	M_r
amount of substance	n
molar mass	M
molar volume	V_m
molar internal energy	U_m
molar enthalpy	H_m
molar heat capacity	C_m
molar entropy	S_m
molar Helmholtz function	A_m
molar Gibbs function	G_m
(molar) gas constant	R
compression factor	Z
mole fraction of substance B	x_B
mass fraction of substance B	w_B
volume fraction of substance B	ϕ_B
molality of solute B	m_B
amount of substance concentration of solute B	c_B
chemical potential of substance B	μ_B
absolute activity of substance B	λ_B
partial pressure of substance B in a gas mixture	p_B
fugacity of substance B in a gas mixture	f_B
relative activity of substance B	α_B
activity coefficient (mole fraction basis)	f_B
activity coefficient (molality basis)	γ_B
activity coefficient (concentration basis)	y_B
osmotic coefficient	ϕ, g
osmotic pressure	Π
surface concentration	Γ

electromotive force	E
Faraday constant	F
charge number of ion i	z_i
ionic strength	I
velocity of ion i	v_i
electric mobility of ion i	u_i
electrolytic conductivity	κ
molar conductance of electrolyte	Λ
transport number of ion i	t_i
molar conductance of ion i	λ_i
overpotential	η
exchange current density	j_0
electrokinetic potential	ζ
intensity of light	I
transmittance	T
absorbance	A
(linear) absorption coefficient	a
molar (linear) absorption coefficient	ε
angle of optical rotation	α
specific optical rotatory power	α_m
molar optical rotatory power	α_n
molar refraction	R_m
stoiciometric coefficient of molecules B	ν_B
extent of reaction	ξ
affinity of a reaction	$\mathbf{A}$
equilibrium constant	K
degree of dissociation	α
rate of reaction	ξ, J
rate constant of a reaction	k
activation energy of a reaction	E

(i) Molecular physics

Avogadro constant	L, N_A
number of molecules	N
number density of molecules	n
molecular mass	m
molecular velocity	$\mathbf{c}$, $\mathbf{u}$
molecular position	$\mathbf{r}$
molecular momentum	$\mathbf{p}$

average velocity	$\langle \mathbf{c} \rangle, \langle \mathbf{u} \rangle, \mathbf{c_0}, \mathbf{u_0}$
average speed	$\langle c \rangle, \langle u \rangle, \bar{c}, \bar{u}$
most probable speed	$\hat{c}, \hat{u}$
mean free path	l, λ
molecular attraction energy	ε
interaction energy between molecules i and j	ϕ_{ij}, V_{ij}
distribution function of speeds	$f(c)$
Boltzmann function	H
generalized co-ordinate	q
generalized momentum	p
volume in phase space	Ω
Boltzmann constant	k
partition function	Q, Z
grand partition function	Ξ
statistical weight	g
symmetrical number	σ, s
dipole moment of molecule	p, μ
quadrupole moment of molecule	Θ
polarizability of molecule	α
Planck constant	h
characteristic temperature	Θ
Debye temperature	Θ_D
Einstein temperature	Θ_E
rotational temperature	Θ_r
vibrational temperature	Θ_v
Stefan-Boltzmann constant	σ
first radiation constant	c_1
second radiation constant	c_2
rotational quantum number	J, K
vibrational quantum number	v

(j) Atomic and nuclear physics

nucleon number; mass number	A
atomic number; proton number	Z
neutron number	N
(rest) mass of atom	m_a
unified atomic mass constant	m_u

(rest) mass of electron	m_e
(rest) mass of proton	m_p
(rest) mass of neutron	m_n
elementary charge (of protons)	e
Planck constant	h
Planck constant divided by 2π	$\hbar$
Bohr radius	a_0
Rydberg constant	R_∞
magnetic moment of particle	μ
Bohr magneton	μ_B
Bohr magneton number, nuclear magneton	μ_N
nuclear gyromagnetic ratio	γ
g-factor	g
Larmor (angular) frequency	ω_L
nuclear angular precession frequency	ω_N
cyclotron angular frequency of electron	ω_c
nuclear quadrupole moment	Q
nuclear radius	R
orbital angular momentum quantum number	L, l_1
spin angular momentum quantum number	S, s_1
total angular momentum quantum number	J, j_1
nuclear spin quantum number	I, J
hyperfine structure quantum number	F
principal quantum number	n, n_1
magnetic quantum number	M, m_1
fine structure constant	α
electron radius	r_e
Compton wavelength	λ_C
mass excess	Δ
packing fraction	f
mean life	τ
level width	Γ
activity	A
specific activity	a

decay constant	λ
half-life	$T_{\frac{1}{2}}, t_{\frac{1}{2}}$
disintegration energy	Q
spin-lattice relaxation time	T_1
spin-spin relaxation time	T_2
indirect spin-spin coupling	J

(k) Nuclear reactions and ionising radiations

reaction energy	Q
cross-section	σ
macroscopic cross-section	Σ
impact parameter	b
scattering angle	θ, ϕ
internal conversion coefficient	α
linear attenuation coefficient	μ, μ_1
atomic attenuation coefficient	μ
mass attenuation coefficient	μ_m
linear stopping power	S, S_1
atomic stopping power	S_a
linear range	R, R_1
recombination coefficient	α

2 Some Algebra Topics

2.1 Polynomial division

Application: Divide $2x^2 + x - 3$ by $x - 1$

$2x^2 + x - 3$ is called the **dividend** and $x - 1$ the **divisor**. The usual layout is shown below with the dividend and divisor both arranged in descending powers of the symbols.

$$\begin{array}{r|l} & 2x + 3 \\ \hline x - 1 & 2x^2 + x - 3 \\ & \underline{2x^2 - 2x} \\ & \quad 3x - 3 \\ & \quad \underline{3x - 3} \\ & \quad \underline{\quad \cdot \quad \cdot \quad} \end{array}$$

Dividing the first term of the dividend by the first term of the divisor, i.e. $2x^2/x$ gives $2x$, which is placed above the first term of the dividend as shown. The divisor is then multiplied by $2x$, i.e. $2x(x - 1) = 2x^2 - 2x$, which is placed under the dividend as shown. Subtracting gives $3x - 3$.

The process is then repeated, i.e. the first term of the divisor, x, is divided into $3x$, giving $+3$, which is placed above the dividend as shown. Then $3(x - 1) = 3x - 3$ which is placed under the $3x - 3$. The remainder, on subtraction, is zero, which completes the process.

Thus, $\mathbf{(2x^2 + x - 3) \div (x - 1) = (2x + 3)}$

Application: Divide $(x^2 + 3x - 2)$ by $(x - 2)$

$$
\begin{array}{r|l}
 & x + 5 \\
\hline
x - 2 & x^2 + 3x - 2 \\
 & \underline{x^2 - 2x} \\
 & \quad 5x - 2 \\
 & \quad \underline{5x - 10} \\
 & \qquad 8 \\
\end{array}
$$

Hence $\dfrac{x^2 + 3x - 2}{x - 2} = \mathbf{x + 5 + \dfrac{8}{x - 2}}$

2.2 The factor theorem

A factor of $(x - a)$ in an equation corresponds to a root of $x = a$

If $x = a$ is a root of the equation $f(x) = 0$, then $(x - a)$ is a factor of $f(x)$

Application: Factorise $x^3 - 7x - 6$ and use it to solve the cubic equation $x^3 - 7x - 6 = 0$

Let $f(x) = x^3 - 7x - 6$

If $x = 1$, then $f(1) = 1^3 - 7(1) - 6 = -12$

If $x = 2$, then $f(2) = 2^3 - 7(2) - 6 = -12$

If $x = 3$, then $f(3) = 3^3 - 7(3) - 6 = 0$

If $f(3) = 0$, then $(x - 3)$ is a factor – from the factor theorem.

We have a choice now. We can divide $x^3 - 7x - 6$ by $(x - 3)$ or we could continue our 'trial and error' by substituting further values for x in the given expression – and hope to arrive at $f(x) = 0$.

Let us do both ways. Firstly, dividing out gives:

$$\begin{array}{r|l} & x^2 + 3x + 2 \\ \hline x - 3 & x^3 + 0 \quad - 7x - 6 \\ & \underline{x^3 - 3x^2} \\ & \quad 3x^2 - 7x - 6 \\ & \quad \underline{3x^2 - 9x} \\ & \qquad\quad 2x - 6 \\ & \qquad\quad \underline{2x - 6} \\ & \qquad\quad \underline{\quad\cdot\quad\cdot} \end{array}$$

Hence, $\dfrac{x^3 - 7x - 6}{x - 3} = x^2 + 3x + 2$

i.e. $x^3 - 7x - 6 = (x - 3)(x^2 + 3x + 2)$

$x^2 + 3x + 2$ factorises 'on sight' as $(x + 1)(x + 2)$

Therefore, $\mathbf{x^3 - 7x - 6 = (x - 3)(x + 1)(x + 2)}$

A second method is to continue to substitute values of x into f(x).

Our expression for f(3) was $3^3 - 7(3) - 6$. We can see that if we continue with positive values of x the first term will predominate such that f(x) will not be zero.

Therefore let us try some negative values for x.

$f(-1) = (-1)^3 - 7(-1) - 6 = 0$; hence $(x + 1)$ is a factor (as shown above).

Also $f(-2) = (-2)^3 - 7(-2) - 6 = 0$; hence $(x + 2)$ is a factor.

To solve $x^3 - 7x - 6 = 0$, we substitute the factors, i.e.

$$(x - 3)(x + 1)(x + 2) = 0$$

from which, $\mathbf{x = 3, x = -1}$ **and** $\mathbf{x = -2}$

Note that the values of x, i.e. 3, −1 and −2, are all factors of the constant term, i.e. the 6. This can give us a clue as to what values of x we should consider.

2.3 The remainder theorem

If $(ax^2 + bx + c)$ is divided by $(x - p)$, the remainder will be $ap^2 + bp + c$

If $(ax^3 + bx^2 + cx + d)$ is divided by $(x - p)$, the remainder will be $ap^3 + bp^2 + cp + d$

Application: When $(3x^2 - 4x + 5)$ is divided by $(x - 2)$ find the remainder

$ap^2 + bp + c$, (where $a = 3$, $b = -4$, $c = 5$ and $p = 2$),

hence the remainder is $3(2)^2 + (-4)(2) + 5 = 12 - 8 + 5 = \mathbf{9}$

We can check this by dividing $(3x^2 - 4x + 5)$ by $(x - 2)$ by long division:

$$\begin{array}{r|l} & 3x + 2 \\ \hline x - 2 & 3x^2 - 4x + 5 \\ & \underline{3x^2 - 6x} \\ & \qquad\quad 2x + 5 \\ & \qquad\quad \underline{2x - 4} \\ & \qquad\qquad\quad 9 \end{array}$$

Application: When $(2x^2 + x - 3)$ is divided by $(x - 1)$, find the remainder

$ap^2 + bp + c$, (where $a = 2$, $b = 1$, $c = -3$ and $p = 1$), hence the **remainder is** $2(1)^2 + 1(1) - 3 = \mathbf{0}$,

which means that $(x - 1)$ is a factor of $(2x^2 + x - 3)$.

In this case, the other factor is $(2x + 3)$,
i.e. $(2x^2 + x - 3) = (x - 1)(2x + 3)$

Application: When $(3x^3 + 2x^2 - x + 4)$ is divided by $(x - 1)$, find the remainder

The remainder is $ap^3 + bp^2 + cp + d$ (where $a = 3$, $b = 2$, $c = -1$, $d = 4$ and $p = 1$), i.e. the remainder is:
$3(1)^3 + 2(1)^2 + (-1)(1) + 4 = 3 + 2 - 1 + 4 = \mathbf{8}$

2.4 Continued fractions

Any fraction may be expressed in the form shown below for the fraction 26/55:

$$\frac{26}{55} = \frac{1}{\frac{55}{26}} = \frac{1}{2+\frac{3}{26}} = \frac{1}{2+\frac{1}{\frac{26}{3}}} = \frac{1}{2+\frac{1}{8+\frac{2}{3}}} = \frac{1}{2+\frac{1}{8+\frac{1}{\frac{3}{2}}}}$$

$$= \frac{1}{2+\frac{1}{8+\frac{1}{1+\frac{1}{2}}}}$$

The latter factor can be expressed as:

$$\cfrac{1}{A + \cfrac{\alpha}{B + \cfrac{\beta}{C + \cfrac{\gamma}{D + \delta}}}}$$

Comparisons show that A, B, C and D are 2, 8, 1 and 2 respectively.

A fraction written in the general form is called a **continued fraction** and the integers A, B, C and D are called the **quotients** of the continued fraction. The quotients may be used to obtain closer and closer approximations, called **convergents**.

A tabular method may be used to determine the convergents of a fraction:

	1	2	3	4	5
a		2	8	1	2
b $\begin{cases} bp \\ bq \end{cases}$	$\frac{0}{1}$	$\frac{1}{2}$	$\frac{8}{17}$	$\frac{9}{19}$	$\frac{26}{55}$

The quotients 2, 8, 1 and 2 are written in cells a2, a3, a4 and a5 with cell a1 being left empty.

The fraction $\frac{0}{1}$ is always written in cell b1.

The reciprocal of the quotient in cell a2 is always written in cell b2, i.e. $\frac{1}{2}$ in this case.

The fraction in cell b3 is given by $\dfrac{(a3 \times b2p) + b1p}{(a3 \times b2q) + b1q}$,

i.e. $\dfrac{(8 \times 1) + 0}{(8 \times 2) + 1} = \dfrac{8}{17}$

The fraction in cell b4 is given by $\dfrac{(a4 \times b3p) + b2p}{(a4 \times b3q) + b2q}$,

i.e. $\dfrac{(1 \times 8) + 1}{(1 \times 17) + 2} = \dfrac{9}{19}$, and so on.

Hence the convergents of $\dfrac{26}{55}$ are $\dfrac{1}{2}, \dfrac{8}{17}, \dfrac{9}{19}$ and $\dfrac{26}{55}$, each value approximating closer and closer to $\dfrac{26}{55}$.

These approximations to fractions are used to obtain practical ratios for **gearwheels** or for a **dividing head** (used to give a required angular displacement).

2.5 Solution of quadratic equations by formula

If $ax^2 + bx + c = 0$ then $\mathbf{x} = \dfrac{-b \pm \sqrt{b^2 - 4ac}}{2a}$

Application: Solve $3x^2 - 11x - 4 = 0$ by using the quadratic formula

Comparing $3x^2 - 11x - 4 = 0$ with $ax^2 + bx + c = 0$ gives $a = 3$, $b = -11$ and $c = -4$

$$\text{Hence, } x = \frac{-(-11) \pm \sqrt{(-11)^2 - 4(3)(-4)}}{2(3)} = \frac{+11 \pm \sqrt{121 + 48}}{6}$$

$$= \frac{11 \pm \sqrt{169}}{6} = \frac{11 \pm 13}{6}$$

$$= \frac{11 + 13}{6} \text{ or } \frac{11 - 13}{6}$$

Hence, $\mathbf{x} = \frac{24}{6} = \mathbf{4}$ or $\frac{-2}{6} = \mathbf{-\frac{1}{3}}$

Application: Solve $4x^2 + 7x + 2 = 0$ giving the roots correct to 2 decimal places

Comparing $4x^2 + 7x + 2 = 0$ with $ax^2 + bx + c$ gives $a = 4$, $b = 7$ and $c = 2$

$$\text{Hence, } x = \frac{-7 \pm \sqrt{7^2 - 4(4)(2)}}{2(4)} = \frac{-7 \pm \sqrt{17}}{8}$$

$$= \frac{-7 \pm 4.123}{8} = \frac{-7 + 4.123}{8} \text{ or } \frac{-7 - 4.123}{8}$$

Hence, $\mathbf{x = -0.36}$ or **−1.39, correct to 2 decimal places.**

Application: The height s metres of a mass projected vertically upwards at time t seconds is $s = ut - \frac{1}{2}gt^2$. Determine how long the mass will take after being projected to reach a height of 16 m (a) on the ascent and (b) on the descent, when $u = 30$ m/s and $g = 9.81$ m/s^2

When height s = 16 m, $16 = 30t - \frac{1}{2}(9.81)t^2$

i.e. $4.905t^2 - 30t + 16 = 0$

Using the quadratic formula:

$$t = \frac{-(-30) \pm \sqrt{(-30)^2 - 4(4.905)(16)}}{2(4.905)}$$

$$= \frac{30 \pm \sqrt{586.1}}{9.81} = \frac{30 \pm 24.21}{9.81} = 5.53 \text{ or } 0.59$$

Hence the mass will reach a height of 16 m after 0.59 s on the ascent and after 5.53 s on the descent.

Application: A shed is 4.0 m long and 2.0 m wide. A concrete path of constant width is laid all the way around the shed and the area of the path is 9.50 m^2. Calculate its width, to the nearest centimetre

Figure 2.1 shows a plan view of the shed with its surrounding path of width t metres

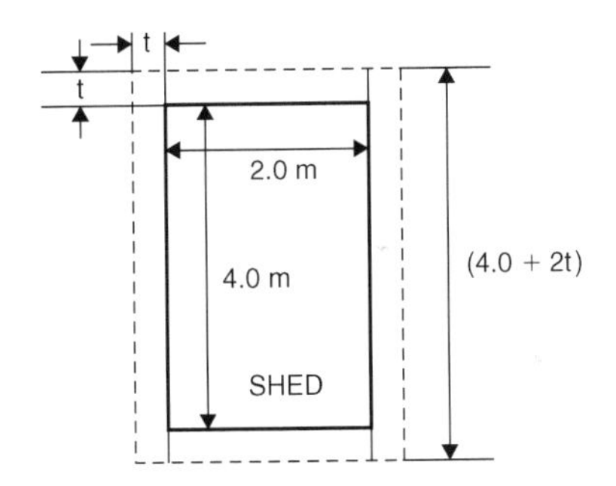

Figure 2.1

Area of path $= 2(2.0 \times t) + 2t(4.0 + 2t)$

i.e. $\quad 9.50 = 4.0t + 8.0t + 4t^2$

or $\quad 4t^2 + 12.0t - 9.50 = 0$

Hence $\quad t = \dfrac{-(12.0) \pm \sqrt{(12.0)^2 - 4(4)(-9.50)}}{2(4)}$

$$= \frac{-12.0 \pm \sqrt{296.0}}{8} = \frac{-12.0 \pm 17.20465}{8}$$

Hence, $t = 0.6506\,m$ or $-3.65058\,m$

Neglecting the negative result which is meaningless, the width of the path, $\mathbf{t = 0.651\,m}$ or **65 cm**, correct to the nearest centimetre.

2.6 Logarithms

Definition of a logarithm: If $y = a^x$ then $x = \log_a y$

Laws of logarithms:

$$\log(A \times B) = \log A + \log B$$

$$\log\left(\frac{A}{B}\right) = \log A - \log B$$

$$\lg A^n = n \log A$$

Application: Evaluate (a) $\log_3 9$ (b) $\log_{16} 8$

(a) Let $x = \log_3 9$ then $3^x = 9$ from the definition of a logarithm,

i.e. $3^x = 3^2$, from which $x = 2$

Hence, $\mathbf{\log_3 9 = 2}$

(b) Let $x = \log_{16} 8$ then $16^x = 8$, from the definition of a logarithm,

i.e. $(2^4)^x = 2^3$, i.e. $2^{4x} = 2^3$ from the laws of indices, from

which, $4x = 3$ and $x = \frac{3}{4}$

Hence, $\mathbf{\log_{16} 8 = \frac{3}{4}}$

Application: Evaluate (a) lg 0.001 (b) ln e (c) $\log_3 \frac{1}{81}$

(a) Let $x = \lg 0.001 = \log_{10} 0.001$ then $10^x = 0.001$, i.e. $10^x = 10^{-3}$, from which $x = -3$

Hence, $\mathbf{\lg 0.001 = -3}$ (which may be checked by a calculator)

(b) Let $x = \ln e = \log_e e$ then $e^x = e$, i.e. $e^x = e^1$ from which $x = 1$.

Hence, $\mathbf{\ln e = 1}$ (which may be checked by a calculator)

(c) Let $x = \log_3 \frac{1}{81}$ then $3^x = \frac{1}{81} = \frac{1}{3^4} = 3^{-4}$ from which $x = -4$

Hence, $\mathbf{\log_3 \frac{1}{81} = -4}$

Application: Solve the equations: (a) $\lg x = 3$ (b) $\log_5 x = -2$

(a) If $\lg x = 3$ then $\log_{10} x = 3$ and $x = 10^3$, i.e. $\mathbf{x = 1000}$

(b) If $\log_5 x = -2$ then $x = 5^{-2} = \frac{1}{5^2} = \mathbf{\frac{1}{25}}$

Application: Solve $3^x = 27$

Logarithms to a base of 10 are taken of both sides, i.e.

$$\log_{10} 3^x = \log_{10} 27$$

and $\log_{10} 3 = \log_{10} 27$ by the third law of logarithms

Rearranging gives: $\mathbf{x} = \frac{\log_{10} 27}{\log_{10} 3} = \frac{1.43136\ldots}{0.4771\ldots} = \mathbf{3}$ which may be readily checked.

Application: Solve the equation $2^{x+1} = 3^{2x-5}$ correct to 2 decimal places

Taking logarithms to base 10 of both sides gives:

$$\log_{10} 2^{x+1} = \log_{10} 3^{2x-5}$$

i.e. $(x+1)\log_{10} 2 = (2x-5)\log_{10} 3$

$$x \log_{10} 2 + \log_{10} 2 = 2x \log_{10} 3 - 5 \log_{10} 3$$

$$x(0.3010) + (0.3010) = 2x(0.4771) - 5(0.4771)$$

i.e. $0.3010x + 0.3010 = 0.9542x - 2.3855$

Hence $2.3855 + 0.3010 = 0.9542x - 0.3010x$

$$2.6865 = 0.6532x$$

from which, $\mathbf{x} = \frac{2.6865}{0.6532} = \mathbf{4.11}$, correct to 2 decimal places.

Application: Solve the equation $x^{3.2} = 41.15$, correct to 4 significant figures

Taking logarithms to base 10 of both sides gives:

$$\log_{10} x^{3.2} = \log_{10} 41.15$$

$$3.2 \log_{10} x = \log_{10} 41.15$$

Hence, $$\log_{10} x = \frac{\log_{10} 41.15}{3.2} = 0.50449$$

Thus, $\mathbf{x} = \text{antilog } 0.50449 = 10^{0.50449} = \mathbf{3.195}$ correct to 4 significant figures.

Graphs of logarithmic functions

A graph of $y = \log_{10} x$ is shown in Figure 2.2 and a graph of $y = \log_e x$ is shown in Figure 2.3. Both are seen to be of similar shape; in fact, the same general shape occurs for a logarithm to any base.

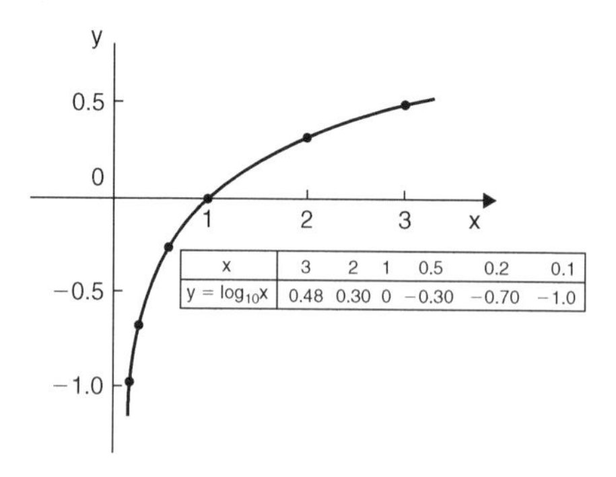

x	3	2	1	0.5	0.2	0.1
$y = \log_{10} x$	0.48	0.30	0	−0.30	−0.70	−1.0

Figure 2.2

In general, with a logarithm to any base a, it is noted that:

1. $\mathbf{\log_a 1 = 0}$
2. $\mathbf{\log_a a = 1}$
3. $\mathbf{\log_a 0 \rightarrow -\infty}$

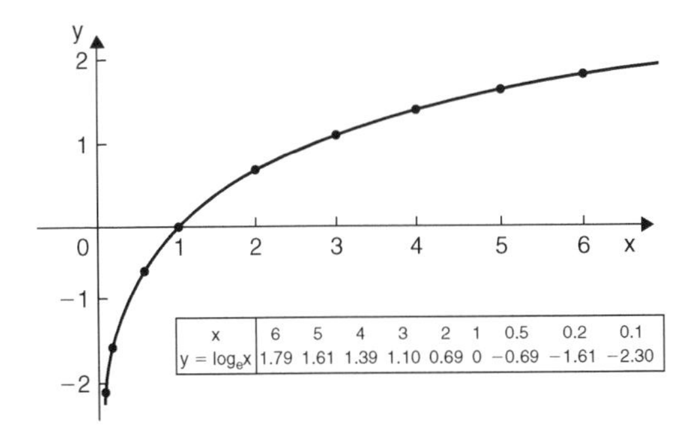

x	6	5	4	3	2	1	0.5	0.2	0.1
$y = \log_e x$	1.79	1.61	1.39	1.10	0.69	0	−0.69	−1.61	−2.30

Figure 2.3

2.7 Exponential functions

The power series for e^x is:

$$e^x = 1 + x + \frac{x^2}{2!} + \frac{x^3}{3!} + \frac{x^4}{4!} + \cdots \quad (1)$$

(where $3! = 3 \times 2 \times 1$ and is called 'factorial 3')

The series is valid for all values of x.

Graphs of exponential functions

Figure 2.4 shows graphs of $y = e^x$ and $y = e^{-x}$

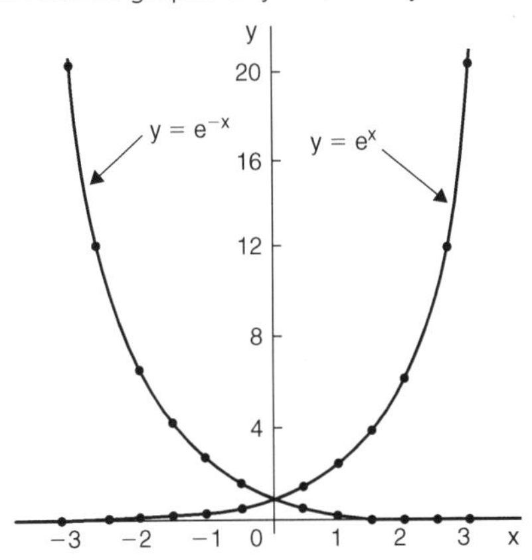

Figure 2.4

Application: The decay of voltage, v volts, across a capacitor at time t seconds is given by $v = 250e^{-t/3}$. Draw a graph showing the natural decay curve over the first 6 seconds. Determine (a) the voltage after 3.4 s, and (b) the time when the voltage is 150 volts

A table of values is drawn up as shown below.

t	0	1	2	3	4	5	6
$e^{-t/3}$	1.00	0.7165	0.5134	0.3679	0.2636	0.1889	0.1353
$v = 250e^{-t/3}$	250.0	179.1	128.4	91.97	65.90	47.22	33.83

The natural decay curve of $v = 250e^{-t/3}$ is shown in Figure 2.5.

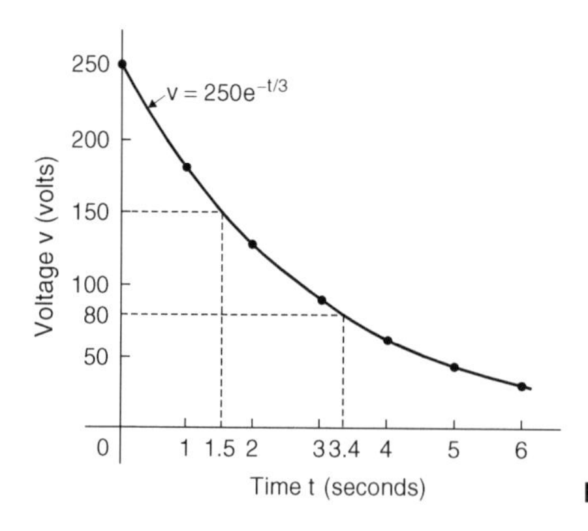

Figure 2.5

From the graph,

(a) when time t = 3.4 s, **voltage v = 80 volts**

(b) when voltage v = 150 volts, **time t = 1.5 seconds**

2.8 Napierian logarithms

$$\log_e y = 2.3026 \log_{10} y$$

$$\log_e e^x = x$$

Application: Solve $e^{3x} = 8$

Taking Napierian logarithms of both sides, gives

$$\ln e^{3x} = \ln 8$$

i.e. $$3x = \ln 8$$

from which $$x = \frac{1}{3}\ln 8 = \mathbf{0.6931}, \text{ correct to 4 decimal places}$$

Application: The work done in an isothermal expansion of a gas from pressure p_1 to p_2 is given by:

$$w = w_0 \ln\left(\frac{p_1}{p_2}\right)$$

If the initial pressure $p_1 = 7.0\,\text{kPa}$, calculate the final pressure p_2 if $w = 3w_0$

If $w = 3w_0$ then $$3\,w_0 = w_0 \ln\left(\frac{p_1}{p_2}\right)$$

i.e. $$3 = \ln\left(\frac{p_1}{p_2}\right)$$

and $$e^3 = \frac{p_1}{p_2} = \frac{7000}{p_2}$$

from which, **final pressure**, $$p_2 = \frac{7000}{e^3} = 7000e^{-3} = \mathbf{348.5\,Pa}$$

Laws of growth and decay

The laws of exponential growth and decay are of the form $y = Ae^{-kx}$ and $y = A(1 - e^{-kx})$, where A and k are constants. When plotted, the form of each of these equations is as shown in Figure 2.6. The laws occur frequently in engineering and science and examples of quantities related by a natural law include

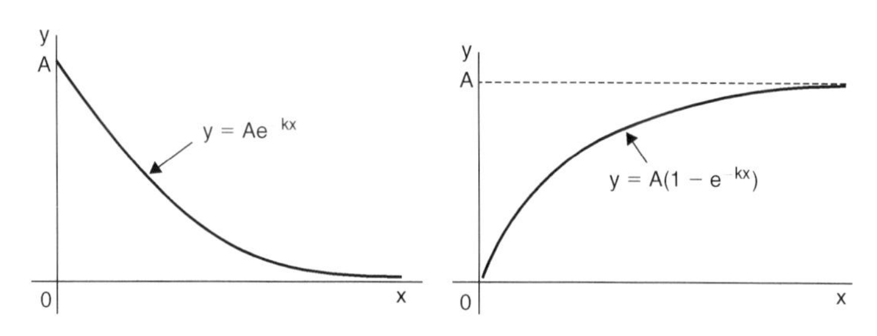

Figure 2.6

(i) Linear expansion	$l = l_0\, e^{\alpha\theta}$
(ii) Change in electrical resistance with temperature	$R_\theta = R_0\, e^{\alpha\theta}$
(iii) Tension in belts	$T_1 = T_0\, e^{\mu\theta}$
(iv) Newton's law of cooling	$\theta = \theta_0\, e^{-kt}$
(v) Biological growth	$y = y_0\, e^{kt}$
(vi) Discharge of a capacitor	$q = Q\, e^{-t/CR}$
(vii) Atmospheric pressure	$p = p_0\, e^{-h/c}$
(viii) Radioactive decay	$N = N_0\, e^{-\lambda t}$
(ix) Decay of current in an inductive circuit	$i = I\, e^{-Rt/L}$
(x) Growth of current in a capacitive circuit	$i = I(1 - e^{-t/CR})$

Application: In an experiment involving Newton's law of cooling, the temperature θ(°C) is given by $\theta = \theta_0 e^{-kt}$. Find the value of constant k when $\theta_0 = 56.6$°C, $\theta = 16.5$°C and $t = 83.0$ seconds

Transposing $\theta = \theta_0\, e^{-kt}$ gives $\frac{\theta}{\theta_0} = e^{-kt}$ from which,

$$\frac{\theta_0}{\theta} = \frac{1}{e^{-kt}} = e^{kt}$$

Taking Napierian logarithms of both sides gives: $\ln \frac{\theta_0}{\theta} = kt$

$$\text{fromwhich, } \mathbf{k} = \frac{1}{t}\ln\frac{\theta_0}{\theta} = \frac{1}{83.0}\ln\left(\frac{56.6}{16.5}\right) = \frac{1}{83.0}(1.2326486\,..)$$

$$= \mathbf{1.485 \times 10^{-2}}$$

Application: The current i amperes flowing in a capacitor at time t seconds is given by $i = 8.0(1 - e^{-t/CR})$, where the circuit resistance R is $25\,k\Omega$ and capacitance C is $16\,\mu F$. Determine (a) the current i after 0.5 seconds and (b) the time, to the nearest ms, for the current to reach 6.0 A

(a) Current $i = 8.0(1 - e^{-t/CR}) = 8.0[1 - e^{-0.5/(16\times10^{-6})(25\times10^{3})}]$

$$= 8.0(1 - e^{-1.25})$$

$$= 8.0(1 - 0.2865047..) = 8.0(0.7134952..)$$

$$= \mathbf{5.71\ amperes}$$

(b) Transposing $i = 8.0(1 - e^{-t/CR})$ gives: $\frac{i}{80} = 1 - e^{-t/CR}$

from which, $e^{-t/CR} = 1 - \frac{i}{8.0} = \frac{8.0 - i}{8.0}$

Taking the reciprocal of both sides gives: $e^{t/CR} = \frac{8.0}{8.0 - i}$

Taking Napierian logarithms of both sides gives:

$$\frac{t}{CR} = \ln\left(\frac{8.0}{8.0 - i}\right)$$

Hence $t = CR\ln\left(\frac{8.0}{8.0 - i}\right)$

$$= (16\times10^{-6})(25\times10^{3})\ln\left(\frac{8.0}{8.0 - 6.0}\right) \text{ when } i = 6.0 \text{ amperes,}$$

i.e.
$$\mathbf{t} = 0.40 \ln\left(\frac{8.0}{2.0}\right) = 0.4 \ln 4.0$$
$$= 0.4(1.3862943..)$$
$$= 0.5545\text{ s}$$
$$= \mathbf{555\ ms}, \text{ to the nearest millisecond.}$$

A graph of current against time is shown in Figure 2.7.

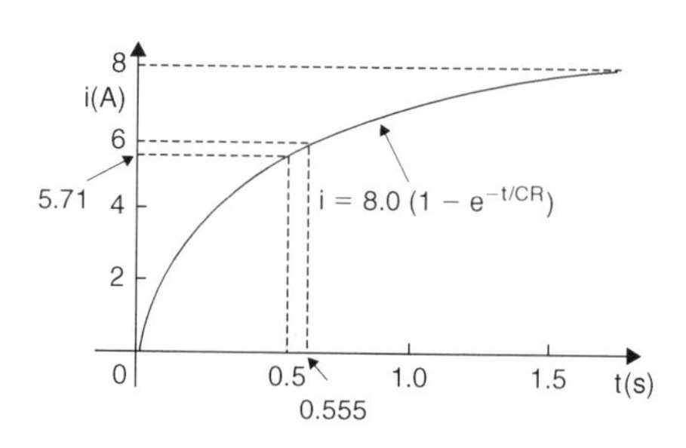

Figure 2.7

2.9 Hyperbolic functions

$$\sinh x = \frac{e^x - e^{-x}}{2} \qquad \text{cosech}\, x = \frac{1}{\sinh x} = \frac{2}{e^x - e^{-x}}$$

$$\cosh x = \frac{e^x + e^{-x}}{2} \qquad \text{sech}\, x = \frac{1}{\cosh x} = \frac{2}{e^x + e^{-x}}$$

$$\tanh x = \frac{\sinh x}{\cosh x} = \frac{e^x - e^{-x}}{e^x + e^{-x}} \qquad \coth x = \frac{1}{\tanh x} = \frac{e^x + e^{-x}}{e^x - e^{-x}}$$

$$\cosh x = 1 + \frac{x^2}{2!} + \frac{x^4}{4!} + \cdots \quad \text{(which is valid for all values of x)}$$

$$\sinh x = x + \frac{x^3}{3!} + \frac{x^5}{5!} + \cdots \quad \text{(which is valid for all values of x)}$$

Graphs of hyperbolic functions

A graph of **y = sinh x** is shown in Figure 2.8. Since the graph is symmetrical about the origin, sinh x is an **odd function**.

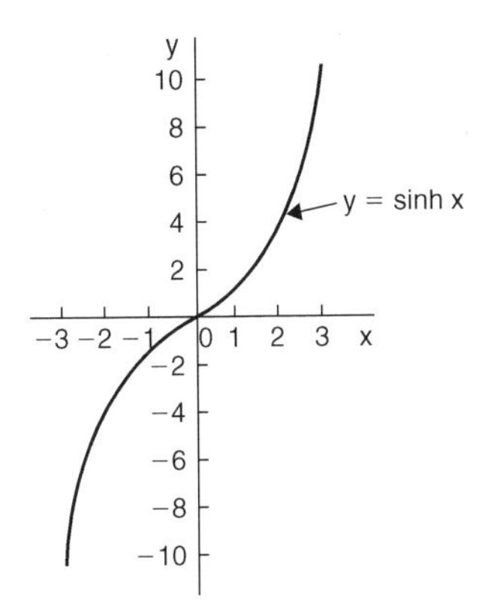

Figure 2.8

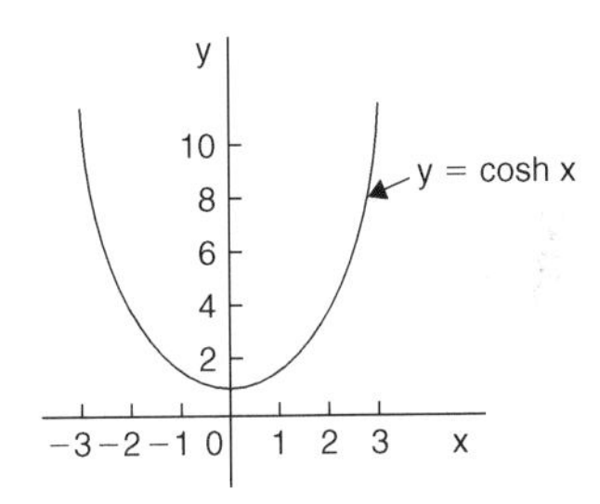

Figure 2.9

A graph of **y = cosh x** is shown in Figure 2.9. Since the graph is symmetrical about the y-axis, cosh x is an **even function**. The shape of y = cosh x is that of a heavy rope or chain hanging freely under gravity and is called a **catenary**. Examples include **transmission lines**, a **telegraph wire** or a **fisherman's line**, and are used in the **design of roofs and arches**. Graphs of y = tanh x, y = coth x, y = cosech x and y = sech x are shown in Figures 2.10 and 2.11.

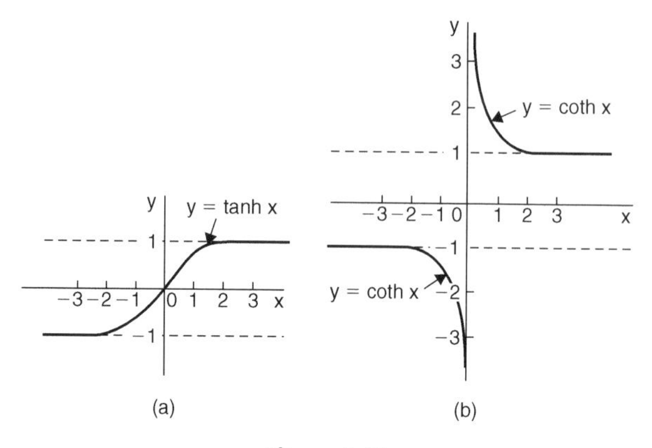

Figure 2.10

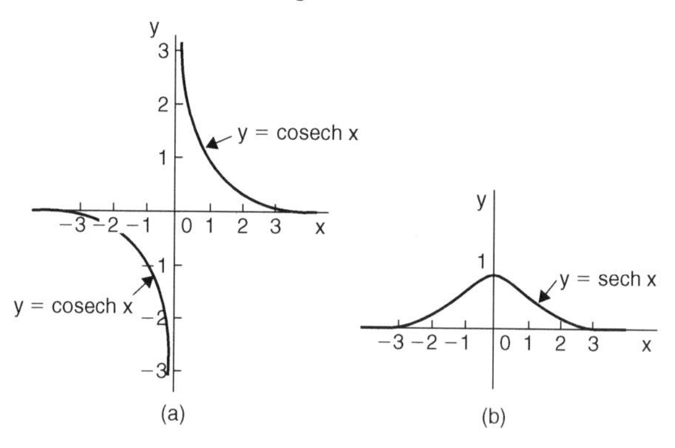

Figure 2.11

Hyperbolic identities

Trigonometric identity	Corresponding hyperbolic identity
$\cos^2 x + \sin^2 x = 1$	$\text{ch}^2 x - \text{sh}^2 x = 1$
$1 + \tan^2 x = \sec^2 x$	$1 - \text{th}^2 x = \text{sech}^2 x$
$\cot^2 x + 1 = \text{cosec}^2 x$	$\text{coth}^2 x - 1 = \text{cosech}^2 x$

Compound angle formulae

$\sin(A \pm B) = \sin A \cos B \pm \cos A \sin B$	$\text{sh}(A \pm B) = \text{sh}\, A\, \text{ch}\, B \pm \text{ch}\, A\, \text{sh}\, B$
$\cos(A \pm B) = \cos A \cos B \mp \sin A \sin B$	$\text{ch}(A \pm B) = \text{ch}\, A\, \text{ch}\, B \pm \text{sh}\, A\, \text{sh}\, B$
$\tan(A \pm B) = \dfrac{\tan A \pm \tan B}{1 \mp \tan A \tan B}$	$\tan(A \pm B) = \dfrac{\text{th}\, A \pm \text{th}\, B}{1 \pm \text{th}\, A\, \text{th}\, B}$

Double angles

$\sin 2x = 2 \sin x \cos x$	$\text{sh}\, 2x = 2\, \text{sh}\, x\, \text{ch}\, x$
$\cos 2x = \cos^2 x - \sin^2 x$	$\text{ch}\, 2x = \text{ch}^2 x + \text{sh}^2 x$
$= 2\cos^2 x - 1$	$= 2\, \text{ch}^2 x - 1$
$= 1 - 2\sin^2 x$	$= 1 + 2\, \text{sh}^2 x$
$\tan 2x = \dfrac{2 \tan x}{1 - \tan^2 x}$	$\text{th}\, 2x = \dfrac{2\, \text{th}\, x}{1 + \text{th}^2 x}$

Solving equations involving hyperbolic functions

Equations of the form **a ch x + b sh x = c**, where a, b and c are constants may be solved either by:

(a) plotting graphs of $y = a\, \text{ch}\, x + b\, \text{sh}\, x$ and $y = c$ and noting the points of intersection, or more accurately,

(b) by adopting the following procedure:

1. Change sh x to $\left(\dfrac{e^x - e^{-x}}{2}\right)$ and ch x to $\left(\dfrac{e^x + e^{-x}}{2}\right)$
2. Rearrange the equation into the form $pe^x + qe^{-x} + r = 0$, where p, q and r are constants.
3. Multiply each term by e^x, which produces an equation of the form $p(e^x)^2 + re^x + q = 0$ (since $(e^{-x})(e^x) = e^0 = 1$)

4. Solve the quadratic equation $p(e^x)^2 + re^x + q = 0$ for e^x by factorising or by using the quadratic formula.
5. Given e^x = a constant (obtained by solving the equation in 4), take Napierian logarithms of both sides to give $x = \ln(\text{constant})$

Application: Solve the equation sh $x = 3$, correct to 4 significant figures

Following the above procedure:

1. $\text{sh}\, x = \left(\dfrac{e^x - e^{-x}}{2}\right) = 3$
2. $e^x - e^{-x} = 6$, i.e. $e^x - e^{-x} - 6 = 0$
3. $(e^x)^2 - (e^{-x})(e^x) - 6e^x = 0$, i.e. $(e^x)^2 - 6e^x - 1 = 0$
4. $e^x = \dfrac{-(-6) \pm \sqrt{[(-6)^2 - 4(1)(-1)]}}{2(1)} = \dfrac{6 \pm \sqrt{40}}{2} = \dfrac{6 \pm 6.3246}{2}$

   Hence, $e^x = 6.1623$ or -0.1623
5. $x = \ln 6.1623$ or $x = \ln(-0.1623)$ which has no solution since it is not possible in real terms to find the logarithm of a negative number.

   Hence $x = \ln 6.1623 = \mathbf{1.818}$, correct to 4 significant figures.

Application: A chain hangs in the form given by $y = 40\,\text{ch}\dfrac{x}{40}$. Determine, correct to 4 significant figures, (a) the value of y when x is 25 and (b) the value of x when $y = 54.30$

(a) $y = 40\,\text{ch}\dfrac{x}{40}$ and when $x = 25$,

$$\mathbf{y} = 40\,\text{ch}\frac{25}{40} = 40\,\text{ch}\,0.625 = 40\left(\frac{e^{0.625} + e^{-0.625}}{2}\right)$$

$$= 20(1.8682 + 0.5353) = \mathbf{48.07}$$

(b) When y = 54.30, $54.30 = 40 \text{ ch} \frac{x}{40}$, from which

$$\text{ch} \frac{x}{40} = \frac{54.30}{40} = 1.3575$$

Following the above procedure:

1. $\frac{e^{x/40} + e^{-x/40}}{2} = 1.3575$
2. $e^{x/40} + e^{-x/40} = 2.715$ i.e. $e^{x/40} + e^{-x/40} - 2.715 = 0$
3. $(e^{x/40})^2 + 1 - 2.715\, e^{x/40} = 0$ i.e. $(e^{x/40})^2 - 2.715\, e^{x/40} + 1 = 0$
4. $$e^{x/40} = \frac{-(-2.715) \pm \sqrt{[(-2.715)^2 - 4(1)(1)]}}{2(1)}$$
   $$= \frac{2.715 \pm \sqrt{(3.3712)}}{2} = \frac{2.715 \pm 1.8361}{2}$$
   Hence $e^{x/40} = 2.2756$ or 0.43945
5. $\frac{x}{40} = \ln 2.2756$ or $\frac{x}{40} = \ln(0.43945)$

   Hence, $\frac{x}{40} = 0.8222$ or $\frac{x}{40} = -0.8222$

   Hence, x = 40(0.8222) or x = 40(−0.8222)

   i.e. **x = ±32.89**, correct to 4 significant figures.

2.10 Partial fractions

Provided that the numerator f(x) is of less degree than the relevant denominator, the following identities are typical examples of the form of partial fraction used:

Linear factors

$$\frac{\mathbf{f(x)}}{\mathbf{(x + a)(x - b)(x + c)}} \equiv \frac{\mathbf{A}}{\mathbf{(x + a)}} + \frac{\mathbf{B}}{\mathbf{(x - b)}} + \frac{\mathbf{C}}{\mathbf{(x + c)}}$$

Repeated linear factors

$$\frac{f(x)}{(x+a)^3} \equiv \frac{A}{(x+a)} + \frac{B}{(x+a)^2} + \frac{C}{(x+a)^3}$$

Quadratic factors

$$\frac{f(x)}{(ax^2+bx+c)(x+d)} \equiv \frac{Ax+B}{(ax^2+bx+c)} + \frac{C}{(x+d)}$$

Application: Resolve $\frac{11-3x}{x^2+2x-3}$ into partial fractions

The denominator factorises as $(x-1)(x+3)$ and the numerator is of less degree than the denominator.

Thus $\frac{11-3x}{x^2+2x-3}$ may be resolved into partial fractions.

Let $\frac{11-3x}{x^2+2x-3} = \frac{11-3x}{(x-1)(x+3)} \equiv \frac{A}{(x-1)} + \frac{B}{(x+3)}$ where A and B are constants to be determined,

i.e. $$\frac{11-3x}{(x-1)(x+3)} \equiv \frac{A(x+3)+B(x-1)}{(x-1)(x+3)}$$ by algebraic addition

Since the denominators are the same on each side of the identity then the numerators are equal to each other.

Thus, $$11-3x = A(x+3) + B(x-1)$$

To determine constants A and B, values of x are chosen to make the term in A or B equal to zero.

When $x = 1$, then $11 - 3(1) = A(1+3) + B(0)$

i.e. $$8 = 4A$$

i.e. $$\mathbf{A = 2}$$

When $x = -3$, then $11 - 3(-3) = A(0) + B(-3-1)$

i.e. $$20 = -4B$$

i.e. $$\mathbf{B = -5}$$

Thus $\frac{\mathbf{11-3x}}{\mathbf{x^2+2x-3}} = \frac{2}{(x-1)} + \frac{-5}{(x+3)} = \frac{\mathbf{2}}{\mathbf{(x-1)}} - \frac{\mathbf{5}}{\mathbf{(x+3)}}$

$$\left[\text{Check: } \frac{2}{(x-1)} - \frac{5}{(x+3)} = \frac{2(x+3)-5(x-1)}{(x-1)(x+3)} = \frac{11-3x}{x^2+2x-3}\right]$$

Application: Express $\frac{x^3-2x^2-4x-4}{x^2+x-2}$ in partial fractions

The numerator is of higher degree than the denominator. Thus dividing out gives:

$$\begin{array}{r|l} & x-3 \\ \hline x^2+x-2 & x^3-2x^2-4x-4 \\ & \underline{x^3+x^2-2x} \\ & \quad -3x^2-2x-4 \\ & \quad \underline{-3x^2-3x+6} \\ & \qquad\qquad x-10 \end{array}$$

Thus $$\frac{x^3-2x^2-4x-4}{x^2+x-2} \equiv x-3+\frac{x-10}{x^2+x-2}$$
$$\equiv x-3+\frac{x-10}{(x-2)(x-1)}$$

Let $\frac{x-10}{(x+2)(x-1)} \equiv \frac{A}{(x+2)} + \frac{B}{(x-1)} = \frac{A(x-1)+B(x+2)}{(x+2)(x-1)}$

Equating the numerators gives: $x-10 = A(x-1)+B(x+2)$

Let $x=-2$, then $-12=-3A$

i.e. $\mathbf{A=4}$

Let $x=1$, then $-9=3B$

i.e. $\mathbf{B=-3}$

Hence $\frac{x-10}{(x+2)(x-1)} = \frac{4}{(x+2)} - \frac{3}{(x-1)}$

Thus $$\frac{\mathbf{x^3 - 2x^2 - 4x - 4}}{\mathbf{x^2 + x - 2}} = \mathbf{x - 3 + \frac{4}{(x + 2)} - \frac{3}{(x - 1)}}$$

Application: Express $\frac{5x^2 - 2x - 19}{(x + 3)(x - 1)^2}$ as the sum of three partial fractions

The denominator is a combination of a linear factor and a repeated linear factor.

Let

$$\frac{5x^2 - 2x - 19}{(x + 3)(x - 1)^2} = \frac{A}{(x + 3)} + \frac{B}{(x - 1)} + \frac{C}{(x - 1)^2}$$

$$= \frac{A(x - 1)^2 + B(x + 3)(x - 1) + C(x + 3)}{(x + 3)(x - 1)^2}$$ by algebraic addition

Equating the numerators gives:

$$5x^2 - 2x - 19 \equiv A(x - 1)^2 + B(x + 3)(x - 1) + C(x + 3) \qquad (1)$$

Let $x = -3$, then $5(-3)^2 - 2(-3) - 19 = A(-4)^2 + B(0)(-4) + C(0)$

i.e. $32 = 16A$

i.e. $\mathbf{A = 2}$

Let $x = 1$, then $5(1)^2 - 2(1) - 19 = A(0)^2 + B(4)(0) + C(4)$

i.e. $-16 = 4C$

i.e. $\mathbf{C = -4}$

Without expanding the RHS of equation (1) it can be seen that equating the coefficients of x^2 gives:

$5 = A + B$, and since $A = 2$, $\mathbf{B = 3}$

Hence $$\frac{\mathbf{5x^2 - 2x - 19}}{\mathbf{(x + 3)(x - 1)^2}} \equiv \mathbf{\frac{2}{(x + 2)} + \frac{3}{(x - 1)} - \frac{4}{(x - 1)^2}}$$

Application: Resolve $\dfrac{3 + 6x + 4x^2 - 2x^3}{x^2(x^2 + 3)}$ into partial fractions

Terms such as x^2 may be treated as $(x + 0)^2$, i.e. they are repeated linear factors.

$(x^2 + 3)$ is a quadratic factor which does not factorise without containing surds and imaginary terms.

$$\text{Let } \frac{3 + 6x + 4x^2 - 2x^3}{x^2(x^2 + 3)} = \frac{A}{x} + \frac{B}{x^2} + \frac{Cx + D}{(x^2 + 3)}$$

$$= \frac{Ax(x^2 + 3) + B(x^2 + 3) + (Cx + D)x^2}{x^2(x^2 + 3)}$$

Equating the numerators gives:

$$3 + 6x + 4x^2 - 2x^3 = Ax(x^2 + 3) + B(x^2 + 3) + (Cx + D)x^2$$

$$= Ax^3 + 3Ax + Bx^2 + 3B + Cx^3 + Dx^2$$

Let $x = 0$, then $3 = 3B$

i.e. $\mathbf{B = 1}$

Equating the coefficients of x^3 terms gives: $-2 = A + C$ (1)

Equating the coefficients of x^2 terms gives: $4 = B + D$

Since $B = 1$, $\mathbf{D = 3}$

Equating the coefficients of x terms gives: $6 = 3A$

i.e. $\mathbf{A = 2}$

From equation (1), since $A = 2$, $\mathbf{C = -4}$

$$\text{Hence } \mathbf{\frac{3 + 6x + 4x^2 - 2x^3}{x^2(x^2 + 3)}} = \frac{2}{x} + \frac{1}{x^2} + \frac{-4x + 3}{x^2 + 3}$$

$$= \mathbf{\frac{2}{x} + \frac{1}{x^2} + \frac{3 - 4x}{x^2 + 3}}$$

3 Some Number Topics

3.1 Arithmetic progressions

If a = first term, d = common difference and n = number of terms, then the arithmetic progression is:

$$\mathbf{a, a + d, a + 2d, \ldots}$$

The n'th term is:

$$\mathbf{a + (n - 1)d}$$

The sum of n terms,

$$\mathbf{S_n = \frac{n}{2}[2a + (n - 1)d]}$$

Application: Find the sum of the first 7 terms of the series 1, 4, 7, 10, 13, ...

The sum of the first 7 terms is given by

$$S_7 = \frac{7}{2}[2(1) + (7 - 1)3] \qquad \text{since } a = 1 \text{ and } d = 3$$

$$= \frac{7}{2}[2 + 18] = \frac{7}{2}[20] = \mathbf{70}$$

Application: Determine (a) the ninth, and (b) the sixteenth term of the series 2, 7, 12, 17, ...

2, 7, 12, 17, ... is an arithmetic progression with a common difference, d, of 5

(a) The n'th term of an AP is given by $a + (n - 1)d$
Since the first term $a = 2$, $d = 5$ and $n = 9$
then the 9th term is: $2 + (9 - 1)5 = 2 + (8)(5) = 2 + 40 = \mathbf{42}$

(b) The 16th term is: $2 + (16 - 1)5 = 2 + (15)(5) = 2 + 75 = \mathbf{77}$

Application: Find the sum of the first 12 terms of the series 5, 9, 13, 17, ...

5, 9, 13, 17, ... is an AP where $a = 5$ and $d = 4$

The sum of n terms of an AP, $S_n = \frac{n}{2}[2a + (n - 1)d]$

Hence the sum of the first 12 terms, $S_{12} = \frac{12}{2}[2(5) + (12 - 1)4]$

$$= 6[10 + 44] = 6(54)$$

$$= \mathbf{324}$$

3.2 Geometric progressions

If a = first term, r = common ratio and n = number of terms, then the geometric progression is:

$$a, ar, ar^2, ar^3, ...$$

The n'th term is: ar^{n-1}

The sum of n terms,

$$S_n = \frac{a(1 - r^n)}{(1 - r)} \quad \text{which is valid when } r < 1$$

or

$$S_n = \frac{a(r^n - 1)}{(r - 1)} \quad \text{which is valid when } r > 1$$

If $-1 < r < 1$, $S_\infty = \frac{a}{(1 - r)}$

Application: Find the sum of the first 8 terms of the GP 1, 2, 4, 8, 16, ...

The sum of the first 8 terms is given by

$$S_8 = \frac{1(2^8 - 1)}{(2 - 1)} \quad \text{since } a = 1 \text{ and } r = 2$$

i.e.
$$S_8 = \frac{1(256 - 1)}{1} = \mathbf{255}$$

Application: Determine the tenth term of the series 3, 6, 12, 24, ...

3, 6, 12, 24, ... is a geometric progression with a common ratio r of 2.

The n'th term of a GP is ar^{n-1}, where a is the first term.

Hence the 10th term is: $(3)(2)^{10-1} = (3)(2)^9 = 3(512) = \mathbf{1536}$

Application: A hire tool firm finds that their net return from hiring tools is decreasing by 10% per annum. Their net gain on a certain tool this year is £400. Find the possible total of all future profits from this tool (assuming the tool lasts for ever)

The net gain forms a series: £400 + £400 × 0.9 + £400 × 0.9^2 + ·····,

which is a GP with a = 400 and r = 0.9

The sum to infinity,

$$S_\infty = \frac{a}{(1 - r)} = \frac{400}{(1 - 0.9)} = \mathbf{£4000} = \textbf{total future profits}$$

Application: A drilling machine is to have 6 speeds ranging from 50 rev/min to 750 rev/min. Determine their values, each correct to the nearest whole number, if the speeds form a geometric progression

Let the GP of n terms be given by a, ar, ar^2, ... ar^{n-1}

The first term a = 50 rev/min

The 6th term is given by ar^{6-1}, which is 750 rev/min,

i.e. $ar^5 = 750$

from which $r^5 = \frac{750}{a} = \frac{750}{50} = 15$

Thus the common ratio, $r = \sqrt[5]{15} = 1.7188$

The first term is a = 50 rev/min

the second term is $ar = (50)(1.7188) = 85.94$,

the third term is $ar^2 = (50)(1.7188)^2 = 147.71$,

the fourth term is $ar^3 = (50)(1.7188)^3 = 253.89$,

the fifth term is $ar^4 = (50)(1.7188)^4 = 436.39$,

the sixth term is $ar^5 = (50)(1.7188)^5 = 750.06$

Hence, correct to the nearest whole number, the 6 speeds of the drilling machine are:

50, 86, 148, 254, 436 and 750 rev/min

3.3 The binomial series

$$(a + x)^n = a^n + na^{n-1}x + \frac{n(n-1)}{2!}a^{n-2}x^2 + \frac{n(n-1)(n-2)}{3!}a^{n-3}x^3 + \cdots + x^n$$

$$(1 + x)^n = 1 + nx + \frac{n(n-1)}{2!}x^2 + \frac{n(n-1)(n-2)}{3!}x^3 + \cdots$$

which is valid for $-1 < x < 1$

The r'th term of the expansion $(a + x)^n$ is:

$$\frac{n(n-1)(n-2)\ldots \text{ to } (r-1) \text{ terms}}{(r-1)!}a^{n-(r-1)}x^{r-1}$$

Application: Using the binomial series, determine the expansion of $(2 + x)^7$

From above, when a = 2 and n = 7:

$$(2+x)^7 = 2^7 + 7(2)^6x + \frac{(7)(6)}{(2)(1)}(2)^5x^2 + \frac{(7)(6)(5)}{(3)(2)(1)}(2)^4x^3$$

$$+ \frac{(7)(6)(5)(4)}{(4)(3)(2)(1)}(2)^3x^4 + \frac{(7)(6)(5)(4)(3)}{(5)(4)(3)(2)(1)}(2)^2x^5$$

$$+ \frac{(7)(6)(5)(4)(3)(2)}{(6)(5)(4)(3)(2)(1)}(2)x^6 + \frac{(7)(6)(5)(4)(3)(2)(1)}{(7)(6)(5)(4)(3)(2)(1)}x^7$$

i.e.

$$\mathbf{(2+x)^7 = 128 + 448x + 672x^2 + 560x^3 + 280x^4 + 84x^5 + 14x^6 + x^7}$$

Application: Determine the fifth term $(3 + x)^7$ without fully expanding

The r'th term of the expansion $(a + x)^n$ is given by:

$$\frac{n(n-1)(n-2)\ldots \text{ to } (r-1) \text{ terms}}{(r-1)!} a^{n-(r-1)} x^{r-1}$$

Substituting n = 7, a = 3 and r − 1 = 5 − 1 = 4 gives:

$$\frac{(7)(6)(5)(4)}{(4)(3)(2)(1)}(3)^{7-4}x^4$$

i.e. the fifth term of $(3 + x)^7 = 35(3)^3 x^4 = \mathbf{945x^4}$

Application: Expand $\dfrac{1}{(1+2x)^3}$ in ascending powers of x as far as the term in x^3, using the binomial series

Using the binomial expansion of $(1 + x)^n$, where $n = -3$ and x is replaced by 2x gives:

$$\frac{1}{(1+2x)^3} = (1+2x)^{-3}$$

$$= 1 + (-3)(2x) + \frac{(-3)(-4)}{2!}(2x)^2 + \frac{(-3)(-4)(-5)}{3!}(2x)^3 + \cdots$$

$$= \mathbf{1 - 6x + 24x^2 - 80x^3 +}$$

The expansion is valid provided $|2x| < 1$

i.e. $\quad \mathbf{|x| < \frac{1}{2}} \quad$ or $\quad \mathbf{-\frac{1}{2} < x < \frac{1}{2}}$

Application: Using the binomial theorem, expand $\sqrt{4+x}$ in ascending powers of x to four terms

$$\sqrt{4+x} = \sqrt{\left[4\left(1+\frac{x}{4}\right)\right]} = \sqrt{4}\sqrt{\left(1+\frac{x}{4}\right)} = 2\left(1+\frac{x}{4}\right)^{1/2}$$

Using the expansion of $(1 + x)^n$,

$$2\left(1+\frac{x}{4}\right)^{1/2}$$

$$= 2\left[1 + \left(\frac{1}{2}\right)\left(\frac{x}{4}\right) + \frac{(1/2)(-1/2)}{2!}\left(\frac{x}{4}\right)^2 + \frac{(1/2)(-1/2)(-3/2)}{3!}\left(\frac{x}{4}\right)^3 + \cdots\right]$$

$$= 2\left(1 + \frac{x}{8} - \frac{x^2}{128} + \frac{x^3}{1024} - \cdots\right)$$

$$= \mathbf{2 + \frac{x}{4} - \frac{x^2}{64} + \frac{x^3}{512} - \cdots\cdots}$$

This is valid when $\left|\frac{x}{4}\right| < 1$, i.e. $\mathbf{|x| < 4}$ or $\mathbf{-4 < x < 4}$

Application: Simplify $\dfrac{\sqrt[3]{(1-3x)}\,\sqrt{(1+x)}}{\left(1+\dfrac{x}{2}\right)^3}$ given that powers of x above the first may be neglected

$$\frac{\sqrt[3]{(1-3x)}\,\sqrt{(1+x)}}{\left(1+\dfrac{x}{2}\right)^3} = (1-3x)^{\frac{1}{3}}(1+x)^{\frac{1}{2}}\left(1+\frac{x}{2}\right)^{-3}$$

$$\approx \left[1+\left(\frac{1}{3}\right)(-3x)\right]\left[1+\left(\frac{1}{2}\right)(x)\right]\left[1+(-3)\left(\frac{x}{2}\right)\right]$$

when expanded by the binomial theorem as far as the x term only,

$$= (1-x)\left(1+\frac{x}{2}\right)\left(1-\frac{3x}{2}\right)$$

$$= \left(1-x+\frac{x}{2}-\frac{3x}{2}\right)$$ when powers of x higher than unity are neglected

$$= \mathbf{(1-2x)}$$

Application: The second moment of area of a rectangle through its centroid is given by $\dfrac{bl^3}{12}$. Determine the approximate change in the second moment of area if b is increased by 3.5% and l is reduced by 2.5%

New values of b and l are $(1 + 0.035)b$ and $(1 - 0.025)l$ respectively.

New second moment of area $= \dfrac{1}{12}[(1+0.035)b][(1-0.025)1]^3$

$$= \frac{bl^3}{12}(1+0.035)(1-0.025)^3$$

$$\approx \frac{bl^3}{12}(1+0.035)(1-0.075)$$

neglecting powers of small terms

$$\approx \frac{bl^3}{12}(1+0.035-0.075)$$

neglecting products of small terms

$$\approx \frac{bl^3}{12}(1-0.040) \text{ or } (0.96)\frac{bl^3}{12}$$

i.e. 96% of the original second moment of area

Hence the second moment of area is reduced by approximately 4%

Application: The resonant frequency of a vibrating shaft is given by: $f = \frac{1}{2\pi}\sqrt{\frac{k}{I}}$, where k is the stiffness and I is the inertia of the shaft. Using the binomial theorem, determine the approximate percentage error in determining the frequency using the measured values of k and I, when the measured value of k is 4% too large and the measured value of I is 2% too small

Let f, k and I be the true values of frequency, stiffness and inertia respectively. Since the measured value of stiffness, k_1, is 4% too large, then $k_1 = \frac{104}{100}k = (1+0.04)k$

The measured value of inertia, I_1, is 2% too small, hence $I_1 = \frac{98}{100}I = (1-0.02)I$

The measured value of frequency,

$$f_1 = \frac{1}{2\pi}\sqrt{\frac{k_1}{I_1}} = \frac{1}{2\pi}k_1^{1/2}\, I_1^{-1/2}$$

$$= \frac{1}{2\pi}[(1+0.04)k]^{1/2}\,[(1-0.02)I]^{-1/2}$$

$$= \frac{1}{2\pi}(1+0.04)^{1/2}\, k^{1/2}\,(1-0.02)^{-1/2}\, l^{-1/2}$$

$$= \frac{1}{2\pi}\, k^{1/2}\, l^{-1/2}\,(1+0.04)^{1/2}\,(1-0.02)^{-1/2}$$

i.e. $$f_1 = f\,(1+0.04)^{1/2}\,(1-0.02)^{-1/2}$$

$$\approx f\,[1+(1/2)(0.04)][(1+(-1/2)(-0.02)]$$

$$\approx f\,(1+0.02)(1+0.01)$$

Neglecting the products of small terms,

$$f_1 \approx (1+0.02+0.01)\,f \approx 1.03\,f$$

Thus the percentage error in f based on the measured values of k and l is approximately **3% too large**.

3.4 Maclaurin's theorem

$$\mathbf{f(x) = f(0) + xf'(0) + \frac{x^2}{2!}f''(0) + \frac{x^3}{3!}f'''(0) + \cdots}$$

Application: Determine the first four terms of the power series for cos x

The values of f(0), f′(0), f″(0), ... in the Maclaurin's series are obtained as follows:

$f(x) = \cos x$	$f(0) = \cos 0 = 1$
$f'(x) = -\sin x$	$f'(0) = -\sin 0 = 0$
$f''(x) = -\cos x$	$f''(0) = -\cos 0 = -1$
$f'''(x) = \sin x$	$f'''(0) = \sin 0 = 0$
$f^{iv}(x) = \cos x$	$f^{iv}(0) = \cos 0 = 1$
$f^{v}(x) = -\sin x$	$f^{v}(0) = -\sin 0 = 0$
$f^{vi}(x) = -\cos x$	$f^{vi}(0) = -\cos 0 = -1$

Substituting these values into the Maclaurin's series gives:

$$f(x) = \cos x = 1 + x(0) + \frac{x^2}{2!}(-1) + \frac{x^3}{3!}(0) + \frac{x^4}{4!}(1) + \frac{x^5}{5!}(0) + \frac{x^6}{6!}(-1) + \cdots$$

i.e. $$\mathbf{\cos x = 1 - \frac{x^2}{2!} + \frac{x^4}{4!} - \frac{x^6}{6!} + \cdots}$$

Application: Determine the power series for $\cos 2\theta$

Replacing x with 2θ in the series obtained in the previous example gives:

$$\cos 2\theta = 1 - \frac{(2\theta)^2}{2!} + \frac{(2\theta)^4}{4!} - \frac{(2\theta)^6}{6!} + \cdots$$

$$= 1 - \frac{4\theta^2}{2} + \frac{16\theta^4}{24} - \frac{64\theta^6}{720} + \cdots$$

i.e. $$\mathbf{\cos 2\theta = 1 - 2\theta^2 + \frac{2}{3}\theta^4 - \frac{4}{45}\theta^6 + \cdots}$$

Application: Expand $\ln(1 + x)$ to five terms

$f(x) = \ln(1 + x)$ $\qquad f(0) = \ln(1 + 0) = 0$

$f'(x) = \dfrac{1}{(1 + x)}$ $\qquad f'(0) = \dfrac{1}{1 + 0} = 1$

$f''(x) = \dfrac{-1}{(1 + x)^2}$ $\qquad f''(0) = \dfrac{-1}{(1 + 0)^2} = -1$

$f'''(x) = \dfrac{2}{(1 + x)^3}$ $\qquad f'''(0) = \dfrac{2}{(1 + 0)^3} = 2$

$f^{iv}(x) = \dfrac{-6}{(1 + x)^4}$ $\qquad f^{iv}(0) = \dfrac{-6}{(1 + 0)^4} = -6$

$f^{v}(x) = \dfrac{24}{(1 + x)^5}$ $\qquad f^{v}(0) = \dfrac{24}{(1 + 0)^5} = 24$

Substituting these values into the Maclaurin's series gives:

$$f(x) = \ln(1 + x) = 0 + x(1) + \frac{x^2}{2!}(-1) + \frac{x^3}{3!}(2) + \frac{x^4}{4!}(-6) + \frac{x^5}{5!}(24)$$

i.e. $$\mathbf{\ln(1 + x) = x - \frac{x^2}{2} + \frac{x^3}{3} - \frac{x^4}{4} + \frac{x^5}{5} - \cdots}$$

Application: Find the expansion of $(2 + x)^4$ using Maclaurin's series

$f(x) = (2 + x)^4 \qquad f(0) = 2^4 = 16$

$f'(x) = 4(2 + x)^3 \qquad f'(0) = 4(2)^3 = 32$

$f''(x) = 12(2 + x)^2 \qquad f''(0) = 12(2)^2 = 48$

$f'''(x) = 24(2 + x)^1 \qquad f'''(0) = 24(2) = 48$

$f^{iv}(x) = 24 \qquad f^{iv}(0) = 24$

Substituting in Maclaurin's series gives:

$$\mathbf{(2 + x)^4} = f(0) + xf'(0) + \frac{x^2}{2!}f''(0) + \frac{x^3}{3!}f'''(0) + \frac{x^4}{4!}f^{iv}(0)$$

$$= 16 + (x)(32) + \frac{x^2}{2!}(48) + \frac{x^3}{3!}(48) + \frac{x^4}{4!}(24)$$

$$\mathbf{= 16 + 32x + 24x^2 + 8x^3 + x^4}$$

Numerical integration using Maclaurin's series

Application: Evaluate $\int_{0.1}^{0.4} 2e^{\sin\theta}\, d\theta$, correct to 3 significant figures

A power series for $e^{\sin\theta}$ is firstly obtained using Maclaurin's series.

$f(\theta) = e^{\sin\theta} \qquad f(0) = e^{\sin 0} = e^0 = 1$

$f'(\theta) = \cos\theta\, e^{\sin\theta} \qquad f'(0) = \cos 0\, e^{\sin 0} = (1)\, e^0 = 1$

$$f''(\theta) = (\cos\theta)(\cos\theta\, e^{\sin\theta}) + (e^{\sin\theta})(-\sin\theta) \text{ by the product rule,}$$

$$= e^{\sin\theta}(\cos^2\theta - \sin\theta) \qquad f''(0) = e^0(\cos^2 0 - \sin 0) = 1$$

$$f'''(\theta) = (e^{\sin\theta})[(2\cos\theta\,(-\sin\theta) - \cos\theta] + (\cos^2\theta - \sin\theta)(\cos\theta\, e^{\sin\theta})$$

$$= e^{\sin\theta}\cos\theta[-2\sin\theta - 1 + \cos^2\theta - \sin\theta]$$

$$f'''(0) = e^0\cos 0[(0 - 1 + 1 - 0)] = 0$$

Hence from the Maclaurin's series:

$$e^{\sin\theta} = f(0) + \theta\, f'(0) + \frac{\theta^2}{2!} f''(0) + \frac{\theta^3}{3!} f'''(0) + \cdots = 1 + \theta + \frac{\theta^2}{2} + 0$$

Thus $\displaystyle\int_{0.1}^{0.4} 2\, e^{\sin\theta}\, d\theta = \int_{0.1}^{0.4} 2\left(1 + \theta + \frac{\theta^2}{2}\right) d\theta$

$$= \int_{0.1}^{0.4} (2 + 2\theta + \theta^2)d\theta = \left[2\theta + \frac{2\theta^2}{2} + \frac{\theta^3}{3}\right]_{0.1}^{0.4}$$

$$= \left(0.8 + (0.4)^2 + \frac{(0.4)^3}{3}\right)$$

$$- \left(0.2 + (0.1)^2 + \frac{(0.1)^3}{3}\right)$$

$$= 0.98133 - 0.21033$$

$= \mathbf{0.771}$, correct to 3 significant figures

3.5 Limiting values

L'Hopital's rule states:

$$\lim_{\delta x \to a} it \left\{\frac{f(x)}{g(x)}\right\} = \lim_{\delta x \to a} it \left\{\frac{f'(x)}{g'(x)}\right\} \text{ provided } g'(a) \neq 0$$

Application: Determine $\displaystyle\lim_{\delta x - 1} it \left\{\frac{x^2 + 3x - 4}{x^2 - 7x + 6}\right\}$

The first step is to substitute x = 1 into both numerator and denominator. In this case we obtain $\frac{0}{0}$.

It is only when we obtain such a result that we then use L'Hopital's rule. Hence applying L'Hopital's rule,

$$\lim_{x \to 1} \text{it} \left\{ \frac{x^2 + 3x - 4}{x^2 - 7x + 6} \right\} = \lim_{x \to 1} \text{it} \left\{ \frac{2x + 3}{2x - 7} \right\}$$ i.e. both numerator and denominator have been differentiated

$$= \frac{5}{-5} = \mathbf{-1}$$

Application: Determine $\lim_{\theta \to 0} \left\{ \frac{\sin\theta - \theta\cos\theta}{\theta^3} \right\}$

$$\lim_{\theta \to 0} \left\{ \frac{\sin\theta - \theta\cos\theta}{\theta^3} \right\} = \lim_{\theta \to 0} \left\{ \frac{\cos\theta - [(\theta)(-\sin\theta) + \cos\theta]}{3\theta^2} \right\}$$

$$= \lim_{\theta \to 0} \left\{ \frac{\theta\sin\theta}{3\theta^2} \right\} = \lim_{\theta \to 0} \left\{ \frac{\theta\cos\theta + \sin\theta}{6\theta} \right\}$$

$$= \lim_{\theta \to 0} \left\{ \frac{\theta(-\sin\theta) + \cos\theta(1) + \cos\theta}{6} \right\}$$

$$= \frac{1+1}{6} = \frac{2}{6} = \mathbf{\frac{1}{3}}$$

3.6 Solving equations by iterative methods

Three iterative methods are

(i) the bisection method
(ii) an algebraic method and
(iii) by using the Newton-Raphson formula.

(i) The bisection method

In the **method of bisection** the mid-point of the interval, i.e. $x_3 = \frac{x_1 + x_2}{2}$, is taken, and from the sign of $f(x_3)$ it can be deduced whether a root lies in the half interval to the left or right of x_3. Whichever half interval is indicated, its mid-point is then taken and the procedure repeated. The method often requires many iterations and is therefore slow, but never fails to eventually produce the root. The procedure stops when two successive values of x are equal, to the required degree of accuracy.

Application: Using the bisection method, determine the positive root of the equation $x + 3 = e^x$, correct to 3 decimal places

Let $f(x) = x + 3 - e^x$ then, using functional notation:

$$\mathbf{f(0)} = 0 + 3 - e^0 = \mathbf{+2}$$

$$\mathbf{f(1)} = 1 + 3 - e^1 = \mathbf{+1.2817..}$$

$$\mathbf{f(2)} = 2 + 3 - e^2 = \mathbf{-2.3890..}$$

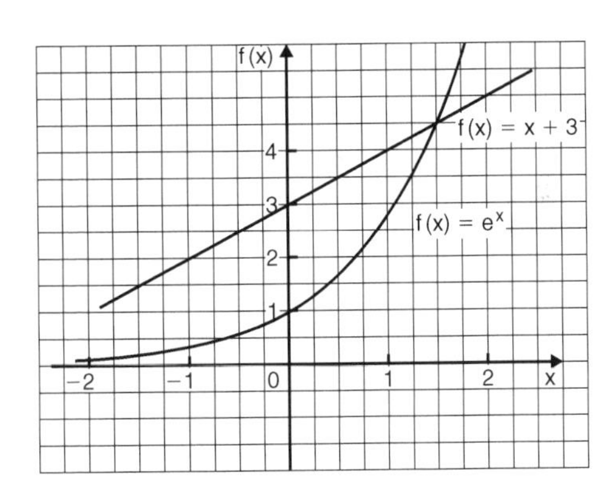

Figure 3.1

Since f(1) is positive and f(2) is negative, a root lies between x = 1 and x = 2. A sketch of $f(x) = x + 3 - e^x$, i.e. $x + 3 = e^x$ is shown in Figure 3.1.

Bisecting the interval between x = 1 and x = 2 gives $\frac{1+2}{2}$ i.e. 1.5

Hence **f(1.5)** $= 1.5 + 3 - e^{1.5} =$ **+0.01831..**

Since f(1.5) is positive and f(2) is negative, a root lies between x = 1.5 and x = 2. Bisecting this interval gives $\frac{1.5+2}{2}$ i.e. 1.75

Hence **f(1.75)** $= 1.75 + 3 - e^{1.75} =$ **−1.00460..**

Since f(1.75) is negative and f(1.5) is positive, a root lies between x = 1.75 and x = 1.5

Bisecting this interval gives $\frac{1.75+1.5}{2}$ i.e. 1.625

Hence **f(1.625)** $= 1.625 + 3 - e^{1.625} =$ **−0.45341..**

Since f(1.625) is negative and f(1.5) is positive, a root lies between x = 1.625 and x = 1.5

Bisecting this interval gives $\frac{1.625+1.5}{2}$ i.e. 1.5625

Hence **f(1.5625)** $= 1.5625 + 3 - e^{1.5625} =$ **−0.20823..**

Since f(1.5625) is negative and f(1.5) is positive, a root lies between x = 1.5625 and x = 1.5.

The iterations are continued and the results are presented in the table shown.

The last two values of x_3 in the table are 1.504882813 and 1.505388282, i.e. both are equal to 1.505, correct to 3 decimal places. The process therefore stops.

Hence the root of x + 3 = e^x is x = 1.505, correct to 3 decimal places.

x_1	x_2	$x_3 = \frac{x_1+x_2}{2}$	$f(x_3)$
		0	+2
		1	+1.2817..
		2	−2.3890..
1	2	1.5	+0.0183..

1.5	2	1.75	−1.0046..
1.5	1.75	1.625	−0.4534..
1.5	1.625	1.5625	−0.2082..
1.5	1.5625	1.53125	−0.0927..
1.5	1.53125	1.515625	−0.0366..
1.5	1.515625	1.5078125	−0.0090..
1.5	1.5078125	1.50390625	+0.0046..
1.50390625	1.5078125	1.505859375	−0.0021..
1.50390625	1.505859375	**1.504882813**	+0.0012..
1.504882813	1.505859375	**1.505388282**	

(ii) An algebraic method of successive approximations

Procedure:

First approximation

(a) Using a graphical or functional notation method, determine an approximate value of the root required, say x_1

Second approximation

(b) Let the true value of the root be $(x_1 + \delta_1)$

(c) Determine x_2 the approximate value of $(x_1 + \delta_1)$ by determining the value of $f(x_1 + \delta_1) = 0$, but neglecting terms containing products of δ_1

Third approximation

(d) Let the true value of the root be $(x_2 + \delta_2)$

(e) Determine x_3, the approximate value of $(x_2 + \delta_2)$ by determining the value of $f(x_2 + \delta_2) = 0$, but neglecting terms containing products of δ_2

(f) The fourth and higher approximations are obtained in a similar way.

Using the techniques given in paragraphs (b) to (f), it is possible to continue getting values nearer and nearer to the required root. The procedure is repeated until the value of the required root does not change on two consecutive approximations, when expressed to the required degree of accuracy.

Application: Determine the value of the smallest positive root of the equation $3x^3 - 10x^2 + 4x + 7 = 0$, correct to 3 significant figures, using an algebraic method of successive approximations

The functional notation method is used to find the value of the first approximation.

$$f(x) = 3x^3 - 10x^2 + 4x + 7$$

$$f(0) = 3(0)^3 - 10(0)^2 + 4(0) + 7 = 7$$

$$f(1) = 3(1)^3 - 10(1)^2 + 4(1) + 7 = 4$$

$$f(2) = 3(2)^3 - 10(2)^2 + 4(2) + 7 = -1$$

Following the above procedure:

First approximation

(a) Let the first approximation be such that it divides the interval 1 to 2 in the ratio of 4 to −1, i.e. let x_1 be 1.8

Second approximation

(b) Let the true value of the root, x_2, be $(x_1 + \delta_1)$

(c) Let $f(x_1 + \delta_1) = 0$, then since $x_1 = 1.8$,
$3(1.8 + \delta_1)^3 - 10(1.8 + \delta_1)^2 + 4(1.8 + \delta_1) + 7 = 0$

Neglecting terms containing products of δ_1 and using the binomial series gives:

$$3[1.8^3 + 3(1.8)^2\delta_1] - 10[1.8^2 + (2)(1.8)\delta_1] + 4(1.8 + \delta_1) + 7 \approx 0$$

$$3(5.832 + 9.720\,\delta_1) - 32.4 - 36\,\delta_1 + 7.2 + 4\,\delta_1 + 7 \approx 0$$

$$17.496 + 29.16\,\delta_1 - 32.4 - 36\,\delta_1 + 7.2 + 4\,\delta_1 + 7 \approx 0$$

$$\delta_1 \approx \frac{-17.496 + 32.4 - 7.2 - 7}{29.16 - 36 + 4} \approx -\frac{0.704}{2.84} \approx -0.2479$$

Thus, $x_2 \approx 1.8 - 0.2479 = 1.5521$

Third approximation

(d) Let the true value of the root, x_3, be $(x_2 + \delta_2)$

(e) Let $f(x_2 + \delta_2) = 0$, then since $x_2 = 1.5521$,

$3(1.5521 + \delta_2)^3 - 10(1.5521 + \delta_2)^2 + 4(1.5521 + \delta_2) + 7 = 0$

Neglecting terms containing products of δ_2 gives:

$$11.217 + 21.681\,\delta_2 - 24.090 - 31.042\,\delta_2 + 6.2084 + 4\,\delta_2 + 7 \approx 0$$

$$\delta_2 \approx \frac{-11.217 + 24.090 - 6.2084 - 7}{21.681 - 31.042 + 4} \approx \frac{-0.3354}{-5.361} \approx 0.06256$$

Thus $x_3 \approx 1.5521 + 0.06256 \approx 1.6147$

(f) Values of x_4 and x_5 are found in a similar way.

$$f(x_3 + \delta_3) = 3(1.6147 + \delta_3)^3 - 10(1.6147 + \delta_3)^2 + 4(1.6147 + \delta_3) + 7 = 0$$

giving $\delta_3 \approx 0.003175$ and $x_4 \approx 1.618$, i.e. 1.62 correct to 3 significant figures

$$f(x_4 + \delta_4) = 3(1.618 + \delta_4)^3 - 10(1.618 + \delta_4)^2 + 4(1.618 + \delta_4) + 7 = 0$$

giving $\delta_4 \approx 0.0000417$, and $x_5 \approx 1.62$, correct to 3 significant figures.

Since x_4 and x_5 are the same when expressed to the required degree of accuracy, then the required root is **1.62**, correct to 3 significant figures.

(iii) The Newton-Raphson method

If r_1 is the approximate value of a real root of the equation $f(x) = 0$, then a closer approximation to the root, r_2, is given by:

$$r_2 = r_1 - \frac{f(r_1)}{f'(r_1)}$$

Application: Using Newton's method, find the positive root of $(x+4)^3 - e^{1.92x} + 5\cos\frac{x}{3} = 9$, correct to 3 significant figures

The functional notational method is used to determine the approximate value of the root.

$$f(x) = (x+4)^3 - e^{1.92x} + 5\cos\frac{x}{3} - 9$$

$$f(0) = (0+4)^3 - e^0 + 5\cos 0 - 9 = 59$$

$$f(1) = 5^3 - e^{1.92} + 5\cos\frac{1}{3} - 9 \approx 114$$

$$f(2) = 6^3 - e^{3.84} + 5\cos\frac{2}{3} - 9 \approx 164$$

$$f(3) = 7^3 - e^{5.76} + 5\cos 1 - 9 \approx 19$$

$$f(4) = 8^3 - e^{7.68} + 5\cos\frac{4}{3} - 9 \approx -1660$$

From these results, let a first approximation to the root be $r_1 = 3$

Newton's method states that a better approximation to the root,

$$r_2 = r_1 - \frac{f(r_1)}{f'(r_1)}$$

$$f(r_1) = f(3) = 7^3 - e^{5.76} + 5\cos 1 - 9 = 19.35$$

$$f'(x) = 3(x+4)^2 - 1.92e^{1.92x} - \frac{5}{3}\sin\frac{x}{3}$$

$$f'(r_1) = f'(3) = 3(7)^2 - 1.92e^{5.76} - \frac{5}{3}\sin 1 = -463.7$$

Thus, $r_2 = 3 - \frac{19.35}{-463.7} = 3 + 0.042 = 3.042 = 3.04$,

correct to 3 significant figures.

Similarly,

$$r_3 = 3.042 - \frac{f(3.042)}{f'(3.042)} = 3.042 - \frac{(-1.146)}{(-513.1)} = 3.042 - 0.0022$$

$$= 3.0398 = 3.04, \text{ correct to 3 significant figures.}$$

Since r_2 and r_3 are the same when expressed to the required degree of accuracy, then the required root is **3.04**, correct to 3 significant figures.

3.7 Computer numbering systems

Conversion of binary to decimal

Application: Change the binary number 1101.1 to its equivalent decimal form

$$1101.1 = 1\times 2^3 + 1\times 2^2 + 0\times 2^1 + 1\times 2^0 + 1\times 2^{-1}$$

$$= 8 + 4 + 0 + 1 + \frac{1}{2}, \text{ that is } 13.5$$

i.e. $\mathbf{1101.1_2 = 13.5_{10}}$, the suffixes 2 and 10 denoting binary and decimal systems of numbers respectively.

Application: Convert 101.0101_2 to a decimal number

$$101.0101_2 = 1\times 2^2 + 0\times 2^1 + 1\times 2^0 + 0\times 2^{-1}$$

$$+ 1\times 2^{-2} + 0\times 2^{-3} + 1\times 2^{-4}$$

$$= 4 + 0 + 1 + 0 + 0.25 + 0 + 0.0625 = \mathbf{5.3125_{10}}$$

Conversion of decimal to binary

An integer decimal number can be converted to a corresponding binary number by repeatedly dividing by 2 and noting the remainder at each stage, as shown below.

Application: Change 39_{10} into binary

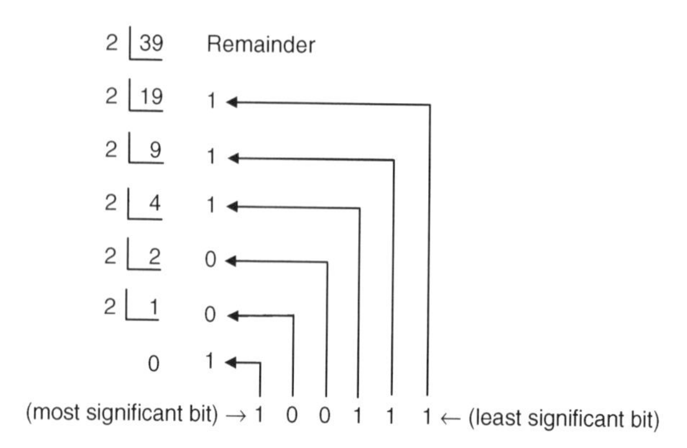

The result is obtained by writing the top digit of the remainder as the least significant bit, (a bit is a **b**inary dig**it** and the least significant bit is the one on the right). The bottom bit of the remainder is the most significant bit, i.e. the bit on the left.

Thus, $39_{10} = 100111_2$

Application: Change 0.625 in decimal into binary form

The fractional part of a denary number can be converted to a binary number by repeatedly multiplying by 2, as shown below for the fraction 0.625

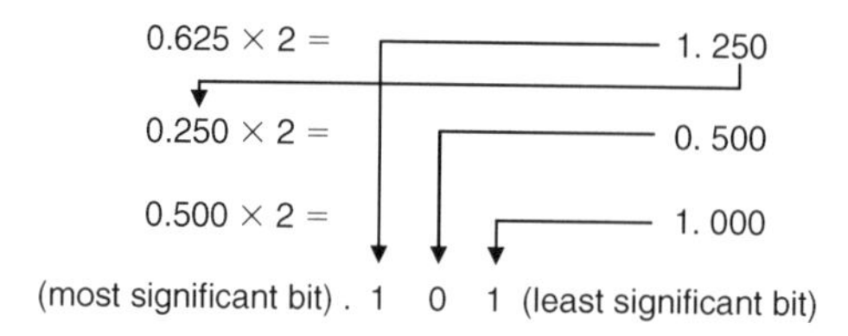

For fractions, the most significant bit of the result is the top bit obtained from the integer part of multiplication by 2. The least significant bit of the result is the bottom bit obtained from the integer part of multiplication by 2.

Thus $0.625_{10} = 0.101_2$

Conversion of decimal to binary via octal

For denary integers containing several digits, repeatedly dividing by 2 can be a lengthy process. In this case, it is usually easier to convert a denary number to a binary number via the octal system of numbers. This system has a radix of 8, using the digits 0, 1, 2, 3, 4, 5, 6 and 7.

Application: Find the decimal number equivalent to the octal number 4317_8

$$4317_8 = 4 \times 8^3 + 3 \times 8^2 + 1 \times 8^1 + 7 \times 8^0$$
$$= 4 \times 512 + 3 \times 64 + 1 \times 8 + 7 \times 1 = 2255_{10}$$

Thus, $\mathbf{4317_8 = 2255_{10}}$

Application: Convert 493_{10} into octal

An integer decimal number can be converted to a corresponding octal number by repeatedly dividing by 8 and noting the remainder at each stage.

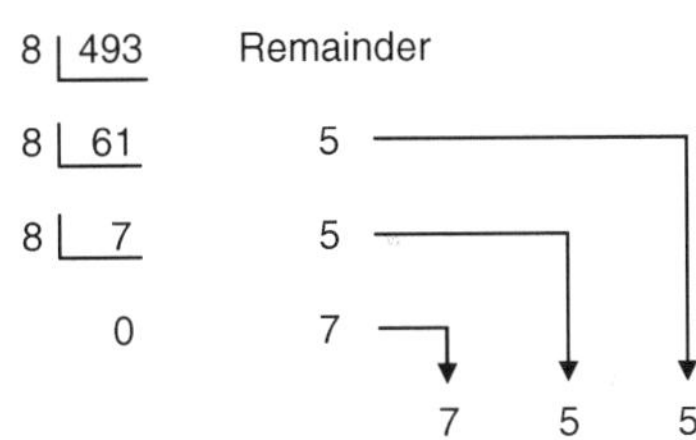

Thus, $\mathbf{493_{10} = 755_8}$

Application: Convert 0.4375_{10} into octal

The fractional part of a denary number can be converted to an octal number by repeatedly multiplying by 8, as shown below.

$0.4375 \times 8 = 3.5$

$0.5 \times 8 = 4.0$

.3 4

For fractions, the most significant bit is the top integer obtained by multiplication of the denary fraction by 8, thus

$\mathbf{0.4375_{10} = 0.34_8}$

Conversion of octal to binary and decimal

The natural binary code for digits 0 to 7 is shown in Table 3.1, and an octal number can be converted to a binary number by writing down the three bits corresponding to the octal digit.

Table 3.1

Octal digit	Natural binary number
0	000
1	001
2	010
3	011
4	100
5	101
6	110
7	111

Application: Change 437_8 into binary

From Table 3.1, $\mathbf{437_8 = 100\ 011\ 111_2}$

Application: Change 26.35_8 into binary

From Table 3.1, $26.35_8 = 010\ 110.011\ 101_2$

The '0' on the extreme left does not signify anything, thus

$$\mathbf{26.35_8 = 10\ 110.011\ 101_2}$$

Application: Convert $11\ 110\ 011.100\ 01_2$ to a decimal number via octal

Grouping the binary number in three's from the binary point gives:

$$011\ 110\ 011.100\ 010_2$$

Using Table 3.1 to convert this binary number to an octal number gives:

363.42_8 and

$363.42_8 = 3 \times 8^2 + 6 \times 8^1 + 3 \times 8^0 + 4 \times 8^{-1} + 2 \times 8^{-2}$

$= 192 + 48 + 3 + 0.5 + 0.03125 = \mathbf{243.53125_{10}}$

Hence, $\mathbf{11\ 110\ 011.100\ 01_2 = 363.42_8 = 243.53125_{10}}$

Hexadecimal numbers

A **hexadecimal numbering system** has a radix of 16 and uses the following 16 distinct digits:

0, 1, 2, 3, 4, 5, 6, 7, 8, 9, A, B, C, D, E and F

'A' corresponds to 10 in the denary system, B to 11, C to 12, and so on.

Table 3.2 compares decimal, binary, octal and hexadecimal numbers.

Table 3.2

Decimal	Binary	Octal	Hexadecimal
0	0000	0	0
1	0001	1	1
2	0010	2	2
3	0011	3	3
4	0100	4	4
5	0101	5	5
6	0110	6	6
7	0111	7	7
8	1000	10	8
9	1001	11	9

Table 3.2 Continued

Decimal	Binary	Octal	Hexadecimal
10	1010	12	A
11	1011	13	B
12	1100	14	C
13	1101	15	D
14	1110	16	E
15	1111	17	F
16	10000	20	10
17	10001	21	11
18	10010	22	12
19	10011	23	13
20	10100	24	14
21	10101	25	15
22	10110	26	16
23	10111	27	17
24	11000	30	18
25	11001	31	19
26	11010	32	1A
27	11011	33	1B
28	11100	34	1C
29	11101	35	1D
30	11110	36	1E
31	11111	37	1F
32	100000	40	20

For example, $\mathbf{23_{10} = 10111_2 = 27_8 = 17_{16}}$

Conversion from hexadecimal to decimal

Application: Change $1A_{16}$ into decimal form

$1A_{16} = 1 \times 16^1 + A \times 16^0 = 1 \times 16^1 + 10 \times 1 = 16 + 10 = 26$

i.e. $\mathbf{1A_{16} = 26_{10}}$

Application: Change $2E_{16}$ into decimal form

$$\mathbf{2E_{16}} = 2 \times 16^1 + E \times 16^0 = 2 \times 16^1 + 14 \times 16^0 = 32 + 14 = \mathbf{46_{10}}$$

Application: Change $1BF_{16}$ into decimal form

$$\begin{aligned}\mathbf{1BF_{16}} &= 1 \times 16^2 + B \times 16^1 + F \times 16^0 \\ &= 1 \times 16^2 + 11 \times 16^1 + 15 \times 16^0 \\ &= 256 + 176 + 15 = \mathbf{447_{10}}\end{aligned}$$

Application: Convert $1A4E_{16}$ into a decimal number

$$\begin{aligned}1A4E_{16} &= 1 \times 16^3 + A \times 16^2 + 4 \times 16^1 + E \times 16^0 \\ &= 1 \times 16^3 + 10 \times 16^2 + 4 \times 16^1 + 14 \times 16^0 \\ &= 1 \times 4096 + 10 \times 256 + 4 \times 16 + 14 \times 1 \\ &= 4096 + 2560 + 64 + 14 = 6734\end{aligned}$$

Thus, $\mathbf{1A4E_{16} = 6734_{10}}$

Conversion from decimal to hexadecimal

This is achieved by repeatedly dividing by 16 and noting the remainder at each stage, as shown below.

Application: Change 26_{10} into hexadecimal

```
16|26   Remainder
16| 1    10 ≡ A16 ──┐
   0      1 ≡ 116 ┐ │
                  ↓ ↓
most significant bit → 1  A  ← least significant bit
```

Hence, $\mathbf{26_{10} = 1A_{16}}$

Application: Change 447_{10} into hexadecimal

$$\begin{array}{r|l l}
16 & 447 & \text{Remainder} \\
16 & 27 & 15 \equiv F_{16} \\
16 & 1 & 11 \equiv B_{16} \\
 & 0 & 1 \equiv 1_{16} \\
 & & \quad 1 \; B \; F
\end{array}$$

Thus, $\mathbf{447_{10} = 1BF_{16}}$

Conversion from binary to hexadecimal

The binary bits are arranged in groups of four, starting from right to left, and a hexadecimal symbol is assigned to each group.

Application: Convert the binary number 1110011110101001 into hexadecimal

The binary number 1110011110101001 is initially grouped in fours as:

1110 0111 1010 1001

and a hexadecimal symbol assigned to each group as:

E 7 A 9

from Table 3.2

Hence, $\mathbf{1110011110101001_2 = E7A9_{16}}$

Conversion from hexadecimal to binary

Application: Convert $6CF3_{16}$ into binary form

$6CF3_{16} = 0110\ 1100\ 1111\ 0011$ from Table 3.2

i.e. $\mathbf{6CF3_{16} = 110110011110011_2}$

---

4 Areas and Volumes

4.1 Areas of plane figures

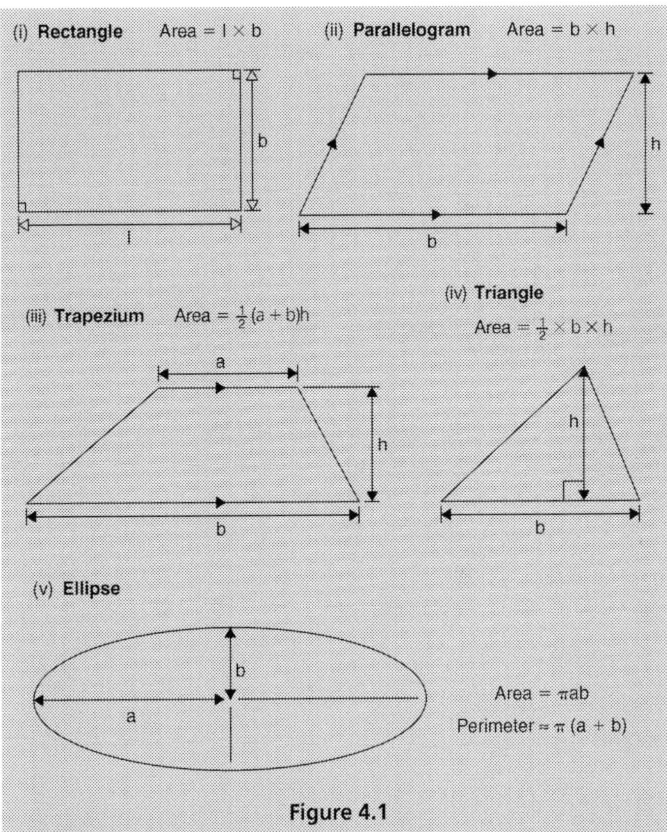

Figure 4.1

A **polygon** is a closed plane figure bounded by straight lines. A polygon, which has:

(i) 3 sides is called a **triangle**
(ii) 4 sides is called a **quadrilateral**

(iii) 5 sides is called a **pentagon**
(iv) 6 sides is called a **hexagon**
(v) 7 sides is called a **heptagon**
(vi) 8 sides is called an **octagon**

Application: Find (a) the cross-sectional area of the girder shown in Figure 4.2(a), and (b) the area of the path shown in Figure 4.2(b)

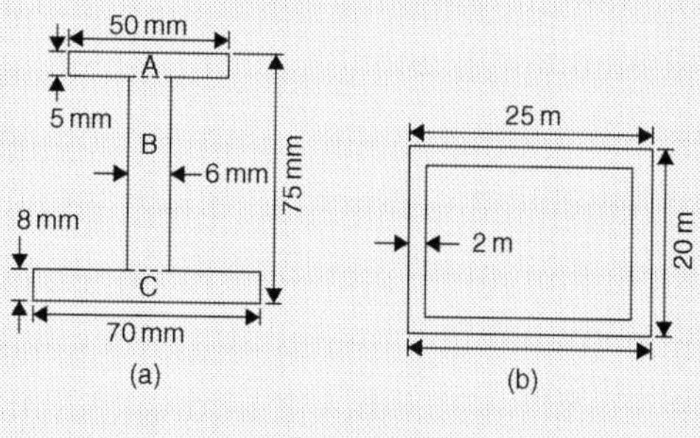

Figure 4.2

(a) The girder may be divided into three separate rectangles as shown.
Area of rectangle A $= 50 \times 5 = 250\,\text{mm}^2$
Area of rectangle B $= (75 - 8-5) \times 6 = 62 \times 6 = 372\,\text{mm}^2$
Area of rectangle C $= 70 \times 8 = 560\,\text{mm}^2$
Total area of girder $= 250 + 372 + 560 = \mathbf{1182\,mm^2}$ or $\mathbf{11.82\,cm^2}$

(b) Area of path = area of large rectangle − area of small rectangle
$= (25 \times 20) - (21 \times 16) = 500 - 336$
$= \mathbf{164\,m^2}$

Application: Figure 4.3 shows the gable end of a building. Determine the area of brickwork in the gable end

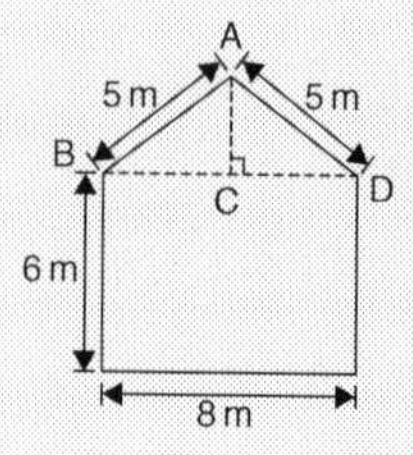

Figure 4.3

The shape is that of a rectangle and a triangle.

Area of rectangle $= 6 \times 8 = 48\,\text{m}^2$

Area of triangle $= \dfrac{1}{2} \times \text{base} \times \text{height}$

CD = 4 m, AD = 5 m, hence AC = 3 m (since it is a 3, 4, 5 triangle)

Hence, area of triangle ABD $= \dfrac{1}{2} \times 8 \times 3 = 12\,\text{m}^2$

Total area of brickwork $= 48 + 12 =$ $\mathbf{60\,m^2}$

Application: Calculate the area of a regular octagon, if each side is 5 cm and the width across the flats is 12 cm

An octagon is an 8-sided polygon. If radii are drawn from the centre of the polygon to the vertices then 8 equal triangles are produced (see Figure 4.4).

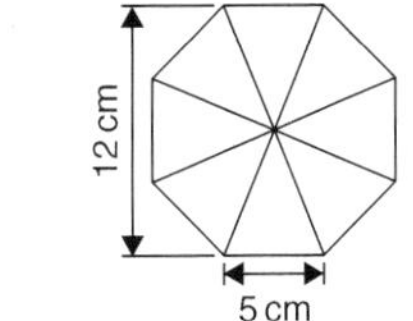

Figure 4.4

$$\text{Area of one triangle} = \frac{1}{2} \times \text{base} \times \text{height}$$

$$= \frac{1}{2} \times 5 \times \frac{12}{2} = 15\,\text{cm}^2$$

Area of octagon $= 8 \times 15 =$ $\mathbf{120\,cm^2}$

Application: Determine the area of a regular hexagon which has sides 8 cm long

A hexagon is a 6-sided polygon that may be divided into 6 equal triangles as shown in Figure 4.5. The angle subtended at the centre of each triangle is $360°/6 = 60°$.

The other two angles in the triangle add up to 120° and are equal to each other.

Hence each of the triangles is equilateral with each angle 60° and each side 8 cm.

Area of one triangle $= \dfrac{1}{2} \times \text{base} \times \text{height} = \dfrac{1}{2} \times 8 \times h$

h is calculated using Pythagoras' theorem:

$$8^2 = h^2 + 4^2$$

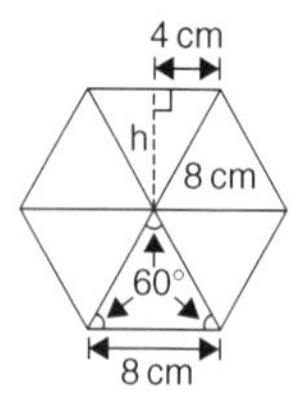

Figure 4.5

from which, $h = \sqrt{8^2 - 4^2} = 6.928\,\text{cm}$

Hence area of one triangle $= \frac{1}{2} \times 8 \times 6.928 = 27.71\,\text{cm}^2$

Area of hexagon $= 6 \times 27.71 =$ **166.3 cm²**

Areas of similar shapes

The areas of similar shapes are proportional to the squares of corresponding linear dimensions.

For example, Figure 4.6 shows two squares, one of which has sides three times as long as the other.

Area of Figure 4.6(a) $= (x)(x) = x^2$

Area of Figure 4.6(b) $= (3x)(3x) = 9x^2$

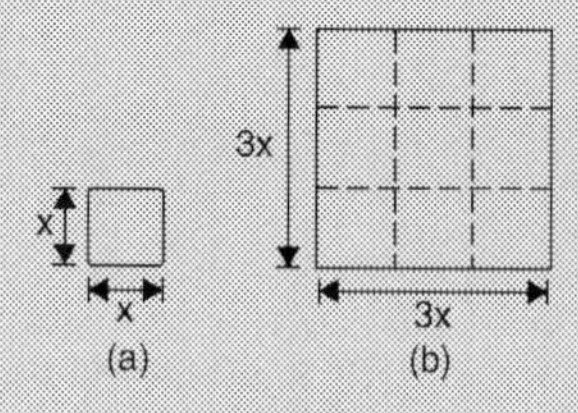

Figure 4.6

Hence Figure 4.6(b) has an area $(3)^2$, i.e. 9 times the area of Figure 4.6(a).

Application: A rectangular garage is shown on a building plan having dimensions 10 mm by 20 mm. If the plan is drawn to a scale of 1 to 250, determine the true area of the garage in square metres

Area of garage on the plan $= 10\,\text{mm} \times 20\,\text{mm} = 200\,\text{mm}^2$

Since the areas of similar shapes are proportional to the squares of corresponding dimensions then

true area of garage $= 200 \times (250)^2 = 12.5 \times 10^6\,\text{mm}^2$

$$= \frac{12.5 \times 10^6}{10^6}\,\text{m}^2 = \mathbf{12.5\,m^2}$$

4.2 Circles

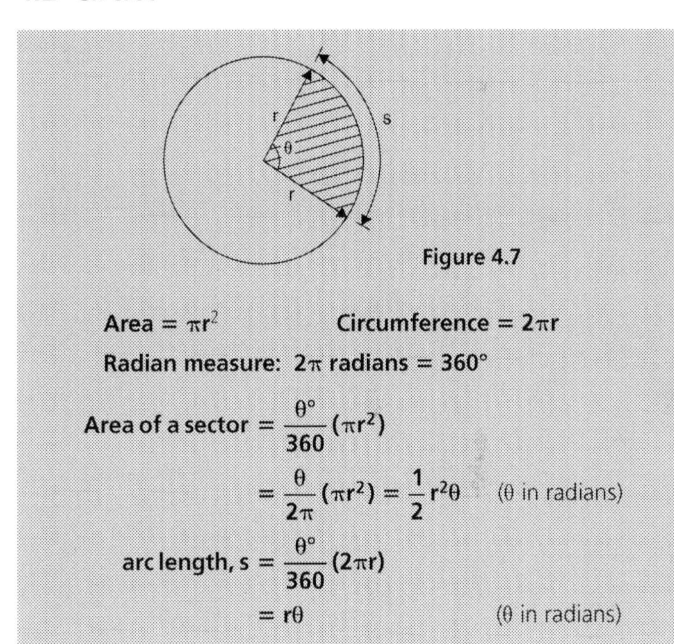

Figure 4.7

Area $= \pi r^2$ **Circumference** $= 2\pi r$

Radian measure: 2π **radians** $= 360°$

$$\textbf{Area of a sector} = \frac{\theta°}{360}(\pi r^2)$$

$$= \frac{\theta}{2\pi}(\pi r^2) = \frac{1}{2}r^2\theta \quad (\theta \text{ in radians})$$

$$\textbf{arc length, } s = \frac{\theta°}{360}(2\pi r)$$

$$= r\theta \quad (\theta \text{ in radians})$$

Application: Find the areas of the circles having (a) a radius of 5 cm, (b) a diameter of 15 mm, (c) a circumference of 70 mm

Area of a circle $= \pi r^2$ or $\dfrac{\pi d^2}{4}$

(a) Area $= \pi r^2 = \pi\,(5)^2 = 25\pi =$ **78.54 cm²**

(b) Area $= \dfrac{\pi d^2}{4} = \dfrac{\pi(15)^2}{4} = \dfrac{225\pi}{4} =$ **176.7 mm²**

(c) Circumference, $c = 2\pi r$, hence $r = \dfrac{c}{2\pi} = \dfrac{70}{2\pi} = \dfrac{35}{\pi}$ mm

$$\text{Area of circle} = \pi r^2 = \pi\left(\frac{35}{\pi}\right)^2 = \frac{35^2}{\pi}$$

$= \mathbf{389.9\ mm^2}$ or $\mathbf{3.899\ cm^2}$

Application: A hollow shaft has an outside diameter of 5.45 cm and an inside diameter of 2.25 cm. Calculate the cross-sectional area of the shaft

The cross-sectional area of the shaft is shown by the shaded part in Figure 4.8 (often called an **annulus**).

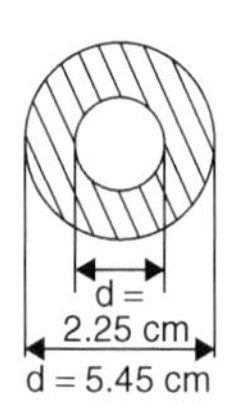

Figure 4.8

Area of shaded part = area of large circle − area of small circle

$$= \frac{\pi D^2}{4} - \frac{\pi d^2}{4} = \frac{\pi}{4}(D^2 - d^2)$$

$$= \frac{\pi}{4}(5.45^2 - 2.25^2) = \mathbf{19.35\ cm^2}$$

Application: Convert (a) 125° (b) 69°47′ to radians

(a) Since 180° = π rad then 1° = $\pi/180$ rad, therefore

$$125° = 125\left(\frac{\pi}{180}\right)\text{rads} = \mathbf{2.182\ radians}$$

(b) $69°47' = 69\frac{47°}{60} = 69.783°$

$$69.783° = 69.783\left(\frac{\pi}{180}\right) = \mathbf{1.218\ radians}$$

Application: Convert (a) 0.749 radians, (b) $3\pi/4$ radians, to degrees and minutes

(a) Since π rad $= 180°$ then 1 rad $= 180°/\pi$, therefore

$$0.749 = 0.749\left(\frac{180}{\pi}\right)^{\circ} = 42.915°$$

$0.915° = (0.915 \times 60)' = 55'$, correct to the nearest minute,

hence **0.749 radians = 42°55′**

(b) Since 1 rad $= \left(\frac{180}{\pi}\right)^{\circ}$

then $\frac{3\pi}{4}$ rad $= \frac{3\pi}{4}\left(\frac{180}{\pi}\right)^{\circ} = \frac{3}{4}(180)^{\circ} = \mathbf{135°}$

Application: Find the length of arc of a circle of radius 5.5 cm when the angle subtended at the centre is 1.20 radians

Length of arc, $s = r\theta$, where θ is in radians, hence

$$s = (5.5)(1.20) = \mathbf{6.60\ cm}$$

Application: Determine the diameter and circumference of a circle if an arc of length 4.75 cm subtends an angle of 0.91 radians

Since $s = r\theta$ then $r = \frac{s}{\theta} = \frac{4.75}{0.91} = 5.22$ cm

Diameter $= 2 \times$ radius $= 2 \times 5.22 =$ **10.44 cm**

Circumference, $c = \pi d = \pi(10.44) =$ **32.80 cm**

Application: Determine the angle, in degrees and minutes, subtended at the centre of a circle of diameter 42 mm by an arc of length 36 mm and the area of the minor sector formed

Since length of arc, $s = r\theta$ then $\theta = s/r$

Radius, $r = \dfrac{\text{diameter}}{2} = \dfrac{42}{2} = 21\,\text{mm}$

hence $\theta = \dfrac{s}{r} = \dfrac{36}{21} = 1.7143\,\text{radians}$

$1.7143\ \text{rad} = 1.7143 \times (180/\pi)^\circ = 98.22^\circ = \mathbf{98^\circ 13'}$

= angle subtended at centre of circle

Area of sector $= \dfrac{1}{2}r^2\theta = \dfrac{1}{2}(21)^2(1.7143) = \mathbf{378\ mm^2}$

Application: A football stadium floodlight can spread its illumination over an angle of 45° to a distance of 55 m. Determine the maximum area that is floodlit

Floodlit area = area of sector $= \dfrac{1}{2}r^2\theta = \dfrac{1}{2}(55)^2\left(45 \times \dfrac{\pi}{180}\right)$

$= \mathbf{1188\ m^2}$

Application: An automatic garden spray produces a spray to a distance of 1.8 m and revolves through an angle α which may be varied. If the desired spray catchment area is to be $2.5\,\text{m}^2$, determine the required angle α, correct to the nearest degree

Area of sector $= \dfrac{1}{2}r^2\theta$, hence $2.5 = \dfrac{1}{2}(1.8)^2\alpha$

from which, $\alpha = \dfrac{2.5 \times 2}{1.8^2} = 1.5432\,\text{radians}$

$1.5432\,\text{rad} = \left(1.5432 \times \dfrac{180}{\pi}\right)^\circ = 88.42^\circ$

Hence **angle α = 88°**, correct to the nearest degree.

Application: The angle of a tapered groove is checked using a 20 mm diameter roller as shown in Figure 4.9. If the roller lies 2.12 mm below the top of the groove, determine the value of angle θ

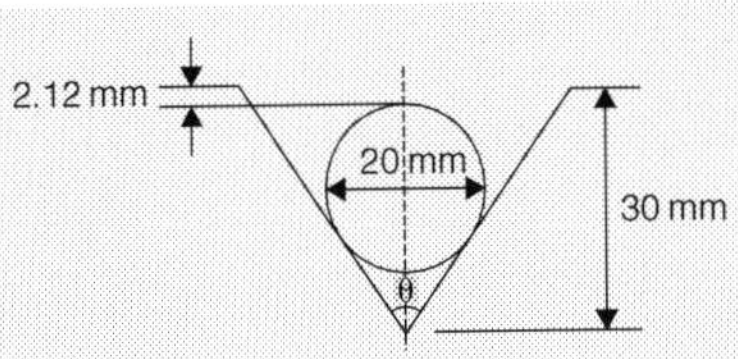

Figure 4.9

In Figure 4.10, triangle ABC is right-angled at C
Length BC = 10 mm (i.e. the radius of the circle),
and AB = 30 − 10 − 2.12 = 17.88 mm from Figure 4.10.

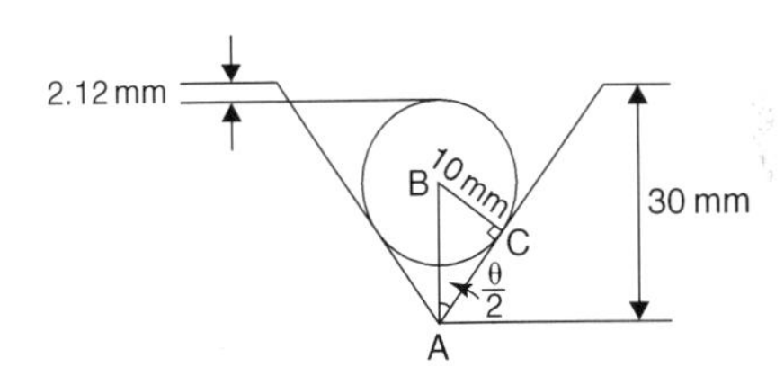

Figure 4.10

Hence,

$$\sin\frac{\theta}{2} = \frac{10}{17.88} \text{ and } \frac{\theta}{2} = \sin^{-1}\left(\frac{10}{17.88}\right) = 34° \text{ and } \textbf{angle } \boldsymbol{\theta} \textbf{ = 68°}$$

The equation of a circle

The equation of a circle, centre at the origin, radius r, is given by:

$$\mathbf{x^2 + y^2 = r^2}$$

The equation of a circle, centre (a, b), radius r, is given by:

$$\mathbf{(x - a)^2 + (y - b)^2 = r^2}$$

Figure 4.11 shows a circle
$(x - 2)^2 + (y - 3)^2 = 4$

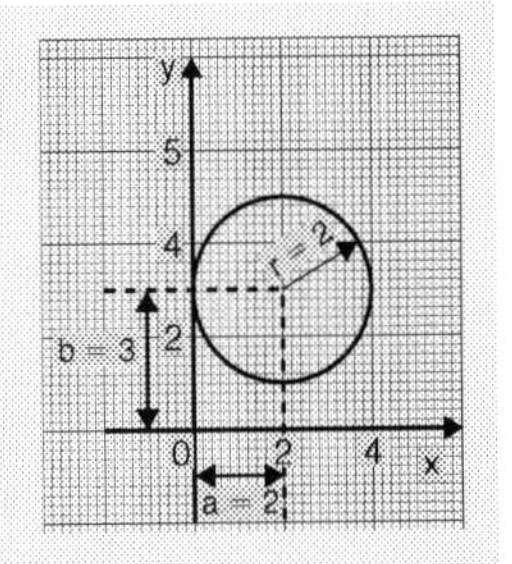

Figure 4.11

Application: Determine the radius and the co-ordinates of the centre of the circle given by the equation $x^2 + y^2 + 8x - 2y + 8 = 0$

$x^2 + y^2 + 8x - 2y + 8 = 0$ may be rearranged as:

$$(x + 4)^2 + (y - 1)^2 - 9 = 0$$

i.e.
$$(x + 4)^2 + (y - 1)^2 = 3^2$$

which represents a circle, **centre (−4, 1)** and **radius 3** as shown in Figure 4.12.

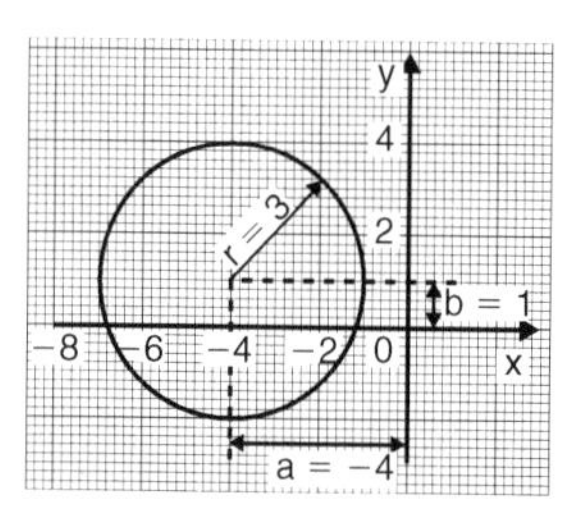

Figure 4.12

4.3 Volumes and surface areas of regular solids

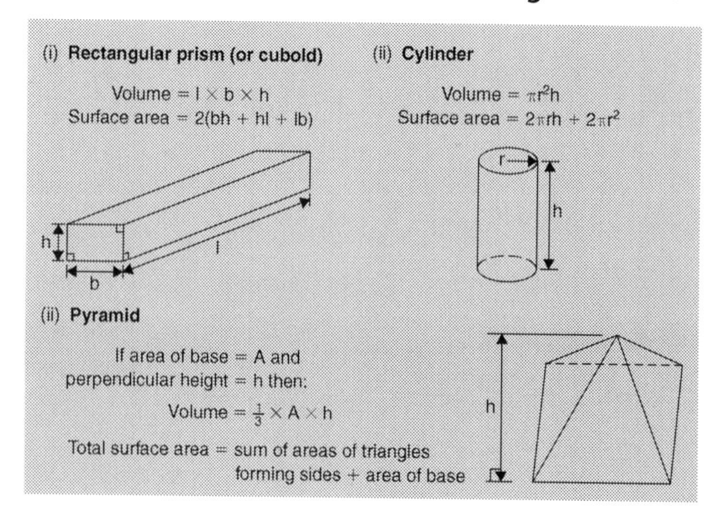

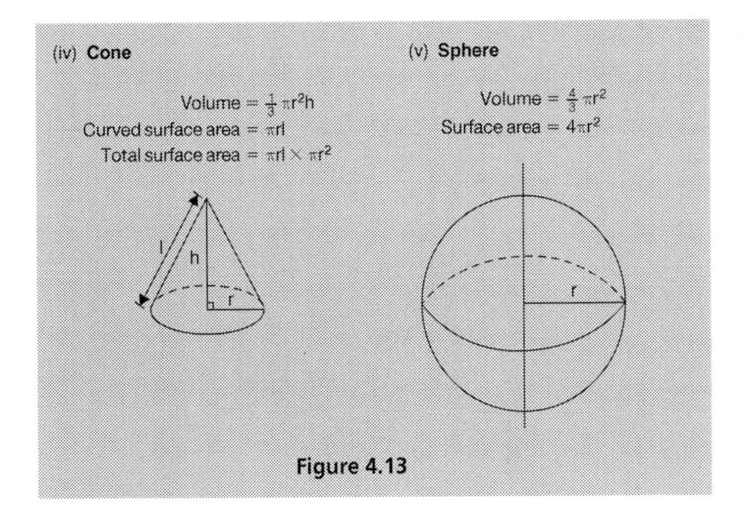

Figure 4.13

Application: A water tank is the shape of a rectangular prism having length 2 m, breadth 75 cm and height 50 cm. Determine the capacity of the tank in (a) m^3 (b) cm^3 (c) litres

Volume of rectangular prism = $l \times b \times h$

(a) **Volume of tank** = $2 \times 0.75 \times 0.5 =$ **0.75 m^3**

(b) $1\,m^3 = 10^6\,cm^3$, hence $0.75\,m^3 = 0.75 \times 10^6\,cm^3 =$ **750000 cm^3**

(c) $1\,\text{litre} = 1000\,cm^3$, hence $750000\,cm^3 = \frac{750000}{1000}$ litres

= **750 litres**

Application: Find the volume and total surface area of a cylinder of length 15 cm and diameter 8 cm

Volume of cylinder = $\pi r^2 h$

Since diameter = 8 cm, then radius r = 4 cm.

Hence, **volume** = $\pi \times 4^2 \times 15 =$ **754 cm^3**

Total surface area = $2\pi rh + 2\pi r^2$ (i.e. including the two ends)

$= (2 \times \pi \times 4 \times 15) + (2 \times \pi \times 4^2) =$ **477.5 cm^2**

Application: Determine the volume and the total surface area of the square pyramid shown in Figure 4.14 if its perpendicular height is 12 cm

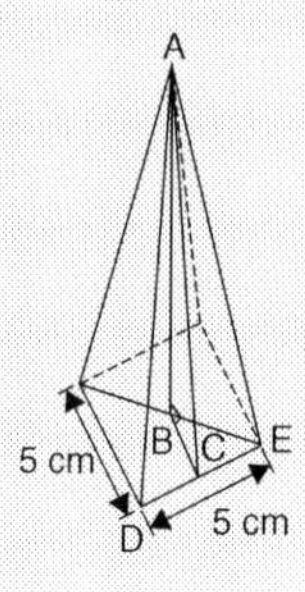

Figure 4.14

Volume of pyramid $= \frac{1}{3}$(area of base) × perpendicular height

$= \frac{1}{3}(5 \times 5) \times 12 = \mathbf{100\ cm^3}$

The total surface area consists of a square base and 4 equal triangles.

Area of triangle ADE $= \frac{1}{2} \times$ base × perpendicular height

$= \frac{1}{2} \times 5 \times AC$

The length AC may be calculated using Pythagoras' theorem on triangle ABC, where

$AB = 12\ cm,\ BC = \frac{1}{2} \times 5 = 2.5\ cm,$

and $AC = \sqrt{AB^2 + BC^2} = \sqrt{12^2 + 2.5^2} = 12.26\ cm$

Hence area of triangle ADE $= \frac{1}{2} \times 5 \times 12.26 = 30.65\ cm^2$

Total surface area of pyramid $= (5 \times 5) + 4(30.65) = \mathbf{147.6\ cm^2}$

Application: Determine the volume and total surface area of a cone of radius 5 cm and perpendicular height 12 cm

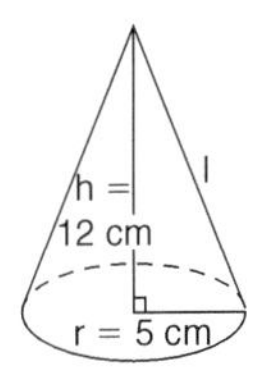

Figure 4.15

The cone is shown in Figure 4.15.

Volume of cone $= \frac{1}{3}\pi r^2 h = \frac{1}{3} \times \pi \times 5^2 \times 12 =$ **314.2 cm³**

Total surface area = curved surface area + area of base $= \pi r l + \pi r^2$

From Figure 4.15, slant height l may be calculated using Pythagoras' theorem: $1 = \sqrt{12^2 + 5^2} = 13$ cm
Hence, **total surface area** $= (\pi \times 5 \times 13) + (\pi \times 5^2) =$ **282.7 cm²**

Application: A wooden section is shown in Figure 4.16. Find (a) its volume (in m³), and (b) its total surface area

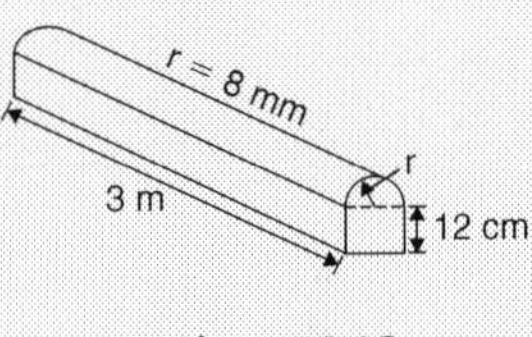

Figure 4.16

The section of wood is a prism whose end comprises a rectangle and a semicircle.

Since the radius of the semicircle is 8 cm, the diameter is 16 cm.

Hence the rectangle has dimensions 12 cm by 16 cm.

Area of end $= (12 \times 16) + \frac{1}{2}\pi 8^2 = 292.5$ cm²

Volume of wooden section = area of end × perpendicular height

$= 292.5 \times 300 = 87750$ cm³

$= \frac{87750 \text{ m}^3}{10^6} =$ **0.08775 m³**

The total surface area comprises the two ends (each of area 292.5 cm^2), three rectangles and a curved surface (which is half a cylinder), hence

$$\text{total surface area} = (2 \times 292.5) + 2(12 \times 300) + (16 \times 300) + \frac{1}{2}(2\pi \times 8 \times 300)$$

$$= 585 + 7200 + 4800 + 2400\pi$$

$$= \mathbf{20125\ cm^2} \text{ or } \mathbf{2.0125\ m^2}$$

Application: A rivet consists of a cylindrical head, of diameter 1 cm and depth 2 mm, and a shaft of diameter 2 mm and length 1.5 cm. Determine the volume of metal in 2000 such rivets

Radius of cylindrical head $= \frac{1}{2}$ cm $= 0.5$ cm

and height of cylindrical head 2 mm $= 0.2$ cm

Hence, volume of cylindrical head $= \pi r^2 h = \pi(0.5)^2(0.2) = 0.1571\ cm^3$

Volume of cylindrical shaft $= \pi r^2 h = \pi\left(\frac{0.2}{2}\right)^2(1.5) = 0.0471\ cm^3$

Total volume of 1 rivet $= 0.1571 + 0.0471 = 0.2042\ cm^3$

Volume of metal in 2000 such rivets $= 2000 \times 0.2042 =$ **408.4 cm^3**

Application: A boiler consists of a cylindrical section of length 8 m and diameter 6 m, on one end of which is surmounted a hemispherical section of diameter 6 m, and on the other end a conical section of height 4 m and base diameter 6 m. Calculate the volume of the boiler and the total surface area

The boiler is shown in Figure 4.17.

Volume of hemisphere, $P = \frac{2}{3}\pi r^3 = \frac{2}{3} \times \pi \times 3^3 = 18\pi\ m^3$

Volume of cylinder, $Q = \pi r^2 h = \pi \times 3^2 \times 8 = 72\pi\ m^3$

Volume of cone, $R = \frac{1}{3}\pi r^2 h = \frac{1}{3} \times \pi \times 3^2 \times 4 = 12\pi\ m^3$

Total volume of boiler $= 18\pi + 72\pi + 12\pi = 102\pi =$ **320.4 m^3**

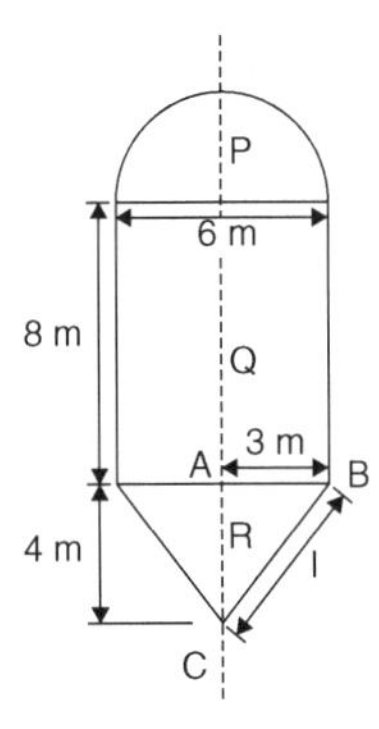

Figure 4.17

Surface area of hemisphere, $P = \frac{1}{2}(4\pi r^2) = 2 \times \pi \times 3^2 = 18\pi\ \text{m}^2$

Curved surface area of cylinder, $Q = 2\pi rh = 2 \times \pi \times 3 \times 8 = 48\pi\ \text{m}^2$

The slant height of the cone, l, is obtained by Pythagoras' theorem on triangle ABC, i.e. $l = \sqrt{(4^2 + 3^2)} = 5$

Curved surface area of cone, $R = \pi rl = \pi \times 3 \times 5 = 15\pi\ \text{m}^2$

Total surface area of boiler $= 18\pi + 48\pi + 15\pi = 81\pi =$ **254.5 m²**

Volumes of similar shapes

The volumes of similar bodies are proportional to the cubes of corresponding linear dimensions.

For example, Figure 4.18 shows two cubes, one of which has sides three times as long as those of the other.

Volume of Figure 4.18(a) $= (x)(x)(x) = x^3$

Volume of Figure 4.18(b) $= (3x)(3x)(3x)$
$= 27x^3$

Hence Figure 4.18(b) has a volume $(3)^3$, i.e. 27 times the volume of Figure 4.18(a).

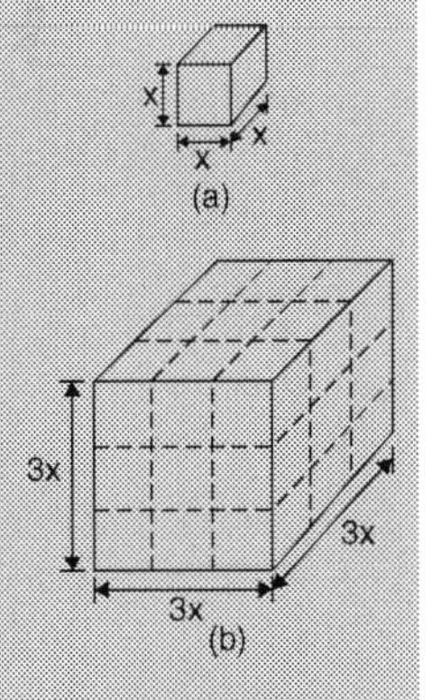

Figure 4.18

Application: A car has a mass of 1000 kg. A model of the car is made to a scale of 1 to 50. Determine the mass of the model if the car and its model are made of the same material

$\dfrac{\text{Volume of model}}{\text{Volume of car}} = \left(\dfrac{1}{50}\right)^3$ since the volume of similar bodies are proportional to the cube of corresponding dimensions.

Mass = density × volume, and since both car and model are made of the same material then:

$$\frac{\text{Mass of model}}{\text{Mass of car}} = \left(\frac{1}{50}\right)^3$$

Hence, **mass of model** $= (\text{mass of car})\left(\dfrac{1}{50}\right)^3 = \dfrac{1000}{50^3}$

$= \mathbf{0.008\ kg}$ or $\mathbf{8\ g}$

4.4 Volumes and surface areas of frusta of pyramids and cones

For the **frustum of a cone** shown in Figure 4.19:

$$\textbf{Volume} = \frac{1}{3}\pi h(R^2 + Rr + r^2)$$

$$\textbf{Curved surface area} = \pi l(R + r)$$

$$\textbf{Total surface area} = \pi l(R + r) + \pi r^2 + \pi R^2$$

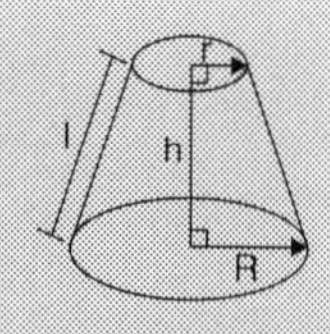

Figure 4.19

Application:

(a) Determine the volume of a frustum of a cone if the diameter of the ends are 6.0 cm and 4.0 cm and its perpendicular height is 3.6 cm.

(b) Find the total surface area of the frustum of the cone

(a) **Method 1**

A section through the vertex of a complete cone is shown in Figure 4.20.

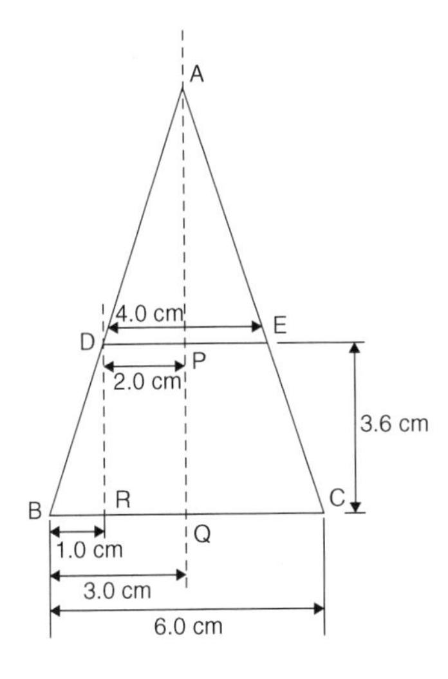

Figure 4.20

Using similar triangles $\frac{AP}{DP} = \frac{DR}{BR}$

Hence $\frac{AP}{2.0} = \frac{3.6}{1.0}$, from which $AP = \frac{(2.0)(3.6)}{1.0} = 7.2\,\text{cm}$

The height of the large cone $= 3.6 + 7.2 = 10.8\,\text{cm}$

Volume of frustum of cone = volume of large cone − volume of small cone cut off

$$= \frac{1}{3}\pi(3.0)^2(10.8) - \frac{1}{3}\pi(2.0)^2(7.2)$$

$$= 101.79 - 30.16 = \mathbf{71.6\,cm^3}$$

Method 2

From above, volume of the frustum of a cone $= \frac{1}{3}\pi h(R^2 + Rr + r^2)$, where $R = 3.0\,\text{cm}$, $r = 2.0\,\text{cm}$ and $h = 3.6\,\text{cm}$

Hence, **volume of frustum** $= \frac{1}{3}\pi(3.6)[(3.0)^2 + (3.0)(2.0) + (2.0)^2]$

$= \frac{1}{3}\pi(3.6)(19.0) = \mathbf{71.6\ cm^3}$

(b) **Method 1**

Curved surface area of frustum = curved surface area of large cone − curved surface area of small cone cut off

From Figure 4.20, using Pythagoras' theorem:

$AB^2 = AQ^2 + BQ^2$, from which, $AB = \sqrt{[10.8^2 + 3.0^2]} = 11.21\,cm$

and $AD^2 = AP^2 + DP^2$, from which, $AD = \sqrt{[7.2^2 + 2.0^2]} = 7.47\,cm$

Curved surface area of large cone $= \pi rl = \pi\,(BQ)(AB)$
$= \pi\,(3.0)(11.21) = 105.65\,cm^2$

and curved surface area of small cone $= \pi\,(DP)(AD)$
$= \pi\,(2.0)(7.47) = 46.94\,cm^2$

Hence, curved surface area of frustum $= 105.65 - 46.94 = 58.71\,cm^2$

Total surface area of frustum = curved surface area + area of two circular ends

$= 58.71 + \pi(2.0)^2 + \pi(3.0)^2$

$= 58.71 + 12.57 + 28.27$

$= \mathbf{99.6\,cm^2}$

Method 2

Total surface area of frustum $= \pi l(R + r) + \pi r^2 + \pi R^2$,

where $l = BD = 11.21 - 7.47 = 3.74\,cm$, $R = 3.0\,cm$ and $r = 2.0\,cm$.

Hence, **total surface area of frustum**

$= \pi(3.74)(3.0 + 2.0) + \pi(2.0)^2 + \pi(3.0)^2 = \mathbf{99.6\,cm^2}$

Application: A lampshade is in the shape of a frustum of a cone. The vertical height of the shade is 25.0 cm and the diameters of the ends are 20.0 cm and 10.0 cm, respectively. Determine the area of the material needed to form the lampshade, correct to 3 significant figures

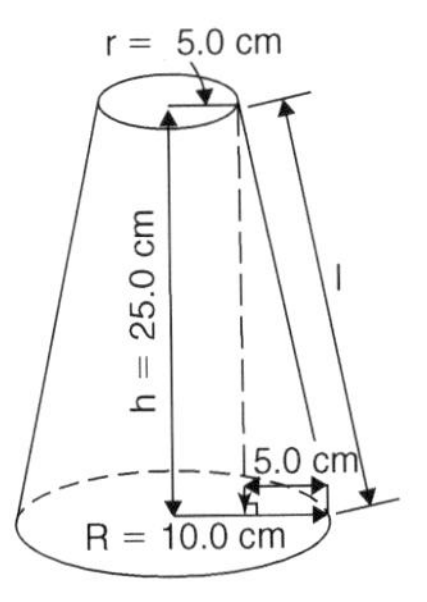

Figure 4.21

The curved surface area of a frustum of a cone $= \pi l(R + r)$

Since the diameters of the ends of the frustum are 20.0 cm and 10.0 cm, then from Figure 4.21, $r = 5.0$ cm, $R = 10.0$ cm and $l = \sqrt{[25.0^2 + 5.0^2]} = 25.50$ cm, from Pythagoras' theorem.

Hence curved surface area $= \pi(25.50)(10.0 + 5.0) = 1201.7\,\text{cm}^2$, i.e. **the area of material needed to form the lampshade is 1200 cm²**, correct to 3 significant figures.

Application: A cooling tower is in the form of a cylinder surmounted by a frustum of a cone as shown in Figure 4.22. Determine the volume of air space in the tower if 40% of the space is used for pipes and other structures

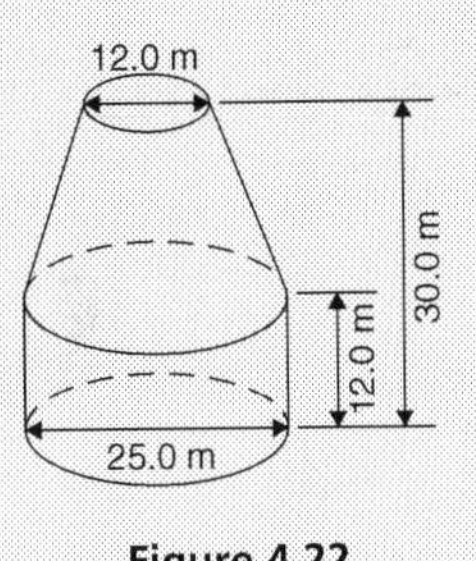

Figure 4.22

Volume of cylindrical protion $= \pi r^2 h = \pi\left(\dfrac{25.0}{2}\right)^2 (12.0) = 5890\,\text{m}^3$

Volume of frustum of cone $= \dfrac{1}{3}\pi h(R^2 + Rr + r^2)$

where $h = 30.0 - 12.0 = 18.0$ m, $R = 25.0/2 = 12.5$ m and $r = 12.0/2 = 6.0$ m.

Hence volume of frustum of cone

$$= \frac{1}{3}\pi(18.0)\,[(12.5)^2 + (12.5)(6.0) + (6.0)^2]$$
$$= 5038\,\text{m}^3$$

Total volume of cooling tower $= 5890 + 5038 = 10928\,\text{m}^3$

If 40% of space is occupied then

volume of air space $= 0.6 \times 10928 =$ **6557 m³**

4.5 The frustum and zone of a sphere

With reference to the **zone of a sphere** shown in Figure 4.23:

Surface area of a zone of a sphere $= 2\pi rh$

Volume of frustum of sphere

$$= \frac{\pi h}{6}(h^2 + 3r_1^2 + 3r_2^2)$$

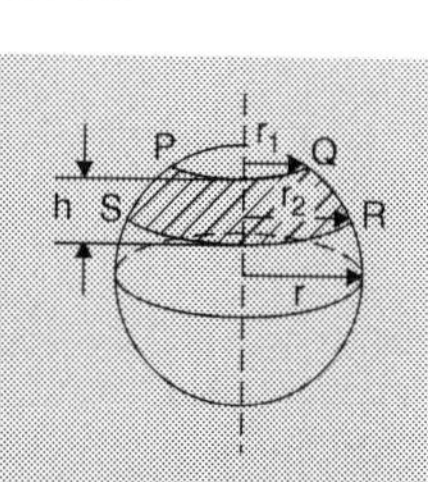

Figure 4.23

Application:

(a) Determine the volume of a frustum of a sphere of diameter 49.74 cm if the diameter of the ends of the frustum are 24.0 cm and 40.0 cm, and the height of the frustum is 7.00 cm.

(b) Determine the curved surface area of the frustum

(a) Volume of frustum of a sphere $= \frac{\pi h}{6}(h^2 + 3r_1^2 + 3r_2^2)$

where $h = 7.00\,\text{cm}$, $r_1 = 24.0/2 = 12.0\,\text{cm}$ and $r_2 = 40.0/2 = 20.0\,\text{cm}$.

Hence,

volume of frustum $= \frac{\pi(7.00)}{6}[(7.00)^2 + 3(12.0)^2 + 3(20.0)^2]$

$=$ **6161 cm³**

(b) The curved surface area of the frustum = surface area of zone = $2\pi rh$ (from above), where r = radius of sphere $= 49.74/2 = 24.87\,\text{cm}$ and $h = 7.00\,\text{cm}$.

Hence, **surface area of zone** $= 2\pi(24.87)(7.00) =$ **1094 cm²**

Application: A frustum of a sphere, of diameter 12.0 cm, is formed by two parallel planes, one through the diameter and the other distance h from the diameter. The curved surface area of the frustum is required to be $\frac{1}{4}$ of the total surface area of the sphere. Determine (a) the volume and surface area of the sphere, (b) the thickness h of the frustum, (c) the volume of the frustum, and (d) the volume of the frustum expressed as a percentage of the sphere

(a) **Volume of sphere**, $V = \frac{4}{3}\pi r^3 = \frac{4}{3}\pi\left(\frac{12.0}{2}\right)^3 = \mathbf{904.8\ cm^3}$

Surface area of sphere $= 4\pi r^2 = 4\pi\left(\frac{12.0}{2}\right)^2 = \mathbf{452.4\ cm^2}$

(b) Curved surface area of frustum $= \frac{1}{4} \times$ surface area of sphere

$$= \frac{1}{4} \times 452.4 = 113.1\,\text{cm}^2$$

From above, $113.1 = 2\pi rh = 2\pi\left(\frac{12.0}{2}\right)h$

Hence, **thickness of frustum**, $h = \frac{113.1}{2\pi(6.0)} = \mathbf{3.0\ cm}$

(c) Volume of frustum, $V = \frac{\pi h}{6}(h^2 + 3r_1^2 + 3r_2^2)$

where $h = 3.0\,\text{cm}$, $r_2 = 6.0\,\text{cm}$ and $r_1 = \sqrt{OQ^2 - OP^2}$, from Figure 4.24,

i.e. $r_1 = \sqrt{6.0^2 - 3.0^2} = 5.196\,\text{cm}$

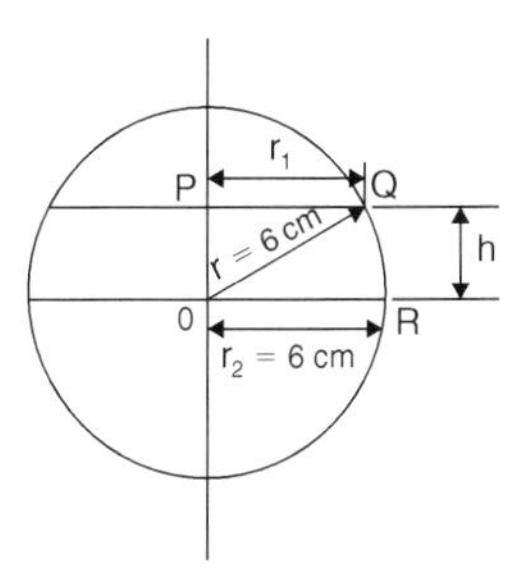

Figure 4.24

Hence,

$$\textbf{volume of frustum} = \frac{\pi(3.0)}{6}[(3.0)^2 + 3(5.196)^2 + 3(6.0)^2]$$

$$= \frac{\pi}{2}[9.0 + 81 + 108.0] = \mathbf{311.0\ cm^3}$$

(d) $\dfrac{\text{Volume of frustum}}{\text{Volume of sphere}} = \dfrac{311.0}{904.8} \times 100\% = \mathbf{34.37\%}$

Application: A spherical storage tank is filled with liquid to a depth of 20 cm. Determine the number of litres of liquid in the container (1 litre = 1000 cm^3), if the internal diameter of the vessel is 30 cm

The liquid is represented by the shaded area in the section shown in Figure 4.25.

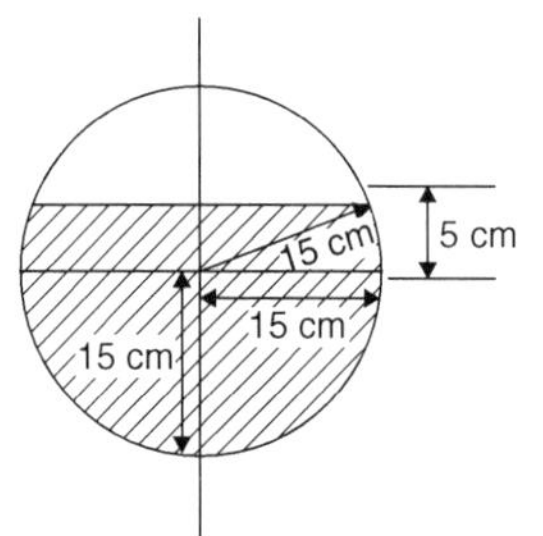

Figure 4.25

The volume of liquid comprises a hemisphere and a frustum of thickness 5 cm.

$$\text{Hence volume of liquid} = \frac{2}{3}\pi r^3 + \frac{\pi h}{6}[h^2 + 3r_1^2 + 3r_2^2]$$

where $r_2 = 30/2 = 15$ cm and $r_1 = \sqrt{15^2 - 5^2} = 14.14$ cm

$$\text{Volume of liquid} = \frac{2}{3}\pi(15)^3 + \frac{\pi(5)}{6}[5^2 + 3(14.14)^2 + 3(15)^2]$$

$$= 7069 + 3403 = 10470\ \text{cm}^3$$

Since 1 litre = 1000 cm^3,

$$\textbf{the number of litres of liquid} = \frac{10470}{1000} = \mathbf{10.47\ litres}$$

4.6 Areas and volumes of irregular figures and solids

Areas of irregular figures

Trapezoidal rule

To determine the areas PQRS in Figure 4.26:

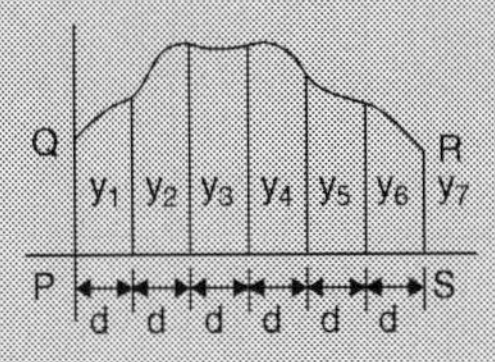

Figure 4.26

(i) Divide base PS into any number of equal intervals, each of width d (the greater the number of intervals, the greater the accuracy)

(ii) Accurately measure ordinates y_1, y_2, y_3, etc.

(iii) $$\text{Area PQRS} = d\left[\frac{y_1 + y_7}{2} + y_2 + y_3 + y_4 + y_5 + y_6\right]$$

In general, the trapezoidal rule states:

$$\textbf{Area} = \textbf{(width of interval)}\left[\frac{1}{2}\textbf{(first + last ordinate) + sum of remaining ordinates}\right]$$

Mid-ordinate rule

To determine the area ABCD of Figure 4.27:

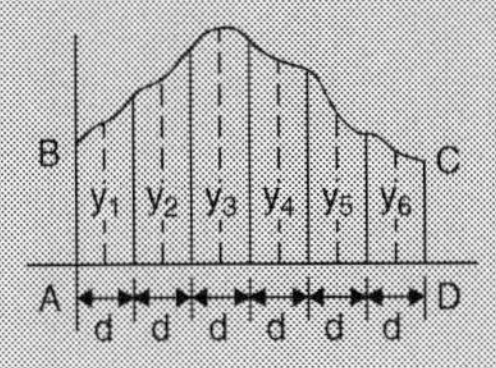

Figure 4.27

(i) Divide base AD into any number of equal intervals, each of width d (the greater the number of intervals, the greater the accuracy)
(ii) Erect ordinates in the middle of each interval (shown by broken lines in Figure 4.27)
(iii) Accurately measure ordinates y_1, y_2, y_3, etc.
(iv) Area ABCD $= d(y_1 + y_2 + y_3 + y_4 + y_5 + y_6)$

In general, the mid-ordinate rule states:

Area = (width of interval)(sum of mid-ordinates)

Simpson's rule

To determine the area PQRS of Figure 4.26:

(i) Divide base PS into an **even** number of intervals, each of width d (the greater the number of intervals, the greater the accuracy)
(ii) Accurately measure ordinates y_1, y_2, y_3, etc.
(iii) Area PQRS $= \frac{d}{3}[(y_1 + y_7) + 4(y_2 + y_4 + y_6) + 2(y_3 + y_5)]$

In general, Simpson's rule states:

$$\textbf{Area} = \frac{1}{3}\begin{pmatrix}\text{width of}\\ \text{interval}\end{pmatrix}\left[\begin{pmatrix}\text{first + last}\\ \text{ordinate}\end{pmatrix} + 4\begin{pmatrix}\text{sum of even}\\ \text{ordinates}\end{pmatrix} + 2\begin{pmatrix}\text{sum of remaining}\\ \text{odd ordinates}\end{pmatrix}\right]$$

Application: A car starts from rest and its speed is measured every second for 6 s:

Time t(s)	0	1	2	3	4	5	6
Speed v (m/s)	0	2.5	5.5	8.75	12.5	17.5	24.0

Determine the distance travelled in 6 seconds (i.e. the area under the v/t graph), by (a) the trapezoidal rule, (b) the mid-ordinate rule, and (c) Simpson's rule

A graph of speed/time is shown in Figure 4.28.

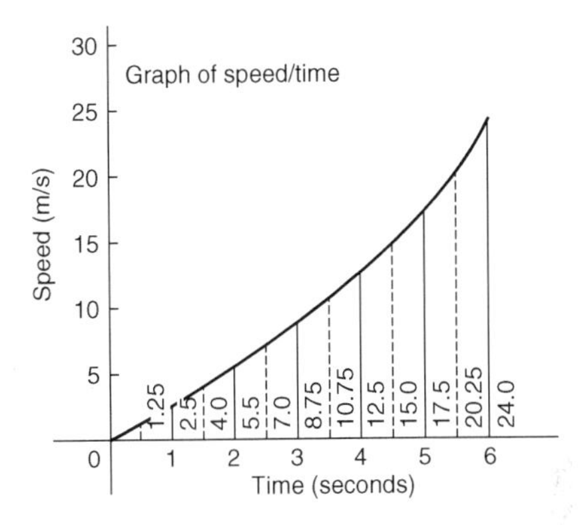

Figure 4.28

(a) **Trapezoidal rule**

The time base is divided into 6 strips each of width 1 s, and the length of the ordinates measured.

Thus

$$\textbf{area} = (1)\left[\left(\frac{0 + 24.0}{2}\right) + 2.5 + 5.5 + 8.75 + 12.5 + 17.5\right]$$

$$= \textbf{58.75 m}$$

(b) **Mid-ordinate rule**

The time base is divided into 6 strips each of width 1 second. Mid-ordinates are erected as shown in Figure 4.28 by the broken lines.

The length of each mid-ordinate is measured. Thus

$$\textbf{area} = (1)[1.25 + 4.0 + 7.0 + 10.75 + 15.0 + 20.25] = \textbf{58.25 m}$$

(c) **Simpson's rule**

The time base is divided into 6 strips each of width 1 s, and the length of the ordinates measured.

Thus,

$$\textbf{area} = \frac{1}{3}(1)[(0 + 24.0) + 4(2.5 + 8.75 + 17.5) + 2(5.5 + 12.5)]$$

$$= \textbf{58.33 m}$$

Application: A river is 15 m wide. Soundings of the depth are made at equal intervals of 3 m across the river and are as shown below.

Depth (m)	0	2.2	3.3	4.5	4.2	2.4	0

Calculate the cross-sectional area of the flow of water at this point using Simpson's rule

From above,

$$\text{Area} = \frac{1}{3}(3)[(0+0)+4(2.2+4.5+2.4)+2(3.3+4.2)]$$
$$= (1)[0+36.4+15] = \mathbf{51.4\ m^2}$$

Volumes of irregular solids using Simpson's rule

If the cross-sectional areas A_1, A_2, A_3, ... of an irregular solid bounded by two parallel planes are known at equal intervals of width d (as shown in Figure 4.29), then by Simpson's rule:

$$\textbf{Volume, } \mathbf{V = \frac{d}{3}[(A_1 + A_7) + 4(A_2 + A_4 + A_6) + 2(A_3 + A_5)]}$$

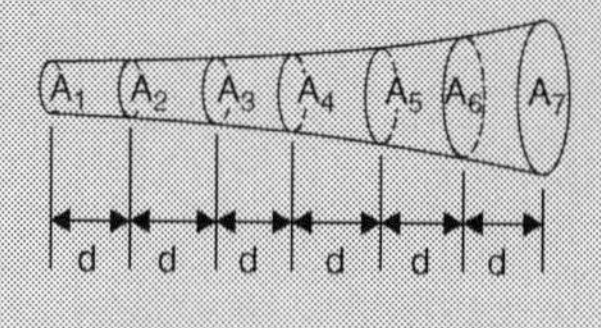

Figure 4.29

Application: A tree trunk is 12 m in length and has a varying cross-section. The cross-sectional areas at intervals of 2 m measured from one end are:

$$0.52, 0.55, 0.59, 0.63, 0.72, 0.84, 0.97\ m^2$$

Estimate the volume of the tree trunk

A sketch of the tree trunk is similar to that shown in Figure 4.29 above, where d = 2 m, $A_1 = 0.52\,m^2$, $A_2 = 0.55\,m^2$, and so on. Using Simpson's rule for volumes gives:

$$\text{Volume} = \frac{2}{3}[(0.52 + 0.97) + 4(0.55 + 0.63 + 0.84) + 2(0.59 + 0.72)]$$

$$= \frac{2}{3}[1.49 + 8.08 + 2.62] = \mathbf{8.13\,m^3}$$

Application: The areas of seven horizontal cross-sections of a water reservoir at intervals of 10 m are: 210, 250, 320, 350, 290, 230, 170 m^2. Calculate the capacity of the reservoir in litres

Using Simpson's rule for volumes gives:

$$\text{Volume} = \frac{10}{3}[(210 + 170) + 4(250 + 350 + 230) + 2(320 + 290)]$$

$$= \frac{10}{3}[380 + 3320 + 1220] = \mathbf{16400\,m^3}$$

$16400\,m^3 = 16400 \times 10^6\,cm^3$

Since 1 litre = 1000 cm^3, capacity of reservoir $= \frac{16400 \times 10^6}{1000}$ litres

$= 16400000$

$= \mathbf{1.64 \times 10^7}$ **litres**

Prismoidal rule for finding volumes

With reference to Figure 4.30,

$$\textbf{Volume, } V = \frac{x}{6}[A_1 + 4A_2 + A_3]$$

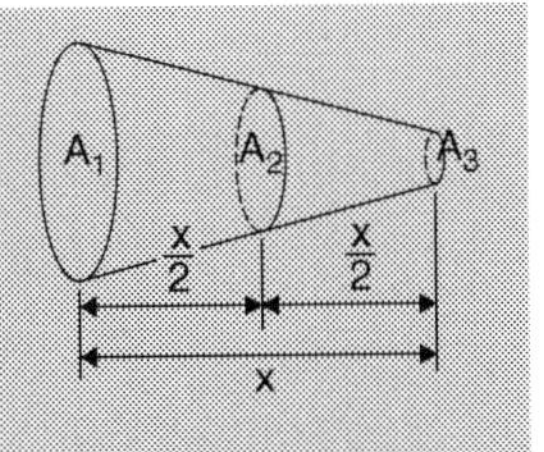

Figure 4.30

Application: A container is in the shape of a frustum of a cone. Its diameter at the bottom is 18 cm and at the top 30 cm. Determine the capacity of the container, correct to the nearest litre, by the prismoidal rule, if the depth is 24 cm.

The container is shown in Figure 4.31. At the mid-point, i.e. at a distance of 12 cm from one end, the radius r_2 is $(9 + 15)/2 = 12$ cm, since the sloping sides change uniformly.

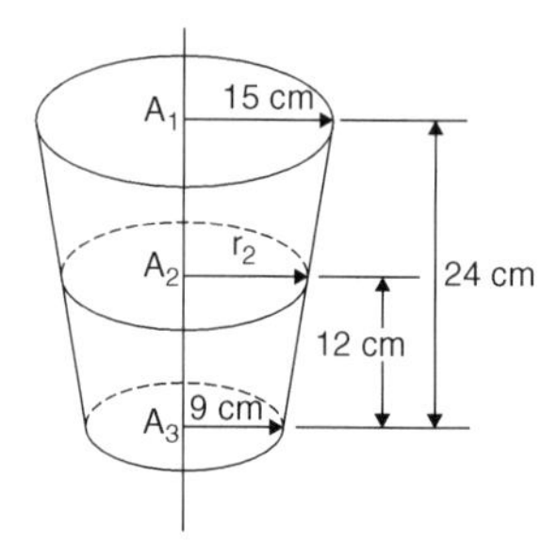

Figure 4.31

Volume of container by the prismoidal rule $= \dfrac{x}{6}[A_1 + 4A_2 + A_3]$ from above, where

$x = 24$ cm, $A_1 = \pi(15)^2$ cm^2, $A_2 = \pi(12)^2$ cm^2 and $A_3 = \pi(9)^2$ cm^2

Hence,

$$\textbf{volume of container} = \frac{24}{6}[\pi(15)^2 + 4\pi(12)^2 + \pi(9)^2]$$
$$= 4[706.86 + 1809.56 + 254.47]$$
$$= 11080\ \text{cm}^3 = \frac{11080}{1000}\ \text{litres}$$
$$= \textbf{11 litres, correct to the nearest litre}$$

Application: The roof of a building is in the form of a frustum of a pyramid with a square base of side 5.0 m. The flat top is a square of side 1.0 m and all the sloping sides are pitched at the same angle. The vertical height of the flat top above the level of the eaves is 4.0 m. Calculate, using the prismoidal rule, the volume enclosed by the roof

Let area of top of frustum be $A_1 = (1.0)^2 = 1.0\,m^2$

Let area of bottom of frustum be $A_3 = (5.0)^2 = 25.0\,m^2$

Let area of section through the middle of the frustum parallel to A_1 and A_3 be A_2. The length of the side of the square forming A_2 is the average of the sides forming A_1 and A_3, i.e. $(1.0 + 5.0)/2 = 3.0\,m$.

Hence $A_2 = (3.0)^2 = 9.0\,m^2$.

Using the prismoidal rule,

$$\text{volume of frustum} = \frac{x}{6}[A_1 + 4A_2 + A_3] = \frac{4.0}{6}[1.0 + 4(9.0) + 25.0]$$

Hence, **volume enclosed by roof = 41.3 m³**

4.7 The mean or average value of a waveform

The mean or average value, y, of the waveform shown in Figure 4.32 is given by:

$$y = \frac{\text{area under curve}}{\text{length of base, b}}$$

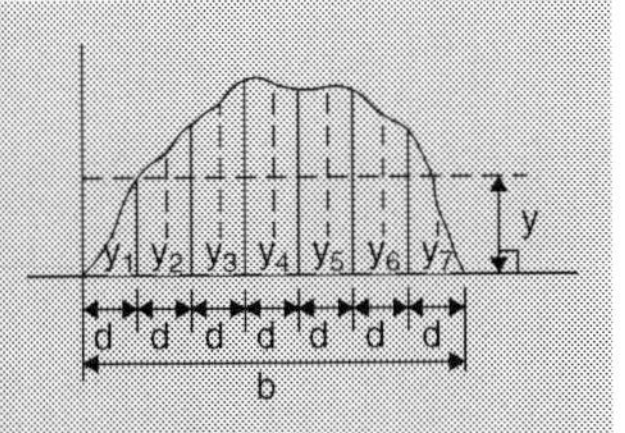

Figure 4.32

If the mid-ordinate rule is used to find the area under the curve, then:

$$\mathbf{y = \frac{sum\ of\ mid\text{-}ordinates}{number\ of\ mid\text{-}ordinates}}$$

$$\left(= \frac{y_1 + y_2 + y_3 + y_4 + y_5 + y_6 + y_7}{7} \text{ for Figure 4.32}\right)$$

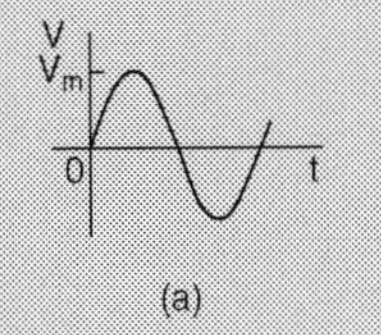

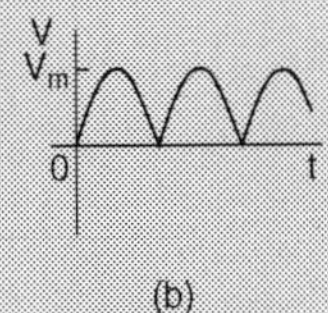

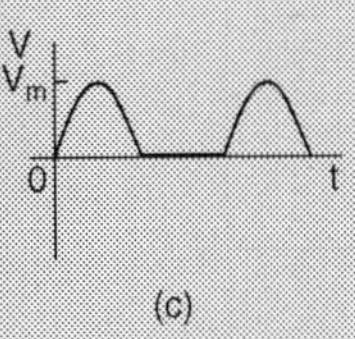

Figure 4.33

For a **sine wave**, the mean or average value:

1. over one complete cycle is zero (see Figure 4.33(a)),
2. over half a cycle is **0.637 × maximum value**, or **2/π × maximum value**,
3. of a full-wave rectified waveform (see Figure 4.33(b)) is **0.637 × maximum value**
4. of a half-wave rectified waveform (see Fig. 4.33(c)) is **0.318 × maximum value** or **1/π × maximum value**

Application: Determine the average values over half a cycle of the periodic waveforms shown in Figure 4.34:

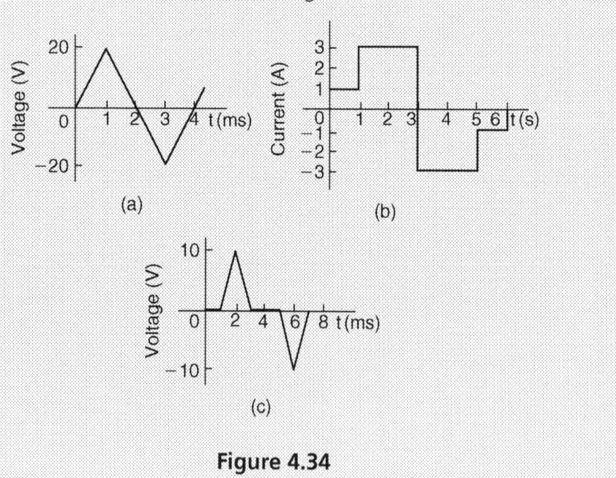

Figure 4.34

(a) Area under triangular waveform (a) for a half cycle is given by:

$$\text{Area} = \frac{1}{2}(\text{base})(\text{perpendicular height}) = \frac{1}{2}(2 \times 10^{-3})(20)$$

$$= 20 \times 10^{-3}\ \text{Vs}$$

$$\text{Average value of waveform} = \frac{\text{area under curve}}{\text{length of base}}$$

$$= \frac{20 \times 10^{-3}\ \text{Vs}}{2 \times 10^{-3}\ \text{s}} = \mathbf{10\ V}$$

(b) Area under waveform (b) for a half cycle $= (1 \times 1) + (3 \times 2) = 7\,\text{As}$

$$\text{Average value of waveform} = \frac{\text{area under curve}}{\text{length of base}} = \frac{7\,\text{As}}{3\,\text{s}}$$

$$= \mathbf{2.33\ A}$$

(c) A half cycle of the voltage waveform (c) is completed in 4 ms.

$$\text{Area under curve} = \frac{1}{2}\{(3-1)10^{-3}\}(10) = 10 \times 10^{-3}\,\text{Vs}$$

$$\text{Average value of waveform} = \frac{\text{area under curve}}{\text{length of base}}$$

$$= \frac{10 \times 10^{-3}\,\text{Vs}}{4 \times 10^{-3}\,\text{s}} = \mathbf{2.5\ V}$$

Application: The power used in a manufacturing process during a 6 hour period is recorded at intervals of 1 hour as shown below.

Time (h)	0	1	2	3	4	5	6
Power (kW)	0	14	29	51	45	23	0

Determine (a) the area under the curve and (b) the average value of the power by plotting a graph of power against time and by using the mid-ordinate rule

The graph of power/time is shown in Figure 4.35.

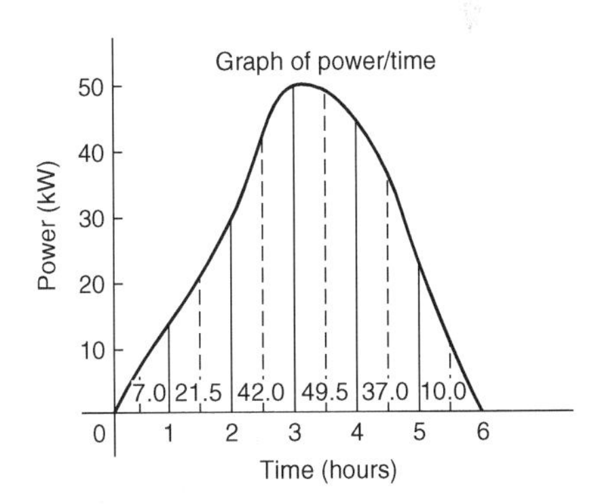

Figure 4.35

(a) The time base is divided into 6 equal intervals, each of width 1 hour.

Mid-ordinates are erected (shown by broken lines in Figure 4.35) and measured. The values are shown in Figure 4.35.

$$\text{Area under curve} = (\text{width of interval})(\text{sum of mid-ordinates})$$
$$= (1)[7.0 + 21.5 + 42.0 + 49.5 + 37.0 + 10.0]$$
$$= \mathbf{167\,kWh} \text{ (i.e. a measure of electrical energy)}$$

(b) $$\text{Average value of waveform} = \frac{\text{area under curve}}{\text{length of base}} = \frac{167\,\text{kWh}}{6\,\text{h}} = \mathbf{27.83\,kW}$$

$$\left(\text{Alternatively, average value} = \frac{\text{sum of mid-ordinates}}{\text{number of mid-ordinates}}\right)$$

Application: An indicator diagram for a steam engine is shown in Figure 4.36. The base line has been divided into 6 equally spaced intervals and the lengths of the 7 ordinates measured with the results shown in centimetres.

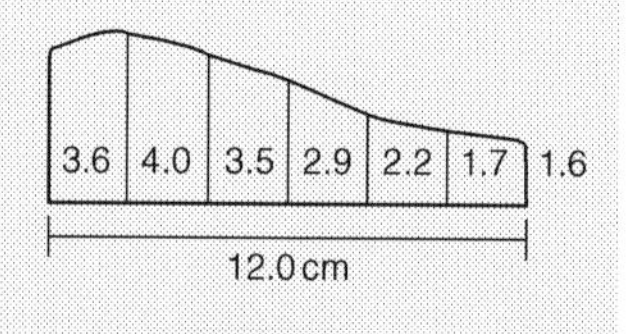

Figure 4.36

Determine (a) the area of the indicator diagram using Simpson's rule, and (b) the mean pressure in the cylinder given that 1 cm represents 100 kPa

(a) The width of each interval is $\frac{12.0}{6} = 2.0$ cm. Using Simpson's rule,

$$\text{area} = \frac{1}{3}(2.0)[(3.6 + 1.6) + 4(4.0 + 2.9 + 1.7) + 2(3.5 + 2.2)]$$
$$= \frac{2}{3}[5.2 + 34.4 + 11.4] = \mathbf{34\ cm^2}$$

(b) $$\text{Mean height of ordinates} = \frac{\text{area of diagram}}{\text{length of base}} = \frac{34}{12} = 2.83\,\text{cm}$$

Since 1 cm represents 100 kPa, the mean pressure in the cylinder

$$= 2.83\,\text{cm} \times 100\,\text{kPa/cm} = \mathbf{283\,kPa}$$

5 Geometry and Trigonometry

5.1 Types and properties of angles

1. Any angle between 0° and 90° is called an **acute angle**.
2. An angle equal to 90° is called a **right angle**.
3. Any angle between 90° and 180° is called an **obtuse angle**.
4. Any angle greater than 180° and less than 360° is called a **reflex angle**.
5. An angle of 180° lies on a straight line.
6. If two angles add up to 90° they are called **complementary angles**.
7. If two angles add up to 180° they are called **supplementary angles**.
8. **Parallel lines** are straight lines that are in the same plane and never meet. (Such lines are denoted by arrows, as in Figure 5.1).
9. A straight line that crosses two parallel lines is called a **transversal** (see MN in Figure. 5.1).
10. With reference to Figure 5.1:
    (i) $a = c$, $b = d$, $e = g$ and $f = h$. Such pairs of angles are called **vertically opposite angles**.
    (ii) $a = e$, $b = f$, $c = g$ and $d = h$. Such pairs of angles are called **corresponding angles**.
    (iii) $c = e$ and $b = h$. Such pairs of angles are called **alternate angles**.
    (iv) $b + e = 180°$ and $c + h = 180°$. Such pairs of angles are called **interior angles**.

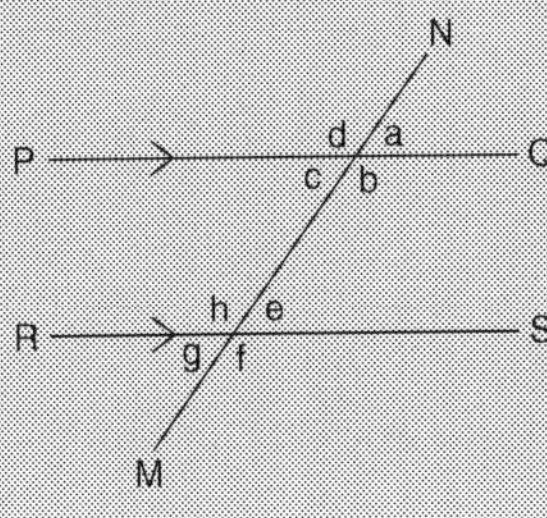

Figure 5.1

5.2 Properties of triangles

1. The sum of the three angles of a triangle is equal to 180°.
2. An **acute-angled triangle** is one in which all the angles are acute, i.e. all the angles less than 90°.
3. A **right-angled triangle** is one that contains a right angle.
4. An **obtuse-angled triangle** is one that contains an obtuse angle, i.e. one angle which lies between 90° and 180°.
5. An **equilateral triangle** is one in which all the sides and all the angles are equal (i.e. each 60°).
6. An **isosceles triangle** is one in which two angles and two sides are equal.
7. A **scalene triangle** is one with unequal angles and therefore unequal sides.
8. With reference to Figure 5.2:
   (i) Angles A, B and C are called **interior angles** of the triangle.
   (ii) Angle θ is called an **exterior angle** of the triangle and is equal to the sum of the two opposite interior angles, i.e. $\theta = A + C$
   (iii) $a + b + c$ is called the **perimeter** of the triangle.

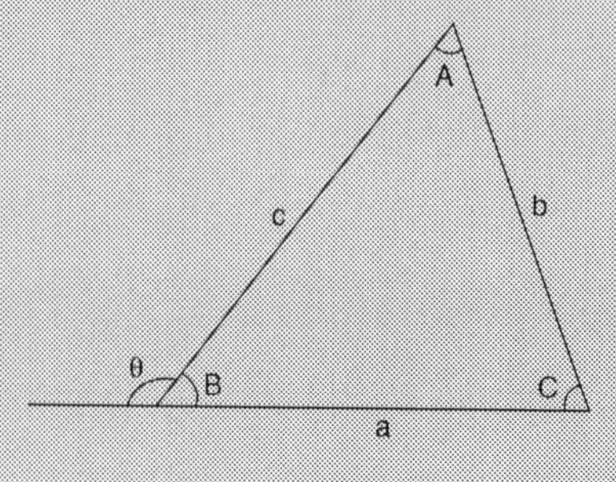

Figure 5.2

9. **Congruent triangles** – two triangles are congruent if:
   (i) the three sides of one are equal to the three sides of the other,
   (ii) they have two sides of the one equal to two sides of the other, and if the angles included by these sides are equal,
   (iii) two angles of the one are equal to two angles of the other and any side of the first is equal to the corresponding side of the other, or
   (iv) their hypotenuses are equal and if one other side of one is equal to the corresponding side of the other.

10. **Similar triangles**
    With reference to Figure 5.3, triangles ABC and PQR are similar and the corresponding sides are in proportion to each other,

    i.e. $$\frac{p}{a} = \frac{q}{b} = \frac{r}{c}$$

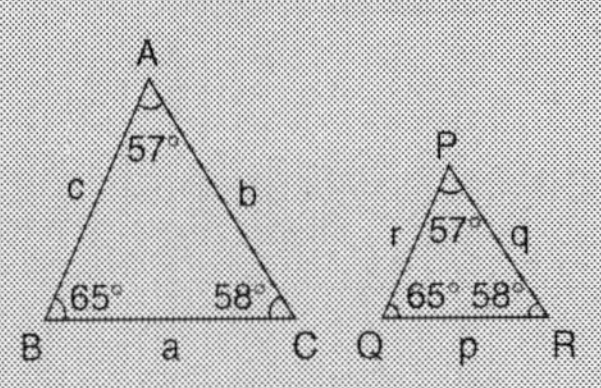

Figure 5.3

Application: A rectangular shed 2 m wide and 3 m high stands against a perpendicular building of height 5.5 m. A ladder is used to gain access to the roof of the building. Determine the minimum distance between the bottom of the ladder and the shed

A side view is shown in Figure 5.4 where AF is the minimum length of ladder. Since BD and CF are parallel, $\angle ADB = \angle DFE$ (corresponding angles between parallel lines). Hence triangles BAD and EDF are similar since their angles are the same.
$AB = AC - BC = AC - DE = 5.5 - 3 = 2.5\,m$

By proportion: $$\frac{AB}{DE} = \frac{BD}{EF} \quad \text{i.e.} \quad \frac{2.5}{3} = \frac{2}{EF}$$

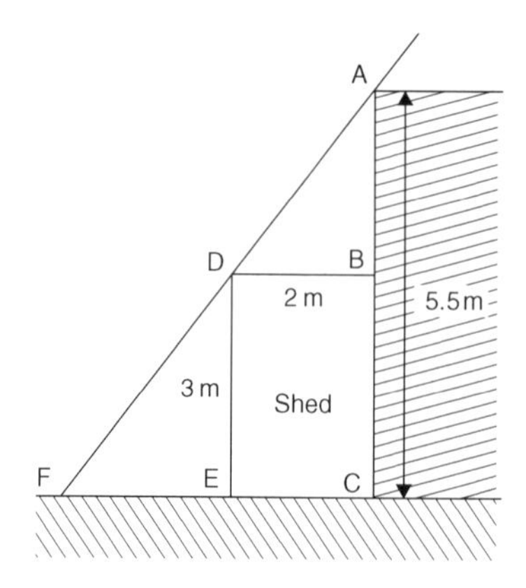

Figure 5.4

Hence $EF = 2\left(\frac{3}{2.5}\right)$ = **2.4 m = minimum distance from bottom of ladder to the shed**

5.3 Introduction to trigonometry

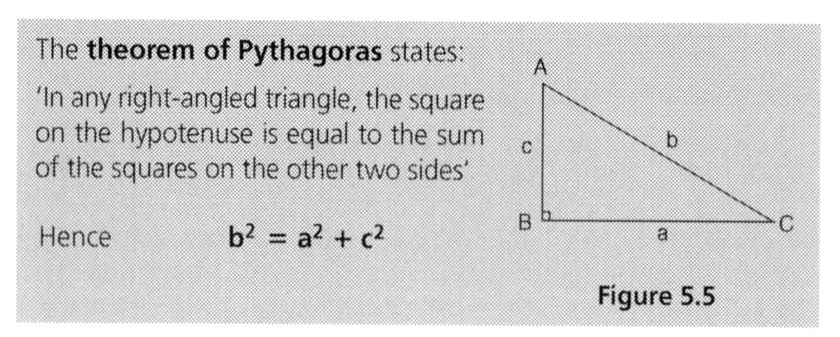

The **theorem of Pythagoras** states:

'In any right-angled triangle, the square on the hypotenuse is equal to the sum of the squares on the other two sides'

Hence $\quad \mathbf{b^2 = a^2 + c^2}$

Figure 5.5

Application: Two aircraft leave an airfield at the same time. One travels due north at an average speed of 300 km/h and the other due west at an average speed of 220 km/h. Calculate their distance apart after 4 hours

After 4 hours, the first aircraft has travelled $4 \times 300 = 1200$ km, due north, and the second aircraft has travelled $4 \times 220 = 880$ km due west, as shown in Figure 5.6.

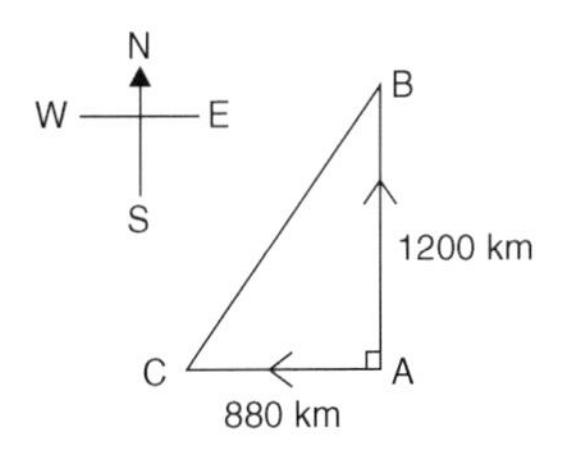

Figure 5.6

Distance apart after 4 hours = BC

From Pythagoras' theorem:

$$BC^2 = 1200^2 + 880^2$$
$$= 1440000 + 774400 \text{ and } BC = \sqrt{2214400}$$

Hence distance apart after 4 hours = 1488 km

5.4 Trigonometric ratios of acute angles

With reference to the right-angled triangle shown in Figure 5.7:

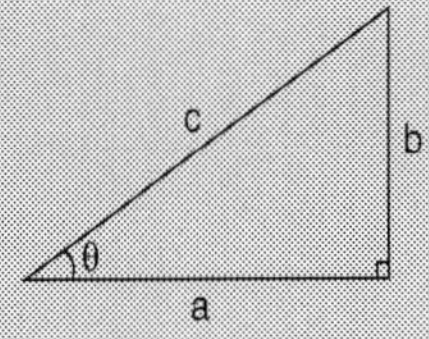

Figure 5.7

1. $\text{sine}\,\theta = \dfrac{\text{opposite side}}{\text{hypotenuse}}$ i.e. $\mathbf{\sin\theta = \dfrac{b}{c}}$
2. $\text{cosine}\,\theta = \dfrac{\text{adjacent side}}{\text{hypotenuse}}$ i.e. $\mathbf{\cos\theta = \dfrac{a}{c}}$
3. $\text{tangent}\,\theta = \dfrac{\text{opposite side}}{\text{adjacent side}}$ i.e. $\mathbf{\tan\theta = \dfrac{b}{a}}$
4. $\text{secant}\,\theta = \dfrac{\text{hypotenuse}}{\text{adjacent side}}$ i.e. $\mathbf{\sec\theta = \dfrac{c}{a}}$
5. $\text{cosecant}\,\theta = \dfrac{\text{hypotenuse}}{\text{opposite side}}$ i.e. $\mathbf{\text{cosec}\,\theta = \dfrac{c}{b}}$

6. cotangent $\theta = \dfrac{\text{adjacent side}}{\text{opposite side}}$ i.e. $\mathbf{\cot\theta = \dfrac{a}{b}}$

From above,

7. $\dfrac{\sin\theta}{\cos\theta} = \dfrac{\frac{b}{c}}{\frac{a}{c}} = \dfrac{b}{a} = \tan\theta$ i.e. $\mathbf{\tan\theta = \dfrac{\sin\theta}{\cos\theta}}$

8. $\dfrac{\cos\theta}{\sin\theta} = \dfrac{\frac{a}{c}}{\frac{b}{c}} = \dfrac{a}{b} = \cot\theta$ i.e. $\mathbf{\cot\theta = \dfrac{\cos\theta}{\sin\theta}}$

9. $\mathbf{\sec\theta = \dfrac{1}{\cos\theta}}$

10. $\mathbf{\text{cosec}\,\theta = \dfrac{1}{\sin\theta}}$ (Note 's' and 'c' go together)

11. $\mathbf{\cot\theta = \dfrac{1}{\tan\theta}}$

Secants, cosecants and cotangents are called the **reciprocal ratios**.

5.5 Evaluating trigonometric ratios

The easiest method of evaluating trigonometric functions of any angle is by using a **calculator**.

The following values, correct to 4 decimal places, may be checked:

sine 18° = 0.3090 cosine 56° = 0.5592
tangent 29° = 0.5543

sine 172° = 0.1392 cosine 115° = −0.4226
tangent 78° = −0.0349

sine 241.63° = −0.8799 cosine 331.78° = 0.8811
tangent 296.42° = −2.0127

Most calculators contain only sine, cosine and tangent functions. Thus to evaluate secants, cosecants and cotangents, reciprocals need to be used.

The following values, correct to 4 decimal places, may be checked:

$$\text{secant } 32° = \frac{1}{\cos 32°} = 1.1792 \qquad \text{secant } 215.12° = \frac{1}{\cos 215.12°} = -1.2226$$

$$\text{cosecant } 75° = \frac{1}{\sin 75°} = 1.0353 \qquad \text{cosecant } 321.62° = \frac{1}{\sin 321.62°} = -1.6106$$

$$\text{cotangent } 41° = \frac{1}{\tan 41°} = 1.1504 \qquad \text{cotangent } 263.59° = \frac{1}{\tan 263.59°} = 0.1123$$

Application: Determine the acute angles:

(a) $\sec^{-1} 2.3164$ (b) $\text{cosec}^{-1} 1.1784$ (c) $\cot^{-1} 2.1273$

(a) $\sec^{-1} 2.3164 = \cos^{-1}\left(\dfrac{1}{2.3164}\right) = \cos^{-1} 0.4317..$

$= \mathbf{64.42°}$ or $\mathbf{64°25'}$ or **1.124 rad**

(b) $\text{cosec}^{-1} 1.1784 = \sin^{-1}\left(\dfrac{1}{1.1784}\right) = \sin^{-1} 0.8486..$

$= \mathbf{58.06°}$ or $\mathbf{58°4'}$ or **1.013 rad**

(c) $\cot^{-1} 2.1273 = \tan^{-1}\left(\dfrac{1}{2.1273}\right) = \tan^{-1} 0.4700..$

$= \mathbf{25.18°}$ or $\mathbf{25°11'}$ or **0.439 rad**

5.6 Fractional and surd forms of trigonometric ratios

In Figure 5.8, ABC is an equilateral triangle of side 2 units. AD bisects angle A and bisects the side BC. Using Pythagoras' theorem on triangle ABD gives:

$$AD = \sqrt{2^2 - 1^2} = \sqrt{3}$$

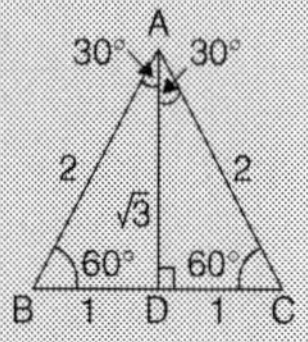

Figure 5.8

Hence,

$$\sin 30° = \frac{BD}{AB} = \frac{1}{2}, \ \cos 30° = \frac{AD}{AB} = \frac{\sqrt{3}}{2}$$

$$\text{and } \tan 30° = \frac{BD}{AD} = \frac{1}{\sqrt{3}}$$

$$\sin 60° = \frac{AD}{AB} = \frac{\sqrt{3}}{2}, \ \cos 60° = \frac{BD}{AB} = \frac{1}{2}$$

$$\text{and } \tan 60° = \frac{AD}{BD} = \sqrt{3}$$

In Figure 5.9, PQR is an isosceles triangle with PQ = QR = 1 unit. By Pythagoras' theorem, $PR = \sqrt{1^2 + 1^2} = \sqrt{2}$

Hence, $\sin 45° = \frac{1}{\sqrt{2}}$, $\cos 45° = \frac{1}{\sqrt{2}}$ and $\tan 45° = 1$

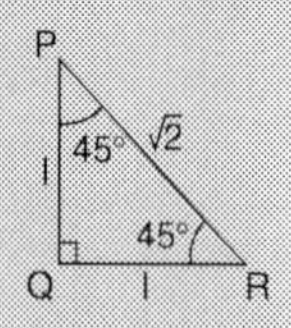

Figure 5.9

A quantity that is not exactly expressible as a rational number is called a **surd**. For example, $\sqrt{2}$ and $\sqrt{3}$ are called surds because they cannot be expressed as a fraction and the decimal part may be continued indefinitely.

From above, sin 30° = cos 60°, sin 45° = cos 45° and sin 60° = cos 30°.

In general, $\mathbf{\sin\theta = \cos(90° - \theta)}$ and $\mathbf{\cos\theta = \sin(90° - \theta)}$

5.7 Solution of right-angled triangles

To 'solve a right-angled triangle' means 'to find the unknown sides and angles'. This is achieved by using (i) the theorem of Pythagoras, and/or (ii) trigonometric ratios.

Application: Find the lengths of PQ and PR in triangle PQR shown in Figure 5.10

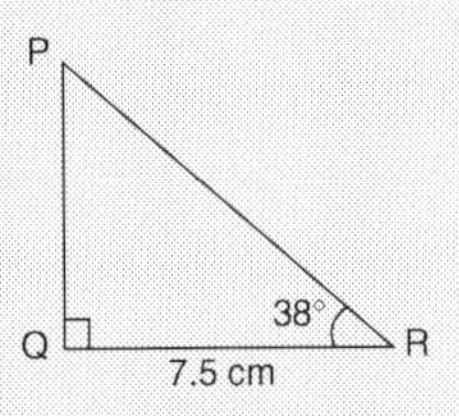

Figure 5.10

$$\tan 38° = \frac{PQ}{QR} = \frac{PQ}{7.5}, \text{ hence } \quad PQ = 7.5 \tan 38°$$

$$= 7.5(0.7813) = \mathbf{5.860\,cm}$$

$$\cos 38° = \frac{QR}{PR} = \frac{7.5}{PR}, \text{ hence } PR = \frac{7.5}{\cos 38°} = \frac{7.5}{0.7880} = \mathbf{9.518\,cm}$$

[Check: Using Pythagoras' theorem $(7.5)^2 + (5.860)^2 = 90.59 = (9.518)^2$]

Angles of elevation and depression

If, in Figure 5.11, BC represents horizontal ground and AB a vertical flagpole, then the **angle of elevation** of the top of the flagpole, A, from the point C is the angle that the imaginary straight line AC must be raised (or elevated) from the horizontal CB, i.e. angle θ.

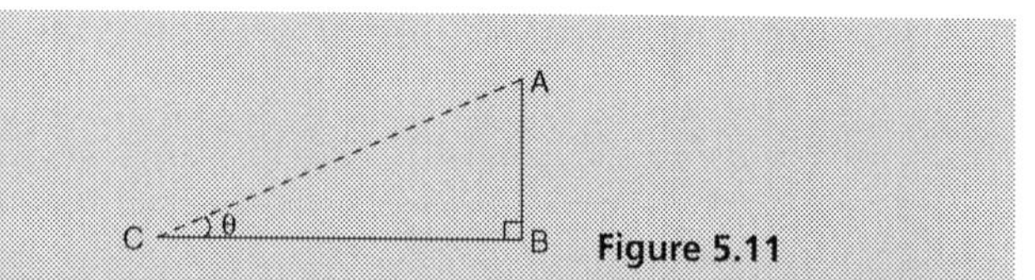

Figure 5.11

If, in Figure 5.12, PQ represents a vertical cliff and R a ship at sea, then the **angle of depression** of the ship from point P is the angle through which the imaginary straight line PR must be lowered (or depressed) from the horizontal to the ship, i.e. angle ϕ.

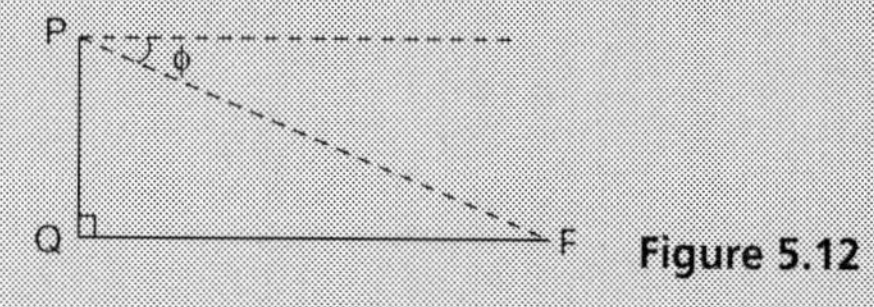

Figure 5.12

(Note, $\angle PRQ$ is also ϕ – alternate angles between parallel lines.)

Application: An electricity pylon stands on horizontal ground. At a point 80 m from the base of the pylon, the angle of elevation of the top of the pylon is 23°. Calculate the height of the pylon to the nearest metre.

Figure 5.13 shows the pylon AB and the angle of elevation of A from point C is 23°.

$$\text{Now } \tan 23° = \frac{AB}{BC} = \frac{AB}{80}$$

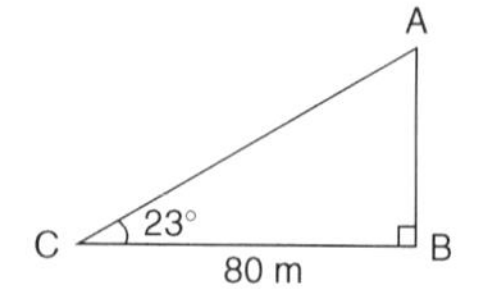

Figure 5.13

Hence, height of pylon, $AB = 80 \tan 23° = 80(0.4245) = 33.96\,\text{m}$

$= \mathbf{34\,m\ to\ the\ nearest\ metre}$

Application: The angle of depression of a ship viewed at a particular instant from the top of a 75 m vertical cliff is 30°. The ship is sailing away from the cliff at constant speed and 1 minute later its angle of depression from the top of the cliff is 20°. Find (a) the initial distance of the ship from the base of the cliff, and (b) the speed of the ship in km/h and in knots

(a) Figure 5.14 shows the cliff AB, the initial position of the ship at C and the final position at D.
Since the angle of depression is initially 30° then $\angle ACB = 30°$ (alternate angles between parallel lines).

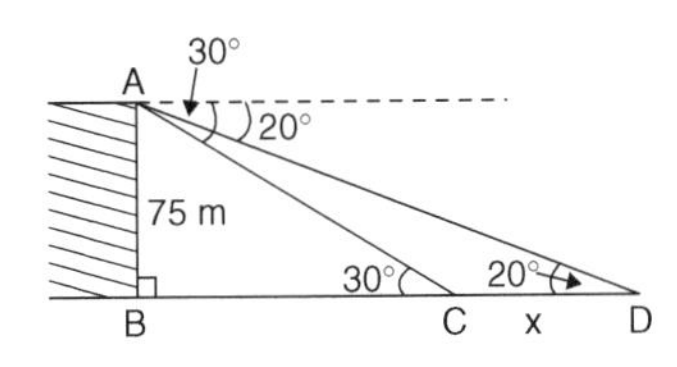

Figure 5.14

$\tan 30° = \frac{AB}{BC} = \frac{75}{BC}$ hence **the initial position of the ship from the base of cliff,**

$$BC = \frac{75}{\tan 30°} = \frac{75}{0.5774} = \mathbf{129.9\,m}$$

(b) In triangle ABD, $\tan 20° = \frac{AB}{BD} = \frac{75}{BC + CD} = \frac{75}{129.9 + x}$

Hence $129.9 + x = \frac{75}{\tan 20°} = \frac{75}{0.3640} = 206.0\,m$

from which $x = 206.0 - 129.9 = 76.1\,m$

Thus the ship sails 76.1 m in 1 minute, i.e. 60 s,

hence, **speed of ship** $= \frac{\text{distance}}{\text{time}} = \frac{76.1}{60}$ m/s

$= \frac{76.1 \times 60 \times 60}{60 \times 1000}$ km/h

$= \mathbf{4.57\,km/h}$

From chapter 1, page 2, 1 km/h = 0.54 knots

Hence, **speed of ship** = 4.57 × 0.54 = **2.47 knots**

5.8 Cartesian and polar co-ordinates

Changing from Cartesian into polar co-ordinates

In Figure 5.15, $\mathbf{r = \sqrt{x^2 + y^2}}$

and $\boldsymbol{\theta = \tan^{-1}\frac{y}{x}}$

The angle θ, which may be expressed in degrees or radians, must **always** be measured from the positive x-axis, i.e. measured from the line OQ in Figure. 5.15.

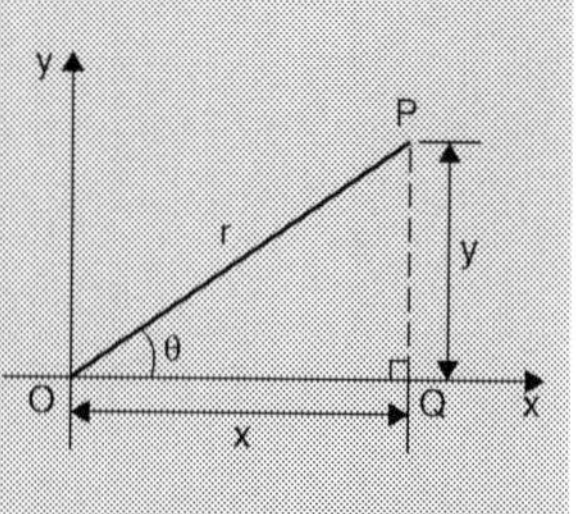

Figure 5.15

Application: Express in polar co-ordinates the position (−4, 3)

A diagram representing the point using the Cartesian co-ordinates (−4, 3) is shown in Figure 5.16.

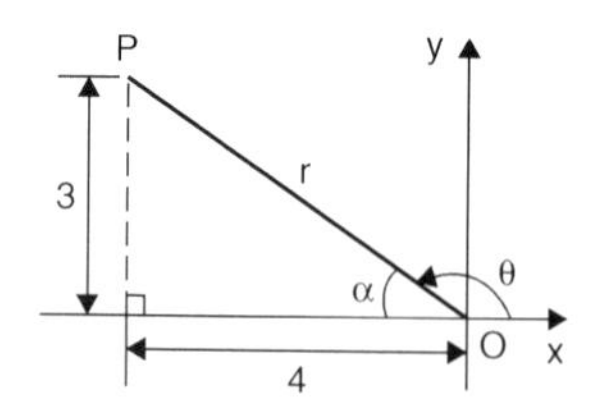

Figure 5.16

From Pythagoras' theorem, $r = \sqrt{4^2 + 3^2} = 5$

By trigonometric ratios, $\alpha = \tan^{-1}\frac{3}{4} = 36.87°$ or 0.644 rad

Hence $\theta = 180° - 36.87° = 143.13°$ or $\theta = \pi - 0.644 = 2.498$ rad

Hence the position of point P in polar co-ordinate form is (5, 143.13°) or (5, 2.498 rad)

Application: Express (−5, −12) in polar co-ordinates

A sketch showing the position (−5, −12) is shown in Figure 5.17.

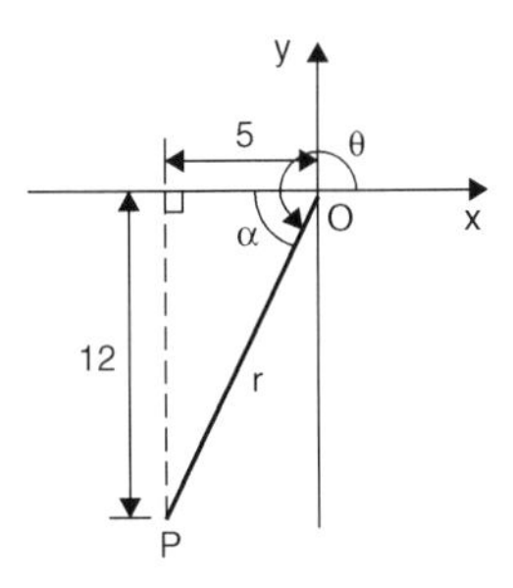

Figure 5.17

$$r = \sqrt{5^2 + 12^2} = 13 \text{ and } \alpha = \tan^{-1}\frac{12}{5} = 67.38° \text{ or } 1.176 \text{ rad}$$

Hence $\theta = 180° + 67.38° = 247.38°$ or $\theta = \pi + 1.176 = 4.318$ rad

Thus (−5, −12) in Cartesian co-ordinates corresponds to (13, 247.38°) or (13, 4.318 rad) in polar co-ordinates.

Changing from polar into Cartesian co-ordinates

From Figure 5.18,

$$\mathbf{x = r\cos\theta} \quad \text{and} \quad \mathbf{y = r\sin\theta}$$

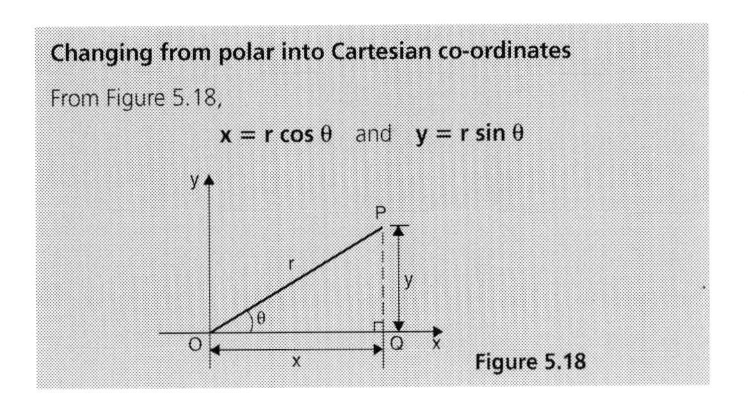

Figure 5.18

Application: Change (4, 32°) into Cartesian co-ordinates

A sketch showing the position (4, 32°) is shown in Figure 5.19.

Now $\quad x = r\cos\theta = 4\cos 32° = 3.392$

and $\quad y = r\sin\theta = 4\sin 32° = 2.120$

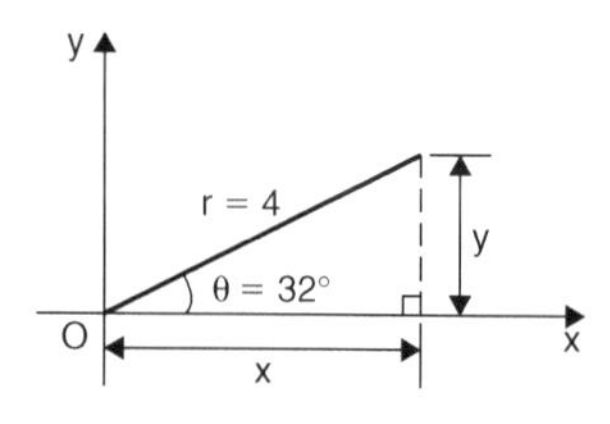

Figure 5.19

Hence, (4, 32°) in polar co-ordinates corresponds to (3.392, 2.120) in Cartesian co-ordinates.

Application: Express (6, 137°) in Cartesian co-ordinates

A sketch showing the position (6, 137°) is shown in Figure 5.20.

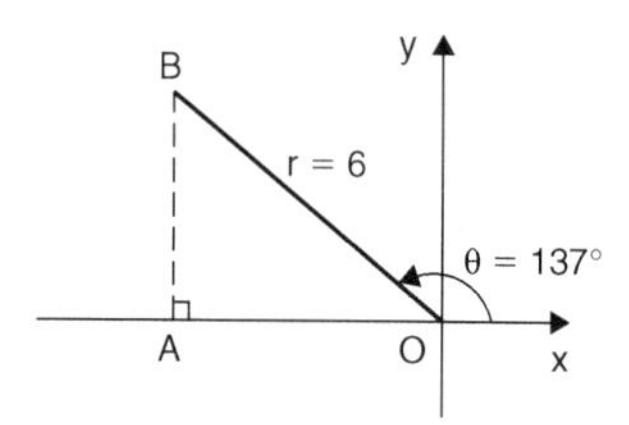

Figure 5.20

$$x = r \cos \theta = 6 \cos 137° = -4.388$$

which corresponds to length OA in Figure 5.20.

$$y = r \sin \theta = 6 \sin 137° = 4.092$$

which corresponds to length AB in Figure 5.20.

Thus, (6, 137°) in polar co-ordinates corresponds to (−4.388, 4.092) in Cartesian co-ordinates.

5.9 Sine and cosine rules and areas of any triangle

Sine rule

With reference to triangle ABC of Figure. 5.21, the **sine rule** states:

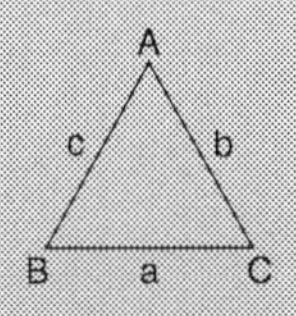

Figure 5.21

$$\frac{a}{\sin A} = \frac{b}{\sin B} = \frac{c}{\sin C}$$

The rule may be used only when:

(i) 1 side and any 2 angles are initially given, or

(ii) 2 sides and an angle (not the included angle) are initially given.

Cosine rule

With reference to triangle ABC of Figure 5.21, the **cosine rule** states:

$$a^2 = b^2 + c^2 - 2bc \cos A$$

or $$b^2 = a^2 + c^2 - 2ac \cos B$$

or $$c^2 = a^2 + b^2 - 2ab \cos C$$

The rule may be used only when:

(i) 2 sides and the included angle are initially given, or

(ii) 3 sides are initially given.

Area of any triangle

The **area of any triangle** such as ABC of Figure. 5.21 is given by:

(i) $\frac{1}{2} \times$ **base** $\times$ **perpendicular height**, or

(ii) $\frac{1}{2}ab \sin C$ or $\frac{1}{2}ac \sin B$ or $\frac{1}{2}bc \sin A$, or

(iii) $\sqrt{[s(s-a)(s-b)(s-c)]}$ where $s = \frac{a+b+c}{2}$

Application: A room 8.0 m wide has a span roof that slopes at 33° on one side and 40° on the other. Find the lengths of the roof slopes, correct to the nearest centimetre

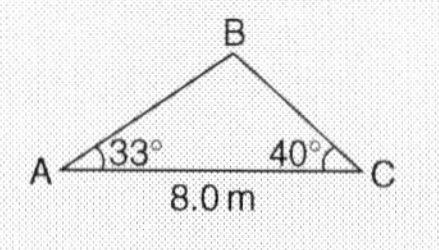

Figure 5.22

A section of the roof is shown in Figure 5.22.

Angle at ridge, $B = 180° - 33° - 40° = 107°$

From the sine rule: $$\frac{8.0}{\sin 107°} = \frac{a}{\sin 33°}$$

from which, $$a = \frac{8.0 \sin 33°}{\sin 107°} = 4.556\,\text{m}$$

Also from the sine rule: $$\frac{8.0}{\sin 107°} = \frac{c}{\sin 40°}$$

from which, $$c = \frac{8.0 \sin 40°}{\sin 107°} = 5.377\,\text{m}$$

Hence the roof slopes are 4.56 m and 5.38 m, correct to the nearest centimetre.

Application: Two voltage phasors are shown in Figure 5.23 where $V_1 = 40$ V and $V_2 = 100$ V. Determine the value of their resultant (i.e. length OA) and the angle the resultant makes with V_1

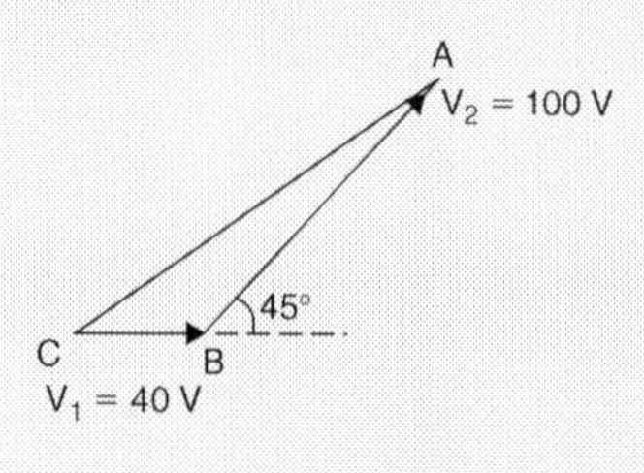

Figure 5.23

Angle OBA = 180° − 45° = 135°

Applying the cosine rule: $OA^2 = V_1^{\,2} + V_2^{\,2} - 2V_1V_2 \cos OBA$

$$= 40^2 + 100^2 - \{2(40)(100)\cos 135°\}$$

$$= 1600 + 10000 - \{-5657\}$$

$$= 1600 + 10000 + 5657 = 17257$$

The resultant $OA = \sqrt{17257} = 131.4\,\text{V}$

Applying the sine rule: $\dfrac{131.4}{\sin 135°} = \dfrac{100}{\sin AOB}$

from which, $\sin AOB = \dfrac{100 \sin 135°}{131.4} = 0.5381$

Hence, angle AOB = $\sin^{-1}0.5381 = 32.55°$ (or 147.45°, which is impossible in this case).

Hence, **the resultant voltage is 131.4 volts at 32.55° to V_1**

Application: In Figure 5.24, PR represents the inclined jib of a crane and is 10.0 m long. PQ is 4.0 m long. Determine the inclination of the jib to the vertical and the length of tie QR

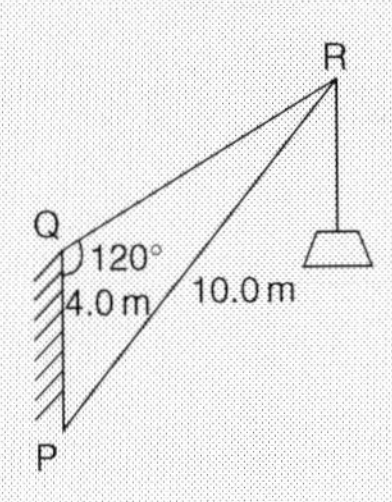

Figure 5.24

Applying the sine rule: $\dfrac{PR}{\sin 120°} = \dfrac{PO}{\sin R}$

from which, $\sin R = \dfrac{PQ \sin 120°}{PR} = \dfrac{(4.0)\sin 120°}{10.0} = 0.3464$

Hence $\angle R = \sin^{-1}0.3464 = 20.27°$ (or 159.73°, which is impossible in this case).

$\angle$**P** $= 180° - 120° - 20.27° =$ **39.73°, which is the inclination of the jib to the vertical.**

Applying the sine rule: $\dfrac{10.0}{\sin 120°} = \dfrac{QR}{\sin 39.73°}$

from which, **length of tie, QR** $= \dfrac{10.0 \sin 39.73°}{\sin 120°} =$ **7.38 m**

Application: A crank mechanism of a petrol engine is shown in Figure 5.25. Arm OA is 10.0 cm long and rotates clockwise about O. The connecting rod AB is 30.0 cm long and end B is constrained to move horizontally. Determine the angle between the connecting rod AB and the horizontal, and the length of OB for the position shown in Figure 5.25

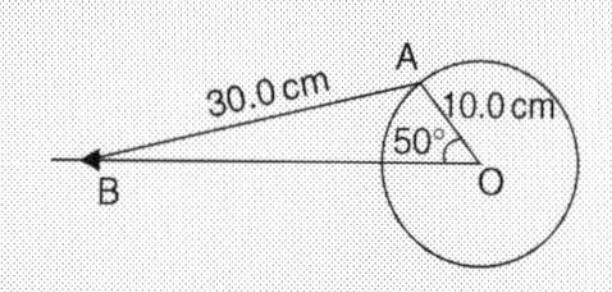

Figure 5.25

Applying the sine rule: $\dfrac{AB}{\sin 50°} = \dfrac{AO}{\sin B}$

from which, $\sin B = \dfrac{AO \sin 50°}{AB} = \dfrac{10.0 \sin 50°}{30.0}$

$= 0.2553$

Hence $B = \sin^{-1} 0.2553 = 14.79°$ (or 165.21°, which is impossible in this case).

Hence, the connecting rod AB makes an angle of 14.79° with the horizontal.

Angle OAB $= 180° - 50° - 14.79° = 115.21°$

Applying the sine rule: $\dfrac{30.0}{\sin 50°} = \dfrac{OB}{\sin 115.21°}$

from which, **OB** $= \dfrac{30.0 \sin 115.21°}{\sin 50°} =$ **35.43 cm**

Application: Determine in Figure 5.25 how far B moves when angle AOB changes from 50° to 120°

Figure 5.26 shows the initial and final positions of the crank mechanism.

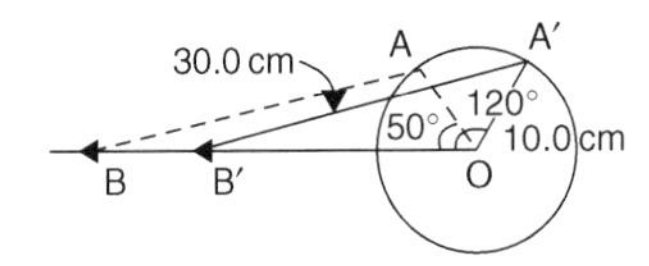

Figure 5.26

In triangle OA′B′, applying the sine rule:

$$\frac{30.0}{\sin 120°} = \frac{10.0}{\sin A'B'O}$$

from which, $\sin A'B'O = \dfrac{10.0 \sin 120°}{30.0} = 0.2887$

Hence $A'B'O = \sin^{-1} 0.2887 = 16.78°$ (or 163.22° which is impossible in this case).

Angle $OA'B' = 180° - 120° - 16.78° = 43.22°$

Applying the sine rule: $\dfrac{30.0}{\sin 120°} = \dfrac{OB'}{\sin 43.22°}$

from which, $OB' = \dfrac{30.0 \sin 43.22°}{\sin 120°} = 23.72\,\text{cm}$

Since OB = 35.43 cm, from the previous example, and OB′ = 23.72 cm then BB′ = 35.43 − 23.72 = 11.71 cm.

Hence, B moves 11.71 cm when angle AOB changes from 50° to 120°

Application: The area of a field is in the form of a quadrilateral ABCD as shown in Figure 5.27. Determine its area

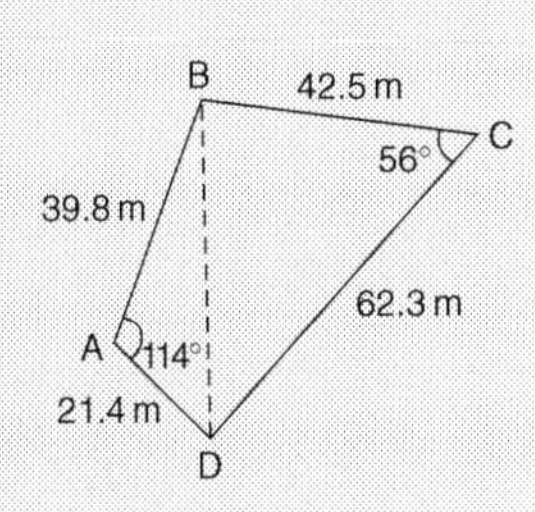

Figure 5.27

A diagonal drawn from B to D divides the quadrilateral into two triangles.

Area of quadrilateral ABCD

$$= \text{area of triangle ABD} + \text{area of triangle BCD}$$

$$= \frac{1}{2}(39.8)(21.4)\sin 114° + \frac{1}{2}(42.5)(62.3)\sin 56°$$

$$= 389.04 + 1097.5 = \mathbf{1487\,m^2}$$

5.10 Graphs of trigonometric functions

Graphs of y = sin A, y = cos A and y = tan A are shown in Figure 5.28.

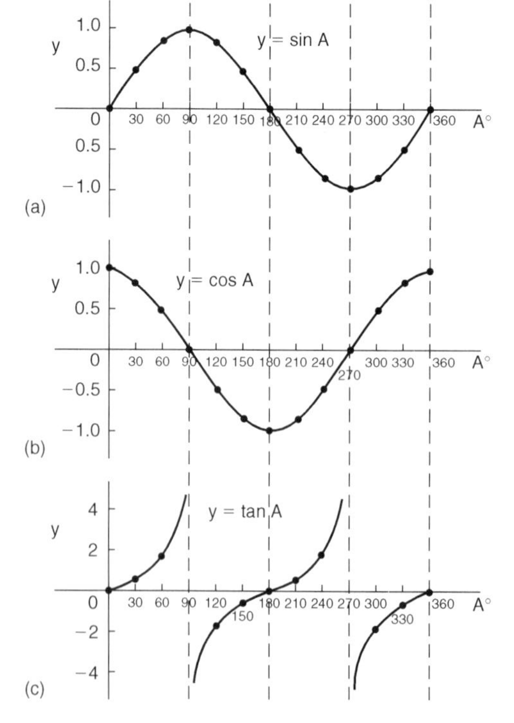

Figure 5.28

5.11 Angles of any magnitude

Figure 5.29 summarises the trigonometric ratios for angles of any magnitude; the letters underlined spell the word CAST when starting in the fourth quadrant and moving in an anticlockwise direction.

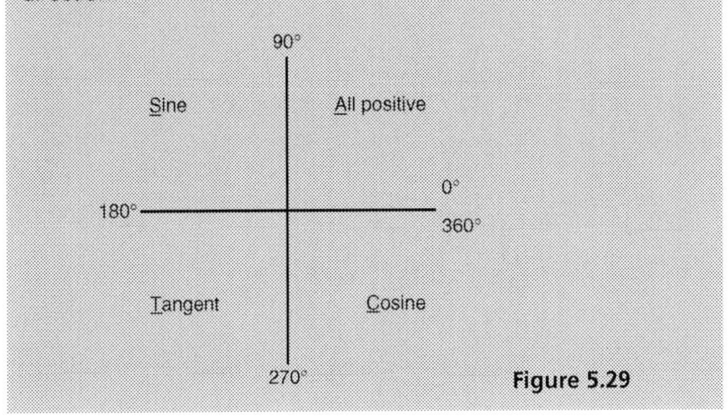

Figure 5.29

Application: Determine all the angles between 0° and 360° whose sine is −0.4638

The angles whose sine is −0.4638 occurs in the third and fourth quadrants since sine is negative in these quadrants – see Figure 5.30.

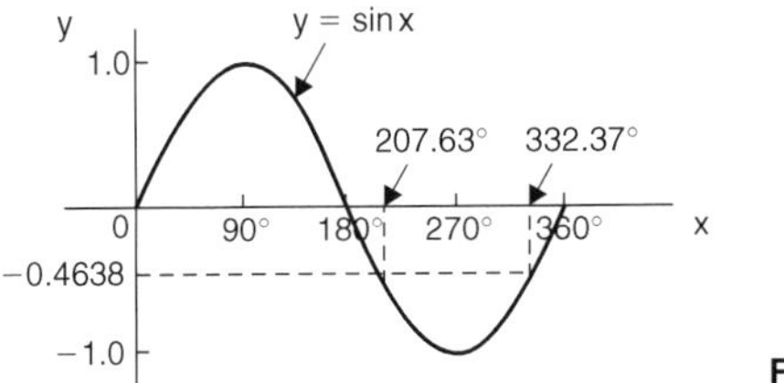

Figure 5.30

From Figure 5.31, $\theta = \sin^{-1} 0.4638 = 27.63°$. Measured from 0°, the two angles between 0° and 360° whose sine is −0.4638 are 180° + 27.63°, i.e. **207.63°** and 360° − 27.63°, i.e. **332.37°**

(Note that a calculator only gives one answer, i.e. −27.632588°)

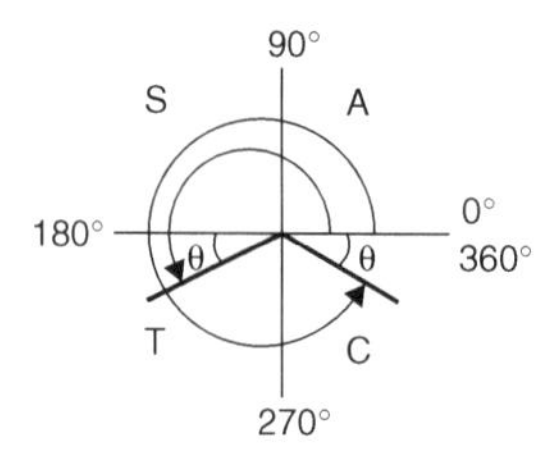

Figure 5.31

Application: Determine all the angles between 0° and 360° whose tangent is 1.7629

A tangent is positive in the first and third quadrants – see Figure 5.32. From Figure 5.33, $\theta = \tan^{-1} 1.7629 = 60.44°$

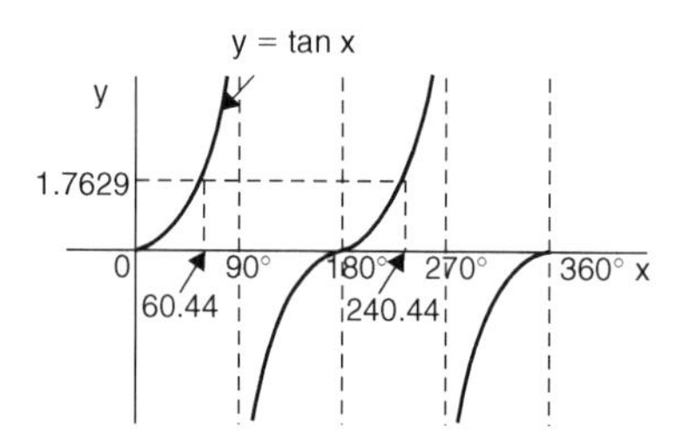

Figure 5.32

Measured from 0°, the two angles between 0° and 360° whose tangent is 1.7629 are **60.44°** and 180° + 60.44°, i.e. **240.44°**

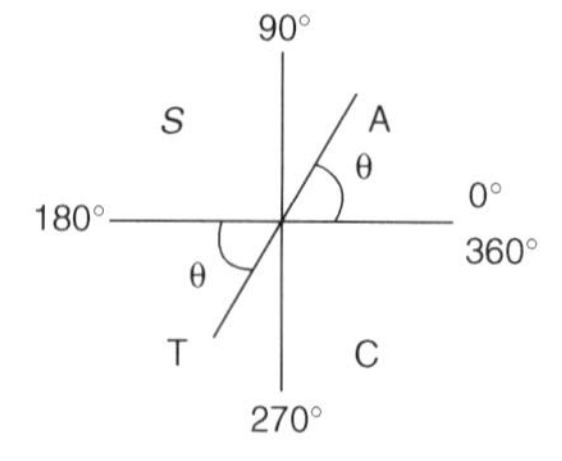

Figure 5.33

Application: Solve the equation $\cos^{-1}(-0.2348) = \alpha$ for angles of α between 0° and 360°

Cosine is positive in the first and fourth quadrants and thus negative in the second and third quadrants – from Figure 5.29 or from Figure 5.28(b).

In Figure 5.34, angle $\theta = \cos^{-1}(0.2348) = 76.42°$

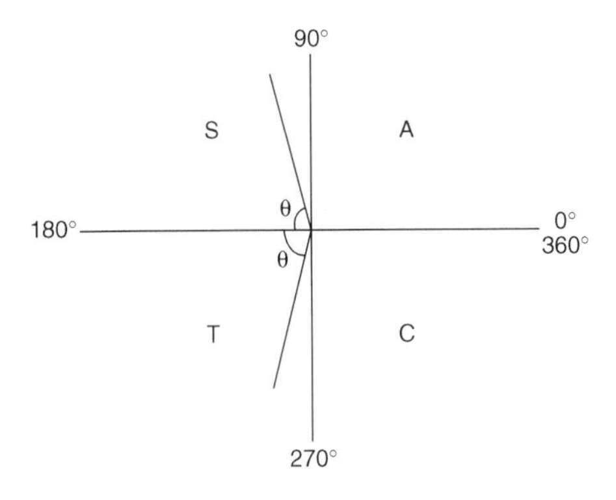

Figure 5.34

Measured from 0°, the two angles whose cosine is -0.2348 are $\alpha = 180° - 76.42°$ i.e. **103.58°** and $\alpha = 180° + 76.42°$ i.e. **256.42°**

5.12 Sine and cosine waveforms

Graphs of $y = \sin A$ and $y = \sin 2A$ are shown in Figure 5.35.

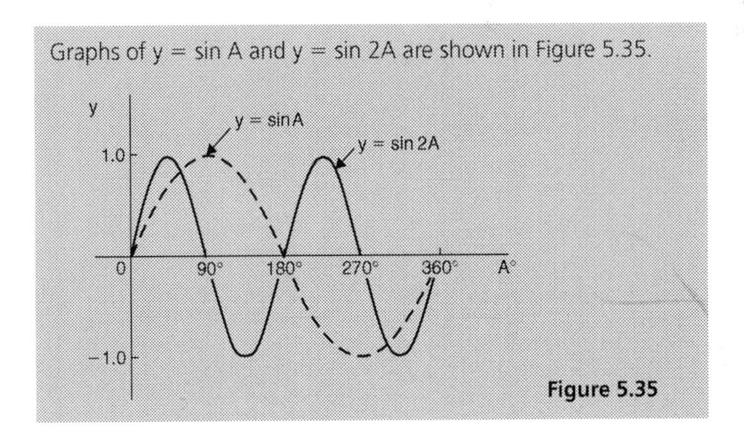

Figure 5.35

A graph of $y = \sin\frac{1}{2}A$ is shown in Figure 5.36.

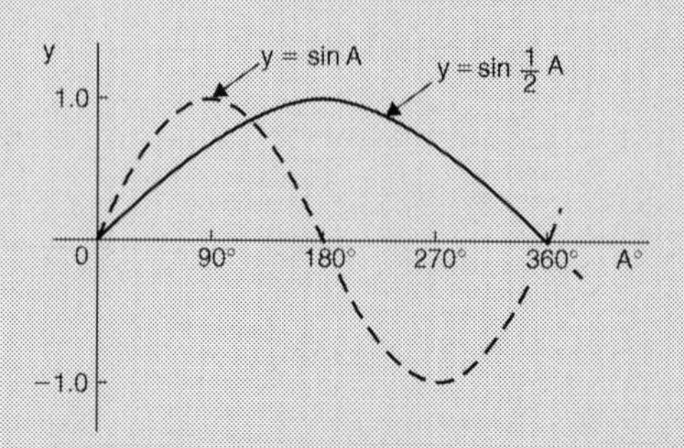

Figure 5.36

Graphs of $y = \cos A$ and $y = \cos 2A$ are shown in Figure 5.37.

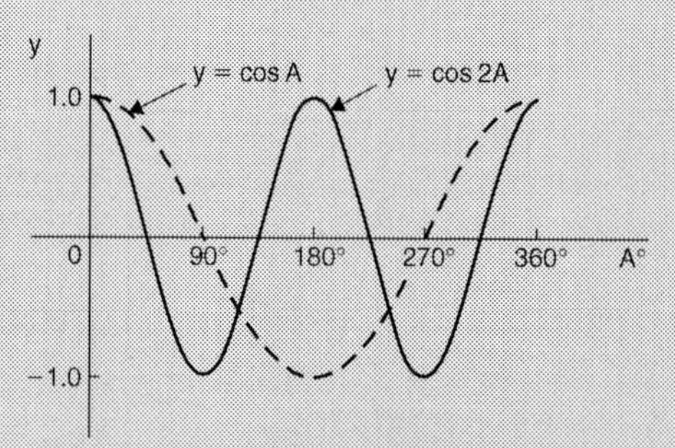

Figure 5.37

A graph of $y = \cos\frac{1}{2}A$ is shown in Figure 5.38.

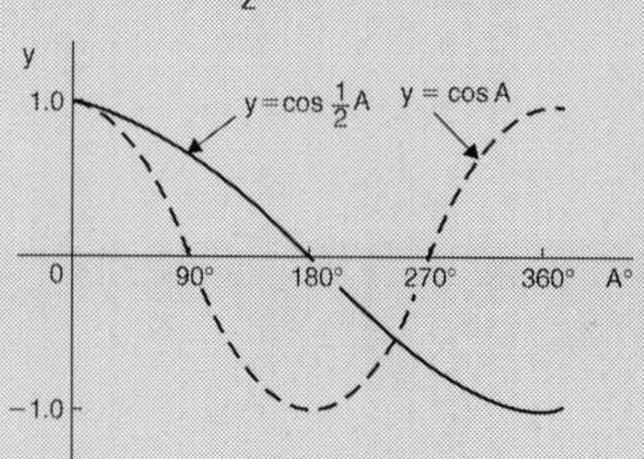

Figure 5.38

Period

If $y = \sin pA$ or $y = \cos pA$ (where p is a constant) then the period of the waveform is 360°/p (or 2π/p rad). Hence if $y = \sin 3A$ then the period is 360/3, i.e. 120°, and if $y = \cos 4A$ then the period is 360/4, i.e. 90°.

Amplitude is the name given to the maximum or peak value of a sine wave. If $y = 4\sin A$ the maximum value, and thus amplitude, is 4. Similarly, if $y = 5\cos 2A$, the amplitude is 5 and the period is 360°/2, i.e. 180°.

Lagging and leading angles

The graph $y = \sin(A - 60°)$ **lags** $y = \sin A$ by 60° as shown in Figure 5.39.

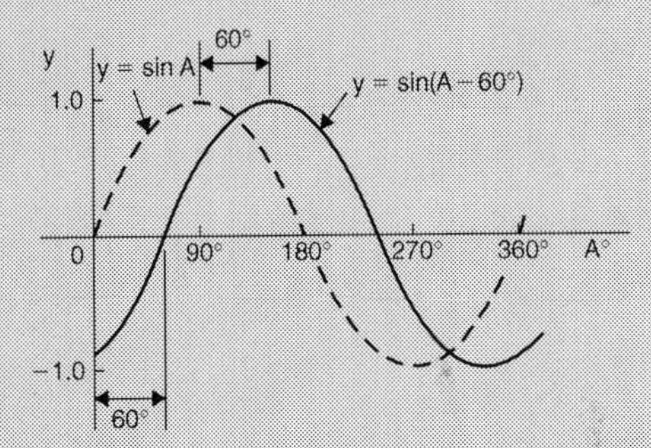

Figure 5.39

The graph of $y = \cos(A + 45°)$ **leads** $y = \cos A$ by 45° as shown in Figure 5.40.

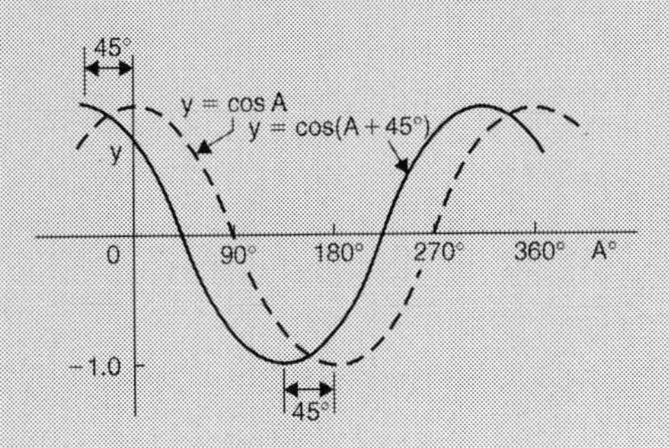

Figure 5.40

Application: Sketch $y = 3 \sin 2A$ from $A = 0$ to $A = 360°$

Amplitude $= 3$ and period $= 360/2 = 180°$

A sketch of $y = 3 \sin 2A$ is shown in Figure 5.41.

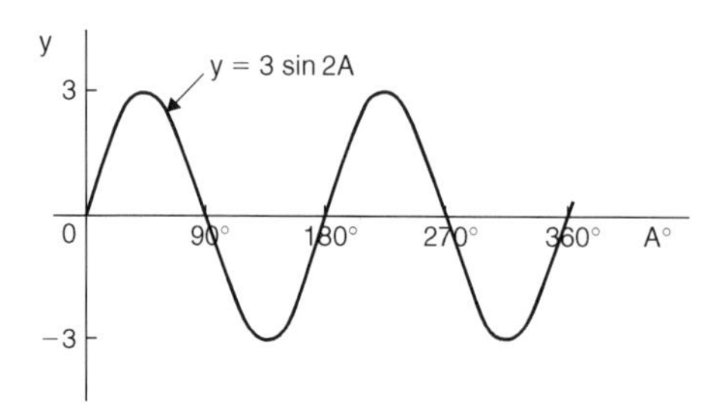

Figure 5.41

Application: Sketch $y = 4 \cos 3x$ from $x = 0°$ to $x = 360°$

Amplitude $= 4$ and period $= 360°/3 = 120°$

A sketch of $y = 4 \cos 3x$ is shown in Figure 5.42.

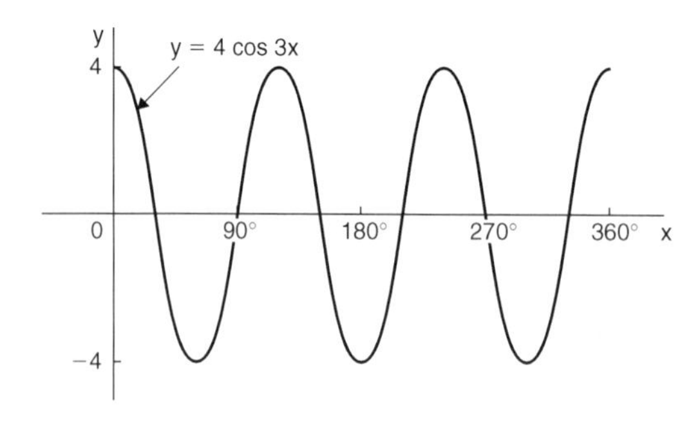

Figure 5.42

Application: Sketch $y = 5 \sin(A + 30°)$ from $A = 0°$ to $A = 360°$

Amplitude $= 5$ and period $= 360°/1 = 360°$

$5 \sin(A + 30°)$ leads $5 \sin A$ by $30°$ (i.e. starts $30°$ earlier)

A sketch of $y = 5 \sin(A + 30°)$ is shown in Figure 5.43.

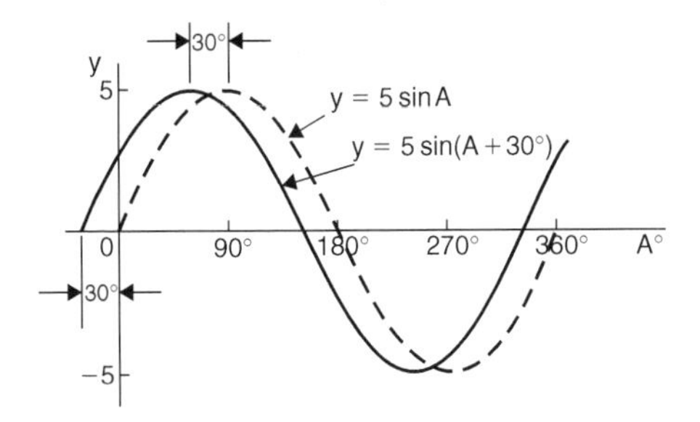

Figure 5.43

Application: Sketch $y = 7\sin(2A - \pi/3)$ in the range $0 \leq A \leq 360°$

Amplitude = 7 and period = $2\pi/2 = \pi$ radians

In general, $\mathbf{y = \sin(pt - \alpha)}$ **lags** $\mathbf{y = \sin pt}$ **by** $\boldsymbol{\alpha/p}$, hence $7\sin(2A - \pi/3)$ lags $7\sin 2A$ by $(\pi/3)/2$, i.e. $\pi/6$ rad or 30°.

A sketch of $y = 7\sin(2A - \pi/3)$ is shown in Figure 5.44.

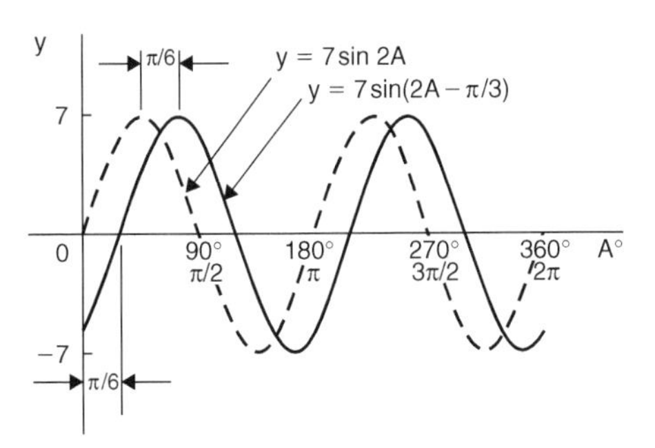

Figure 5.44

Sinusoidal form $A\sin(\omega t \pm \alpha)$

Given a general sinusoidal function $\mathbf{y = A\sin(\omega t \pm \alpha)}$, then

(i) A = amplitude

(ii) ω = angular velocity = $2\pi f$ rad/s

(iii) $\dfrac{\omega}{2\pi}$ = frequency, f hertz

(iv) $\frac{1}{f}$ = periodic time T seconds

(v) α = angle of lead or lag (compared with $y = A \sin \omega t$), in radians.

Application: An alternating current is given by $i = 30 \sin(100\pi t + 0.27)$ amperes. Find the amplitude, frequency, periodic time and phase angle (in degrees and minutes)

$i = 30 \sin(100\pi t + 0.27)$A, hence **amplitude = 30 A**

Angular velocity $\omega = 100\pi$, hence

Frequency, f $= \frac{\omega}{2\pi} = \frac{100\pi}{2\pi} =$ **50 Hz**

Periodic time, T $= \frac{1}{f} = \frac{1}{50} =$ **0.02s or 20 ms**

Phase angle, $\alpha = 0.27 \text{ rad} = \left(0.27 \times \frac{180}{\pi}\right)^\circ$

$= \mathbf{15.47^\circ}$ **leading** $\mathbf{i = 30 \sin(100\pi t)}$

Application: An oscillating mechanism has a maximum displacement of 2.5 m and a frequency of 60 Hz. At time t = 0 the displacement is 90 cm. Express the displacement in the general form $A \sin(\omega t \pm \alpha)$

Amplitude = maximum displacement = 2.5 m

Angular velocity, $\omega = 2\pi f = 2\pi(60) = 120\pi$ rad/s

Hence, displacement $= 2.5 \sin(120\pi t + \alpha)$ m

When t = 0, displacement = 90 cm = 0.90 m

Hence $0.90 = 2.5 \sin(0 + \alpha)$ i.e. $\sin \alpha = \frac{0.90}{2.5} = 0.36$

Hence $\alpha = \sin^{-1} 0.36 = 21.10^\circ = 0.368$ rad

Thus, **displacement = 2.5 sin(120πt + 0.368) m**

Application: The instantaneous value of voltage in an a.c. circuit at any time t seconds is given by $v = 340\sin(50\pi t - 0.541)$ volts. Determine

(a) the amplitude, frequency, periodic time and phase angle (in degrees),
(b) the value of the voltage when $t = 0$,
(c) the value of the voltage when $t = 10$ ms,
(d) the time when the voltage first reaches 200 V, and
(e) the time when the voltage is a maximum

(a) **Amplitude = 340 V**

Angular velocity, $\omega = 50\pi = 2\pi f$

Frequency, f $= \dfrac{\omega}{2\pi} = \dfrac{50\pi}{2\pi} = \mathbf{25\,Hz}$

Periodic time, T $= \dfrac{1}{f} = \dfrac{1}{25} = \mathbf{0.04\,s}$ or $\mathbf{40\,ms}$

Phase angle $= 0.541 \text{ rad} = \left(0.541 \times \dfrac{180}{\pi}\right)^{\circ}$

$= \mathbf{31^{\circ}\ lagging}\ v = 340\sin(50\pi t)$

(b) **When t = 0, v** $= 340\sin(0 - 0.541) = 340\sin(-31^{\circ})$

$= \mathbf{-175.1\,V}$

(c) **When t = 10 ms** then **v** $= 340\sin(50\pi \times 10 \times 10^{-3} - 0.541)$

$= 340\sin(1.0298) = 340\sin 59^{\circ}$

$= \mathbf{291.4\ volts}$

(d) When v = 200 volts then $200 = 340\sin(50\pi t - 0.541)$

$$\frac{200}{340} = \sin(50\pi t - 0.541)$$

Hence $(50\pi t - 0.541) = \sin^{-1}\dfrac{200}{340} = 36.03^{\circ}$ or 0.6288 rad

$$50\pi t = 0.6288 + 0.541 = 1.1698$$

Hence when v = 200 V, **time, t** $= \dfrac{1.1698}{50\pi} = \mathbf{7.447\,ms}$

(e) When the voltage is a maximum, $v = 340\,V$

Hence $340 = 340\sin(50\pi t - 0.541)$

$1 = \sin(50\pi t - 0.541)$

$50\pi t - 0.541 = \sin^{-1}1 = 90°$ or 1.5708 rad

$50\pi t = 1.5708 + 0.541 = 2.1118$

Hence, **time, t** $= \dfrac{2.1118}{50\pi} = $ **13.44 ms**

A sketch of $v = 340\sin(50\pi t - 0.541)$ volts is shown in Figure 5.45.

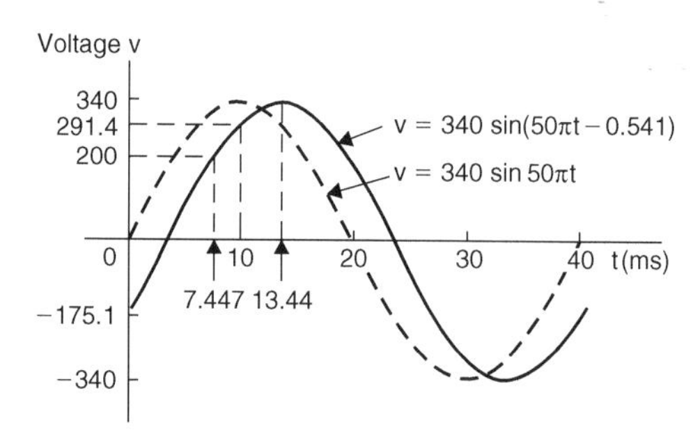

Figure 5.45

5.13 Trigonometric identities and equations

$$\tan\theta = \frac{\sin\theta}{\cos\theta} \quad \cot\theta = \frac{\cos\theta}{\sin\theta} \quad \sec\theta = \frac{1}{\cos\theta}$$

$$\text{cosec}\,\theta = \frac{1}{\sin\theta} \quad \cot\theta = \frac{1}{\tan\theta}$$

$$\cos^2\theta + \sin^2\theta = 1 \quad 1 + \tan^2\theta = \sec^2\theta \quad \cot^2\theta + 1 = \text{cosec}^2\theta$$

Equations of the type $a\sin^2 A + b\sin A + c = 0$

(i) **When $a = 0$**, $b\sin A + c = 0$, hence

$\sin A = -\dfrac{c}{b}$ and $\mathbf{A = \sin^{-1}\left(-\dfrac{c}{b}\right)}$

There are two values of A between 0° and 360° that satisfy such an equation, provided $-1 \le \dfrac{c}{b} \le 1$

(ii) **When b = 0**, $a\sin^2 A + c = 0$, hence

$$\sin^2 A = -\frac{c}{a},\ \sin A = \sqrt{\left(-\frac{c}{a}\right)} \text{ and } \mathbf{A = \sin^{-1}\sqrt{\left(-\frac{c}{a}\right)}}$$

If either a or c is a negative number, then the value within the square root sign is positive. Since when a square root is taken there is a positive and negative answer there are four values of A between 0° and 360° which satisfy such an equation, provided $-1 \leq \frac{c}{a} \leq 1$

(iii) **When a, b and c are all non-zero:**
$a\sin^2 A + b\sin A + c = 0$ is a quadratic equation in which the unknown is sin A. The solution of a quadratic equation is obtained either by factorising (if possible) or by using the quadratic formula:

$$\mathbf{\sin A = \frac{-b \pm \sqrt{(b^2 - 4ac)}}{2a}}$$

(iv) Often the trigonometric identities $\cos^2 A + \sin^2 A = 1$, $1 + \tan^2 A = \sec^2 A$ and $\cot^2 A + 1 = \text{cosec}^2 A$ need to be used to reduce equations to one of the above forms.

Application: Solve the trigonometric equation $5\sin\theta + 3 = 0$ for values of θ from 0° to 360°

$5\sin\theta + 3 = 0$, from which $\sin\theta = -3/5 = -0.6000$

Hence, $\theta = \sin^{-1}(-0.6000)$. Sine is negative in the third and fourth quadrants (see Figure 5.46). The acute angle $\sin^{-1}(0.6000) = 36.87°$ (shown as α in Figure 5.46(b)).

Hence $\theta = 180° + 36.87°$ i.e. **216.87°** or $\theta = 360° - 36.87°$ i.e. **323.13°**

Application: Solve $4\sec t = 5$ for values of t between 0° and 360°

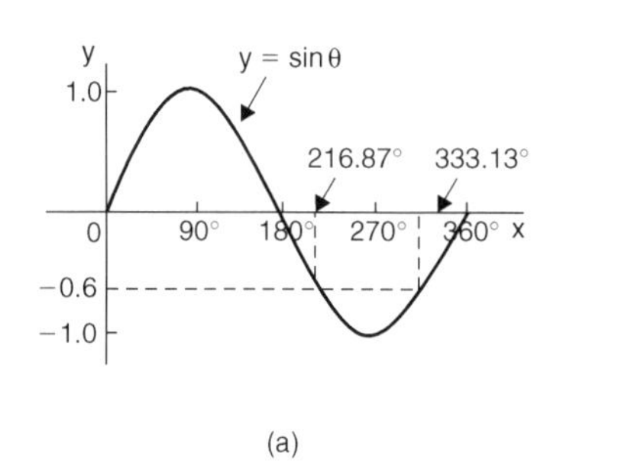

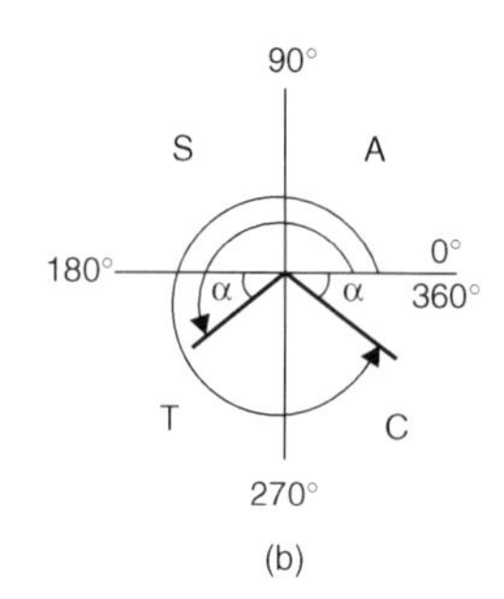

Figure 5.46

$4 \sec t = 5$, from which $\sec t = \dfrac{5}{4} = 1.2500$ and $t = \sec^{-1} 1.2500$

Secant $= \dfrac{1}{\text{cosine}}$ is positive in the first and fourth quadrants (see Figure 5.47).

The acute angle, $\sec^{-1} 1.2500 = \cos^{-1}\left(\dfrac{1}{1.2500}\right) = \cos^{-1} 0.8 = 36.87°$.

Hence, $\mathbf{t = 36.87°}$ or $360° - 36.87° = \mathbf{323.13°}$

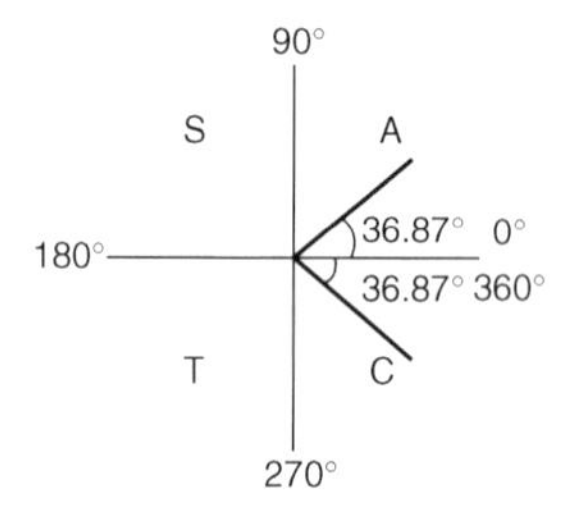

Figure 5.47

Application: Solve $2 - 4\cos^2 A = 0$ for values of A in the range $0° < A < 360°$

$2 - 4\cos^2 A = 0$, from which $\cos^2 A = \dfrac{2}{4} = 0.5000$

Hence $\cos A = \sqrt{0.5000} = \pm 0.7071$ and $A = \cos^{-1}(\pm 0.7071)$

Cosine is positive in quadrants one and four and negative in quadrants two and three. Thus in this case there are four solutions, one in each quadrant (see Figure 5.48).

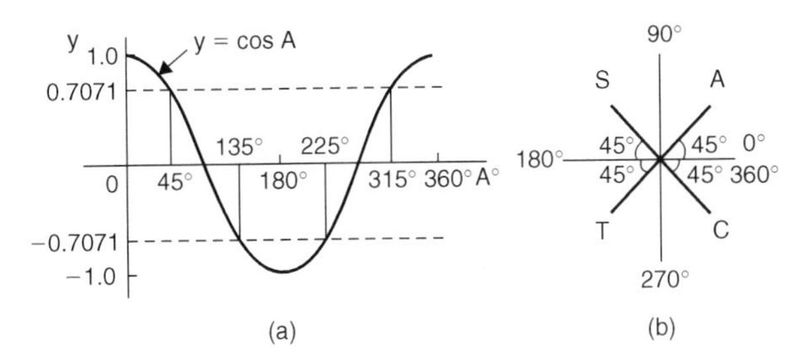

Figure 5.48

The acute angle $\cos^{-1}0.7071 = 45°$

Hence, **A = 45°, 135°, 225° or 315°**

Application: Solve $\frac{1}{2}\cot^2 y = 1.3$ for $0° < y < 360°$

$\frac{1}{2}\cot^2 y = 1.3$, from which, $\cot^2 y = 2(1.3) = 2.6$

Hence $\cot y = \sqrt{2.6} = \pm 1.6125$, and $y = \cot^{-1}(\pm 1.6125)$. There are four solutions, one in each quadrant. The acute angle $\cot^{-1}1.6125 = \tan^{-1}\left(\frac{1}{1.6125}\right) = 31.81°$

Hence, **y = 31.81°, 148.19°, 211.81° or 328.19°**

Application: Solve the equation $8\sin^2\theta + 2\sin\theta - 1 = 0$, for all values of θ between 0° and 360°

Factorising $8\sin^2\theta + 2\sin\theta - 1 = 0$ gives $(4\sin\theta - 1)(2\sin\theta + 1) = 0$

Hence $4\sin\theta - 1 = 0$, from which, $\sin\theta = \frac{1}{4} = 0.2500$

or $2\sin\theta + 1 = 0$, from which, $\sin\theta = -\frac{1}{2} = -0.5000$

(Instead of factorising, the quadratic formula can, of course, be used).

$\theta = \sin^{-1}0.250 = 14.48°$ or $165.52°$, since sine is positive in the first and second quadrants, or

$\theta = \sin^{-1}(-0.5000) = 210°$ or $330°$, since sine is negative in the third and fourth quadrants.

Hence, $\boldsymbol{\theta = 14.48°, 165.52°, 210°}$ **or** $\boldsymbol{330°}$

Application: Solve $5\cos^2 t + 3\sin t - 3 = 0$ for values of t from 0° to 360°

Since $\cos^2 t + \sin^2 t = 1$, then $\cos^2 t = 1 - \sin^2 t$

Substituting for $\cos^2 t$ in $5\cos^2 t + 3\sin t - 3 = 0$ gives

$$5(1 - \sin^2 t) + 3\sin t - 3 = 0$$
$$5 - 5\sin^2 t + 3\sin t - 3 = 0$$
$$-5\sin^2 t + 3\sin t + 2 = 0$$
$$-5\sin^2 t - 3\sin t - 2 = 0$$

Factorising gives $(5\sin t + 2)(\sin t - 1) = 0$

Hence, $5\sin t + 2 = 0$, from which, $\sin t = -\dfrac{2}{5} = -0.4000$ or $\sin t - 1 = 0$, from which, $\sin t = 1$.

$t = \sin^{-1}(-0.4000) = 203.58°$ or $336.42°$, since sine is negative in the third and fourth quadrants, or $t = \sin^{-1}1 = 90°$

Hence, $\mathbf{t = 90°, 203.58°}$ **or** $\mathbf{336.42°}$ as shown in Figure 5.49.

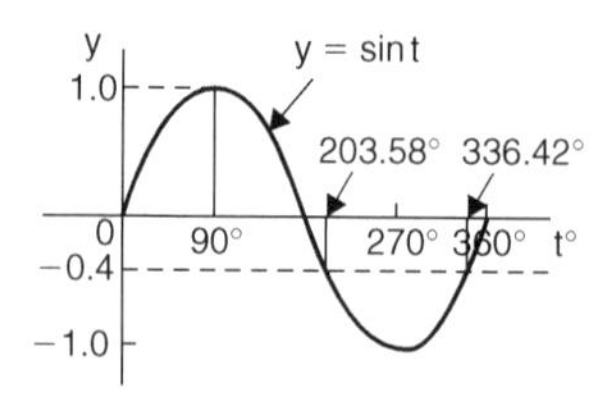

Figure 5.49

Application: Solve $18\sec^2 A - 3\tan A = 21$ for values of A between 0° and 360°

$1 + \tan^2 A = \sec^2 A$. Substituting for $\sec^2 A$ in $18\sec^2 A - 3\tan A = 21$ gives

$$18(1 + \tan^2 A) - 3\tan A = 21$$

i.e. $$18 + 18\tan^2 A - 3\tan A - 21 = 0$$

$$18\tan^2 A - 3\tan A - 3 = 0$$

Factorising gives: $(6\tan A - 3)(3\tan A + 1) = 0$

Hence, $6\tan A - 3 = 0$, from which, $\tan A = \frac{3}{6} = 0.5000$

or $3\tan A + 1 = 0$, from which, $\tan A = -\frac{1}{3} = -0.3333$

Thus, $A = \tan^{-1}(0.5000) = 26.57°$ or $206.57°$, since tangent is positive in the first and third quadrants, or $A = \tan^{-1}(-0.3333) = 161.57°$ or $341.57°$, since tangent is negative in the second and fourth quadrants. Hence, **A = 26.57°, 161.57°, 206.57° or 341.57°**

5.14 The relationship between trigonometric and hyperbolic functions

$$\cos\theta = \frac{1}{2}(e^{j\theta} + e^{-j\theta})$$

$$\sin\theta = \frac{1}{2j}(e^{j\theta} - e^{-j\theta})$$

$$\cos j\theta = \cosh\theta \quad (1)$$

$$\sin j\theta = j\sinh\theta \quad (2)$$

$$\cosh j\theta = \cos\theta$$

$$\sinh j\theta = j\sin\theta$$

$$\tan j\theta = j\tanh\theta$$

$$\tanh j\theta = j\tan\theta$$

Application: Verify that $\cos^2 j\theta + \sin^2 j\theta = 1$

From equation (3), $\cos j\theta = \cosh \theta$, and from equation (4), $\sin j\theta = j\sinh \theta$

Thus, $\cos^2 j\theta + \sin^2 j\theta = \cosh^2\theta + j^2 \sinh^2\theta$, and since $j^2 = -1$ (from chapter 8),

$$\cos^2 j\theta + \sin^2 j\theta = \cosh^2\theta - \sinh^2\theta$$

But, $\cosh^2\theta - \sinh^2\theta = 1$, from page 38,

hence $$\mathbf{\cos^2 j\theta + \sin^2 j\theta = 1}$$

Application: Determine the corresponding hyperbolic identity by writing jA for θ in $\cot^2\theta + 1 = \text{cosec}^2\theta$

Substituting jA for θ gives:

$\cot^2 jA + 1 = \text{cosec}^2 jA$, i.e. $\dfrac{\cos^2 jA}{\sin^2 jA} + 1 = \dfrac{1}{\sin^2 jA}$

But from equation (3), $\cos jA = \cosh A$

and from equation (4), $\sin jA = j\sinh A$

Hence $\dfrac{\cosh^2 A}{j^2 \sinh^2 A} + 1 = \dfrac{1}{j^2 \sinh^2 A}$

and since $j^2 = -1$, $-\dfrac{\cosh^2 A}{\sinh^2 A} + 1 = -\dfrac{1}{\sinh^2 A}$

Multiplying throughout by -1, gives:

$\dfrac{\cosh^2 A}{\sinh^2 A} - 1 = \dfrac{1}{\sinh^2 A}$ i.e. $\mathbf{\coth^2 A - 1 = \text{cosech}^2 A}$

Application: Develop the hyperbolic identity corresponding to $\sin 3\theta = 3\sin\theta - 4\sin^3\theta$ by writing jA for θ

Substituting jA for θ gives: $\sin 3jA = 3\sin jA - 4\sin^3 jA$

and since from equation (4), $\sin jA = j\sinh A$,

$$j\sinh 3A = 3j\sinh A - 4j^3\sinh^3 A$$

Dividing throughout by j gives:

$$\sinh 3A = 3\sinh A - j^2 4\sinh^3 A$$

But $j^2 = -1$, hence $$\mathbf{\sinh 3A = 3\sinh A + 4\sinh^3 A}$$

5.15 Compound angles

Compound angle addition and subtraction formulae

$\sin(A+B) = \sin A \cos B + \cos A \sin B$

$\sin(A-B) = \sin A \cos B - \cos A \sin B$

$\cos(A+B) = \cos A \cos B - \sin A \sin B$

$\cos(A-B) = \cos A \cos B + \sin A \sin B$

$$\tan(A+B) = \frac{\tan A + \tan B}{1 - \tan A \tan B}$$

$$\tan(A-B) = \frac{\tan A - \tan B}{1 + \tan A \tan B}$$

If $R\sin(\omega t + \alpha) = a\sin \omega t + b\cos \omega$ then:

$a = R\cos\alpha$, $b = R\sin\alpha$, $R = \sqrt{a^2 + b^2}$ and $\alpha = \tan^{-1} b/a$

Application: Solve the equation $4\sin(x - 20°) = 5\cos x$ for values of x between 0° and 90°

$4\sin(x - 20°) = 4[\sin x \cos 20° - \cos x \sin 20°]$,
from the formula for $\sin(A - B)$

$= 4[\sin x\,(0.9397) - \cos x\,(0.3420)]$

$= 3.7588 \sin x - 1.3680 \cos x$

Since $\quad 4\sin(x - 20°) = 5\cos x$

then $\quad 3.7588\sin x - 1.3680\cos x = 5\cos x$

Rearranging gives: $3.7588\sin x = 5\cos x + 1.3680\cos x$

$= 6.3680\cos x$

and $$\frac{\sin x}{\cos x} = \frac{6.3680}{3.7588} = 1.6942$$

i.e. $\quad \tan x = 1.6942$, and $x = \tan^{-1}1.6942 =$ **59.45°**

[Check: LHS $= 4\sin(59.45° - 20°) = 4\sin 39.45° = 2.54$

RHS $= 5\cos x = 5\cos 59.45° = 2.54$]

Application: Find an expression for $3\sin\omega t + 4\cos\omega t$ in the form $R\sin(\omega t + \alpha)$ and sketch graphs of $3\sin\omega t$, $4\cos\omega t$ and $R\sin(\omega t + \alpha)$ on the same axes

Let $\quad 3\sin\omega t + 4\cos\omega t = R\sin(\omega t + \alpha)$

then $\quad 3\sin\omega t + 4\cos\omega t = R[\sin\omega t\cos\alpha + \cos\omega t\sin\alpha]$

$$= (R\cos\alpha)\sin\omega t + (R\sin\alpha)\cos\omega t$$

Equating coefficients of $\sin\omega t$ gives:

$$3 = R\cos\alpha, \text{ from which, } \cos\alpha = \frac{3}{R}$$

Equating coefficients of $\cos\omega t$ gives:

$$4 = R\sin\alpha, \text{ from which, } \sin\alpha = \frac{4}{R}$$

There is only one quadrant where both $\sin\alpha$ **and** $\cos\alpha$ are positive, and this is the first, as shown in Figure 5.50. From Figure 5.50, by Pythagoras' theorem:

$$R = \sqrt{3^2 + 4^2} = 5$$

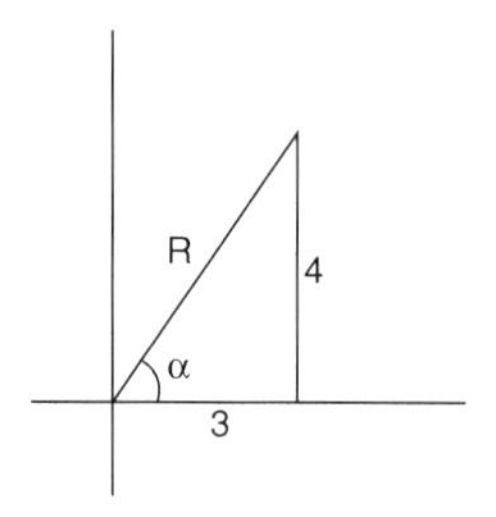

Figure 5.50

From trigonometric ratios: $\alpha = \tan^{-1}\dfrac{4}{3} = 53.13°$ or 0.927 radians

Hence $3\sin\omega t + 4\cos\omega t = 5\sin(\omega t + 0.927)$

A sketch of $3\sin\omega t$, $4\cos\omega t$ and $5\sin(\omega t + 0.927)$ is shown in Figure 5.51.

Application: Express $4.6\sin\omega t - 7.3\cos\omega t$ in the form $R\sin(\omega t + \alpha)$

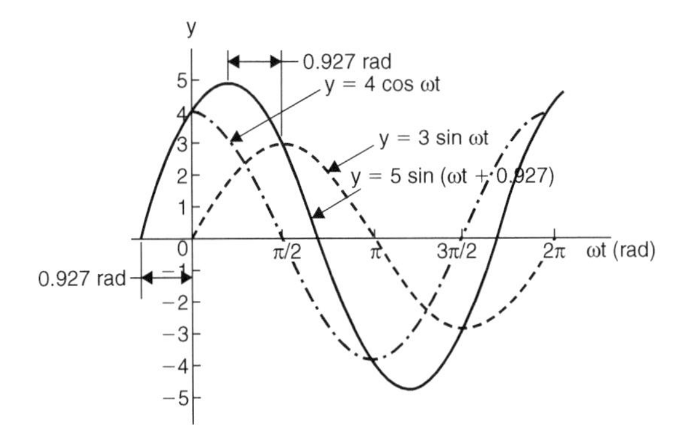

Figure 5.51

Let $4.6 \sin \omega t - 7.3 \cos \omega t = R \sin(\omega t + \alpha)$

then $4.6 \sin \omega t - 7.3 \cos \omega t = R[\sin \omega t \cos \alpha + \cos \omega t \sin \alpha]$

$$= (R \cos \alpha) \sin \omega t + (R \sin \alpha) \cos \omega t$$

Equating coefficients of $\sin \omega t$ gives:

$$4.6 = R \cos \alpha, \text{ from which, } \cos \alpha = \frac{4.6}{R}$$

Equating coefficients of $\cos \omega t$ gives:

$$-7.3 = R \sin \alpha, \text{ from which } \sin \alpha = \frac{-7.3}{R}$$

There is only one quadrant where cosine is positive **and** sine is negative, i.e. the fourth quadrant, as shown in Figure 5.52. By Pythagoras' theorem:

$$R = \sqrt{4.6^2 + (-7.3)^2} = 8.628$$

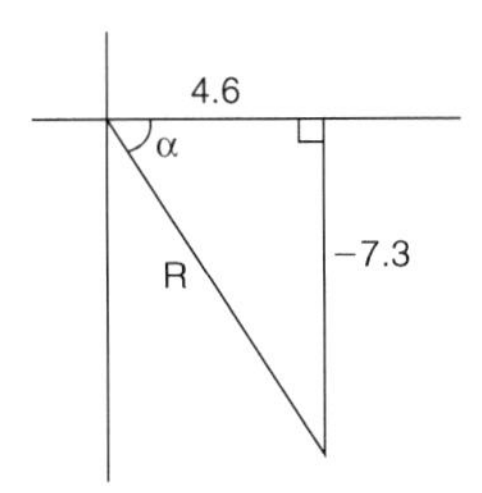

Figure 5.52

By trigonometric ratios:

$$\alpha = \tan^{-1}\left(\frac{-7.3}{4.6}\right) = -57.78° \text{ or } -1.008 \text{ radians}$$

Hence, $4.6 \sin \omega t - 7.3 \cos \omega t = 8.628 \sin(\omega t - 1.008)$

Application: Express $-2.7 \sin \omega t - 4.1 \cos \omega t$ in the form $R \sin(\omega t + \alpha)$

Let
$$-2.7 \sin \omega t - 4.1 \cos \omega t = R \sin(\omega t + \alpha)$$
$$= R\{\sin \omega t \cos \alpha + \cos \omega t \sin \alpha]$$
$$= (R \cos \alpha) \sin \omega t + (R \sin \alpha) \cos \omega t$$

Equating coefficients gives:

$$-2.7 = R \cos \alpha, \text{ from which, } \cos \alpha = \frac{-2.7}{R}$$

and
$$-4.1 = R \sin \alpha, \text{ from which, } \sin \alpha = \frac{-4.1}{R}$$

There is only one quadrant in which both cosine **and** sine are negative, i.e. the third quadrant, as shown in Figure 5.53. From Figure 5.53,

$$R = \sqrt{(-2.7)^2 + (-4.1)^2} = 4.909$$

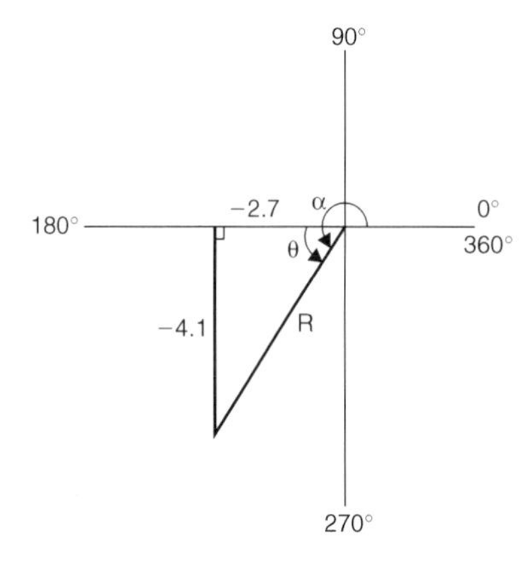

Figure 5.53

and $\theta = \tan^{-1}\dfrac{4.1}{2.7} = 56.63°$

Hence $\alpha = 180° + 56.63° = 236.63°$ or 4.130 radians

Thus, $-2.7\sin\omega t - 4.1\cos\omega t = 4.909\sin(\omega t + 4.130)$

An angle of 236.63° is the same as −123.37° or −2.153 radians

Hence, $-2.7\sin\omega t - 4.1\cos\omega t$ may also be expressed as $\mathbf{4.909\sin(\omega t - 2.153)}$, which is preferred since it is the **principal value** (i.e. $-\pi \le \alpha \le \pi$).

Double angles

$$\sin 2A = 2\sin A\cos A$$

$$\cos 2A = \cos^2 A - \sin^2 A = 1 - 2\sin^2 A = 2\cos^2 A - 1$$

$$\tan 2A = \frac{2\tan A}{1 - \tan^2 A}$$

Application: $I_3 \sin 3\theta$ is the third harmonic of a waveform. Express the third harmonic in terms of the first harmonic $\sin\theta$, when $I_3 = 1$

When $I_3 = 1$, $I_3 \sin 3\theta = \sin 3\theta = \sin(2\theta + \theta)$

$= \sin 2\theta\cos\theta + \cos 2\theta\sin\theta$, from the sin(A + B) formula

$= (2\sin\theta\cos\theta)\cos\theta + (1 - 2\sin^2\theta)\sin\theta$, from the double angle expansions

$= 2\sin\theta\cos^2\theta + \sin\theta - 2\sin^3\theta$

$= 2\sin\theta(1 - \sin^2\theta) + \sin\theta - 2\sin^3\theta$, (since $\cos^2\theta = 1 - \sin^2\theta$)

$= 2\sin\theta - 2\sin^3\theta + \sin\theta - 2\sin^3\theta$

i.e. $\mathbf{\sin 3\theta = 3\sin\theta - 4\sin^3\theta}$

Changing products of sines and cosines into sums or differences

$$\sin A \cos B = \frac{1}{2}[\sin(A+B) + \sin(A-B)] \quad (3)$$

$$\cos A \sin B = \frac{1}{2}[\sin(A+B) - \sin(A-B)] \quad (4)$$

$$\cos A \cos B = \frac{1}{2}[\cos(A+B) + \cos(A-B)] \quad (5)$$

$$\sin A \sin B = -\frac{1}{2}[\cos(A+B) - \cos(A-B)] \quad (6)$$

Application: Express sin 4x cos 3x as a sum or difference of sines and cosines

From equation (3), $\sin 4x \cos 3x = \frac{1}{2}[\sin(4x+3x) + \sin(4x-3x)]$

$$= \mathbf{\frac{1}{2}(\sin 7x + \sin x)}$$

Application: Express $2\cos 5\theta \sin 2\theta$ as a sum or difference of sines or cosines

From equation (4),

$$2\cos 5\theta \sin 2\theta = 2\left\{\frac{1}{2}[\sin(5\theta+2\theta) - \sin(5\theta-2\theta)]\right\}$$

$$= \mathbf{\sin 7\theta - \sin 3\theta}$$

Application: In an alternating current circuit, voltage $v = 5\sin\omega t$ and current $i = 10\sin(\omega t - \pi/6)$. Find an expression for the instantaneous power p at time t given that $p = vi$, expressing the answer as a sum or difference of sines and cosines

$p = vi = (5 \sin \omega t)[10 \sin(\omega t - \pi/6)] = 50 \sin \omega t \sin(\omega t - \pi/6)$

From equation (6),

$$50 \sin \omega t \sin(\omega t - \pi/6) = (50) - \frac{1}{2}\{\cos(\omega t + \omega t - \pi/6) - \cos[\omega t - (\omega t - \pi/6)]\}$$

$$= -25\{\cos(2\omega t - \pi/6) - \cos \pi/6\}$$

i.e. instantaneous power, $p = 25[\cos \pi/6 - \cos(2\omega t - \pi/6)]$

Changing sums or differences of sines and cosines into products

$$\sin X + \sin Y = 2 \sin\left(\frac{X+Y}{2}\right) \cos\left(\frac{X-Y}{2}\right) \quad (7)$$

$$\sin X - \sin Y = 2 \cos\left(\frac{X+Y}{2}\right) \sin\left(\frac{X-Y}{2}\right) \quad (8)$$

$$\cos X + \cos Y = 2 \cos\left(\frac{X+Y}{2}\right) \cos\left(\frac{X-Y}{2}\right) \quad (9)$$

$$\cos X - \cos Y = -2 \sin\left(\frac{X+Y}{2}\right) \sin\left(\frac{X-Y}{2}\right) \quad (10)$$

Application: Express $\sin 5\theta + \sin 3\theta$ as a product

From equation (7),

$$\sin 5\theta + \sin 3\theta = 2 \sin\left(\frac{5\theta + 3\theta}{2}\right) \cos\left(\frac{5\theta - 3\theta}{2}\right) = \mathbf{2 \sin 4\theta \cos \theta}$$

Application: Express sin 7x − sin x as a product

From equation (9),

$$\sin 7x - \sin x = 2\cos\left(\frac{7x + x}{2}\right)\sin\left(\frac{7x - x}{2}\right) = \mathbf{2\cos 4x \sin 3x}$$

Application: Express cos 2t − cos 5t as a product

From equation (10),

$$\cos 2t - \cos 5t = -2\sin\left(\frac{2t + 5t}{2}\right)\sin\left(\frac{2t - 5t}{2}\right)$$

$$= -2\sin\frac{7}{2}t\sin\left(-\frac{3}{2}t\right) = \mathbf{2\sin\frac{7}{2}t\sin\frac{3}{2}t}$$

$$\left[\text{since } \sin\left(-\frac{3}{2}t\right) = -\sin\frac{3}{2}t\right]$$

6 Graphs

6.1 The straight line graph

The equation of a straight line graph is: $\mathbf{y = mx + c}$
where m is the gradient and c the y-axis intercept.

With reference to Figure 6.1, **gradient** $\mathbf{m = \frac{y_2 - y_1}{x_2 - x_1}}$

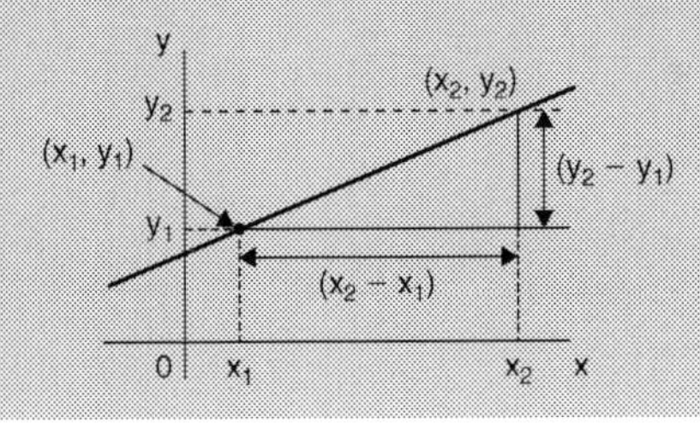

Figure 6.1

Application: Determine the gradient of the straight-line graph passing through the co-ordinates (−2, 5) and (3, 4)

A straight line graph passing through co-ordinates (x_1, y_1) and (x_2, y_2) has a gradient given by:

$$m = \frac{y_2 - y_1}{x_2 - x_1} \qquad \text{(see Figure 6.1)}$$

A straight line passes through (−2, 5) and (3, 4), from which, $x_1 = -2$, $y_1 = 5$, $x_2 = 3$ and $y_2 = 4$, hence

$$\text{gradient, } m = \frac{y_2 - y_1}{x_2 - x_1} = \frac{4 - 5}{3 - (-2)} = -\frac{\mathbf{1}}{\mathbf{5}}$$

Application: The temperature in degrees Celsius and the corresponding values in degrees Fahrenheit are shown in the table below.

°C	10	20	40	60	80	100
°F	50	68	104	140	176	212

Plot a graph of degrees Celsius (horizontally) against degrees Fahrenheit (vertically). From the graph find (a) the temperature in degrees Fahrenheit at 55°C, (b) the temperature in degrees Celsius at 167°F, (c) the Fahrenheit temperature at 0°C, and (d) the Celsius temperature at 230°F

Axes with suitable scales are shown in Figure 6.2. The co-ordinates (10, 50), (20, 68), (40, 104), and so on are plotted as shown. When the co-ordinates are joined, a straight line is produced. Since a straight line results there is a linear relationship between degrees Celsius and degrees Fahrenheit.

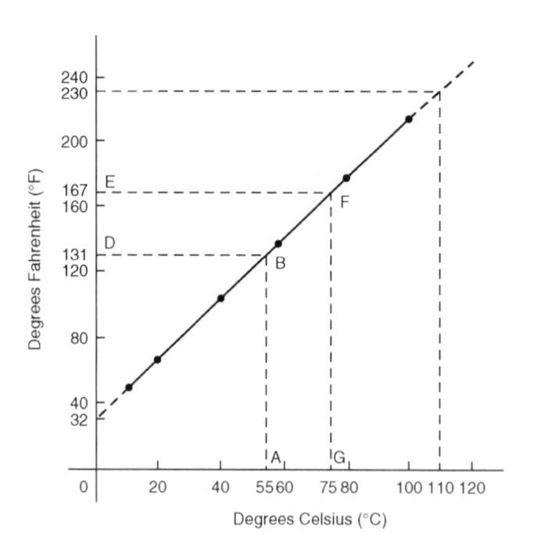

Figure 6.2

(a) To find the Fahrenheit temperature at 55°C, a vertical line AB is constructed from the horizontal axis to meet the straight line at B. The point where the horizontal line BD meets the vertical axis indicates the equivalent Fahrenheit temperature. **Hence 55°C is equivalent to 131°F.** This process of finding an equivalent value in between the given information in the above table is called **interpolation**.

(b) To find the Celsius temperature at 167°F, a horizontal line EF is constructed as shown in Figure 6.2. The point where the vertical line FG cuts the horizontal axis indicates the equivalent Celsius temperature. **Hence 167°F is equivalent to 75°C.**

(c) If the graph is assumed to be linear even outside of the given data, then the graph may be extended at both ends (shown by broken lines in Figure 6.2). From Figure 6.2, it is seen that **0°C corresponds to 32°F.**

(d) **230°F is seen to correspond to 110°C.**

The process of finding equivalent values outside of the given range is called **extrapolation**.

Application: Experimental tests to determine the breaking stress σ of rolled copper at various temperatures t gave the following results.

Stress σ N/cm^2	8.46	8.04	7.78	7.37	7.08	6.63
Temperature t°C	70	200	280	410	500	640

Show that the values obey the law $\sigma = at + b$, where a and b are constants and determine approximate values for a and b. Use the law to determine the stress at 250°C and the temperature when the stress is 7.54 N/cm^2

The co-ordinates (70, 8.46), (200, 8.04), and so on, are plotted as shown in Figure 6.3. Since the graph is a straight line then the values obey the law $\sigma = at + b$, and the gradient of the straight line, is:

$$a = \frac{AB}{BC} = \frac{8.36 - 6.76}{100 - 600} = \frac{1.60}{-500} = \mathbf{-0.0032}$$

Vertical axis intercept, **b = 8.68**

Hence the law of the graph is: $\boldsymbol{\sigma = -0.0032t + 8.68}$

When the temperature is 250°C, stress σ is given by

$$\sigma = -0.0032(250) + 8.68 = \mathbf{7.88\ N/cm^2}$$

Rearranging $\sigma = -0.0032t + 8.68$ gives:

$$0.0032t = 8.68 - \sigma, \quad \text{i.e.} \quad t = \frac{8.68 - \sigma}{0.0032}$$

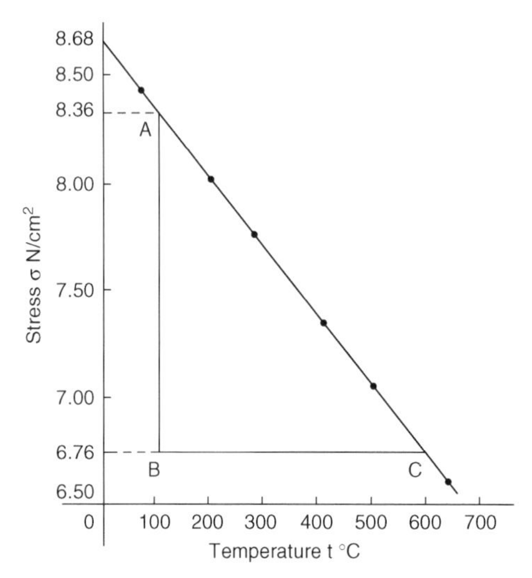

Figure 6.3

Hence, when the stress, $\sigma = 7.54$ N/cm^2,

$$\text{temperature } t = \frac{8.68 - 7.54}{0.0032} = \mathbf{356.3°C}$$

6.2 Determination of law

Some examples of the reduction of equations to linear form include:

1. $y = ax^2 + b$ compares with $Y = mX + c$, where $m = a$, $c = b$ and $X = x^2$.
   Hence y is plotted vertically against x^2 horizontally to produce a straight line graph of gradient 'a' and y-axis intercept 'b'

2. $y + \frac{a}{x} + b$

   y is plotted vertically against $\frac{1}{x}$ horizontally to produce a straight line graph of gradient 'a' and y-axis intercept 'b'

3. $y = ax^2 + bx$

   Dividing both sides by x gives $\frac{y}{x} = ax + b$

   Comparing with $Y = mX + c$ shows that $\frac{y}{x}$ is plotted vertically against x horizontally to produce a straight line graph of gradient 'a' and $\frac{y}{x}$ axis intercept 'b'.

Determination of law involving logarithms

4. If $\mathbf{y = ax^n}$ then $\lg y = \lg(ax^n) = \lg a + \lg x^n$
   i.e. $\mathbf{\lg y = n \lg x + \lg a}$

5. If $\mathbf{y = ab^x}$ then $\lg y = \lg(ab^x) = \lg a + \lg b^x = \lg a + x \lg b$
   i.e. $\mathbf{\lg y = (\lg b)x + \lg a}$

6. If $\mathbf{y = ae^{bx}}$ then $\ln y = \ln(ae^{bx}) = \ln a + \ln(e^{bx}) = \ln a + bx$
   i.e. $\mathbf{\ln y = bx + \ln a}$

Application: Values of load L newtons and distance d metres obtained experimentally are shown in the following table.

Load, L (N)	32.3	29.6	27.0	23.2	18.3	12.8	10.0	6.4
distance, d (m)	0.75	0.37	0.24	0.17	0.12	0.09	0.08	0.07

Verify that the load and distance are related by a law of the form $L = \frac{a}{d} + b$ and determine approximate values of a and b. Hence calculate the load when the distance is 0.20 m and the distance when the load is 20 N

Comparing $L = \frac{a}{d} + b$ i.e. $L = a\left(\frac{1}{d}\right) + b$ with $Y = mX + c$ shows that L is to be plotted vertically against $\frac{1}{d}$ horizontally. Another table of values is drawn up as shown below.

L	32.3	29.6	27.0	23.2	18.3	12.8	10.0	6.4
d	0.75	0.37	0.24	0.17	0.12	0.09	0.08	0.07
$\frac{1}{d}$	1.33	2.70	4.17	5.88	8.33	11.11	12.50	14.29

A graph of L against $\frac{1}{d}$ is shown in Figure 6.4. A straight line can be drawn through the points, which verifies that load and distance are related by a law of the form $L = \frac{a}{d} + b$

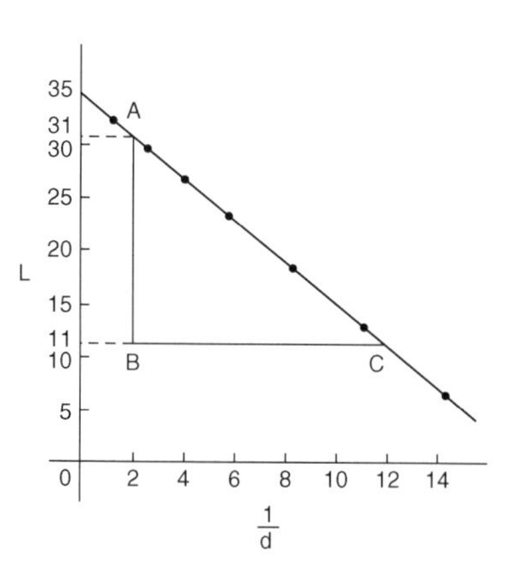

Figure 6.4

Gradient of straight line, $\mathbf{a} = \frac{AB}{BC} = \frac{31 - 11}{2 - 12} = \frac{20}{-10} = \mathbf{-2}$

L-axis intercept, **b = 35**

Hence, the law of the graph is: $\mathbf{L = -\frac{2}{d} + 35}$

When the distance d is 0.20 m, load $L = \frac{-2}{0.20} + 35 =$ **25.0 N**

Rearranging $L = -\frac{2}{d} + 35$ gives $\frac{2}{d} = 35 - L$ and $d = \frac{2}{35 - L}$

Hence, when the load L is 20 N,

$$\text{distance } d = \frac{2}{35 - 20} = \frac{2}{15} = \mathbf{0.13\,m}$$

Application: The current flowing in, and the power dissipated by a resistor are measured experimentally for various values and the results are as shown below.

Current, I amperes	2.2	3.6	4.1	5.6	6.8
Power, P watts	116	311	403	753	1110

Show that the law relating current and power is of the form $P = RI^n$, where R and n are constants, and determine the law

Taking logarithms to a base of 10 of both sides of $P = RI^n$ gives:

$\lg P = \lg(RI^n) = \lg R + \lg I^n = \lg R + n \lg I$ by the laws of logarithms

i.e. $\lg P = n \lg I + \lg R$, which is of the form $Y = mX + c$,

showing that lg P is to be plotted vertically against lg I horizontally.

A table of values for lg I and lg P is drawn up as shown below.

I	2.2	3.6	4.1	5.6	6.8
lg I	0.342	0.556	0.613	0.748	0.833
P	116	311	403	753	1110
lg P	2.064	2.493	2.605	2.877	3.045

A graph of lg P against lg I is shown in Figure 6.5 and since a straight line results the law $P = RI^n$ is verified.

$$\text{Gradient of straight line, } n = \frac{AB}{BC} = \frac{2.98 - 2.18}{0.80 - 0.40} = \frac{0.80}{0.40} = \mathbf{2}$$

It is not possible to determine the vertical axis intercept on sight since the horizontal axis scale does not start at zero. Selecting any point from the graph, say point D, where lg I = 0.70 and lg P = 2.78, and substituting values into

$$\lg P = n \lg I + \lg R$$

gives: $$2.78 = (2)(0.70) + \lg R$$

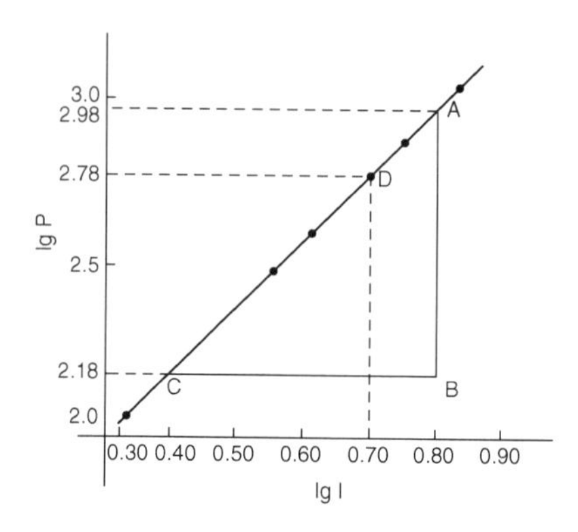

Figure 6.5

from which $\lg R = 2.78 - 1.40 = 1.38$

Hence $R = \text{antilog } 1.38\ (=10^{1.38}) = \mathbf{24.0}$

Hence the law of the graph is: $P = 24.0\, I^2$

Application: The current i mA flowing in a capacitor which is being discharged varies with time t ms as shown below.

i mA	203	61.14	22.49	6.13	2.49	0.615
t ms	100	160	210	275	320	390

Show that these results are related by a law of the form $I = Ie^{t/T}$, where I and T are constants. Determine the approximate values of I and T.

Taking Napierian logarithms of both sides of $i = Ie^{t/T}$ gives

$$\ln i = \ln(Ie^{t/T}) = \ln I + \ln e^{t/T}$$

i.e. $$\ln i = \ln I + \frac{t}{T} \text{ (since } \ln e^x = x)$$

or $$\ln i = \left(\frac{1}{T}\right)t + \ln I$$

which compares with $y = mx + c$, showing that ln i is plotted vertically against t horizontally. Another table of values is drawn up as shown below.

t	100	160	210	275	320	390
i	203	61.14	22.49	6.13	2.49	0.615
ln i	5.31	4.11	3.11	1.81	0.91	−0.49

A graph of ln i against t is shown in Figure 6.6 and since a straight line results the law $i = Ie^{t/T}$ is verified.

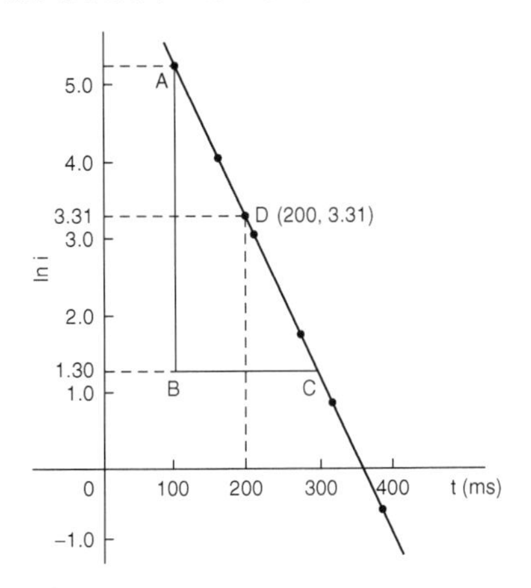

Figure 6.6

Gradient of straight line,

$$\frac{1}{T} = \frac{AB}{BC} = \frac{5.30 - 1.30}{100 - 300} = \frac{4.0}{-200} = -0.02$$

Hence, $\quad \mathbf{T} = \dfrac{1}{-0.02} = \mathbf{-50}$

Selecting any point on the graph, say point D, where t = 200 and ln i = 3.31,

and substituting into $\quad \ln i = \left(\dfrac{1}{T}\right)t + \ln I$

gives: $\quad 3.31 = -\dfrac{1}{50}(200) + \ln I$

from which, $\quad \ln I = 3.31 + 4.0 = 7.31$

and $\quad I$ = antilog 7.31 ($=e^{7.31}$) = 1495 or **1500** correct to 3 significant figures

Hence the law of the graph is $i = 1500e^{-t/50}$

6.3 Logarithmic scales

Application: Experimental values of two related quantities x and y are shown below:

x	0.41	0.63	0.92	1.36	2.17	3.95
y	0.45	1.21	2.89	7.10	20.79	82.46

The law relating x and y is believed to be $y = ax^b$, where a and b are constants.

Verify that this law is true and determine the approximate values of a and b

If $y = ax^b$ then $\lg y = b \lg x + \lg a$, from page 153, which is of the form $Y = mX + c$, showing that to produce a straight line graph lg y is plotted vertically against lg x horizontally. x and y may be plotted directly on to log-log graph paper as shown in Figure 6.7. The values of y range from 0.45 to 82.46 and 3 cycles are needed (i.e. 0.1 to 1, 1 to 10 and 10 to 100). The values of x range from 0.41 to 3.95 and 2 cycles are needed (i.e. 0.1 to 1 and 1 to 10). Hence 'log 3 cycle × 2 cycle' is used as shown in Figure 6.7 where the axes are marked and the points plotted. Since the points lie on a straight line the law $y = ax^b$ is verified.

To evaluate constants a and b:

Method 1. Any two points on the straight line, say points A and C, are selected, and AB and BC are measured (say in centimetres). Then, gradient,

$$\mathbf{b} = \frac{AB}{BC} = \frac{11.5\text{ units}}{5\text{ units}} = \mathbf{2.3}$$

Since $\lg y = b \lg x + \lg a$, when $x = 1$, $\lg x = 0$ and $\lg y = \lg a$.

The straight line crosses the ordinate $x = 1.0$ at $y = 3.5$.

Hence, $\lg a = \lg 3.5$, i.e. $\mathbf{a = 3.5}$

Method 2. Any two points on the straight line, say points A and C, are selected. A has co-ordinates (2, 17.25) and C has co-ordinates (0.5, 0.7).

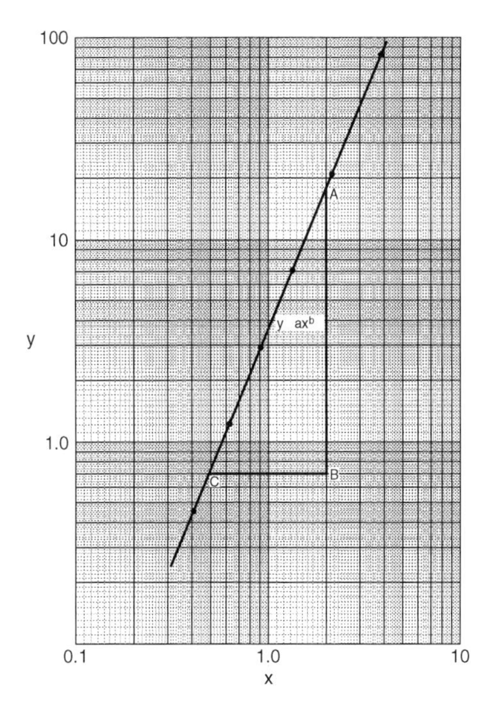

Figure 6.7

Since $y = ax^b$ then $17.25 = a(2)^b$ (1)

and $0.7 = a(0.5)^b$ (2)

i.e. two simultaneous equations are produced and may be solved for a and b.

Dividing equation (1) by equation (2) to eliminate a gives:

$$\frac{17.25}{0.7} = \frac{(2)^b}{(0.5)^b} = \left(\frac{2}{0.5}\right)^b$$

i.e. $24.643 = (4)^b$

Taking logarithms of both sides gives $\lg 24.643 = b \lg 4$,

i.e. $b = \dfrac{\lg 24.643}{\lg 4} = 2.3$, correct to 2 significant figures.

Substituting b = 2.3 in equation (1) gives: $17.25 = a(2)^{2.3}$, i.e.

$$a = \frac{17.25}{(2)^{2.3}} = \frac{17.25}{4.925} = 3.5 \text{ correct to 2 significant figures.}$$

Hence the law of the graph is: $y = 3.5x^{2.3}$

Application: The pressure p and volume v of a gas are believed to be related by a law of the form $p = cv^n$, where c and n are constants. Experimental values of p and corresponding values of v obtained in a laboratory are:

p (Pascals)	2.28×10^5	8.04×10^5	2.03×10^6	5.05×10^6	1.82×10^7
v (m^3)	3.2×10^{-2}	1.3×10^{-2}	6.7×10^{-3}	3.5×10^{-3}	1.4×10^{-3}

Verify that the law is true and determine approximate values of c and n

Since $p = cv^n$, then $\lg p = n \lg v + \lg c$, which is of the form $Y = mX + c$, showing that to produce a straight line graph, lg p is plotted vertically against lg v horizontally. The co-ordinates are plotted on 'log 3 cycle × 2 cycle' graph paper as shown in Figure 6.8. With the data expressed in standard form, the axes are marked in standard form also. Since a straight line results the law $p = cv^n$ is verified.

The straight line has a negative gradient and the value of the gradient is given

by: $\frac{AB}{BC} = \frac{14 \text{ units}}{10 \text{ units}} = 1.4.$ Hence $n = \mathbf{-1.4}$

Selecting any point on the straight line, say point C, having co-ordinates $(2.63 \times 10^{-2}, 3 \times 10^5)$, and substituting these values in $p = cv^n$ gives: $3 \times 10^5 = c(2.63 \times 10^{-2})^{-1.4}$

Hence,

$$c = \frac{3 \times 10^5}{(2.63 \times 10^{-2})^{-1.4}} = \frac{3 \times 10^5}{(0.0263)^{-1.4}}$$

$$= \frac{3 \times 10^5}{1.63 \times 10^2} = \mathbf{1840}\text{, correct to 3 significant figures.}$$

Hence the law of the graph is: $p = 1840v^{-1.4}$ or **$pv^{1.4} = 1840$**

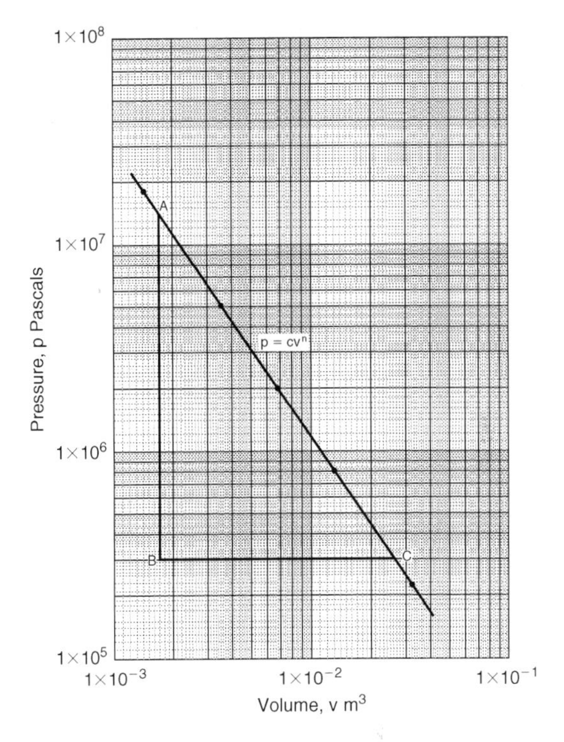

Figure 6.8

Application: The voltage, v volts, across an inductor is believed to be related to time, t ms, by the law $v = Ve^{t/T}$, where V and T are constants. Experimental results obtained are:

v volts	883	347	90	55.5	18.6	5.2
t ms	10.4	21.6	37.8	43.6	56.7	72.0

Show that the law relating voltage and time is as stated and determine the approximate values of V and T. Find also the value of voltage after 25 ms and the time when the voltage is 30.0 V

Since $\quad v = Ve^{t/T}$ then $\ln v = \dfrac{1}{T}t + \ln V$,

which is of the form $Y = mX + c$

Using 'log 3 cycle × linear' graph paper, the points are plotted as shown in Figure 6.9. Since the points are joined by a straight line the law $v = Ve^{t/T}$ is verified.

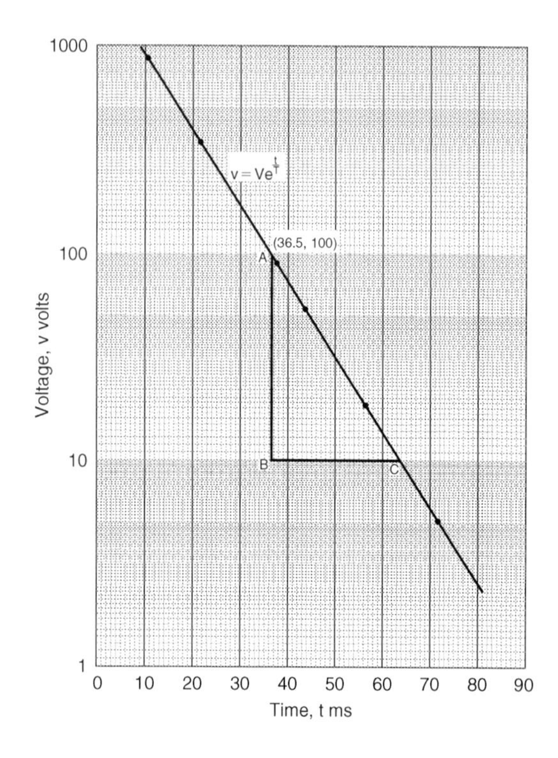

Figure 6.9

Gradient of straight line, $\dfrac{1}{T} = \dfrac{AB}{BC} = \dfrac{\ln 100 - \ln 10}{36.5 - 64.2} = \dfrac{2.3026}{-27.7}$

Hence $\quad \mathbf{T} = \dfrac{-27.7}{2.3026} = \mathbf{-12.0}$, correct to 3 significant figures.

Since the straight line does not cross the vertical axis at $t = 0$ in Figure 6.9, the value of V is determined by selecting any point, say A, having co-ordinates (36.5, 100) and substituting these values into $v = Ve^{t/T}$.

Thus $\quad 100 = Ve^{36.5/-12.0}$

i.e. $\quad V = \dfrac{100}{e^{-36.5/12.0}} =$ **2090 volts**, correct to 3 significant figures.

Hence the law of the graph is: $\mathbf{v = 2090e^{-t/12.0}}$

When time t = 25 ms, voltage v $= 2090e^{-25/12.0} = \mathbf{260\,V}$

When the voltage is 30.0 volts, $30.0 = 2090e^{-t/12.0}$

hence $e^{-t/12.0} = \dfrac{30.0}{2090}$ and $e^{t/12.0} = \dfrac{2090}{30.0} = 69.67$

Taking Napierian logarithms gives: $\dfrac{t}{12.0} = \ln 69.67 = 4.2438$

from which, **time, t**$= (12.0)(4.2438) = \mathbf{50.9\,ms}$

6.4 Graphical solution of simultaneous equations

Linear simultaneous equations in two unknowns may be solved graphically by:

1. plotting the two straight lines on the same axes, and
2. noting their point of intersection.

The co-ordinates of the point of intersection give the required solution.

Application: Solve graphically the simultaneous equations:

$$2x - y = 4$$
$$x + y = 5$$

Rearranging each equation into $y = mx + c$ form gives:

$$y = 2x - 4 \qquad (1)$$

$$y = -x + 5 \qquad (2)$$

Only three co-ordinates need be calculated for each graph since both are straight lines.

x	0	1	2
$y = 2x - 4$	-4	-2	0

x	0	1	2
$y = -x + 5$	5	4	3

Each of the graphs is plotted as shown in Figure 6.10. The point of intersection is at (3, 2) and since this is the only point which lies simultaneously on both lines then **x = 3, y = 2** is the solution of the simultaneous equations.

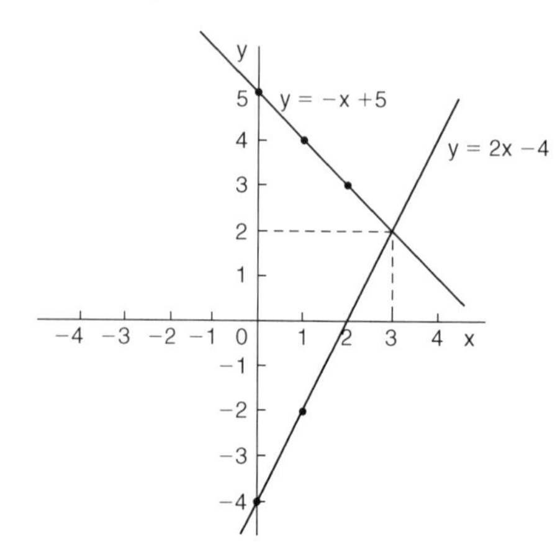

Figure 6.10

6.5 Quadratic graphs

(i) $\mathbf{y = ax^2}$

Graphs of $y = x^2$, $y = 3x^2$ and $y = \frac{1}{2}x^2$ are shown in Figure 6.11.

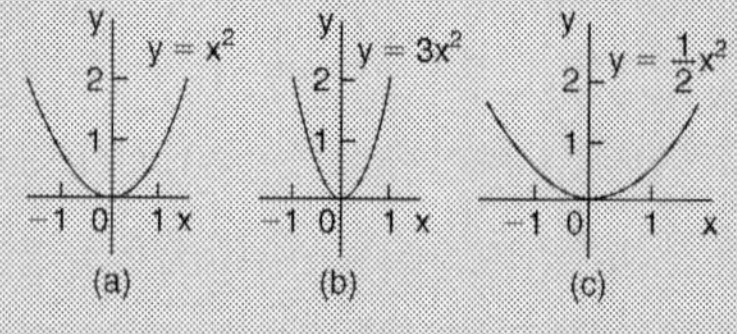

Figure 6.11

All have minimum values at the origin (0, 0).

Graphs of $y = -x^2$, $y = -3x^2$ and $y = -\frac{1}{2}x^2$ are shown in Figure 6.12.

All have maximum values at the origin (0, 0).

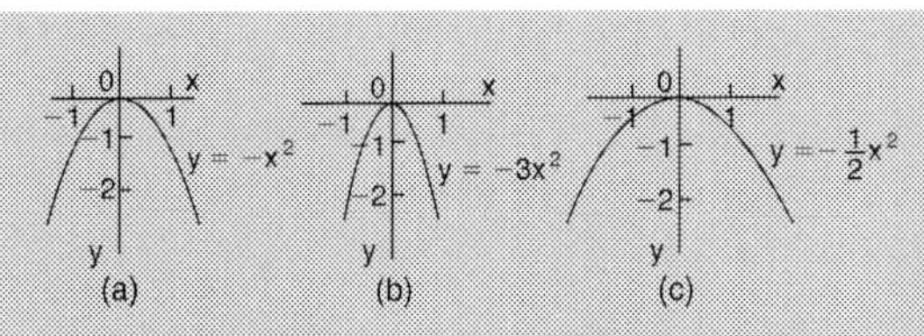

Figure 6.12

When $y = ax^2$,

(a) curves are symmetrical about the y-axis,
(b) the magnitude of 'a' affects the gradient of the curve, and
(c) the sign of 'a' determines whether it has a maximum or minimum value

(ii) $\mathbf{y = ax^2 + c}$

Graphs of $y = x^2 + 3$, $y = x^2 - 2$, $y = -x^2 + 2$ and $y = -2x^2 - 1$ are shown in Figure 6.13.

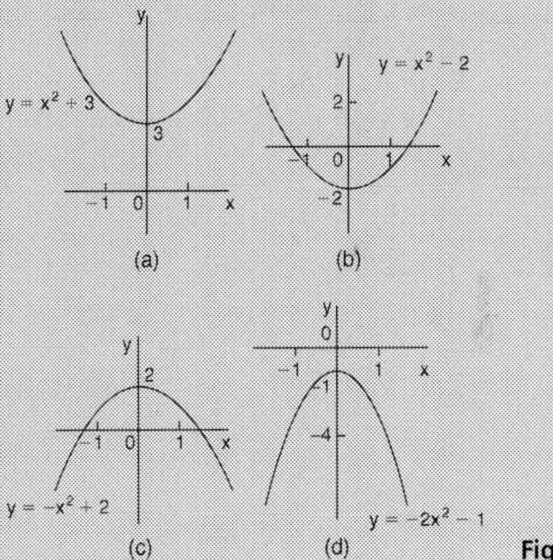

Figure 6.13

When $y = ax^2 + c$:

(a) curves are symmetrical about the y-axis,
(b) the magnitude of 'a' affects the gradient of the curve, and
(c) the constant 'c' is the y-axis intercept

(iii) $\mathbf{y = ax^2 + bx + c}$

Whenever 'b' has a value other than zero the curve is displaced to the right or left of the y-axis. When b/a is positive, the curve is displaced b/2a to the left of the y-axis, as shown in Figure 6.14(a). When b/a is negative the curve is displaced b/2a to the right of the y-axis, as shown in Figure 6.14(b).

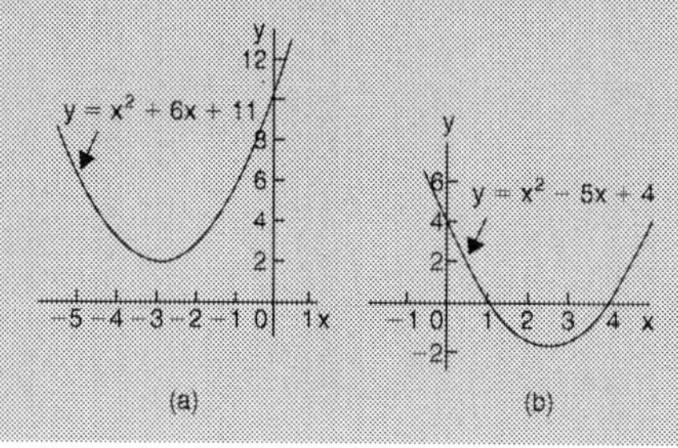

Figure 6.14

Graphical solutions of quadratic equations

Quadratic equations of the form $ax^2 + bx + c = 0$ may be solved graphically by:

(i) plotting the graph $y = ax^2 + bx + c$, and
(ii) noting the points of intersection on the x-axis (i.e. where $y = 0$).

The number of solutions, or roots of a quadratic equation, depends on how many times the curve cuts the x-axis and there can be no real roots (as in Figure 6.14(a)) or one root (as in Figures 6.11 and 6.12) or two roots (as in Figure 6.14(b)).

Application: Solve the quadratic equation $4x^2 + 4x - 15 = 0$ graphically given that the solutions lie in the range $x = -3$ to $x = 2$

Let $y = 4x^2 + 4x - 15$. A table of values is drawn up as shown below.

x	−3	−2	−1	0	1	2
$y = 4x^2 + 4x - 15$	9	−7	−15	−15	−7	9

A graph of $y = 4x^2 + 4x - 15$ is shown in Figure 6.15. The only points where $y = 4x^2 + 4x - 15$ and $y = 0$, are the points marked A and B. This occurs at **$x = -2.5$ and $x = 1.5$** and these are the solutions of the quadratic equation $4x^2 + 4x - 15 = 0$. (By substituting $x = -2.5$ and $x = 1.5$ into the original equation the solutions may be checked). The curve has a turning point at $(-0.5, -16)$ and the nature of the point is a **minimum**.

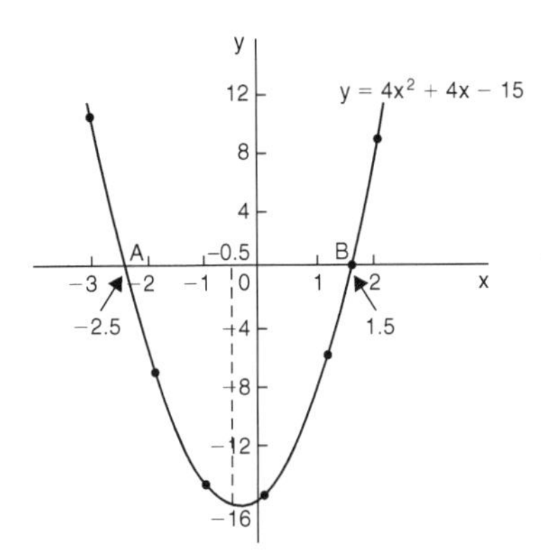

Figure 6.15

An alternative graphical method of solving $4x^2 + 4x - 15 = 0$ is to rearrange the equation as $4x^2 = -4x + 15$ and then plot two separate graphs – in this case $y = 4x^2$ and $y = -4x + 15$. Their points of intersection give the roots of equation $4x^2 = -4x + 15$, i.e. $4x^2 + 4x - 15 = 0$. This is shown in Figure 6.16, where the roots are $x = -2.5$ and $x = 1.5$ as before.

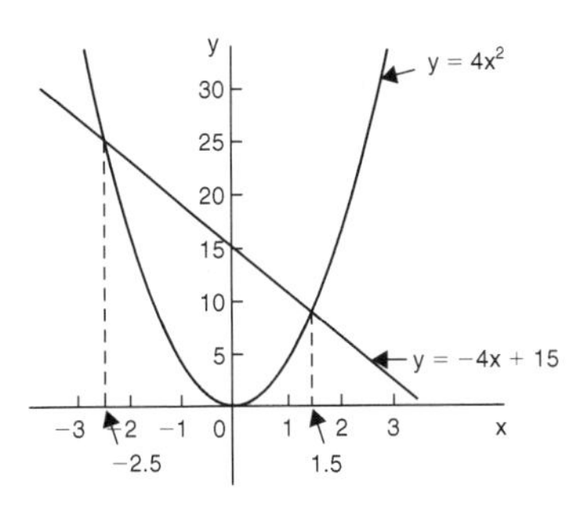

Figure 6.16

Application: Plot a graph of $y = 2x^2$ and hence solve the equations

$$\text{(a) } 2x^2 - 8 = 0 \quad \text{and} \quad \text{(b) } 2x^2 - x - 3 = 0$$

A graph of $y = 2x^2$ is shown in Figure 6.17.

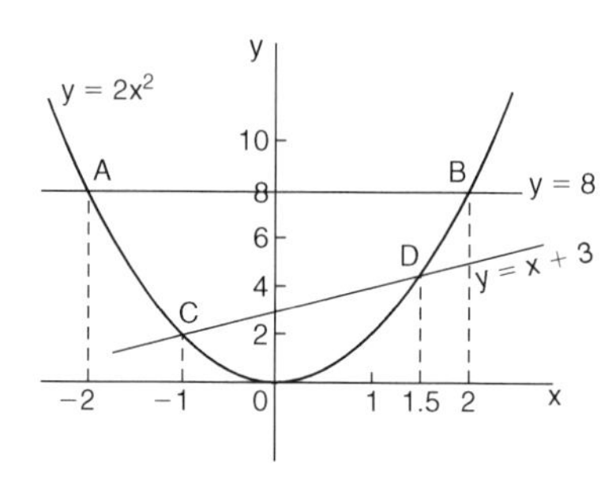

Figure 6.17

(a) Rearranging $2x^2 - 8 = 0$ gives $2x^2 = 8$ and the solution of this equation is obtained from the points of intersection of $y = 2x^2$ and $y = 8$, i.e. at co-ordinates $(-2, 8)$ and $(2, 8)$, shown as A and B, respectively, in Figure 6.17.

Hence the solutions of $2x^2 - 8 = 0$ are $\mathbf{x = -2}$ **and** $\mathbf{x = +2}$

(b) Rearranging $2x^2 - x - 3 = 0$ gives $2x^2 = x + 3$ and the solution of this equation is obtained from the points of intersection of $y = 2x^2$ and $y = x + 3$, i.e. at C and D in Figure 6.17. Hence the solutions of $2x^2 - x - 3 = 0$ are $\mathbf{x = -1}$ **and** $\mathbf{x = 1.5}$

Application: Plot the graph of $y = -2x^2 + 3x + 6$ for values of x from $x = -2$ to $x = 4$ and to use the graph to find the roots of the following equations

(a) $-2x^2 + 3x + 6 = 0$ (b) $-2x^2 + 3x + 2 = 0$

(c) $-2x^2 + 3x + 9 = 0$ (d) $-2x^2 + x + 5 = 0$

A table of values is drawn up as shown below.

x	−2	−1	0	1	2	3	4
$y = -2x^2 + 3x + 6$	−8	1	6	7	4	−3	−14

A graph of $-2x^2 + 3x + 6$ is shown in Figure 6.18.

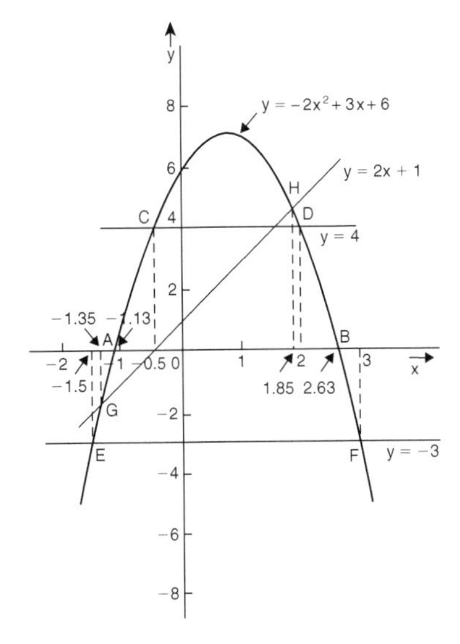

Figure 6.18

(a) The parabola $y = -2x^2 + 3x + 6$ and the straight line $y = 0$ intersect at A and B, where $\mathbf{x = -1.13}$ **and** $\mathbf{x = 2.63}$ and these are the roots of the equation $-2x^2 + 3x + 6 = 0$

(b) Comparing $\qquad y = -2x^2 + 3x + 6 \qquad (1)$

with $\qquad 0 = -2x^2 + 3x + 2 \qquad (2)$

shows that if 4 is added to both sides of equation (2), the right-hand side of both equations will be the same. Hence $4 = -2x^2 + 3x + 6$. The solution of this equation is found from the points of intersection of the line $y = 4$ and the parabola $y = -2x^2 + 3x + 6$, i.e. points C and D in Figure 6.18.

Hence the roots of $-2x^2 + 3x + 2 = 0$ are $\mathbf{x = -0.5}$ **and** $\mathbf{x = 2}$

(c) $-2x^2 + 3x + 9 = 0$ may be rearranged as $-2x^2 + 3x + 6 = -3$, and the solution of this equation is obtained from the points of intersection of the line $y = -3$ and the parabola $y = -2x^2 + 3x + 6$, i.e. at points E and F in Figure 6.18. Hence the roots of $-2x^2 + 3x + 9 = 0$ are $\mathbf{x = -1.5}$ **and** $\mathbf{x = 3}$

(d) Comparing $y = -2x^2 + 3x + 6$ (3)

with $0 = -2x^2 + x + 5$ (4)

shows that if $2x + 1$ is added to both sides of equation (4) the right-hand side of both equations will be the same. Hence equation (4) may be written as $2x + 1 = -2x^2 + 3x + 6$. The solution of this equation is found from the points of intersection of the line $y = 2x + 1$ and the parabola $y = -2x^2 + 3x + 6$, i.e. points G and H in Figure 6.18. Hence the roots of $-2x^2 + x + 5 = 0$ are **$x = -1.35$ and $x = 1.85$**

6.6 Graphical solution of cubic equations

A **cubic equation** of the form $ax^3 + bx^2 + cx + d = 0$ may be solved graphically by:

(i) plotting the graph $y = ax^3 + bx^2 + cx + d$,

and (ii) noting the points of intersection on the x-axis (i.e. where $y = 0$).

The number of solutions, or roots of a cubic equation depends on how many times the curve cuts the x-axis and there can be one, two or three possible roots, as shown in Figure 6.19.

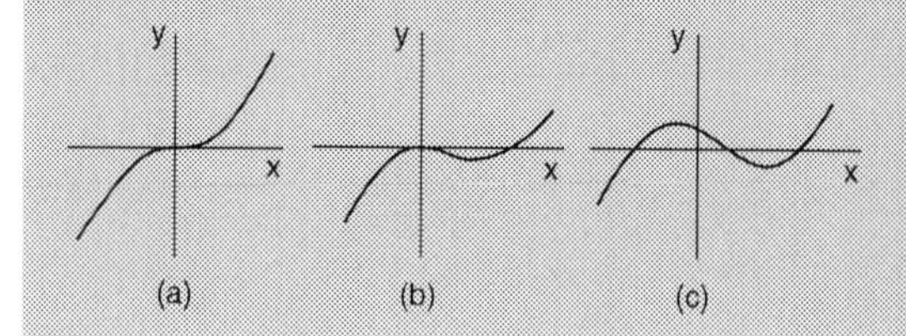

Figure 6.19

Application: Solve graphically the cubic equation $4x^3 - 8x^2 - 15x + 9 = 0$ given that the roots lie between $x = -2$ and $x = 3$. Find also the co-ordinates of the turning points on the curve.

Let $y = 4x^3 - 8x^2 - 15x + 9$. A table of values is drawn up as shown below.

x	−2	−1	0	1	2	3
y	−25	12	9	−10	−21	0

A graph of $y = 4x^3 - 8x^2 - 15x + 9$ is shown in Figure 6.20.

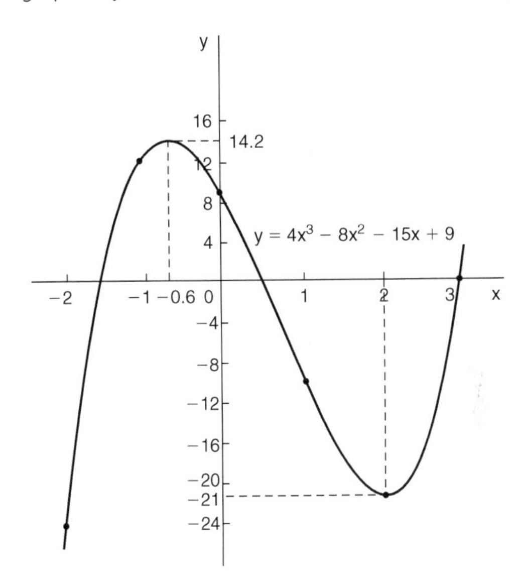

Figure 6.20

The graph crosses the x-axis (where $y = 0$) at $\mathbf{x = -1.5}$, $\mathbf{x = 0.5}$ **and** $\mathbf{x = 3}$ and these are the solutions to the cubic equation $4x^3 - 8x^2 - 15x + 9 = 0$.

The turning points occur at **(−0.6, 14.2)**, which is a **maximum**, and **(2, −21)**, which is a **minimum**.

6.7 Polar curves

Application: Plot the polar graph of $r = 5\sin\theta$ between $\theta = 0°$ and $\theta = 360°$ using increments of 30°

A table of values at 30° intervals is produced as shown below.

θ	0	30°	60°	90°	120°	150°	180°
$r = 5\sin\theta$	0	2.50	4.33	5.00	4.33	2.50	0

θ	210°	240°	270°	300°	330°	360°
$r = 5\sin\theta$	−2.50	−4.33	−5.00	−4.33	−2.50	0

The graph is plotted as shown in Figure 6.21.

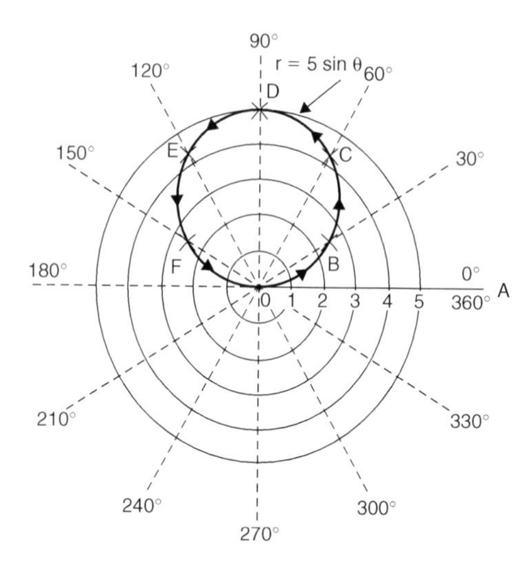

Figure 6.21

Initially the zero line 0A is constructed and then the broken lines in Figure 6.21 at 30° intervals are produced. The maximum value of r is 5.00 hence 0A is scaled and circles drawn as shown with the largest at a radius of 5 units. The polar co-ordinates (0, 0°), (2.50, 30°), (4.33, 60°), (5.00, 90°).... are plotted and shown as points 0, B, C, D, ... in Figure 6.21. When polar co-ordinate (0, 180°) is plotted and the points joined with a smooth curve a complete circle is seen to have been produced. When plotting the next point, (−2.50, 210°), since r is negative it is plotted in the opposite direction to 210°, i.e. 2.50 units long on the 30° axis. Hence the point (−2.50, 210°) is equivalent to the point (2.50, 30°). Similarly, (−4.33, 240°) is the same point as (4.33, 60°).

When all the co-ordinates are plotted the graph $r = 5\sin\theta$ appears as a single circle; it is, in fact, two circles, one on top of the other.

In general, a polar curve $\mathbf{r = a \sin \theta}$ is as shown in Figure 6.22.

In a similar manner to that explained above, it may be shown that the polar curve $\mathbf{r = a \cos \theta}$ is as sketched in Figure 6.23.

Figure 6.22

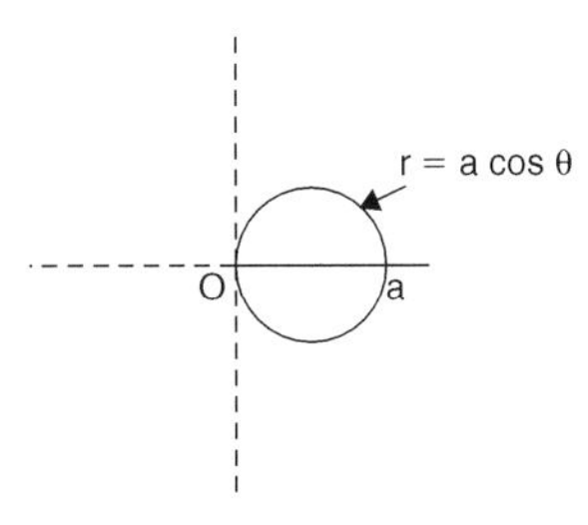

Figure 6.23

Application: Plot the polar graph of $r = 4\sin^2\theta$ between $\theta = 0$ and $\theta = 2\pi$ radians using intervals of $\frac{\pi}{6}$

A table of values is produced as shown below.

θ	0	$\frac{\pi}{6}$	$\frac{\pi}{3}$	$\frac{\pi}{2}$	$\frac{2\pi}{3}$	$\frac{5\pi}{6}$	π	$\frac{7\pi}{6}$	$\frac{4\pi}{3}$	$\frac{3\pi}{2}$	$\frac{5\pi}{3}$	$\frac{11\pi}{6}$	2π
$r = 4\sin^2\theta$	0	1	3	4	3	1	0	1	3	4	3	1	0

The zero line 0A is firstly constructed and then the broken lines at intervals of $\frac{\pi}{6}$ rad (or 30°) are produced. The maximum value of r is 4 hence 0A is scaled and circles produced as shown with the largest at a radius of 4 units.

The polar co-ordinates (0, 0), $(1, \frac{\pi}{6})$, $(3, \frac{\pi}{3})$, … $(0, \pi)$ are plotted and shown as points 0, B, C, D, E, F, 0, respectively. Then $(1, \frac{7\pi}{6})$, $(3, \frac{4\pi}{3})$,⋯ (0, 0) are plotted as shown by points G, H, I, J, K, 0 respectively. Thus two distinct loops are produced as shown in Figure 6.24.

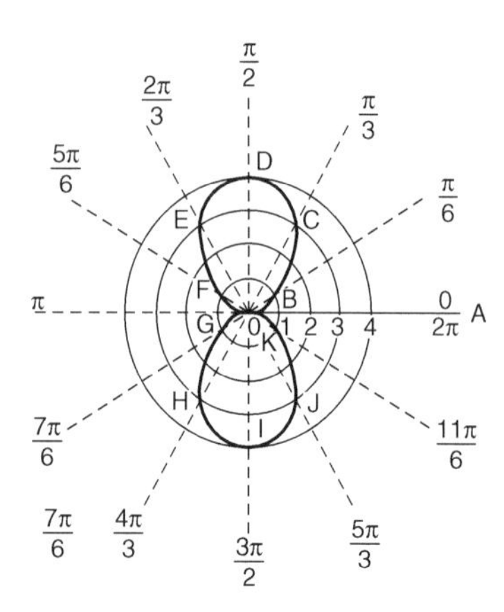

Figure 6.24

In general, a polar curve $r = a\sin^2\theta$ is as shown in Figure 6.25.

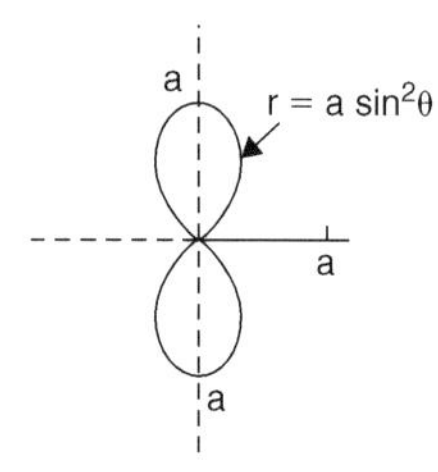

Figure 6.25

In a similar manner it may be shown that the polar curve $r = a\cos^2\theta$ is as sketched in Figure 6.26.

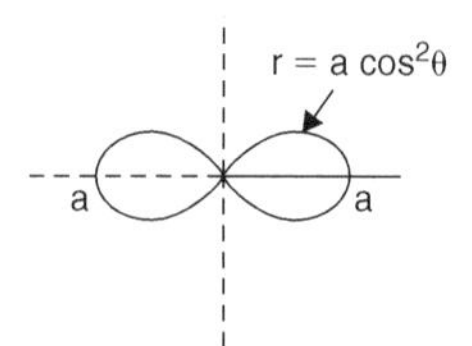

Figure 6.26

Application: Plot the polar graph of $r = 3\sin 2\theta$ between $\theta = 0°$ and $\theta = 360°$, using 15° intervals

As in previous applications a table of values may be produced.

The polar graph $r = 3\sin 2\theta$ is plotted as shown in Figure 6.27 and is seen to contain four similar shaped loops displaced at 90° from each other.

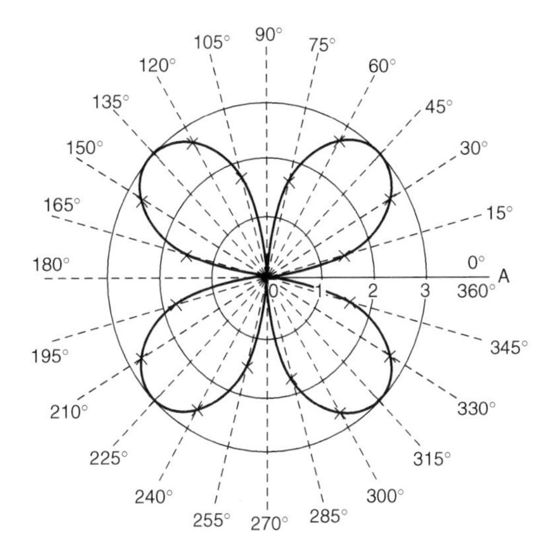

Figure 6.27

In general, a polar curve $r = a\sin 2\theta$ is as shown in Figure 6.28.

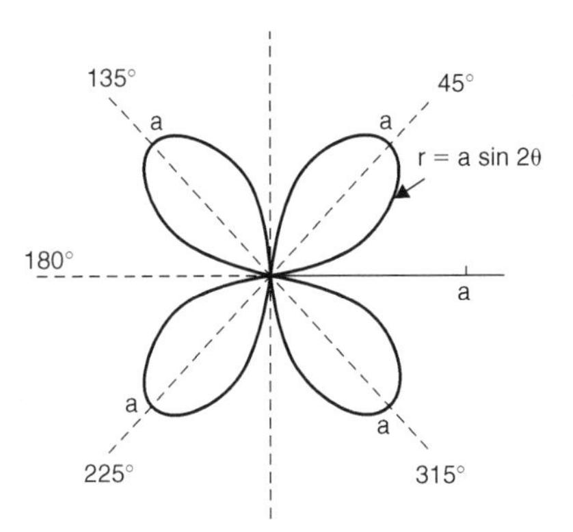

Figure 6.28

In a similar manner it may be shown that polar curves of $r = a\cos 2\theta$, $r = a\sin 3\theta$ and $r = a\cos 3\theta$ are as sketched in Figure 6.29.

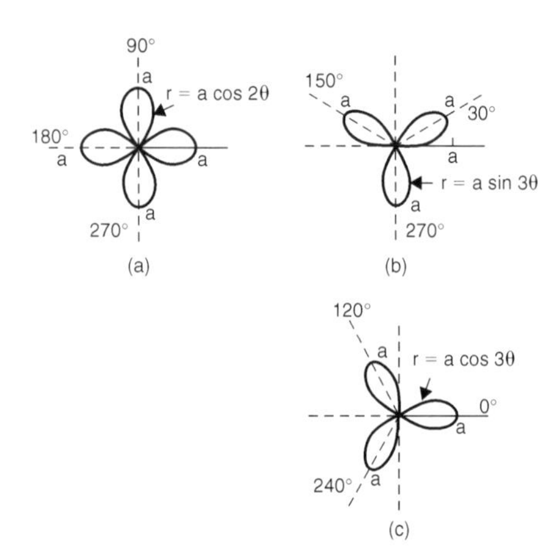

Figure 6.29

Application: Sketch the polar curve r = 2θ between θ = 0 and $\theta = \frac{5\pi}{2}$ rad at intervals of $\frac{\pi}{6}$

A table of values may be produced and the polar graph of r = 2θ is shown in Figure 6.30 and is seen to be an ever-increasing spiral.

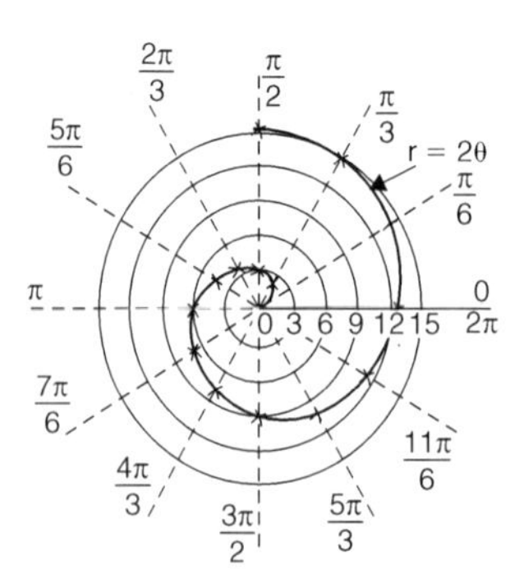

Figure 6.30

Application: Plot the polar curve r = 5(1 + cos θ) from θ = 0° to θ = 360°, using 30° intervals

A table of values may be produced and the polar curve $r = 5(1 + \cos\theta)$ is shown in Figure 6.31.

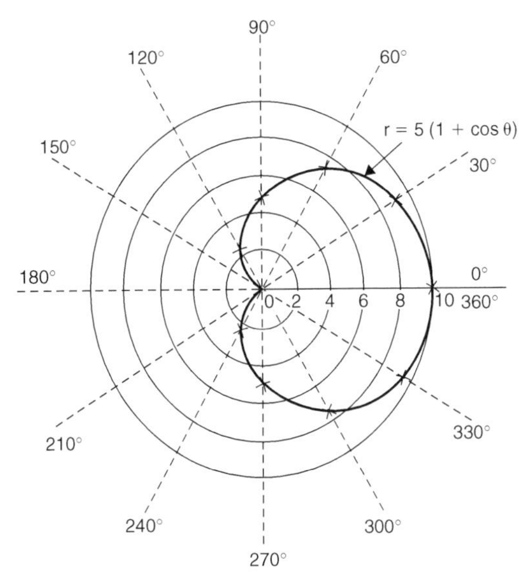

Figure 6.31

In general, a polar curve $r = a(1 + \cos\theta)$ is as shown in Figure 6.32 and the shape is called a **cardioid**.

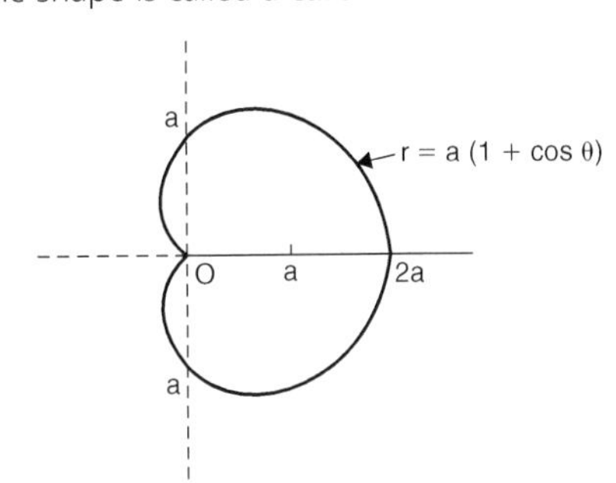

Figure 6.32

In a similar manner it may be shown that the polar curve $r = a + b\cos\theta$ varies in shape according to the relative values of a and b. When $a = b$ the polar curve shown in Figure 6.32 results.

When $a < b$ the general shape shown in Figure 6.33(a) results and when $a > b$ the general shape shown in Figure 6.33(b) results.

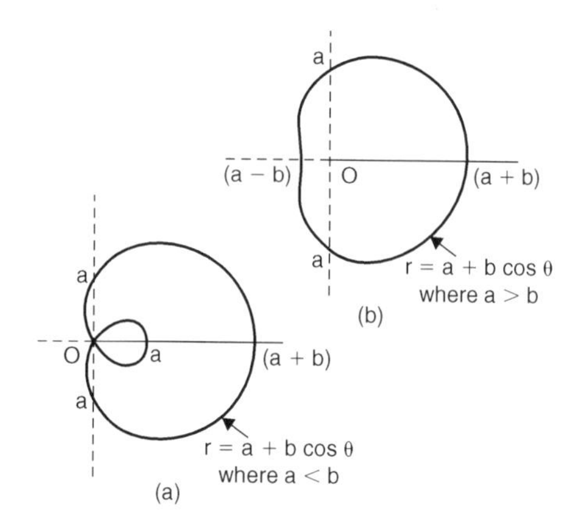

Figure 6.33

6.8 The ellipse and hyperbola

Ellipse

The equation of an ellipse is $\frac{x^2}{a^2} + \frac{y^2}{b^2} = 1$ and the general shape is as shown in Figure 6.34.

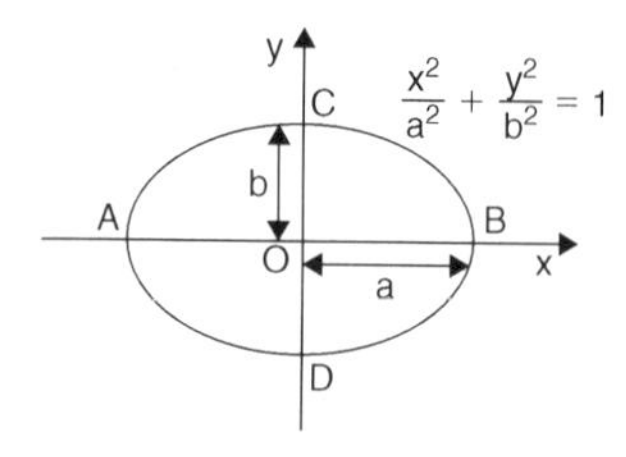

Figure 6.34

The length AB is called the **major axis** and CD the **minor axis**.

In the above equation, 'a' is the semi-major axis and 'b' is the semi-minor axis.

(Note that if $b = a$, the equation becomes $\frac{x^2}{a^2} + \frac{y^2}{a^2} = 1$, i.e. $x^2 + y^2 = a^2$, which is a circle of radius a.)

Hyperbola

The equation of a hyperbola is $\dfrac{x^2}{a^2} - \dfrac{y^2}{b^2} = 1$ and the general shape is shown in Figure 6.35. The curve is seen to be symmetrical about both the x- and y-axes. The distance AB in Figure 6.35 is given by 2a.

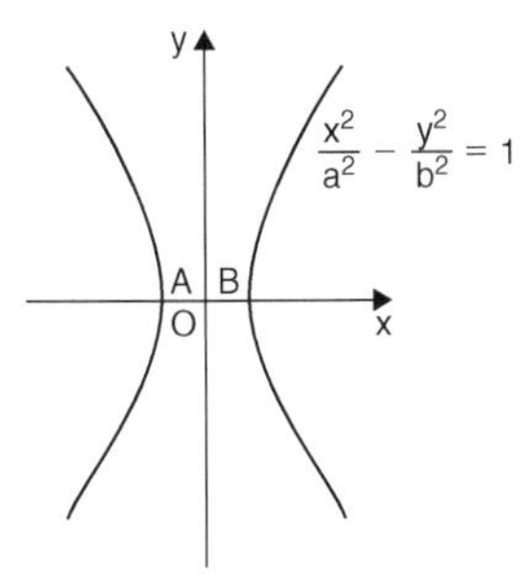

Figure 6.35

Rectangular hyperbola

The equation of a rectangular hyperbola is $xy = c$ or $y = \dfrac{c}{x}$ and the general shape is shown in Figure 6.36.

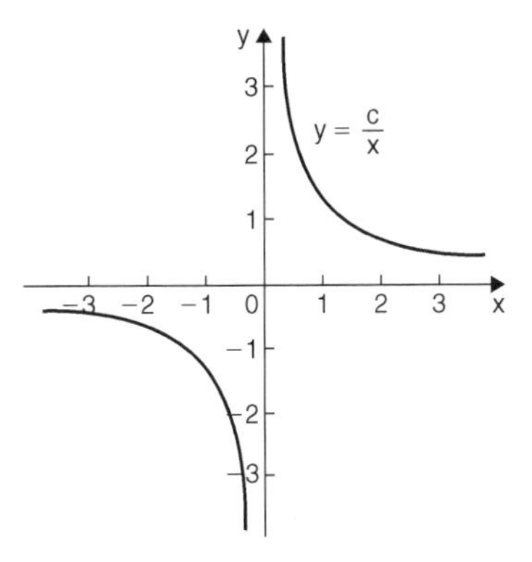

Figure 6.36

6.9 Graphical functions

Periodic functions

A function f(x) is said to be **periodic** if $f(x + T) = f(x)$ for all values of x, where T is some positive number. T is the interval between two successive repetitions and is called the **period** of the function f(x). For example, $y = \sin x$ is periodic in x with period 2π since $\sin x = \sin(x + 2\pi) = \sin(x + 4\pi)$, and so on. Similarly, $y = \cos x$ is a periodic function with period 2π since $\cos x = \cos(x + 2\pi) = \cos(x + 4\pi)$, and so on. In general, if $y = \sin \omega t$ or $y = \cos \omega t$ then the period of the waveform is $2\pi/\omega$. The function shown in Figure 6.37 is also periodic of period 2π and is defined by:

$$f(x) = \begin{cases} -1, \text{ when } -\pi \leq x \leq 0 \\ 1, \text{ when } \quad 0 \leq x \leq \pi \end{cases}$$

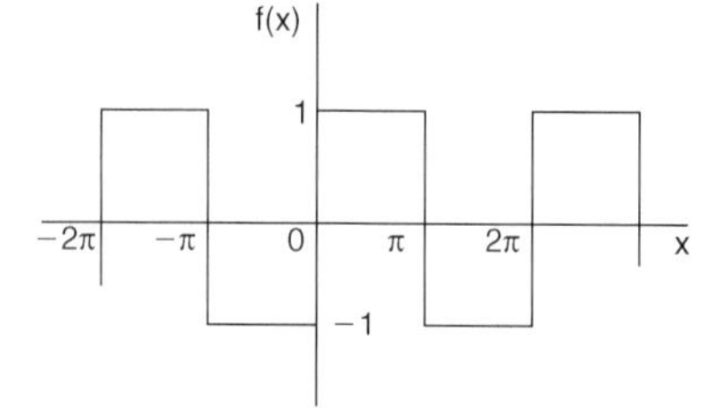

Figure 6.37

Continuous and discontinuous functions

If a graph of a function has no sudden jumps or breaks it is called a **continuous function**, examples being the graphs of sine and cosine functions. However, other graphs make finite jumps at a point or points in the interval. The square wave shown in Figure 6.37 has **finite discontinuities** as $x = \pi, 2\pi, 3\pi$, and so on, and is therefore a discontinuous function. $y = \tan x$ is another example of a discontinuous function.

Even and odd functions

A function $y = f(x)$ is said to be **even** if $f(-x) = f(x)$ for all values of x. Graphs of even functions are always **symmetrical about the**

y-axis (i.e. is a mirror image). Two examples of even functions are $y = x^2$ and $y = \cos x$ as shown in Figure 6.38.

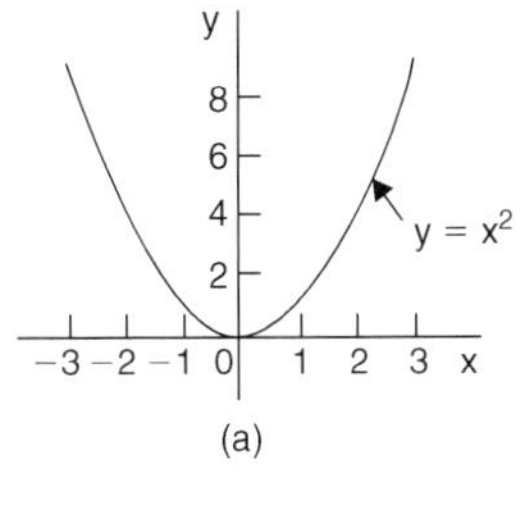

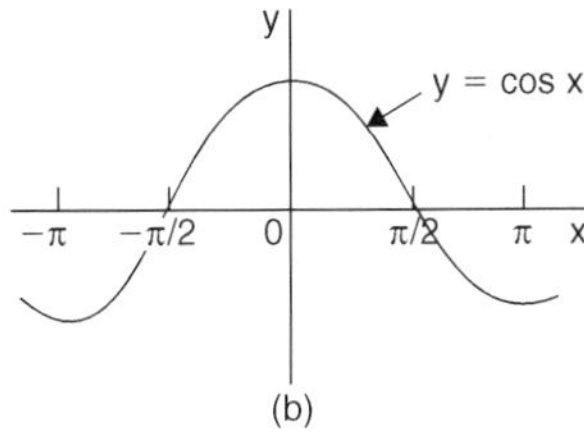

Figure 6.38

A function $y = f(x)$ is said to be **odd** if $f(-x) = -f(x)$ for all values of x. Graphs of odd functions are always **symmetrical about the origin**. Two examples of odd functions are $y = x^3$ and $y = \sin x$ as shown in Figure 6.39.

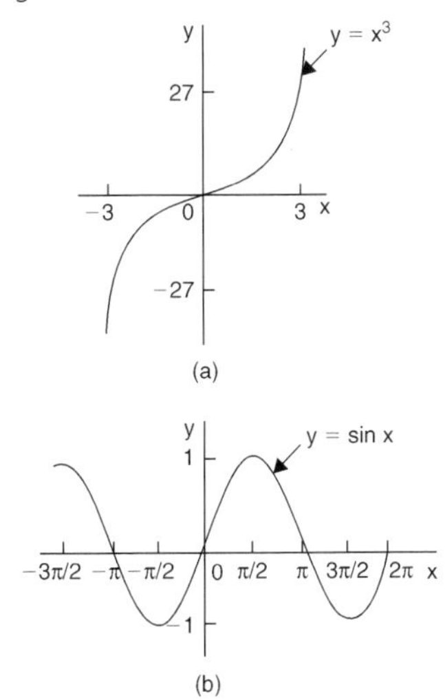

Figure 6.39

Inverse functions

Given a function $y = f(x)$, its inverse may be obtained by interchanging the roles of x and y and then transposing for y. The inverse function is denoted by $y = f^{-1}(x)$.

Application: Find the inverse of $y = 2x + 1$

(i) Transposing for x, i.e. $x = \frac{y-1}{2} = \frac{y}{2} - \frac{1}{2}$

and (ii) interchanging x and y, gives the inverse as $y = \frac{x}{2} - \frac{1}{2}$

Thus if $f(x) = 2x + 1$, then $\mathbf{f^{-1}(x) = \frac{x}{2} - \frac{1}{2}}$

A graph of $f(x) = 2x + 1$ and its inverse $f^{-1}(x) = \frac{x}{2} - \frac{1}{2}$ is shown in Figure 6.40 and $f^{-1}(x)$ is seen to be a reflection of f(x) in the line $y = x$.

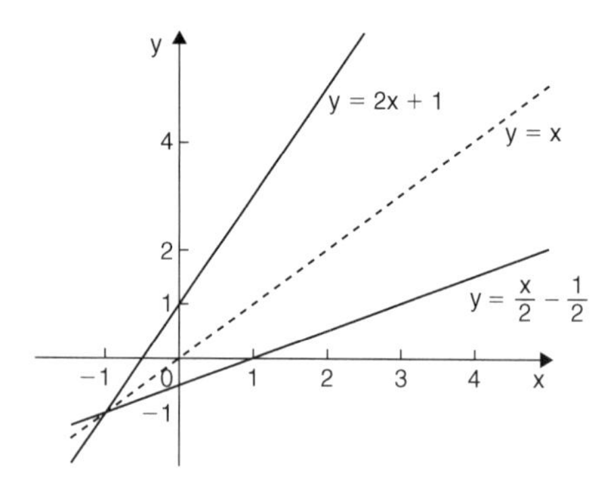

Figure 6.40

Application: Find the inverse of $y = x^2$

(i) Transposing for x, i.e. $x = \pm\sqrt{y}$

and (ii) interchanging x and y, gives the inverse $y = \pm\sqrt{x}$

Hence the inverse has two values for every value of x. Thus $f(x) = x^2$ does not have a single inverse. In such a case the domain of the original function may be restricted to $y = x^2$ for $x > 0$. Thus the inverse is then $\mathbf{f^{-1}(x) = +\sqrt{x}}$

A graph of $f(x) = x^2$ and its inverse $f^{-1}(x) = \sqrt{x}$ for $x > 0$ is shown in Figure 6.41 and, again, $f^{-1}(x)$ is seen to be a reflection of $f(x)$ in the line $y = x$.

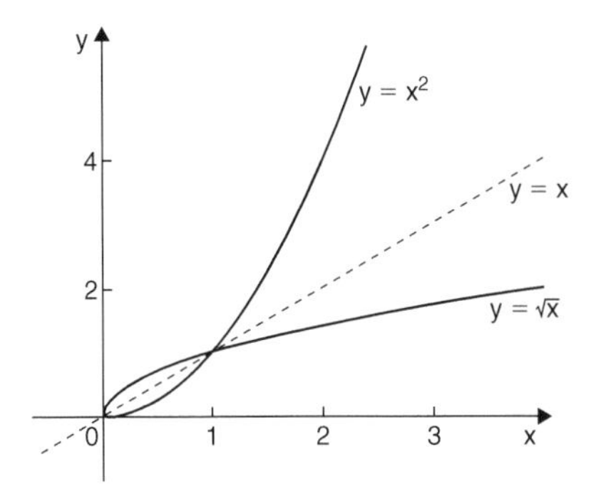

Figure 6.41

Inverse trigonometric functions

If $y = \sin x$, then x is the angle whose sine is y. Inverse trigonometric functions are denoted either by prefixing the function with 'arc' or more commonly^{-1}. Hence, transposing $y = \sin x$ for x gives $x = \sin^{-1} y$. Interchanging x and y gives the inverse $y = \sin^{-1} x$. Similarly, $y = \cos^{-1} x$, $y = \tan^{-1} x$, $y = \sec^{-1} x$, $y = \text{cosec}^{-1} x$ and $y = \cot^{-1} x$ are all inverse trigonometric functions. The angle is always expressed in radians.

Inverse trigonometric functions are periodic so it is necessary to specify the smallest or principal value of the angle. For $\sin^{-1} x$, $\tan^{-1} x$, $\text{cosec}^{-1} x$ and $\cot^{-1} x$, the principal value is in the range $-\frac{\pi}{2} < y < \frac{\pi}{2}$. For $\cos^{-1} x$ and $\sec^{-1} x$ the principal value is in the range $0 < y < \pi$.

Graphs of the six inverse trigonometric functions are shown in Figure 11.6, page 282.

Application: Determine the principal values of

(a) arcsin 0.5 (b) arctan(−1) (c) $\arccos\left(-\frac{\sqrt{3}}{2}\right)$ (d) $\text{arccosec}(\sqrt{2})$

Using a calculator,

(a) $\arcsin 0.5 \equiv \sin^{-1} 0.5 = 30° = \dfrac{\pi}{6}$ **rad** or **0.5236 rad**

(b) $\arctan(-1) \equiv \tan^{-1}(-1) = -45° = -\dfrac{\pi}{4}$ **rad** or **−0.7854 rad**

(c) $\arccos\left(-\dfrac{\sqrt{3}}{2}\right) \equiv \cos^{-1}\left(-\dfrac{\sqrt{3}}{2}\right) = 150° = \dfrac{5\pi}{6}$ **rad** or **2.6180 rad**

(d) $\text{arccosec}\left(\sqrt{2}\right) = \arcsin\left(\dfrac{1}{\sqrt{2}}\right) \equiv \sin^{-1}\left(\dfrac{1}{\sqrt{2}}\right) = 45° = \dfrac{\pi}{4}$ **rad** or **0.7854 rad**

Asymptotes

If a table of values for the function $y = \dfrac{x+2}{x+1}$ is drawn up for various values of x and then y plotted against x, the graph would be as shown in Figure 6.42. The straight lines AB, i.e. $x = -1$, and CD, i.e. $y = 1$, are known as **asymptotes**.

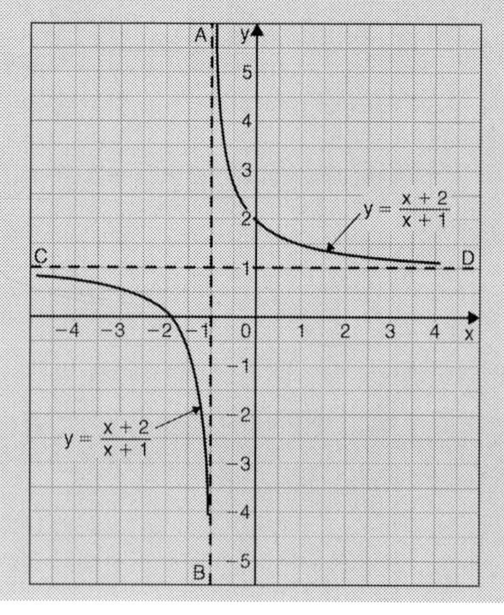

Figure 6.42

An asymptote to a curve is defined as a straight line to which the curve approaches as the distance from the origin increases. Alternatively, an asymptote can be considered as a tangent to the curve at infinity.

Asymptotes parallel to the x- and y-axes

For a curve $y = f(x)$:

(i) the asymptotes parallel to the x-axis are found by equating the coefficient of the highest power of x to zero

(ii) the asymptotes parallel to the y-axis are found by equating the coefficient of the highest power of y to zero

Other asymptotes

To determine asymptotes other than those parallel to x- and y-axes a simple procedure is:

(i) substitute $y = mx + c$ in the given equation

(ii) simplify the expression

(iii) equate the coefficients of the two highest powers of x to zero and determine the values of m and c. $y = mx + c$ gives the asymptote.

Application: Show that asymptotes occur at $y = 1$ and $x = -1$ for the curve $y = \frac{x+2}{x+1}$

Rearranging $y = \frac{x+2}{x+1}$ gives:

$$y(x + 1) = x + 2$$

i.e. $$yx + y - x - 2 = 0 \qquad (1)$$

and $$x(y - 1) + y - 2 = 0$$

The coefficient of the highest power of x (in this case x^1) is $(y - 1)$.

Equating to zero gives: $y - 1 = 0$

from which, $\mathbf{y = 1}$, which is an asymptote of $y = \frac{x+2}{x+1}$ as shown in Figure 6.42.

Returning to equation (1): $yx + y - x - 2 = 0$

from which, $y(x + 1) - x - 2 = 0$

The coefficient of the highest power of y (in this case y^1) is $(x + 1)$.

Equating to zero gives: $x + 1 = 0$

from which, $\mathbf{x = -1}$, which is another asymptote of $y = \dfrac{x+2}{x+1}$ as shown in Figure 6.42.

Application: Determine the asymptotes parallel to the x- and y-axes for the function $x^2y^2 = 9(x^2 + y^2)$

Asymptotes parallel to the x-axis:

Rearranging $x^2y^2 = 9(x^2 + y^2)$ gives $x^2y^2 - 9x^2 - 9y^2 = 0$

Hence $x^2(y^2 - 9) - 9y^2 = 0$

Equating the coefficient of the highest power of x to zero gives:

$y^2 - 9 = 0$ from which, $y^2 = 9$ and $\mathbf{y = \pm 3}$

Asymptotes parallel to the y-axis:

Since $x^2y^2 - 9x^2 - 9y^2 = 0$ then $y^2(x^2 - 9) - 9y^2 = 0$

Equating the coefficient of the highest power of y to zero gives:

$x^2 - 9 = 0$ from which, $x^2 = 9$ and $\mathbf{x = \pm 3}$

Hence, asymptotes occur at $\mathbf{y = \pm 3}$ and $\mathbf{x = \pm 3}$

Application: Determine the asymptotes for the function:

$$y(x + 1) = (x - 3)(x + 2)$$

(i) Substituting $y = mx + c$ into $y(x + 1) = (x - 3)(x + 2)$ gives $(mx + c)(x + 1) = (x - 3)(x + 2)$

(ii) Simplifying gives $mx^2 + mx + cx + c = x^2 - x - 6$ and $(m - 1)x^2 + (m + c + 1)x + c + 6 = 0$

(iii) Equating the coefficient of the highest power of x to zero gives $m - 1 = 0$ from which, $\mathbf{m = 1}$

Equating the coefficient of the next highest power of x to zero gives $m + c + 1 = 0$

and since $m = 1$, $1 + c + 1 = 0$ from which, $\mathbf{c = -2}$

Hence $y = mx + c = 1x - 2$

i.e. **$\mathbf{y = x - 2}$ is an asymptote**

To determine any asymptotes parallel to the x-axis:

Rearranging $y(x + 1) = (x - 3)(x + 2)$

gives $yx + y = x^2 - x - 6$

The coefficient of the highest power of x (i.e. x^2) is 1. Equating this to zero gives $1 = 0$, which is not an equation of a line. Hence there is no asymptote parallel to the x-axis

To determine any asymptotes parallel to the y-axis:

Since $y(x + 1) = (x - 3)(x + 2)$ the coefficient of the highest power of y is $x + 1$. Equating this to zero gives $x + 1 = 0$, from which, $x = -1$. Hence, $\mathbf{x = -1}$ **is an asymptote**.

When $\mathbf{x = 0}$, $\quad y(1) = (-3)(2)$, i.e. $\mathbf{y = -6}$

When $\mathbf{y = 0}$, $\quad 0 = (x - 3)(x + 2)$, i.e. $\mathbf{x = 3}$ and $\mathbf{x = -2}$

A sketch of the function $y(x + 1) = (x - 3)(x + 2)$ is shown in Figure 6.43.

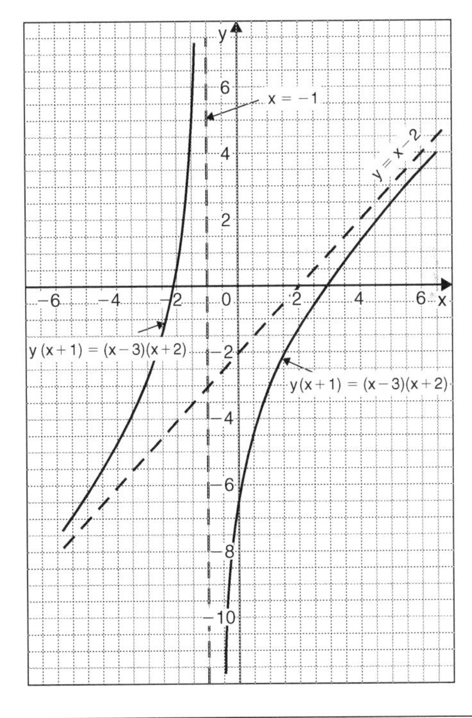

Figure 6.43

7 Vectors

7.1 Scalars and vectors

Some physical quantities are entirely defined by a numerical value and are called **scalar quantities** or **scalars**. Examples of scalars include time, mass, temperature, energy and volume. Other physical quantities are defined by both a numerical value **and** a direction in space and these are called **vector quantities** or **vectors**. Examples of vectors include force, velocity, moment and displacement.

Various ways of showing vector quantities include:

1. **bold print**.
2. two capital letters with an arrow above them to denote the sense of direction, e.g. $\overrightarrow{AB}$, where A is the starting point and B the end point of the vector,
3. a line over the top of letters, e.g. $\overline{AB}$ or $\bar{a}$
4. letters with an arrow above, e.g. $\vec{a}$, $\vec{A}$
5. underlined letters, e.g. $\underline{a}$
6. xi + jy, where i and j are axes at right-angles to each other; for example, 3i + 4j means 3 units in the i direction and 4 units in the j direction, as shown in Figure 7.1
7. a column matrix $\begin{pmatrix} a \\ b \end{pmatrix}$; for example, the vector **OA** of Figure 7.1 could be represented by $\begin{pmatrix} 3 \\ 4 \end{pmatrix}$

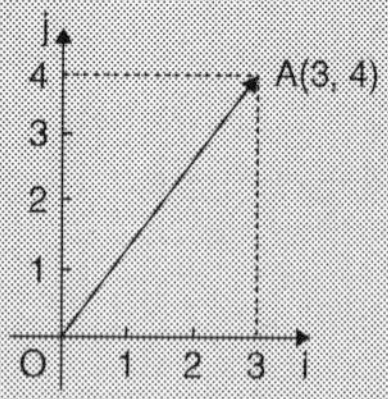

Figure 7.1

Thus, in Figure 7.1, $\mathbf{OA} = \overrightarrow{OA} = \overline{OA} = 3i + 4j = \begin{pmatrix} 3 \\ 4 \end{pmatrix}$

The one adopted in this text is to denote vector quantities in **bold print**.

7.2 Vector addition

The resultant of adding two vectors together, say $\mathbf{V_1}$ at an angle θ_1 and $\mathbf{V_2}$ at angle $(-\theta_2)$, as shown in Figure 7.2(a), can be obtained by drawing **oa** to represent $\mathbf{V_1}$ and then drawing **ar** to represent $\mathbf{V_2}$. The resultant of $\mathbf{V_1} + \mathbf{V_2}$ is given by **or**. This is shown in Figure 7.2(b), the vector equation being **oa** + **ar** = **or**. This is called the **'nose-to-tail' method** of vector addition.

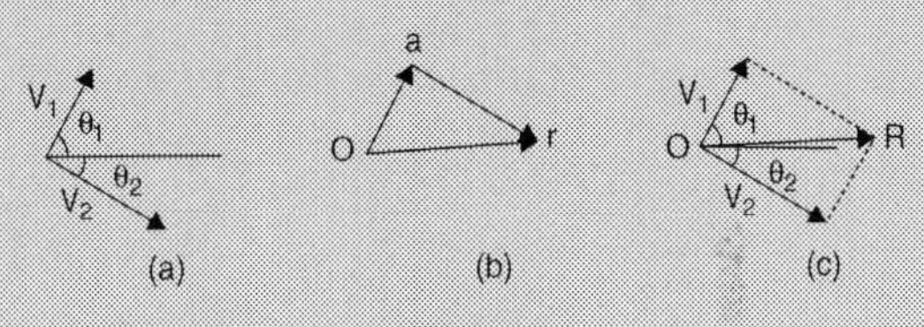

Figure 7.2

Alternatively, by drawing lines parallel to $\mathbf{V_1}$ and $\mathbf{V_2}$ from the noses of $\mathbf{V_2}$ and $\mathbf{V_1}$, respectively, and letting the point of intersection of these parallel lines be R, gives **OR** as the magnitude and direction of the resultant of adding $\mathbf{V_1}$ and $\mathbf{V_2}$, as shown in Figure 7.2(c). This is called the **'parallelogram' method** of vector addition.

Application: A force of 4 N is inclined at an angle of 45° to a second force of 7 N, both forces acting at a point. Find the magnitude of the resultant of these two forces and the direction of the resultant with respect to the 7 N force by both the 'triangle' and the 'parallelogram' methods

The forces are shown in Figure 7.3(a). Although the 7 N force is shown as a horizontal line, it could have been drawn in any direction.

Using the **'nose-to-tail' method**, a line 7 units long is drawn horizontally to give vector **oa** in Figure 7.3(b). To the nose of this vector **ar** is drawn 4 units long at an angle of 45° to **oa**. The resultant of vector addition is **or** and by measurement is **10.2 units long and at an angle of 16° to the 7 N force**. Figure 7.3(c) uses the **'parallelogram' method** in which lines are drawn parallel to the

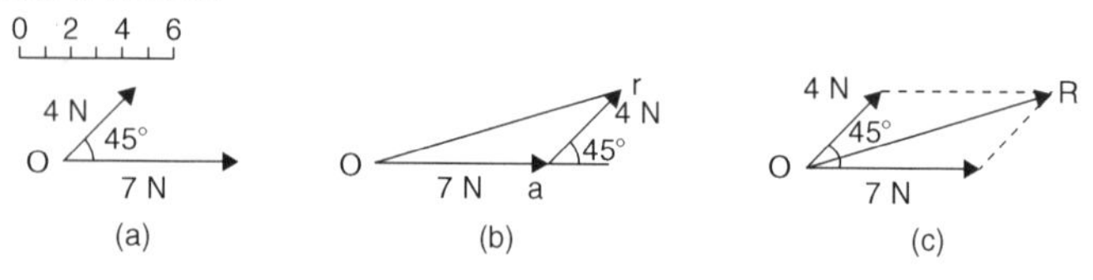

Figure 7.3

7 N and 4 N forces from the noses of the 4 N and 7 N forces, respectively. These intersect at R. Vector **OR** give the magnitude and direction of the resultant of vector addition and, as obtained by the 'nose-to-tail' method, is **10.2 units long at an angle of 16° to the 7 N force**.

Application: Use a graphical method to determine the magnitude and direction of the resultant of the three velocities shown in Figure 7.4

Figure 7.4

It is easier to use the 'nose-to-tail' method when more than two vectors are being added. The order in which the vectors are added is immaterial. In this case the order taken is v_1, then v_2, then v_3 but just the same result would have been obtained if the order had been, say, v_1, v_3 and finally v_2.

v_1 is drawn 10 units long at an angle of 20° to the horizontal, shown by **oa** in Figure 7.5. v_2 is added to v_1 by drawing a line 15 units

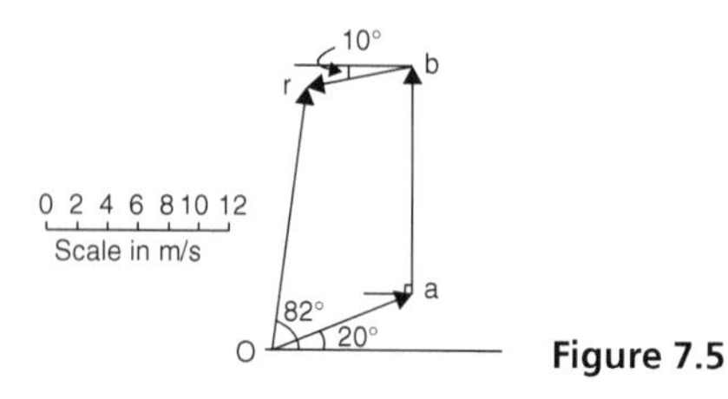

Figure 7.5

long vertically upwards from a, shown as **ab**. Finally, v_3 is added to $v_1 + v_2$ by drawing a line 7 units long at an angle at 190° from b, shown as **br**. The resultant of vector addition is **or** and by measurement is 17.5 units long at an angle of 82° to the horizontal.

Thus, $\mathbf{v_1 + v_2 + v_3 = 17.5\,m/s}$ **at 82° to the horizontal.**

7.3 Resolution of vectors

A vector can be resolved into horizontal component and vertical components. For the vector shown as **F** in Figure 7.6, the horizontal component is $F\cos\theta$ and the vertical component is $F\sin\theta$.

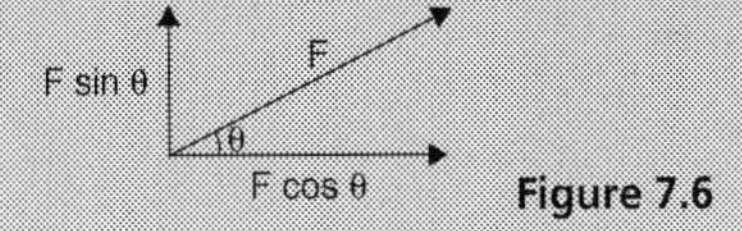

Figure 7.6

Application: Calculate the resultant of the two forces shown in Figure 7.3(a)

Horizontal component of force,

$$H = 7\cos 0° + 4\cos 45° = 7 + 2.828 = \mathbf{9.828\,N}$$

Vertical component of force,

$$V = 7\sin 0° + 4\sin 45° = 0 + 2.828 = \mathbf{2.828\,N}$$

The magnitude of the resultant of vector addition

$$= \sqrt{H^2 + V^2} = \sqrt{9.828^2 + 2.828^2} = \sqrt{104.59} = \mathbf{10.23\,N}$$

The direction of the resultant of vector addition $= \tan^{-1}\left(\dfrac{V}{H}\right)$

$$= \tan^{-1}\left(\frac{2.828}{9.828}\right)$$

$$= \mathbf{16.05°}$$

Thus, the resultant of the two forces is a single vector of 10.23 N at 16.05° to the 7 N vector

Application: Calculate the resultant velocity of the three velocities shown in Figure 7.4

Horizontal component of the velocity,

$$H = 10\cos 20^\circ + 15\cos 90^\circ + 7\cos 190^\circ$$
$$= 9.397 + 0 + (-6.894) = \mathbf{2.503\,m/s}$$

Vertical component of the velocity,

$$V = 10\sin 20^\circ + 15\sin 90^\circ + 7\sin 190^\circ$$
$$= 3.420 + 15 + (-1.216) = \mathbf{17.205\,m/s}$$

Magnitude of the resultant of vector addition

$$= \sqrt{H^2 + V^2} = \sqrt{2.503^2 + 17.205^2} = \sqrt{302.28} = \mathbf{17.39\,m/s}$$

Direction of the resultant of vector addition

$$= \tan^{-1}\left(\frac{V}{H}\right) = \tan^{-1}\left(\frac{17.205}{2.503}\right) = \tan^{-1} 6.8738 = 81.72^\circ$$

Thus, the resultant of the three velocities is a single vector of 17.39 m/s at 81.72° to the horizontal.

7.4 Vector subtraction

In Figure 7.7, a force vector **F** is represented by **oa**. The vector **(−oa)** can be obtained by drawing a vector from o in the opposite sense to **oa** but having the same magnitude, shown as **ob** in Figure 7.7, i.e. **ob = (−oa)**

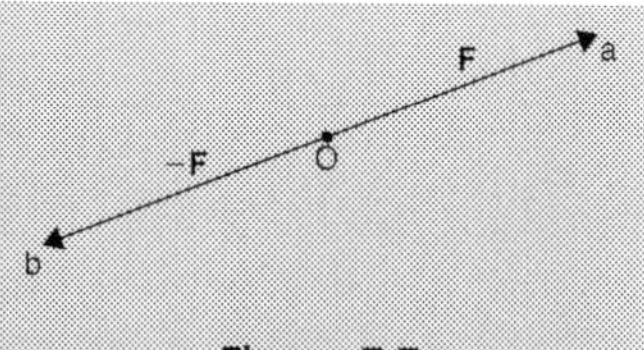

Figure 7.7

For two vectors acting at a point, as shown in Figure 7.8(a), the resultant of vector addition is **os = oa + ob**. Figure 7.8(b) shows vectors **ob + (−oa)**, that is, **ob − oa** and the vector equation is **ob − oa = od**. Comparing **od** in Figure 7.8(b) with the broken line ab in Figure 7.8(a) shows that the second diagonal of the 'parallelogram' method of vector addition gives the magnitude and direction of vector subtraction of **oa** from **ob**.

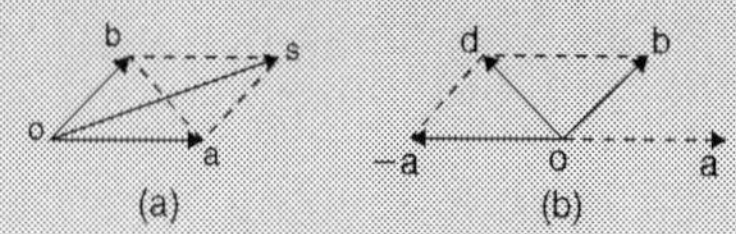

(a) (b)

Figure 7.8

Application: Accelerations of $a_1 = 1.5\,\text{m/s}^2$ at 90° and $a_2 = 2.6\,\text{m/s}^2$ at 145° act at a point. Find $\mathbf{a_1} + \mathbf{a_2}$ and $\mathbf{a_1} - \mathbf{a_2}$ by (i) drawing a scale vector diagram and (ii) by calculation

(i) The scale vector diagram is shown in Figure 7.9. By measurement,

$$\mathbf{a_1} + \mathbf{a_2} = \mathbf{3.7\,m/s^2 \text{ at } 126°}$$

and

$$\mathbf{a_1} - \mathbf{a_2} = \mathbf{2.1\,m/s^2 \text{ at } 0°}$$

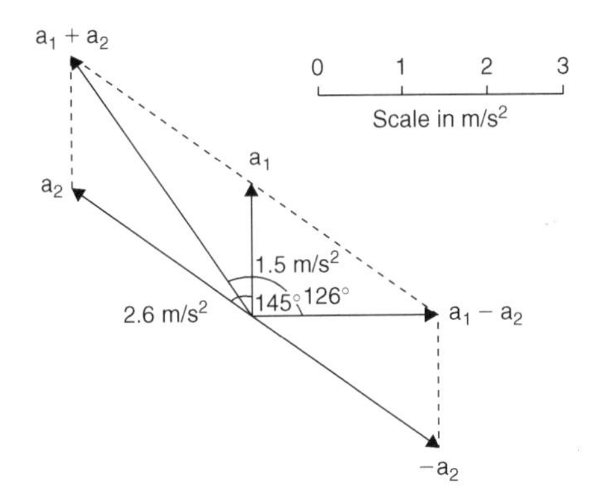

Figure 7.9

(ii) Resolving horizontally and vertically gives:

Horizontal component of $\mathbf{a_1} + \mathbf{a_2}$,

$$\mathbf{H} = 1.5\cos 90° + 2.6\cos 145° = -2.13$$

Vertical component of $\mathbf{a_1} + \mathbf{a_2}$,

$$\mathbf{V} = 1.5\sin 90° + 2.6\sin 145° = 2.99$$

Magnitude of $\mathbf{a_1} + \mathbf{a_2} = \sqrt{(-2.13)^2 + 2.99^2} = \mathbf{3.67\,m/s^2}$

Direction of $\mathbf{a_1} + \mathbf{a_2} = \tan^{-1}\left(\dfrac{2.99}{-2.13}\right)$ and must lie in the second quadrant since H is negative and V is positive.

$\tan^{-1}\left(\dfrac{2.99}{-2.13}\right) = -54.53°$, and for this to be in the second quadrant, the true angle is 180° displaced, i.e. 180° − 54.53° or 125.47°. Thus $\mathbf{a_1} + \mathbf{a_2} = \mathbf{3.67\,m/s^2 \text{ at } 125.47°}$

Horizontal component of $\mathbf{a_1 - a_2}$, that is, $\mathbf{a_1 + (-a_2)}$

$$= 1.5\cos 90° + 2.6\cos(145° - 180°)$$
$$= 2.6\cos(-35°) = 2.13$$

Vertical component of $\mathbf{a_1 - a_2}$, that is,

$$\mathbf{a_1 + (-a_2)} = 1.5\sin 90° + 2.6\sin(-35°) = 0$$

Magnitude of $\mathbf{a_1 - a_2} = \sqrt{2.13^2 + 0^2} = 2.13\,\text{m/s}^2$

Direction of $\mathbf{a_1 - a_2} = \tan^{-1}\left(\frac{0}{2.13}\right) = 0°$

Thus, $\mathbf{a_1 - a_2 = 2.13\,m/s^2\ at\ 0°}$

Application: Calculate the resultant of $\mathbf{v_1 - v_2 + v_3}$ when $\mathbf{v_1} = 22$ units at 140°, $\mathbf{v_2} = 40$ units at 190° and $\mathbf{v_3} = 15$ units at 290°

The vectors are shown in Figure 7.10.

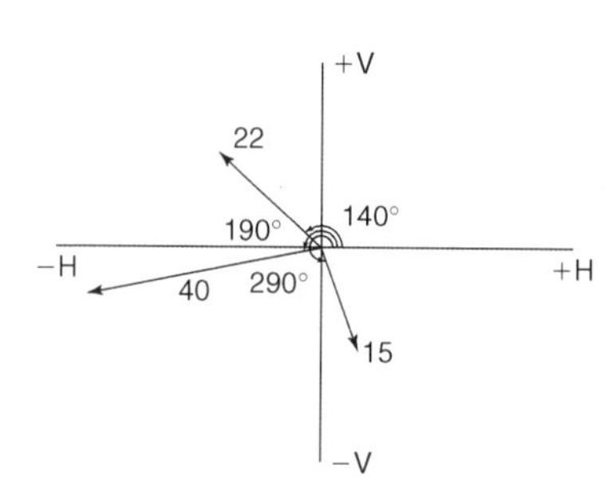

Figure 7.10

The horizontal component of

$$\mathbf{v_1 - v_2 + v_3} = (22\cos 140°) - (40\cos 190°) + (15\cos 290°)$$
$$= (-16.85) - (-39.39) + (5.13) = \mathbf{27.67\,units}$$

The vertical component of

$$\mathbf{v_1 - v_2 + v_3} = (22\sin 140°) - (40\sin 190°) + (15\sin 290°)$$
$$= (14.14) - (-6.95) + (-14.10) = \mathbf{6.99\,units}$$

The magnitude of the resultant, R, is given by:

$$|R| = \sqrt{27.67^2 + 6.99^2} = 28.54\text{ units}$$

The direction of the resultant, **R**, is given by:

$$\arg \mathbf{R} = \tan^{-1}\left(\frac{6.99}{27.67}\right) = 14.18°$$

Thus, $\mathbf{v_1} - \mathbf{v_2} + \mathbf{v_3}$ **= 28.54 units at 14.18°**

7.5 Relative velocity

For relative velocity problems, some fixed datum point needs to be selected. This is often a fixed point on the earth's surface. In any vector equation, only the start and finish points affect the resultant vector of a system. Two different systems are shown in Figure 7.11, but in each of the systems, the resultant vector is **ad**.

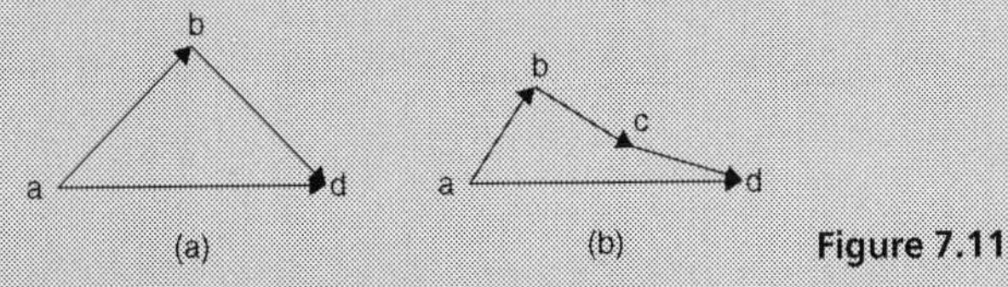

Figure 7.11

The vector equation of the system shown in Figure 7.11(a) is:

$$\mathbf{ad} = \mathbf{ab} + \mathbf{bd}$$

and that for the system shown in Figure 7.11(b) is:

$$\mathbf{ad} = \mathbf{ab} + \mathbf{bc} + \mathbf{cd}$$

Thus in vector equations of this form, only the first and last letters, a and d, respectively, fix the magnitude and direction of the resultant vector.

Application: Two cars, P and Q, are travelling towards the junction of two roads which are at right angles to one another. Car P has a velocity of 45 km/h due east and car Q a velocity of 55 km/h due south. Calculate (i) the velocity of car P relative to car Q, and (ii) the velocity of car Q relative to car P

(i) The directions of the cars are shown in Figure 7.12(a), called a **space diagram**. The velocity diagram is shown in Figure 7.12(b), in which **pe** is taken as the velocity of car P relative to point e on the earth's surface. The velocity of P relative to Q is vector **pq** and the vector equation is **pq** = **pe** + **eq**. Hence the vector directions are as shown, **eq** being in the opposite direction to **qe**.

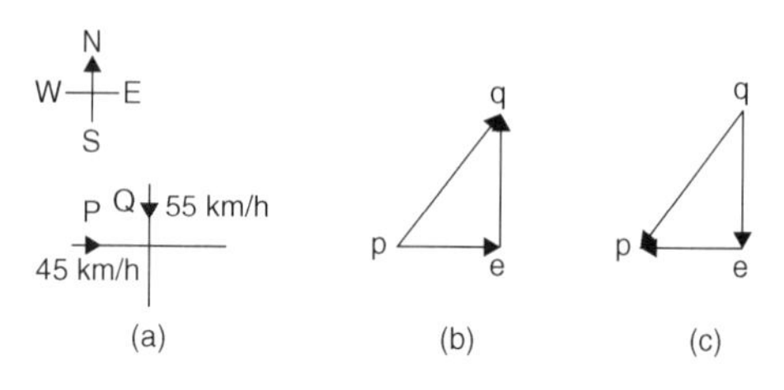

Figure 7.12

From the geometry of the vector triangle,

$$|\mathbf{pq}| = \sqrt{45^2 + 55^2} = 71.06\,\text{km/h}$$

and
$$\arg \mathbf{pq} = \tan^{-1}\left(\frac{55}{45}\right) = 50.71°$$

i.e. **the velocity of car P relative to car Q is 71.06 km/h at 50.71°**

(ii) The velocity of car Q relative to car P is given by the vector equation **qp** = **qe** + **ep** and the vector diagram is as shown in Figure 7.12(c), having **ep** opposite in direction to **pe**. From the geometry of this vector triangle:

$$|\mathbf{qp}| = \sqrt{45^2 + 55^2} = 71.06\,\text{m/s}$$

and
$$\arg \mathbf{qp} = \tan^{-1}\left(\frac{55}{45}\right) = 50.71°$$

but must lie in the third quadrant, i.e. the required angle is $180° + 50.71° = 230.71°$

Thus the velocity of car Q relative to car P is 71.06 m/s at 230.71°

7.6 Combination of two periodic functions

In many engineering situations waveforms have to be combined. There are a number of methods of determining the resultant waveform. These include by:

1. drawing the waveforms and adding graphically
2. drawing the phasors and measuring the resultant
3. using the cosine and sine rules
4. using horizontal and vertical components
5. using complex numbers

Application: Sketch graphs of $y_1 = 4\sin\omega t$ and $y_2 = 3\sin(\omega t - \pi/3)$ on the same axes, over one cycle. Adding ordinates at intervals, obtain a sinusoidal expression for the resultant waveform $y_R = y_1 + y_2$

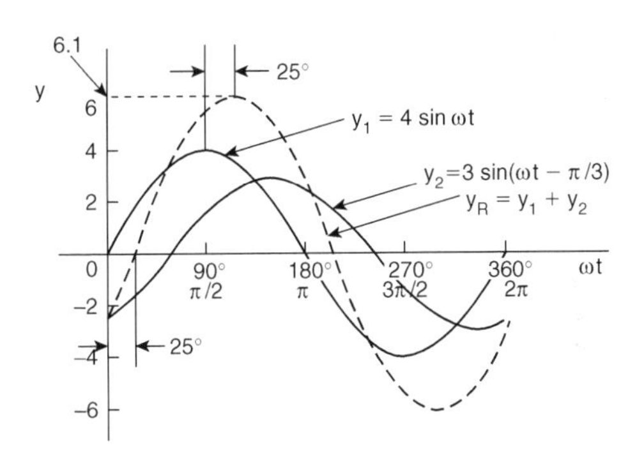

Figure 7.13

$y_1 = 4\sin\omega t$ and $y_2 = 3\sin(\omega t - \pi/3)$ are shown plotted in Figure 7.13.

Ordinates are added at 15° intervals and the resultant is shown by the broken line. The amplitude of the resultant is 6.1 and it **lags** y_1 by 25° or 0.436 rad.

Hence the sinusoidal expression for the resultant waveform is: $\mathbf{y_R = 6.1\sin(\omega t - 0.436)}$

Application: Determine $4\sin\omega t + 3\sin(\omega t - \pi/3)$ by drawing phasors

The resultant of two periodic functions may be found from their relative positions when the time is zero. $4 \sin \omega t$ and $3 \sin(\omega t - \pi/3)$ may each be represented as phasors as shown in Figure 7.14, y_1 being 4 units long and drawn horizontally and y_2 being 3 units long, lagging y_1 by $\pi/3$ radians or $60°$. To determine the resultant of $y_1 + y_2$, y_1 is drawn horizontally as shown in Figure 7.15 and y_2 is joined to the end of y_1 at $60°$ to the horizontal. The resultant is given by y_R. This is the same as the diagonal of a parallelogram that is shown completed in Figure 7.16.

The resultant is measured as 6.1 and angle ϕ as $25°$ or 0.436 rad.

Hence, $\mathbf{4 \sin \omega t + 3 \sin(\omega t - \pi/3) = 6.1 \sin(\omega t - 0.436)}$

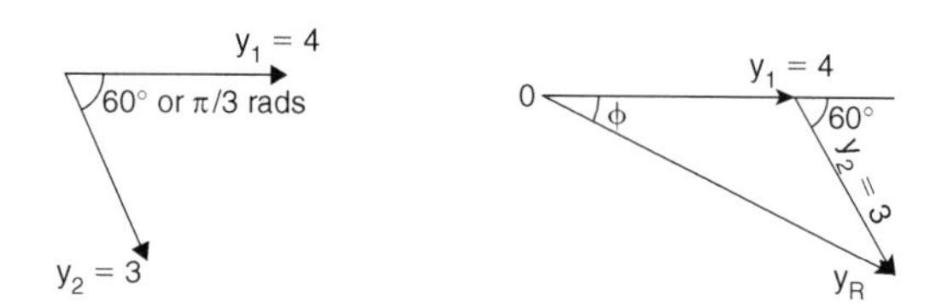

Figure 7.14 **Figure 7.15**

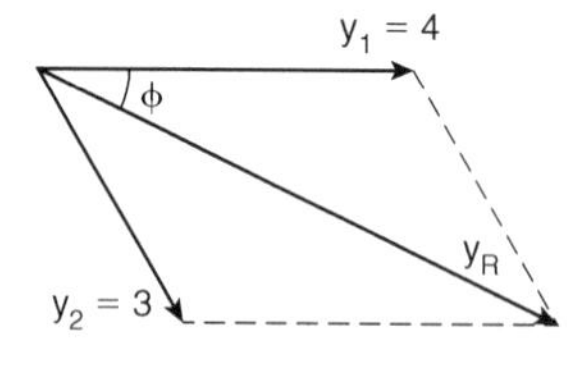

Figure 7.16

Application: Determine $4 \sin \omega t + 3 \sin(\omega t - \pi/3)$ using the cosine and sine rules

From the phasor diagram of Figure 7.15, and using the cosine rule:

$$y_R^2 = 4^2 + 3^2 - 2(4)(3)\cos 120° = 37 \text{ and } y_R = \sqrt{37} = \mathbf{6.083}$$

Using the sine rule gives:

$$\frac{3}{\sin \phi} = \frac{6.083}{\sin 120°} \text{ from which, } \sin \phi = \frac{3 \sin 120°}{6.083} = 0.4271044$$

and $\phi = \sin^{-1} 0.4271044 = \mathbf{25.28°} = 25.28 \times \frac{\pi}{180} = \mathbf{0.441\,rad}$

Hence, by cosine and sine rules,

$$\mathbf{y_R = y_1 + y_2 = 6.083 \sin(\omega t - 0.441)}$$

Application: Determine $4\sin\omega t + 3\sin(\omega t - \pi/3)$ using horizontal and vertical components

From the phasors shown in Figure 7.14:

Total horizontal component $= 4\cos 0° + 3\cos 300° = 5.5$

Total vertical component $= 4\sin 0° + 3\sin 300° = -2.598$

By Pythagoras, the resultant, $i_R = \sqrt{[5.5^2 + 2.598^2]} = \mathbf{6.083}$

Phase angle, $\phi = \tan^{-1}\left(\frac{2.598}{5.5}\right) = \mathbf{25.28°}$ or $\mathbf{0.441\,rad}$ (ϕ being in the 4th quadrant)

Hence, by using horizontal and vertical components,

$$\mathbf{y_R = y_1 + y_2 = 6.083\sin(\omega t - 0.441)}$$

Application: Determine $4\sin\omega t + 3\sin(\omega t - \pi/3)$ using complex numbers

From the phasors shown in Figure 7.14, the resultant may be expressed in polar form (see page 209)

as: $y_R = 4\angle 0° + 3\angle -60°$

i.e. $y_R = (4 + j0) + (1.5 - j2.598)$

$= (5.5 - j2.598) = \mathbf{6.083\angle -25.28°\ A}$ or $\mathbf{6.083\angle -0.441\,rad\ A}$

Hence, by using complex numbers, the resultant is:

$$\mathbf{y_R = y_1 + y_2 = 6.083 \sin(\omega t - 0.441)}$$

7.7 The scalar product of two vectors

If $\mathbf{a} = a_1\mathbf{i} + a_2\mathbf{j} + a_3\mathbf{k}$ and $\mathbf{b} = b_1\mathbf{i} + b_2\mathbf{j} + b_3\mathbf{k}$

scalar or dot product: $\mathbf{a} \bullet \mathbf{b} = a_1b_1 + a_2b_2 + a_3b_3$ (1)

$|a| = \sqrt{(a_1^2 + a_2^2 + a_3^2)}$ and $|b| = \sqrt{(b_1^2 + b_2^2 + b_3^2)}$ (2)

$$\cos\theta = \frac{a \bullet b}{|a||b|} = \frac{a_1b_1 + a_2b_2 + a_3b_3}{\sqrt{(a_1^2 + a_2^2 + a_3^2)}\sqrt{(b_1^2 + b_2^2 + b_3^2)}} \quad (3)$$

Application: Find vector **a** joining points P and Q where point P has co-ordinates (4, −1, 3) and point Q has co-ordinates (2, 5, 0) and find |a|, the magnitude or norm of **a**

Let O be the origin, i.e. its co-ordinates are (0, 0, 0)

The position vector of P and Q are given by $\mathbf{OP} = 4\mathbf{i} - \mathbf{j} + 3\mathbf{k}$ and $\mathbf{OQ} = 2\mathbf{i} + 5\mathbf{j}$

By the addition law of vectors $\mathbf{OP} + \mathbf{PQ} = \mathbf{OQ}$

Hence $\mathbf{a} = \mathbf{PQ} = \mathbf{OQ} - \mathbf{OP}$

i.e. $\mathbf{a} = \mathbf{PQ} = (2\mathbf{i} + 5\mathbf{j}) - (4\mathbf{i} - \mathbf{j} + 3\mathbf{k})$

$= \mathbf{-2i + 6j - 3k}$

From equation (2), the **magnitude** or **norm** of **a**,

$$|\mathbf{a}| = \sqrt{(a^2 + b^2 + c^2)} = \sqrt{[(-2)^2 + 6^2 + (-3)^2]} = \sqrt{49} = \mathbf{7}$$

Application: Determine: (i) $\mathbf{p} \bullet \mathbf{q}$ (ii) $\mathbf{p} + \mathbf{q}$ (iii) $|\mathbf{p} + \mathbf{q}|$ and (iv) $|\mathbf{p}| + |\mathbf{q}|$ if $\mathbf{p} = 2\mathbf{i} + \mathbf{j} - \mathbf{k}$ and $\mathbf{q} = \mathbf{i} - 3\mathbf{j} + 2\mathbf{k}$

(i) From equation (1), if $\mathbf{p} = a_1\mathbf{i} + a_2\mathbf{j} + a_3\mathbf{k}$ and $\mathbf{q} = b_1\mathbf{i} + b_2\mathbf{j} + b_3\mathbf{k}$
then $\mathbf{p} \bullet \mathbf{q} = a_1b_1 + a_2b_2 + a_3b_3$
When $\mathbf{p} = 2\mathbf{i} + \mathbf{j} - \mathbf{k}$, $a_1 = 2$, $a_2 = 1$ and $a_3 = -1$
and when $\mathbf{q} = \mathbf{i} - 3\mathbf{j} + 2\mathbf{k}$, $b_1 = 1$, $b_2 = -3$ and $b_3 = 2$
Hence $\mathbf{p} \bullet \mathbf{q} = (2)(1) + (1)(-3) + (-1)(2)$
i.e. $\mathbf{p} \bullet \mathbf{q} = \mathbf{-3}$

(ii) $\mathbf{p} + \mathbf{q} = (2\mathbf{i} + \mathbf{j} - \mathbf{k}) + (\mathbf{i} - 3\mathbf{j} + 2\mathbf{k}) = \mathbf{3i - 2j + k}$

(iii) $|\mathbf{p}+\mathbf{q}| = |3i - 2j + k|$

From equation (2), $|\mathbf{p}+\mathbf{q}| = \sqrt{[3^2 + (-2)^2 + 1^2)} = \mathbf{\sqrt{14}}$

(iv) From equation (2), $|\mathbf{p}| = |2i + j - k| = \sqrt{[2^2 + 1^2 + (-1)^2]} = \sqrt{6}$

Similarly, $|\mathbf{q}| = |i - 3j + 2k| = \sqrt{[1^2 + (-3)^2 + 2^2]} = \sqrt{14}$

Hence $|\mathbf{p}| + |\mathbf{q}| = \sqrt{6} + \sqrt{14} = \mathbf{6.191}$, correct to 3 decimal places

Application: Determine the angle between vectors **oa** and **ob** when **oa** = **i** + 2**j** − 3**k** and **ob** = 2**i** − **j** + 4**k**

From equation (3), $\cos\theta = \dfrac{a_1b_1 + a_2b_2 + a_3b_3}{\sqrt{(a_1^2 + a_2^2 + a_3^2)}\sqrt{(b_1^2 + b_2^2 + b_3^2)}}$

Since **oa** = **i** + 2**j** − 3**k**, $a_1 = 1$, $a_2 = 2$ and $a_3 = -3$

Since **ob** = 2**i** − **j** + 4**k**, $b_1 = 2$, $b_2 = -1$ and $b_3 = 4$

Thus, $$\cos\theta = \frac{(1\times 2) + (2\times -1) + (-3\times 4)}{\sqrt{(1^2 + 2^2 + (-3)^2)}\sqrt{(2^2 + (-1)^2 + 4^2)}}$$

$$= \frac{-12}{\sqrt{14}\sqrt{21}} = -0.6999$$

i.e. $$\theta = \cos^{-1}\theta = 134.4° \text{ or } 225.6°$$

By sketching the position of the two vectors, it will be seen that 225.6° is not an acceptable answer.

Thus, the angle between the vectors **oa** and **ob**, $\theta = \mathbf{134.4°}$

Application: A constant force of **F** = 10**i** + 2**j** − **k** Newton's displaces an object from **A** = **i** + **j** + **k** to **B** = 2**i** − **j** + 3**k** (in metres). Find the work done in Newton metres

The work done is the product of the applied force and the distance moved in the direction of the force,

i.e. **work done = F • d**

From the sketch shown in Figure 7.17, **AB** = **AO** + **OB** = **OB** − **OA** that is **AB** = (2**i** − **j** + 3**k**) − (**i** + **j** + **k**) = **i** − 2**j** + 2**k**

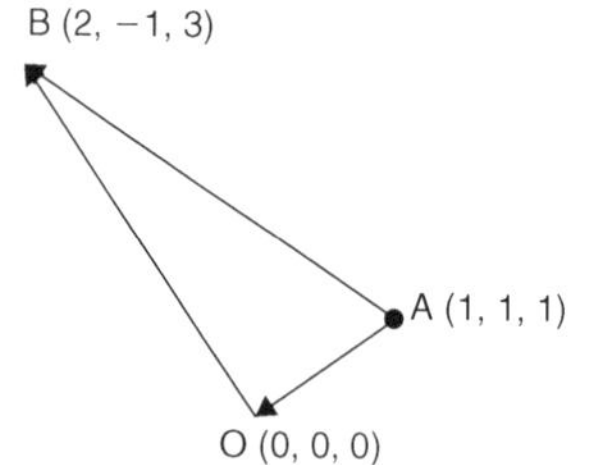

Figure 7.17

The work done is **F • d**, that is **F • AB** in this case

i.e. **work done** $= (10\mathbf{i} + 2\mathbf{j} - \mathbf{k}) \bullet (\mathbf{i} - 2\mathbf{j} + 2\mathbf{k})$

From equation (1), $\mathbf{a} \bullet \mathbf{b} = a_1b_1 + a_2b_2 + a_3b_3$

Hence, **work done** $= (10 \times 1) + (2 \times (-2)) + ((-1) \times 2) =$ **4 Nm**

Direction cosines

Let **or** $= x\mathbf{i} + y\mathbf{j} + z\mathbf{k}$ and from equation (2),

$$|\mathbf{or}| = \sqrt{x^2 + y^2 + z^2}$$

If **or** makes angles of α, β and γ with the co-ordinate axes i, j and k respectively, then:

$$\cos\alpha = \frac{x}{\sqrt{x^2 + y^2 + z^2}}, \quad \cos\beta = \frac{y}{\sqrt{x^2 + y^2 + z^2}}$$

$$\text{and } \cos\gamma = \frac{y}{\sqrt{x^2 + y^2 + z^2}}$$

The values of $\cos\alpha$, $\cos\beta$ and $\cos\gamma$ are called the **direction cosines** of **or**

Application: Find the direction cosines of $3\mathbf{i} + 2\mathbf{j} + \mathbf{k}$

$$\sqrt{x^2 + y^2 + z^2} = \sqrt{3^2 + 2^2 + 1^2} = \sqrt{14}$$

The direction cosines are: $\cos\alpha = \dfrac{x}{\sqrt{x^2 + y^2 + z^2}} = \dfrac{3}{\sqrt{14}} =$ **0.802**

$$\cos\beta = \frac{y}{\sqrt{x^2 + y^2 + z^2}} = \frac{2}{\sqrt{14}} = \mathbf{0.535}$$

and
$$\cos\gamma = \frac{y}{\sqrt{x^2 + y^2 + z^2}} = \frac{1}{\sqrt{14}} = \mathbf{0.267}$$

7.8 Vector products

Let $\mathbf{a} = a_1\mathbf{i} + a_2\mathbf{j} + a_3\mathbf{k}$ and $\mathbf{b} = b_1\mathbf{i} + b_2\mathbf{j} + b_3\mathbf{k}$

Vector or cross product: $$\mathbf{a} \times \mathbf{b} = \begin{vmatrix} \mathbf{i} & \mathbf{j} & \mathbf{k} \\ a_1 & a_2 & a_3 \\ b_1 & b_2 & b_3 \end{vmatrix} \quad (4)$$

$$|\mathbf{a} \times \mathbf{b}| = \sqrt{[(\mathbf{a} \bullet \mathbf{a})(\mathbf{b} \bullet \mathbf{b}) - (\mathbf{a} \bullet \mathbf{b})^2]} \quad (5)$$

Application: Find (i) $\mathbf{a} \times \mathbf{b}$ and (ii) $|\mathbf{a} \times \mathbf{b}|$ for the vectors $\mathbf{a} = \mathbf{i} + 4\mathbf{j} - 2\mathbf{k}$ and $\mathbf{b} = 2\mathbf{i} - \mathbf{j} + 3\mathbf{k}$

(i) From equation (4),

$$\mathbf{a} \times \mathbf{b} = \begin{vmatrix} i & j & k \\ 1 & 4 & -2 \\ 2 & -1 & 3 \end{vmatrix} = \mathbf{i}\begin{vmatrix} 4 & -2 \\ -1 & 3 \end{vmatrix} - \mathbf{j}\begin{vmatrix} 1 & -2 \\ 2 & 3 \end{vmatrix} + \mathbf{k}\begin{vmatrix} 1 & 4 \\ 2 & -1 \end{vmatrix}$$

$$= \mathbf{i}(12 - 2) - \mathbf{j}(3 + 4) + \mathbf{k}(-1 - 8)$$

$$= \mathbf{10i - 7j - 9k}$$

(ii) From equation (5) $|\mathbf{a} \times \mathbf{b}| = \sqrt{[(\mathbf{a} \bullet \mathbf{a})(\mathbf{b} \bullet \mathbf{b}) - (\mathbf{a} \bullet \mathbf{b})^2]}$

Now $\mathbf{a} \bullet \mathbf{a} = (1)(1) + (4)(4) + (-2)(-2) = 21$

$\mathbf{b} \bullet \mathbf{b} = (2)(2) + (-1)(-1) + (3)(3) = 14$

and $\mathbf{a} \bullet \mathbf{b} = (1)(2) + (4)(-1) + (-2)(3) = -8$

Thus, $|\mathbf{a} \times \mathbf{b}| = \sqrt{(21 \times 14 - 64)} = \sqrt{230} = \mathbf{15.17}$

Application: Find (a) $(\mathbf{p} - 2\mathbf{q}) \times \mathbf{r}$ (b) $\mathbf{p} \times (2\mathbf{r} \times 3\mathbf{q})$ if $\mathbf{p} = 4\mathbf{i} + \mathbf{j} - 2\mathbf{k}$, $\mathbf{q} = 3\mathbf{i} - 2\mathbf{j} + \mathbf{k}$ and $\mathbf{r} = \mathbf{i} - 2\mathbf{k}$

(a) $(\mathbf{p} - 2\mathbf{q}) \times \mathbf{r} = [4\mathbf{i} + \mathbf{j} - 2\mathbf{k} - 2(3\mathbf{i} - 2\mathbf{j} + \mathbf{k})] \times (\mathbf{i} - 2\mathbf{k})$

$$= (-2\mathbf{i} + 5\mathbf{j} - 4\mathbf{k}) \times (\mathbf{i} - 2\mathbf{k})$$

$$= \begin{vmatrix} i & j & k \\ -2 & 5 & -4 \\ 1 & 0 & -2 \end{vmatrix} \quad \text{from equation (4)}$$

$$= \mathbf{i}\begin{vmatrix} 5 & -4 \\ 0 & -2 \end{vmatrix} - \mathbf{j}\begin{vmatrix} -2 & -4 \\ 1 & -2 \end{vmatrix} + \mathbf{k}\begin{vmatrix} -2 & 5 \\ 1 & 0 \end{vmatrix}$$

$$= \mathbf{i}(-10 - 0) - \mathbf{j}(4 + 4) + \mathbf{k}(0 - 5)$$

i.e. $(\mathbf{p} - 2\mathbf{q}) \times \mathbf{r} = \mathbf{-10i - 8j - 5k}$

(b) $(2\mathbf{r} \times 3\mathbf{q}) = (2\mathbf{i} - 4\mathbf{k}) \times (9\mathbf{i} - 6\mathbf{j} + 3\mathbf{k})$

$$= \begin{vmatrix} i & j & k \\ 2 & 0 & -4 \\ 9 & -6 & 3 \end{vmatrix} = \mathbf{i}(0 - 24) - \mathbf{j}(6 + 36) + \mathbf{k}(-12 - 0)$$

$$= \mathbf{-24i - 42j - 12k}$$

Hence, $\mathbf{p} \times (2\mathbf{r} \times 3\mathbf{q}) = (4\mathbf{i} + \mathbf{j} - 2\mathbf{k}) \times (-24\mathbf{i} - 42\mathbf{j} - 12\mathbf{k})$

$$= \begin{vmatrix} i & j & k \\ 4 & 1 & -2 \\ -24 & -42 & -12 \end{vmatrix}$$

$$= \mathbf{i}(-12 - 84) - \mathbf{j}(-48 - 48) + \mathbf{k}(-168 + 24)$$

$$= \mathbf{-96i + 96j - 144k} \quad \text{or} \quad \mathbf{-48(2i - 2j + 3k)}$$

Application: Find the moment and the magnitude of the moment of a force of $(\mathbf{i} + 2\mathbf{j} - 3\mathbf{k})$ Newton's about point B having co-ordinates (0, 1, 1), when the force acts on a line through A whose co-ordinates are (1, 3, 4)

The moment **M** about point B of a force vector **F** that has a position vector of **r** from A is given by:

$$\mathbf{M} = \mathbf{r} \times \mathbf{F}$$

r is the vector from B to A, i.e. $\mathbf{r} = \mathbf{BA}$

But $\mathbf{BA} = \mathbf{BO} + \mathbf{OA} = \mathbf{OA} - \mathbf{OB}$

i.e. $\mathbf{r} = (\mathbf{i} + 3\mathbf{j} + 4\mathbf{k}) - (\mathbf{j} + \mathbf{k}) = \mathbf{i} + 2\mathbf{j} + 3\mathbf{k}$

Moment, $\mathbf{M} = \mathbf{r} \times \mathbf{F} = (\mathbf{i} + 2\mathbf{j} + 3\mathbf{k}) \times (\mathbf{i} + 2\mathbf{j} - 3\mathbf{k})$

$$= \begin{vmatrix} i & j & k \\ 1 & 2 & 3 \\ 1 & 2 & -3 \end{vmatrix} = \mathbf{i}(-6 - 6) - \mathbf{j}(-3 - 3) + \mathbf{k}(2 - 2) = \mathbf{-12i + 6j\ Nm}$$

The magnitude of **M**, $|M| = |\mathbf{r} \times \mathbf{F}| = \sqrt{[(\mathbf{r} \bullet \mathbf{r})(\mathbf{F} \bullet \mathbf{F}) - (\mathbf{r} \bullet \mathbf{F})^2]}$

$$\mathbf{r} \bullet \mathbf{r} = (1)(1) + (2)(2) + (3)(3) = 14$$
$$\mathbf{F} \bullet \mathbf{F} = (1)(1) + (2)(2) + (-3)(-3) = 14$$
$$\mathbf{r} \bullet \mathbf{F} = (1)(1) + (2)(2) + (3)(-3) = -4$$

i.e. **magnitude,** $|\mathbf{M}| = \sqrt{[14 \times 14 - (-4)^2]} = \sqrt{180}$ Nm

$= \mathbf{13.42\ Nm}$

Application: The axis of a circular cylinder coincides with the z-axis and it rotates with an angular velocity of $(2\mathbf{i} - 5\mathbf{j} + 7\mathbf{k})$ rad/s. Determine the tangential velocity at a point P on the cylinder, whose co-ordinates are $(\mathbf{j} + 3\mathbf{k})$ metres, and the magnitude of the tangential velocity

The velocity v of point P on a body rotating with angular velocity ω about a fixed axis is given by:

$\mathbf{v} = \omega \times \mathbf{r}$ where r is the point on vector P.

Thus, **velocity,** $\mathbf{v} = (2\mathbf{i} - 5\mathbf{j} + 7\mathbf{k}) \times (\mathbf{j} + 3\mathbf{k})$

$$= \begin{vmatrix} i & j & k \\ 2 & -5 & 7 \\ 0 & 1 & 3 \end{vmatrix} = \mathbf{i}(-15 - 7) - \mathbf{j}(6 - 0) + \mathbf{k}(2 - 0)$$

$= \mathbf{(-22i - 6j + 2k)\ m/s}$

The magnitude of **v**, $|\mathbf{v}| = \sqrt{[(\omega \bullet \omega)(\mathbf{r} \bullet \mathbf{r}) - (\mathbf{r} \bullet \omega)^2]}$

$$\omega \bullet \omega = (2)(2) + (-5)(-5) + (7)(7) = 78$$
$$\mathbf{r} \bullet \mathbf{r} = (0)(0) + (1)(1) + (3)(3) = 10$$
$$\omega \bullet \mathbf{r} = (2)(0) + (-5)(1) + (7)(3) = 16$$

Hence, magnitude, $|\mathbf{v}| = \sqrt{(78 \times 10 - 16^2)} = \sqrt{524}$ m/s

$= \mathbf{22.89\ m/s}$

8 Complex Numbers

8.1 General formulae

$z = a + jb = r(\cos\theta + j\sin\theta) = r\angle\theta = r\,e^{j\theta}$ where $j^2 = -1$

Modulus, $r = |z| = \sqrt{(a^2 + b^2)}$ Argument, $\theta = \arg z = \tan^{-1}\frac{b}{a}$

Addition: $(a + jb) + (c + jd) = (a + c) + j(b + d)$

Subtraction: $(a + jb) - (c + jd) = (a - c) + j(b - d)$

Complex equations: If $(m + jn) = (p + jq)$ then $m = p$ and $n = q$

Multiplication: $z_1 z_2 = r_1 r_2 \angle(\theta_1 + \theta_2)$

Division: $\frac{z_1}{z_2} = \frac{r_1}{r_2}\angle(\theta_1 - \theta_2)$

De Moivre's theorem: $[r\angle\theta]^n = r^n\angle n\theta = r^n(\cos n\theta + j\sin n\theta)$

8.2 Cartesian form

$(-1 + j2)$ and $(3 - j4)$ are examples of **Cartesian** (or **rectangular**) **complex numbers**. They are each of the form $a + jb$, 'a' being termed the **real part** and jb the **imaginary part**.

Application: Solve the quadratic equation $2x^2 + 3x + 5 = 0$

Using the quadratic formula,

$$x = \frac{-3 \pm \sqrt{[(3)^2 - 4(2)(5)]}}{2(2)} = \frac{-3 \pm \sqrt{-31}}{4}$$

$$= \frac{-3 \pm \sqrt{(-1)}\sqrt{31}}{4} = \frac{-3 \pm j\sqrt{31}}{4}$$

Hence, $\mathbf{x = -\frac{3}{4} + j\frac{\sqrt{31}}{4}}$ or $\mathbf{-0.750 \pm j1.392}$, correct to 3 decimal places.

Application: Determine $(2 + j3) + (3 - j4) - (-5 + j)$

$$\begin{aligned}(2 + j3) + (3 - j4) - (-5 + j) &= 2 + j3 + 3 - j4 + 5 - j \\ &= (2 + 3 + 5) + j(3 - 4 - 1) = \mathbf{10 - j2}\end{aligned}$$

Application: Determine $(3 + j2)(4 - j5)$

$$\begin{aligned}(3 + j2)(4 - j5) &= 12 - j15 + j8 - j^2 10 \\ &= (12 - -10) + j(-15 + 8) \quad \text{where } j^2 = -1 \\ &= \mathbf{22 - j7}\end{aligned}$$

Application: Solve the complex equation:

$$(1 + j2)(-2 - j3) = a + jb$$

$$(1 + j2)(-2 - j3) = a + jb$$

i.e. $\quad -2 - j3 - j4 - j^2 6 = a + jb$

Hence, $\quad 4 - j7 = a + jb$

Equating real and imaginary terms gives: $\mathbf{a = 4}$ and $\mathbf{b = -7}$

Application: Solve the equation $(x - j2y) + (y - j3x) = 2 + j3$

Since $(x - j2y) + (y - j3x) = 2 + j3$

then $(x + y) + j(-2y - 3x) = 2 + j3$

Equating real and imaginary parts gives: $x + y = 2$ (1)

and $-3x - 2y = 3$ (2)

Multiplying equation (1) by 2 gives: $2x + 2y = 4$ (3)

Adding equations (2) and (3) gives: $-x = 7$ i.e. $\mathbf{x = -7}$

From equation (1), $\mathbf{y = 9}$, which may be checked in equation (2).

Application: Determine $(3 + j4)(3 - j4)$

$(3 + j4)(3 - j4) = 9 - j12 + j12 - j^2 16 = 9 + 16 = \mathbf{25}$

[$(3 - j4)$ is called the **complex conjugate** of $(3 + j4)$; whenever a complex number is multiplied by its conjugate, a real number results. In general, $(a + jb)(a - jb)$ may be evaluated 'on sight' as $a^2 + b^2$]

Application: Determine $\dfrac{2-j5}{3+j4}$

$$\frac{2-j5}{3+j4}=\frac{2-j5}{3+j4}\times\frac{(3-j4)}{(3-j4)}=\frac{6-j8-j15+j^2 20}{3^2+4^2}$$

$$=\frac{-14-j23}{25}$$

$$=\mathbf{\frac{-14}{25}-j\frac{23}{25}}\quad\text{or}\quad\mathbf{-0.56-j0.92}$$

Application: If $Z_1 = 1 - j3$ and $Z_2 = -2 + j5$ determine $\dfrac{Z_1 Z_2}{Z_1 + Z_2}$ in $(a + jb)$ form

$$\mathbf{\frac{Z_1Z_2}{Z_1+Z_2}}=\frac{(1-j3)(-2+j5)}{(1-j3)+(-2+j5)}=\frac{-2+j5+j6-j^2 15}{1-j3-2+j5}$$

$$=\frac{-2+j5+j6+15}{-1+j2}=\frac{13+j11}{-1+j2}$$

$$=\frac{13+j11}{-1+j2}\times\frac{-1-j2}{-1-j2}$$

$$=\frac{-13-j26-j11-j^2 22}{1^2+2^2}$$

$$=\frac{9-j37}{5}$$

$$=\mathbf{\frac{9}{5}-j\frac{37}{5}}\quad\text{or}\quad\mathbf{1.8-j7.4}$$

Application: Show the following complex numbers on an Argand diagram $(3 + j2)$, $(-2 + j4)$, $(-3 - j5)$ and $(1 - j3)$

In Figure 8.1, the point A represents the complex number $(3 + j2)$ and is obtained by plotting the co-ordinates $(3, j2)$ as in graphical work. The Argand points B, C and D represent the complex numbers $(-2 + j4)$, $(-3 - j5)$ and $(1 - j3)$ respectively.

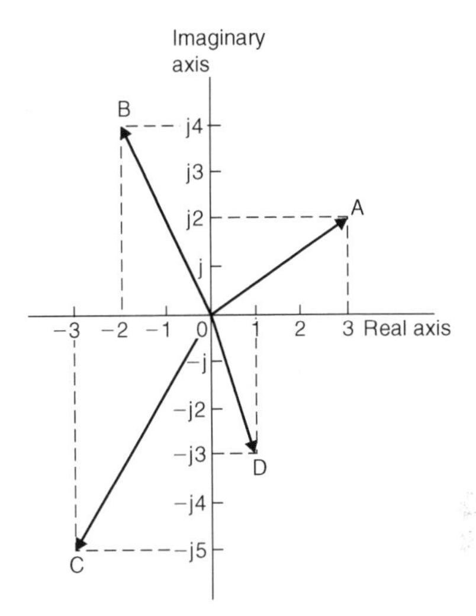

Figure 8.1

8.3 Polar form

A number written in the form $Z = r\angle\theta$ is known as the **polar form** of a complex number.

Application: Express (a) $3 + j4$ and (b) $-3 + j4$ in polar form

(a) $3 + j4$ is shown in Figure 8.2 and lies in the first quadrant.

Modulus, $r = \sqrt{3^2 + 4^2} = 5$

and argument $\theta = \tan^{-1}\dfrac{4}{3} = 53.13°$

Hence, $\mathbf{3 + j4 = 5\angle 53.13°}$

(b) $-3 + j4$ is shown in Figure 8.2 and lies in the second quadrant.

Modulus, $r = 5$ and angle $\alpha = 53.13°$, from part (a).

Argument $= 180° - 53.13° = 126.87°$ (i.e. the argument must be measured from the positive real axis)

Hence $\mathbf{-3 + j4 = 5\angle 126.87°}$

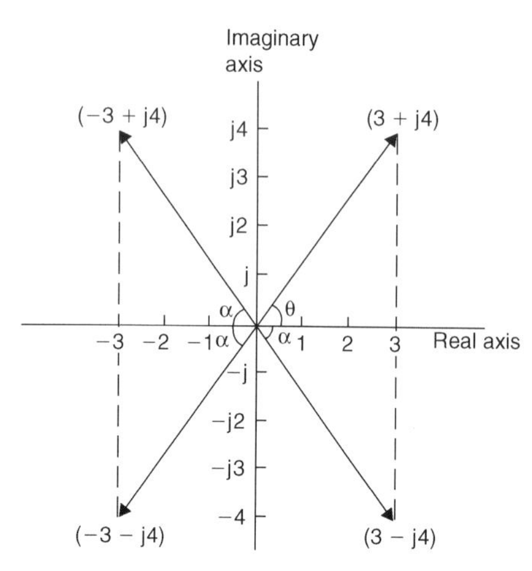

Figure 8.2

Similarly it may be shown that $\mathbf{(-3 - j4) = 5\angle 233.13°}$ **or** $\mathbf{5\angle -126.87°}$**,** (by convention the **principal value** is normally used, i.e. the numerically least value, such that $-\pi < \theta < \pi$), and $\mathbf{(3 - j4) = 5\angle -53.13°}$

Application: Change $7\angle -145°$ into a + jb form:

$7\angle -145°$ is shown in Figure 8.3 and lies in the third quadrant.

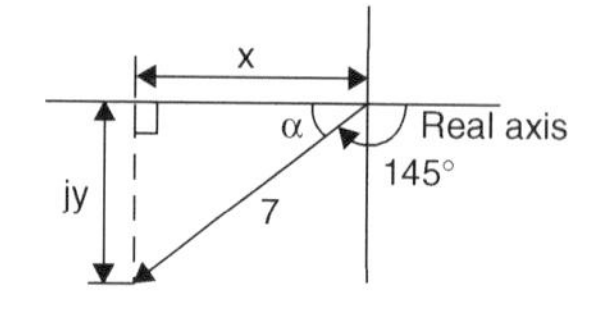

Figure 8.3

$\mathbf{7\angle -145°} = 7\cos(-145°) + j7\sin(-145°) = \mathbf{-5.734 - j4.015}$

Application: Determine $3\angle 16° \times 5\angle -44° \times 2\angle 80°$ in polar form

$$3\angle 16° \times 5\angle -44° \times 2\angle 80° = (3 \times 5 \times 2)\ \angle[16° + (-44°) + 80]° = \mathbf{30\angle 52°}$$

Application: Determine $\dfrac{16\angle 75°}{2\angle 15°}$ in polar form

$$\frac{16\angle 75^\circ}{2\angle 15^\circ} = \frac{16}{2}\angle(75^\circ - 15^\circ) = \mathbf{8\angle 60^\circ}$$

Application: Evaluate, in polar form, $2\angle 30^\circ + 5\angle -45^\circ - 4\angle 120^\circ$

$$2\angle 30^\circ = 2(\cos 30^\circ + j\sin 30^\circ) = 2\cos 30^\circ + j2\sin 30^\circ = 1.732 + j1.000$$

$$5\angle -45^\circ = 5(\cos(-45^\circ) + j\sin(-45^\circ)) = 5\cos(-45^\circ) + j5\sin(-45^\circ) = 3.536 - j3.536$$

$$4\angle 120^\circ = 4(\cos 120^\circ + j\sin 120^\circ) = 4\cos 120^\circ + j4\sin 120^\circ = -2.000 + j3.464$$

Hence, $2\angle 30^\circ + 5\angle -45^\circ - 4\angle 120^\circ$

$$= (1.732 + j1.000) + (3.536 - j3.536) - (-2.000 + j3.464)$$

$= 7.268 - j6.000$, which lies in the fourth quadrant

$$= \sqrt{7.268^2 + 6.000^2}\ \angle\tan^{-1}\left(\frac{-6.000}{7.268}\right)$$

$$= \mathbf{9.425\angle -39.54^\circ}$$

8.4 Applications of complex numbers

There are several applications of complex numbers in science and engineering, in particular in electrical alternating current theory and in mechanical vector analysis.

Application: Determine the value of current I and its phase relative to the 240V supply for the parallel circuit shown in Figure 8.4

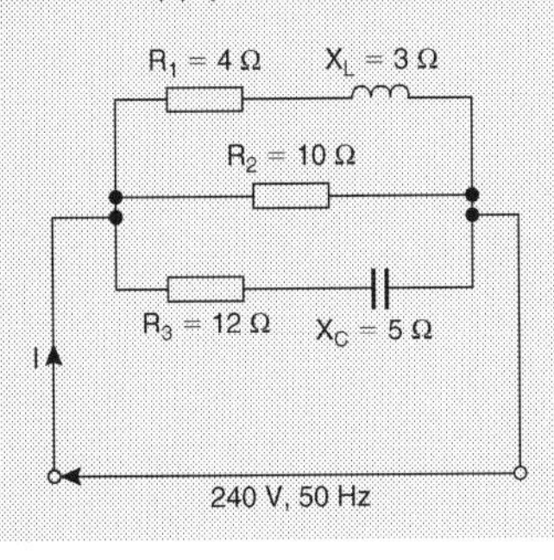

Figure 8.4

Current, $I = \dfrac{V}{Z}$. Impedance Z for the three-branch parallel circuit is given by:

$$\frac{1}{Z} = \frac{1}{Z_1} + \frac{1}{Z_2} + \frac{1}{Z_3},$$

where $Z_1 = 4 + j3\,\Omega$, $Z_2 = 10\,\Omega$ and $Z_3 = 12 - j5\,\Omega$

$$\text{Admittance, } Y_1 = \frac{1}{Z_1} = \frac{1}{4 + j3} = \frac{1}{4 + j3} \times \frac{4 - j3}{4 - j3} = \frac{4 - j3}{4^2 + 3^2}$$
$$= 0.160 - j0.120 \text{ siemens}$$

$$\text{Admittance, } Y_2 = \frac{1}{Z_2} = \frac{1}{10} = 0.10 \text{ siemens}$$

$$\text{Admittance, } Y_3 = \frac{1}{Z_3} = \frac{1}{12 - j5} = \frac{1}{12 - j5} \times \frac{12 + j5}{12 + j5} = \frac{12 + j5}{12^2 + 5^2}$$
$$= 0.0710 + j0.0296 \text{ siemens}$$

Total admittance,

$$\begin{aligned} Y &= Y_1 + Y_2 + Y_3 \\ &= (0.160 - j0.120) + (0.10) + (0.0710 + j0.0296) \\ &= 0.331 - j0.0904 = 0.343\angle -15.28^\circ \text{ siemens} \end{aligned}$$

$$\text{Current, } I = \frac{V}{Z} = VY = (240\angle 0^\circ)(0.343\angle -15.28^\circ)$$
$$= \mathbf{82.32\angle -15.28^\circ\ A}$$

Application: Determine the magnitude and direction of the resultant of the three coplanar forces shown in Figure 8.5

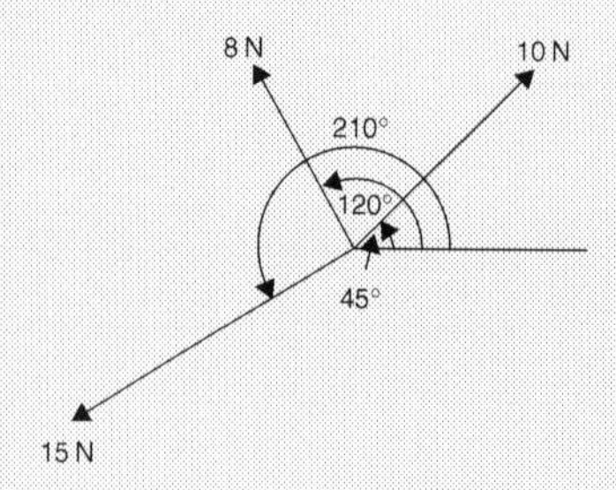

Figure 8.5

Force A, $f_A = 10\angle 45°$ N, force B, $f_B = 8\angle 120°$ N and force C, $f_C = 15\angle 210°$ N

The resultant force

$$\begin{aligned} &= f_A + f_B + f_C \\ &= 10\angle 45° + 8\angle 120° + 15\angle 210° \\ &= 10(\cos 45° + j\sin 45°) + 8(\cos 120° + j\sin 120°) \\ &\quad + 15(\cos 210° + j\sin 210°) \\ &= (7.071 + j7.071) + (-4.00 + j6.928) + (-12.99 - j7.50) \\ &= -9.919 + j6.499 \end{aligned}$$

Magnitude of resultant force $= \sqrt{(-9.919)^2 + 6.499^2} = \mathbf{11.86\ N}$

Direction of resultant force $= \tan^{-1}\left(\dfrac{6.499}{-9.919}\right) = \mathbf{146.77°}$

(since $-9.919 + j6.499$ lies in the second quadrant).

8.5 De Moivre's theorem

De Moivre's theorem states: $\mathbf{[r\angle\theta]^n = r^n\angle n\theta}$

The theorem is used to determine powers and roots of complex numbers.

In general, **when finding the nth root of a complex number, there are n solutions.** For example, there are three solutions to a cube root, five solutions to a fifth root, and so on. In the solutions to the roots of a complex number, the modulus, r, is always the same, but the arguments, θ, are different. Arguments are symmetrically spaced on an Argand diagram and are 360°/n apart, where n is the number of the roots required. Thus if one of the solutions to the cube root of a complex number is, say, $5\angle 20°$, the other two roots are symmetrically spaced 360°/3, i.e. 120° from this root, and the three roots are $5\angle 20°$, $5\angle 140°$ and $5\angle 260°$.

Application: Determine $[3\angle 20°]^4$

$[3\angle 20°]^4 = 3^4\angle(4 \times 20°) = \mathbf{81\angle 80°}$ by de Moivre's theorem.

Application: Determine $(-2 + j3)^6$ in polar form

$$(-2 + j3) = \sqrt{(-2)^2 + 3^2}\angle\tan^{-1}\left(\frac{3}{-2}\right)$$

$$= \sqrt{13}\angle 123.69° \text{ since } -2 + j3 \text{ lies in the second quadrant}$$

$$(-2 + j3)^6 = [\sqrt{13}\angle 123.69°]^6$$

$$= (\sqrt{13})^6\angle(6 \times 123.69°) \text{ by De Moivre's theorem}$$

$$= 2197\angle 742.14°$$

$$= 2197\angle 382.14°$$

$$(\text{since } 742.14° \equiv 742.14° - 360° = 382.14°)$$

$$= \mathbf{2197\angle 22.14°}$$

$$(\text{since } 382.14° \equiv 382.14° - 360° = 22.14°)$$

Application: Determine the two square roots of the complex number $(5 + j12)$ in polar and Cartesian forms

$$(5 + j12) = \sqrt{5^2 + 12^2}\ \angle\tan^{-1}\left(\frac{12}{5}\right) = 13\angle 67.38°$$

When determining square roots two solutions result. To obtain the second solution one way is to express $13\angle 67.38°$ also as $13\angle(67.38° + 360°)$, i.e. $13\angle 427.38°$. When the angle is divided by 2 an angle less than 360° is obtained.

$$\text{Hence } \sqrt{5^2 + 12^2} = \sqrt{13\angle 67.38°} \text{ and } \sqrt{13\angle 427.38°}$$

$$= [13\angle 67.38°]^{1/2} \text{ and } [13\angle 427.38°]^{1/2}$$

$$= 13^{1/2}\angle\left(\frac{1}{2} \times 67.38°\right) \text{and } 13^{1/2}\angle\left(\frac{1}{2} \times 427.38°\right)$$

$$= \sqrt{13}\ \angle 33.69° \text{ and } \sqrt{13}\ \angle 213.69°$$

$$= 3.61\angle 33.69° \text{ and } 3.61\angle 213.69°$$

Thus, in polar form, the two roots are $3.61\angle 33.69°$ and $3.61\angle -146.31°$

$$\sqrt{13}\angle 33.69° = \sqrt{13}\ (\cos 33.69° + j\sin 33.69°) = 3.0 + j2.0$$

$$\sqrt{13}\angle 213.69° = \sqrt{13}\ (\cos 213.69° + j\sin 213.69°) = -3.0 - j2.0$$

Thus, in Cartesian form, the two roots are $\pm(3.0 + j2.0)$

From the Argand diagram shown in Figure 8.6 the two roots are seen to be 180° apart, which is always true when finding square roots of complex numbers.

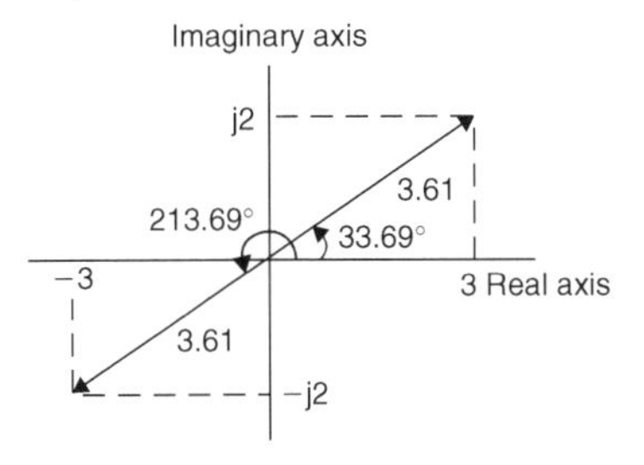

Figure 8.6

Application: Express the roots of $(-14 + j3)^{-2/5}$ in polar form

$$(-14 + j3) = \sqrt{205}\angle 167.905°$$

$$(-14 + j3)^{-2/5} = \sqrt{205}^{-2/5}\angle\left[\left(-\frac{2}{5}\right)\times 167.905°\right]$$

$$= 0.3449\angle -67.16°$$

There are five roots to this complex number, $\left(x^{-2/5} = \frac{1}{x^{2/5}} = \frac{1}{\sqrt[5]{x^2}}\right)$

The roots are symmetrically displaced from one another $\frac{360°}{5}$, i.e. 72° apart around an Argand diagram.

Thus, **the required roots are 0.3449∠−67.16°, 0.3449∠4.84°, 0.3449∠76.84°, 0.3449∠148.84° and 0.3449∠220.84°**

8.6 Exponential form

There are therefore three ways of expressing a complex number:

1. $z = (a + jb)$, called **Cartesian** or **rectangular form**,
2. $z = r(\cos\theta + j\sin\theta)$ or $r\angle\theta$, called **polar form**, and
3. $z = r\,e^{j\theta}$ called **exponential form**.

The **exponential form** is obtained from the polar form. For example, 4∠30° becomes $4e^{j\pi/6}$ in exponential form. (Note that in $r\,e^{j\theta}$, θ must be in radians).

Application: Express $(3 - j4)$ in polar and exponential forms

$(3 - j4) = \mathbf{5\angle -53.13^\circ} = \mathbf{5\angle -0.927}$ in polar form
$= \mathbf{5e^{-j0.927}}$ in exponential form

Application: Express $7.2e^{j1.5}$ in rectangular form

$7.2e^{j1.5} = 7.2\angle 1.5\,\text{rad}\ (=7.2\angle 85.944^\circ)$ in polar form
$= 7.2\cos 1.5 + j7.2\sin 1.5$
$= \mathbf{(0.509 + j7.182)}$ in rectangular form

Application: If $z = 2e^{1+j\pi/3}$ express z in Cartesian form

$z = 2e^{1+j\pi/3} = (2e^1)(e^{j\pi/3})$ by the laws of indices
$= (2e^1)\angle\frac{\pi}{3}$ (or $2e\angle 60^\circ$) in polar form
$= 2e\left(\cos\frac{\pi}{3} + j\sin\frac{\pi}{3}\right) = \mathbf{(2.718 + j4.708)}$ in Cartesian form

Application: If $z = 4e^{j1.3}$ find ln z in polar form

If $z = 4e^{j1.3}$ then $\ln z = \ln(4e^{j1.3})$
$= \mathbf{\ln 4 + j1.3}$ (or $\mathbf{1.386 + j1.300}$) in Cartesian form (by the laws of logarithms)
$= \mathbf{1.90\angle 43.17^\circ}$ or $\mathbf{1.90\angle 0.753}$ in polar form.

Application: Determine $\ln(3 + j4)$

$\ln(3 + j4) = \ln[5\angle 0.927] = \ln[5e^{j0.927}] = \ln 5 + \ln(e^{j0.927})$
$= \ln 5 + j0.927 = 1.609 + j0.927$
$= \mathbf{1.857\angle 29.95^\circ}$ or $\mathbf{1.857\angle 0.523}$

9 Matrices and Determinants

9.1 Addition, subtraction and multiplication of matrices

If $A = \begin{pmatrix} a & b \\ c & d \end{pmatrix}$ and $B = \begin{pmatrix} e & f \\ f & h \end{pmatrix}$

then $\mathbf{A + B} = \begin{pmatrix} \mathbf{a+e} & \mathbf{b+f} \\ \mathbf{c+g} & \mathbf{d+h} \end{pmatrix}$ and $\mathbf{A - B} = \begin{pmatrix} \mathbf{a-e} & \mathbf{b-f} \\ \mathbf{c-g} & \mathbf{d-h} \end{pmatrix}$

and $\mathbf{A \times B} = \begin{pmatrix} \mathbf{ae+bg} & \mathbf{af+bh} \\ \mathbf{ce+dg} & \mathbf{cf+dh} \end{pmatrix}$

Application: Determine $\begin{pmatrix} 2 & -1 \\ -7 & 4 \end{pmatrix} + \begin{pmatrix} -3 & 0 \\ 7 & -4 \end{pmatrix}$

$$\begin{pmatrix} 2 & -1 \\ -7 & 4 \end{pmatrix} + \begin{pmatrix} -3 & 0 \\ 7 & -4 \end{pmatrix} = \begin{pmatrix} 2+(-3) & -1+0 \\ -7+7 & 4+(-4) \end{pmatrix} = \begin{pmatrix} \mathbf{-1} & \mathbf{-1} \\ \mathbf{0} & \mathbf{0} \end{pmatrix}$$

Application: Determine $\begin{pmatrix} 2 & -1 \\ -7 & 4 \end{pmatrix} - \begin{pmatrix} -3 & 0 \\ 7 & -4 \end{pmatrix}$

$$\begin{pmatrix} 2 & -1 \\ -7 & 4 \end{pmatrix} - \begin{pmatrix} -3 & 0 \\ 7 & -4 \end{pmatrix} = \begin{pmatrix} 2-(-3) & -1-0 \\ -7-7 & 4-(-4) \end{pmatrix} = \begin{pmatrix} \mathbf{5} & \mathbf{-1} \\ \mathbf{-14} & \mathbf{8} \end{pmatrix}$$

Application:

If $A = \begin{pmatrix} -3 & 0 \\ 7 & -4 \end{pmatrix}$ and $B = \begin{pmatrix} 2 & -1 \\ -7 & 4 \end{pmatrix}$ determine $2A - 3B$

$$2A - 3B = 2\begin{pmatrix} -3 & 0 \\ 7 & -4 \end{pmatrix} - 3\begin{pmatrix} 2 & -1 \\ -7 & 4 \end{pmatrix} = \begin{pmatrix} -6 & 0 \\ 14 & -8 \end{pmatrix} - \begin{pmatrix} 6 & -3 \\ -21 & 12 \end{pmatrix}$$

$$= \begin{pmatrix} -6-6 & 0-(-3) \\ 14-(-21) & -8-12 \end{pmatrix}$$

$$= \begin{pmatrix} \mathbf{-12} & \mathbf{3} \\ \mathbf{35} & \mathbf{-20} \end{pmatrix}$$

Application: If $A = \begin{pmatrix} 2 & 3 \\ 1 & -4 \end{pmatrix}$ and $B = \begin{pmatrix} -5 & 7 \\ -3 & 4 \end{pmatrix}$ determine $A \times B$

$$A \times B = \begin{pmatrix} [2 \times -5 + 3 \times -3] & [2 \times 7 + 3 \times 4] \\ [1 \times -5 + -4 \times -3] & [1 \times 7 + -4 \times 4] \end{pmatrix} = \begin{pmatrix} \mathbf{-19} & \mathbf{26} \\ \mathbf{7} & \mathbf{-9} \end{pmatrix}$$

Application: Determine $\begin{pmatrix} 3 & 4 & 0 \\ -2 & 6 & -3 \\ 7 & -4 & 1 \end{pmatrix} \times \begin{pmatrix} 2 \\ 5 \\ -1 \end{pmatrix}$

$$\begin{pmatrix} 3 & 4 & 0 \\ -2 & 6 & -3 \\ 7 & -4 & 1 \end{pmatrix} \times \begin{pmatrix} 2 \\ 5 \\ -1 \end{pmatrix} = \begin{pmatrix} (3 \times 2) + (4 \times 5) + (0 \times -1) \\ (-2 \times 2) + (6 \times 5) + (-3 \times -1) \\ (7 \times 2) + (-4 \times 5) + (1 \times -1) \end{pmatrix} = \begin{pmatrix} \mathbf{26} \\ \mathbf{29} \\ \mathbf{-7} \end{pmatrix}$$

9.2 The determinant and inverse of a 2 by 2 matrix

If $A = \begin{pmatrix} a & b \\ c & d \end{pmatrix}$ then

the **determinant of A,** $\begin{vmatrix} a & b \\ c & d \end{vmatrix} = \mathbf{a \times d - b \times c}$

and the **inverse of A,** $\mathbf{A^{-1} = \dfrac{1}{ad - bc}\begin{vmatrix} d & -b \\ -c & a \end{vmatrix}}$

Application: Find the determinant of $\begin{pmatrix} 3 & -4 \\ 1 & 6 \end{pmatrix}$

$$\begin{vmatrix} 3 & -4 \\ 1 & 6 \end{vmatrix} = (3 \times 6) - (-4 \times 1) = 18 - (-4) = \mathbf{22}$$

Application: Find the inverse of $\begin{pmatrix} 3 & -4 \\ 1 & 6 \end{pmatrix}$

$$\text{Inverse of matrix } \begin{pmatrix} 3 & -4 \\ 1 & 6 \end{pmatrix} = \frac{1}{18 - -4}\begin{pmatrix} 6 & 4 \\ -1 & 3 \end{pmatrix} = \mathbf{\frac{1}{22}\begin{pmatrix} 6 & 4 \\ -1 & 3 \end{pmatrix}}$$

Application: If $A = \begin{pmatrix} 3 & -4 \\ 1 & 6 \end{pmatrix}$ determine $A \times A^{-1}$

From above:

$$A \times A^{-1} = \begin{pmatrix} 3 & -4 \\ 1 & 6 \end{pmatrix} \times \frac{1}{22}\begin{pmatrix} 6 & 4 \\ -1 & 3 \end{pmatrix}$$

$$= \frac{1}{22}\begin{pmatrix} 3 & -4 \\ 1 & 6 \end{pmatrix} \times \begin{pmatrix} 6 & 4 \\ -1 & 3 \end{pmatrix}$$

$$= \frac{1}{22}\begin{pmatrix} 18 + 4 & 12 - 12 \\ 6 - 6 & 4 + 18 \end{pmatrix}$$

$$= \frac{1}{22}\begin{pmatrix} 22 & 0 \\ 0 & 22 \end{pmatrix} = \begin{pmatrix} \mathbf{1} & \mathbf{0} \\ \mathbf{0} & \mathbf{1} \end{pmatrix}$$

$\begin{pmatrix} 1 & 0 \\ 0 & 1 \end{pmatrix}$ is called the **unit matrix**; such a matrix has all leading diagonal elements equal to 1 and all other elements equal to 0

9.3 The determinant of a 3 by 3 matrix

(i) The **minor** of an element of a 3 by 3 matrix is the value of the 2 by 2 determinant obtained by covering up the row and column containing that element.

Thus for the matrix $\begin{pmatrix} 1 & 2 & 3 \\ 4 & 5 & 6 \\ 7 & 8 & 9 \end{pmatrix}$ the minor of element 4 is obtained by covering up the row (4 5 6) and the column $\begin{pmatrix} 1 \\ 4 \\ 7 \end{pmatrix}$, leaving the 2 by 2 determinant $\begin{vmatrix} 2 & 3 \\ 8 & 9 \end{vmatrix}$ i.e. the minor of element 4 is $(2 \times 9) - (3 \times 8) = -6$

(ii) The sign of a minor depends on its position within the matrix, the sign pattern being $\begin{pmatrix} + & - & + \\ - & + & - \\ + & - & + \end{pmatrix}$

Thus the signed-minor of element 4 in the matrix $\begin{pmatrix} 1 & 2 & 3 \\ 4 & 5 & 6 \\ 7 & 8 & 9 \end{pmatrix}$ is $-\begin{vmatrix} 2 & 3 \\ 8 & 9 \end{vmatrix} = -(-6) = 6$

The signed-minor of an element is called the **cofactor** of the element.

(iii) **The value of a 3 by 3 determinant is the sum of the products of the elements and their cofactors of any row or any column of the corresponding 3 by 3 matrix.**

There are thus six different ways of evaluating a 3 × 3 determinant – and all should give the same value.

Using the first row:

$$\begin{vmatrix} a_1 & b_1 & c_1 \\ a_2 & b_2 & c_2 \\ a_3 & b_3 & c_3 \end{vmatrix} = a_1 \begin{vmatrix} b_2 & c_2 \\ b_3 & c_3 \end{vmatrix} - b_1 \begin{vmatrix} a_2 & c_2 \\ a_3 & c_3 \end{vmatrix} + c_1 \begin{vmatrix} a_2 & b_2 \\ a_3 & b_3 \end{vmatrix}$$

Application: Evaluate $\begin{vmatrix} 1 & 4 & -3 \\ -5 & 2 & 6 \\ -1 & -4 & 2 \end{vmatrix}$ using (a) the first row, and (b) the second column

(a) Using the first row:

$$\begin{vmatrix} 1 & 4 & -3 \\ -5 & 2 & 6 \\ -1 & -4 & 2 \end{vmatrix} = 1\begin{vmatrix} 2 & 6 \\ -4 & 2 \end{vmatrix} - 4\begin{vmatrix} -5 & 6 \\ -1 & 2 \end{vmatrix} + (-3)\begin{vmatrix} -5 & 2 \\ -1 & -4 \end{vmatrix}$$

$$= (4 + 24) - 4(-10 + 6) - 3(20 + 2)$$

$$= 28 + 16 - 66 = \mathbf{-22}$$

(b) Using the second column:

$$\begin{vmatrix} 1 & 4 & -3 \\ -5 & 2 & 6 \\ -1 & -4 & 2 \end{vmatrix} = -4\begin{vmatrix} -5 & 6 \\ -1 & 2 \end{vmatrix} + 2\begin{vmatrix} 1 & -3 \\ -1 & 2 \end{vmatrix} - (-4)\begin{vmatrix} 1 & -3 \\ -5 & 6 \end{vmatrix}$$

$$= -4(-10 + 6) + 2(2 - 3) + 4(6 - 15)$$

$$= 16 - 2 - 36 = \mathbf{-22}$$

9.4 The inverse of a 3 by 3 matrix

If $A = \begin{pmatrix} a_1 & b_1 & c_1 \\ a_2 & b_2 & c_2 \\ a_3 & b_3 & c_3 \end{pmatrix}$ then the **inverse of matrix A**,

$$\mathbf{A^{-1} = \frac{adj\ A}{|A|}} \quad \text{where adj A is the adjoint}$$

The **adjoint** of a matrix A is obtained by:

(i) forming a matrix B of the cofactors of A, and

(ii) **transposing** matrix B to give B^T, where B^T is the matrix obtained by writing the rows of B as the columns of B^T. Then **adj A =** $\mathbf{B^T}$

Application: Find the inverse of $\begin{pmatrix} 1 & 5 & -2 \\ 3 & -1 & 4 \\ -3 & 6 & -7 \end{pmatrix}$

$$\text{Inverse} = \frac{\text{adjoint}}{\text{determinant}}$$

The matrix of cofactors is $\begin{pmatrix} -17 & 9 & 15 \\ 23 & -13 & -21 \\ 18 & -10 & -16 \end{pmatrix}$

The transpose of the matrix of cofactors

(i.e. the adjoint) is $\begin{pmatrix} -17 & 23 & 18 \\ 9 & -13 & -10 \\ 15 & -21 & -16 \end{pmatrix}$

The determinant of $\begin{pmatrix} 1 & 5 & -2 \\ 3 & -1 & 4 \\ -3 & 6 & -7 \end{pmatrix}$

$= 1(7 - 24) - 5(-21 + 12) - 2(18 - 3)$ using the first row

$= -17 + 45 - 30 = -2$

Hence the inverse of $\begin{pmatrix} 1 & 5 & -2 \\ 3 & -1 & 4 \\ -3 & 6 & -7 \end{pmatrix} = \dfrac{\begin{pmatrix} -17 & 23 & 18 \\ 9 & -13 & -10 \\ 15 & -21 & -16 \end{pmatrix}}{-2}$

$$= \begin{pmatrix} \mathbf{8.5} & \mathbf{-11.5} & \mathbf{-9} \\ \mathbf{-4.5} & \mathbf{6.5} & \mathbf{5} \\ \mathbf{-7.5} & \mathbf{10.5} & \mathbf{8} \end{pmatrix}$$

9.5 Solution of simultaneous equations by matrices

Two unknowns

The procedure for solving linear simultaneous equations in **two unknowns using matrices** is:

(i) write the equations in the form

$$a_1x + b_1y = c_1$$
$$a_2x + b_2y = c_2$$

(ii) write the matrix equation corresponding to these equations,

i.e. $$\begin{pmatrix} a_1 & b_1 \\ a_2 & b_2 \end{pmatrix} \times \begin{pmatrix} x \\ y \end{pmatrix} = \begin{pmatrix} c_1 \\ c_2 \end{pmatrix}$$

(iii) determine the inverse matrix of $\begin{pmatrix} a_1 & b_1 \\ a_2 & b_2 \end{pmatrix}$

i.e. $$\frac{1}{a_1b_2 - b_1a_2}\begin{pmatrix} b_2 & -b_1 \\ -a_2 & a_1 \end{pmatrix}$$

(iv) multiply each side of (ii) by the inverse matrix, and

(v) solve for x and y by equating corresponding elements

Applications: Use matrices to solve the simultaneous equations:

$$3x + 5y - 7 = 0 \qquad (1)$$
$$4x - 3y - 19 = 0 \qquad (2)$$

(i) Writing the equations in the $a_1x + b_1y = c$ form gives:

$$3x + 5y = 7$$
$$4x - 3y = 19$$

(ii) The matrix equation is $\begin{pmatrix} 3 & 5 \\ 4 & -3 \end{pmatrix} \times \begin{pmatrix} x \\ y \end{pmatrix} = \begin{pmatrix} 7 \\ 19 \end{pmatrix}$

(iii) The inverse of matrix $\begin{pmatrix} 3 & 5 \\ 4 & -3 \end{pmatrix}$ is:

$$\frac{1}{3 \times (-3) - 5 \times 4}\begin{pmatrix} -3 & -5 \\ -4 & 3 \end{pmatrix} = \begin{pmatrix} \frac{3}{29} & \frac{5}{29} \\ \frac{4}{29} & \frac{-3}{29} \end{pmatrix}$$

(iv) Multiplying each side of (ii) by (iii) and remembering that $A \times A^{-1} = I$, the unit matrix, gives:

$$\begin{pmatrix} 1 & 0 \\ 0 & 1 \end{pmatrix}\begin{pmatrix} x \\ y \end{pmatrix} = \begin{pmatrix} \frac{3}{29} & \frac{5}{29} \\ \frac{4}{29} & \frac{-3}{29} \end{pmatrix} \times \begin{pmatrix} 7 \\ 19 \end{pmatrix}$$

Thus $$\begin{pmatrix} x \\ y \end{pmatrix} = \begin{pmatrix} \frac{21}{29} + \frac{95}{29} \\ \frac{28}{29} - \frac{57}{29} \end{pmatrix} \text{ i.e. } \begin{pmatrix} x \\ y \end{pmatrix} = \begin{pmatrix} 4 \\ -1 \end{pmatrix}$$

(v) By comparing corresponding elements: $\mathbf{x = 4}$ **and** $\mathbf{y = -1}$**,** which can be checked in the original equations.

Three unknowns

The procedure for solving linear simultaneous equations in **three unknowns using matrices** is:

(i) write the equations in the form

$$\begin{aligned} a_1x + b_1y + c_1z &= d_1 \\ a_2x + b_2y + c_2z &= d_2 \\ a_3x + b_3y + c_3z &= d_3 \end{aligned}$$

(ii) write the matrix equation corresponding to these equations, i.e.

$$\begin{pmatrix} a_1 & b_1 & c_1 \\ a_2 & b_2 & c_2 \\ a_3 & b_3 & c_3 \end{pmatrix} \times \begin{pmatrix} x \\ y \\ z \end{pmatrix} = \begin{pmatrix} d_1 \\ d_2 \\ d_3 \end{pmatrix}$$

(iii) determine the inverse matrix of $\begin{pmatrix} a_1 & b_1 & c_1 \\ a_2 & b_2 & c_2 \\ a_3 & b_3 & c_3 \end{pmatrix}$

(iv) multiply each side of (ii) by the inverse matrix, and

(v) solve for x, y and z by equating the corresponding elements

Application: Use matrices to solve the simultaneous equations:

$$x + y + z - 4 = 0 \qquad (1)$$

$$2x - 3y + 4z - 33 = 0 \qquad (2)$$

$$3x - 2y - 2z - 2 = 0 \qquad (3)$$

(i) Writing the equations in the $a_1x + b_1y + c_1z = d_1$ form gives:

$$x + y + z = 4$$

$$2x - 3y + 4z = 33$$

$$3x - 2y - 2z = 2$$

(ii) The matrix equation is: $\begin{pmatrix} 1 & 1 & 1 \\ 2 & -3 & 4 \\ 3 & -2 & -2 \end{pmatrix} \times \begin{pmatrix} x \\ y \\ z \end{pmatrix} = \begin{pmatrix} 4 \\ 33 \\ 2 \end{pmatrix}$

(iii) The inverse matrix of $A = \begin{pmatrix} 1 & 1 & 1 \\ 2 & -3 & 4 \\ 3 & -2 & -2 \end{pmatrix}$ is given by $A^{-1} = \dfrac{\text{adj } A}{|A|}$

The adjoint of A is the transpose of the matrix of the cofactors of the elements. The matrix of cofactors is $\begin{pmatrix} 14 & 16 & 5 \\ 0 & -5 & 5 \\ 7 & -2 & -5 \end{pmatrix}$ and the transpose of this matrix gives: $\text{adj } A = \begin{pmatrix} 14 & 0 & 7 \\ 16 & -5 & -2 \\ 5 & 5 & -5 \end{pmatrix}$

The determinant of A, i.e. the sum of the products of elements and their cofactors, using a first row expansion is

$$1\begin{vmatrix}-3 & 4\\ -2 & -2\end{vmatrix} - 1\begin{vmatrix}2 & 4\\ 3 & -2\end{vmatrix} + 1\begin{vmatrix}2 & -3\\ 3 & -2\end{vmatrix} = (1\times 14) - (1\times -16) + (1\times 5)$$

$$= 35$$

Hence the inverse of A, $A^{-1} = \dfrac{1}{35}\begin{pmatrix}14 & 0 & 7\\ 16 & -5 & -2\\ 5 & 5 & -5\end{pmatrix}$

(iv) Multiplying each side of (ii) by (iii), and remembering that $A \times A^{-1} = I$, the unit matrix, gives:

$$\begin{pmatrix}1 & 0 & 0\\ 0 & 1 & 0\\ 0 & 0 & 1\end{pmatrix} \times \begin{pmatrix}x\\ y\\ z\end{pmatrix} = \frac{1}{35}\begin{pmatrix}14 & 0 & 7\\ 16 & -5 & -2\\ 5 & 5 & -5\end{pmatrix} \times \begin{pmatrix}4\\ 33\\ 2\end{pmatrix}$$

$$\begin{pmatrix}x\\ y\\ z\end{pmatrix} = \frac{1}{35}\begin{pmatrix}(14\times 4)+(0\times 33)+(7\times 2)\\ (16\times 4)+(-5\times 33)+(-2\times 2)\\ (5\times 4)+(5\times 33)+(-5\times 2)\end{pmatrix} = \frac{1}{35}\begin{pmatrix}70\\ -105\\ 175\end{pmatrix} = \begin{pmatrix}2\\ -3\\ 5\end{pmatrix}$$

(v) By comparing corresponding elements, **x = 2, y = −3, z = 5**, which can be checked in the original equations.

9.6 Solution of simultaneous equations by determinants

Two unknowns

When solving linear simultaneous equations in **two unknowns using determinants:**

(i) write the equations in the form

$$a_1x + b_1y + c_1 = 0$$

$$a_2x + b_2y + c_2 = 0$$

(ii) the solution is given by: $\frac{x}{D_x} = \frac{-y}{D_y} = \frac{1}{D}$

where $D_x = \begin{vmatrix} b_1 & c_1 \\ b_2 & c_2 \end{vmatrix}$ i.e. the determinant of the coefficients left when the x-column is covered up,

$D_y = \begin{vmatrix} a_1 & c_1 \\ a_2 & c_2 \end{vmatrix}$ i.e. the determinant of the coefficients left when the y-column is covered up,

and $D = \begin{vmatrix} a_1 & b_1 \\ a_2 & b_2 \end{vmatrix}$ i.e. the determinant of the coefficients left when the constants-column is covered up

Application: Solve the following simultaneous equations using determinants:

$$3x - 4y = 12$$
$$7x + 5y = 6.5$$

Following the above procedure:

(i) $3x - 4y - 12 = 0$

$7x + 5y - 6.5 = 0$

(ii) $$\frac{x}{\begin{vmatrix} -4 & -12 \\ 5 & -6.5 \end{vmatrix}} = \frac{-y}{\begin{vmatrix} 3 & -12 \\ 7 & -6.5 \end{vmatrix}} = \frac{1}{\begin{vmatrix} 3 & -4 \\ 7 & 5 \end{vmatrix}}$$

i.e. $$\frac{x}{(-4)(-6.5) - (-12)(5)} = \frac{-y}{(3)(-6.5) - (-12)(7)} = \frac{1}{(3)(5) - (-4)(7)}$$

i.e. $$\frac{x}{26 + 60} = \frac{-y}{-19.5 + 84} = \frac{1}{15 + 28}$$

i.e. $$\frac{x}{86} = \frac{-y}{64.5} = \frac{1}{43}$$

Since $\frac{x}{86} = \frac{1}{43}$ then $\mathbf{x} = \frac{86}{43} = \mathbf{2}$

and since $\frac{-y}{64.5} = \frac{1}{43}$ then $\mathbf{y} = -\frac{64.5}{43} = \mathbf{-1.5}$

Three unknowns

When solving simultaneous equations in **three unknowns using determinants**:

(i) write the equations in the form

$$a_1x + b_1y + c_1z + d_1 = 0$$
$$a_2x + b_2y + c_2z + d_2 = 0$$
$$a_3x + b_3y + c_3z + d_3 = 0$$

(ii) the solution is given by: $\frac{x}{D_x} = \frac{-y}{D_y} = \frac{z}{D_z} = \frac{-1}{D}$

where $D_x = \begin{vmatrix} b_1 & c_1 & d_1 \\ b_2 & c_2 & d_2 \\ b_3 & c_3 & d_3 \end{vmatrix}$ i.e. the determinant of the coefficients obtained by covering up the x-column

$D_y = \begin{vmatrix} a_1 & c_1 & d_1 \\ a_2 & c_2 & d_2 \\ a_3 & c_3 & d_3 \end{vmatrix}$ i.e. the determinant of the coefficients obtained by covering up the y-column

$D_z = \begin{vmatrix} a_1 & b_1 & d_1 \\ a_2 & b_2 & d_2 \\ a_3 & b_3 & d_3 \end{vmatrix}$ i.e. the determinant of the coefficients obtained by covering up the z-column

and $D = \begin{vmatrix} a_1 & b_1 & c_1 \\ a_2 & b_2 & c_2 \\ a_3 & b_3 & c_3 \end{vmatrix}$ i.e. the determinant of the coefficients obtained by covering up the constants-column

Application: A d.c. circuit comprises three closed loops. Applying Kirchhoff's laws to the closed loops gives the following equations for current flow in milliamperes:

$$2I_1 + 3I_2 - 4I_3 = 26$$
$$I_1 - 5I_2 - 3I_3 = -87$$
$$-7I_1 + 2I_2 + 6I_3 = 12$$

Use determinants to solve for I_1, I_2 and I_3

Following the above procedure:

(i) $2I_1 + 3I_2 - 4I_3 - 26 = 0$
$I_1 - 5I_2 - 3I_3 + 87 = 0$
$-7I_1 + 2I_2 + 6I_3 - 12 = 0$

(ii) The solution is given by: $\dfrac{I_1}{D_{I_1}} = \dfrac{-I_2}{D_{I_2}} = \dfrac{I_3}{D_{I_3}} = \dfrac{-1}{D}$, where

$$DI_1 = \begin{vmatrix} 3 & -4 & -26 \\ -5 & -3 & 87 \\ 2 & 6 & -12 \end{vmatrix}$$

$$= (3)\begin{vmatrix} -3 & 87 \\ 6 & -12 \end{vmatrix} - (-4)\begin{vmatrix} -5 & 87 \\ 2 & -12 \end{vmatrix} + (-26)\begin{vmatrix} -5 & -3 \\ 2 & 6 \end{vmatrix}$$

$$= 3(-486) + 4(-114) - 26(-24) = \mathbf{-1290}$$

$$DI_2 = \begin{vmatrix} 2 & -4 & -26 \\ 1 & -3 & 87 \\ -7 & 6 & -12 \end{vmatrix}$$

$$= (2)(36 - 522) - (-4)(-12 + 609) + (-26)(6 - 21)$$

$$= -972 + 2388 + 390 = \mathbf{1806}$$

$$DI_3 = \begin{vmatrix} 2 & 3 & -26 \\ 1 & -5 & 87 \\ -7 & 2 & -12 \end{vmatrix}$$

$$= (2)(60 - 174) - (3)(-12 + 609) + (-26)(2 - 35)$$

$$= -228 - 1791 + 858 = \mathbf{-1161}$$

and $D = \begin{vmatrix} 2 & 3 & -4 \\ 1 & -5 & -3 \\ -7 & 2 & 6 \end{vmatrix}$

$= (2)(-30+6) - (3)(6-21) + (-4)(2-35)$

$= -48 + 45 + 132 = \mathbf{129}$

Thus $\dfrac{I_1}{-1290} = \dfrac{-I_2}{1806} = \dfrac{I_3}{-1161} = \dfrac{-1}{129}$

giving: $\mathbf{I_1} = -\dfrac{-1290}{129} = \mathbf{10\,mA}$, $\mathbf{I_2} = \dfrac{1806}{129} = \mathbf{14\,mA}$ and $\mathbf{I_3} = -\dfrac{-1161}{129} = \mathbf{9\,mA}$

9.7 Solution of simultaneous equations using Cramer's rule

Cramer's rule states that if

$$a_{11}x + a_{12}y + a_{13}z = b_1$$
$$a_{21}x + a_{22}y + a_{23}z = b_2$$
$$a_{31}x + a_{32}y + a_{33}z = b_3$$

then $\mathbf{x = \dfrac{D_x}{D}}$, $\mathbf{y = \dfrac{D_y}{D}}$ and $\mathbf{z = \dfrac{D_z}{D}}$

where $D = \begin{vmatrix} a_{11} & a_{12} & a_{13} \\ a_{21} & a_{22} & a_{23} \\ a_{31} & a_{32} & a_{33} \end{vmatrix}$

$D_x = \begin{vmatrix} b_1 & a_{12} & a_{13} \\ b_2 & a_{22} & a_{23} \\ b_3 & a_{32} & a_{33} \end{vmatrix}$ i.e. the x-column has been replaced by the R.H.S. b column

$D_y = \begin{vmatrix} a_{11} & b_1 & a_{13} \\ a_{21} & b_2 & a_{23} \\ a_{31} & b_3 & a_{33} \end{vmatrix}$ i.e. the y-column has been replaced by the R.H.S. b column

$$D_z = \begin{vmatrix} a_{11} & a_{12} & b_1 \\ a_{21} & a_{22} & b_2 \\ a_{31} & a_{32} & b_3 \end{vmatrix}$$ i.e. the z-column has been replaced by the R.H.S. b column

Application: Solve the following simultaneous equations using Cramer's rule

$$x + y + z = 4$$
$$2x - 3y + 4z = 33$$
$$3x - 2y - 2z = 2$$

Following the above method:

$$\mathbf{D} = \begin{vmatrix} 1 & 1 & 1 \\ 2 & -3 & 4 \\ 3 & -2 & -2 \end{vmatrix} = 1(6 - -8) - 1(-4 - 12) + 1(-4 - -9)$$
$$= 14 + 16 + 5 = \mathbf{35}$$

$$\mathbf{D_x} = \begin{vmatrix} 4 & 1 & 1 \\ 33 & -3 & 4 \\ 2 & -2 & -2 \end{vmatrix} = 4(6 - -8) - 1(-66 - 8) + 1(-66 - -6)$$
$$= 56 + 74 - 60 = \mathbf{70}$$

$$\mathbf{D_y} = \begin{vmatrix} 1 & 4 & 1 \\ 2 & 33 & 4 \\ 3 & 2 & -2 \end{vmatrix} = 1(-66 - 8) - 4(-4 - 12) + 1(4 - 99)$$
$$= -74 + 64 - 95 = \mathbf{-105}$$

$$\mathbf{D_z} = \begin{vmatrix} 1 & 1 & 4 \\ 2 & -3 & 33 \\ 3 & -2 & 2 \end{vmatrix} = 1(-6 - -66) - 1(4 - 99) + 4(-4 - -9)$$
$$= 60 + 95 + 20 = \mathbf{175}$$

Hence, $\mathbf{x} = \frac{D_x}{D} = \frac{70}{35} = \mathbf{2}$, $\mathbf{y} = \frac{D_y}{D} = \frac{-105}{35} = \mathbf{-3}$ and

$$\mathbf{z} = \frac{D_z}{D} = \frac{175}{35} = \mathbf{5}$$

9.8 Solution of simultaneous equations using Gaussian elimination

If
$$a_{11}x + a_{12}y + a_{13}z = b_1 \quad (1)$$
$$a_{21}x + a_{22}y + a_{23}z = b_2 \quad (2)$$
$$a_{31}x + a_{32}y + a_{33}z = b_3 \quad (3)$$

the three-step **procedure** to solve simultaneous equations in three unknowns using the **Gaussian elimination method** is:

(i) Equation (2) $-\dfrac{a_{21}}{a_{11}} \times$ equation (1) to form equation (2′)

and equation (3) $-\dfrac{a_{31}}{a_{11}} \times$ equation (1) to form equation (3′)

(ii) Equation (3′) $-\dfrac{a_{32} \text{ (of 3')}}{a_{22} \text{ (of 2')}} \times$ equation (2′) to form equation (3″)

(iii) Determine z from equation (3″), then y from equation (2′) and finally, x from equation (1)

Application: A d.c. circuit comprises three closed loops. Applying Kirchhoff's laws to the closed loops gives the following equations for current flow in milliamperes:

$$2I_1 + 3I_2 - 4I_3 = 26 \quad (1)$$
$$I_1 - 5I_2 - 3I_3 = -87 \quad (2)$$
$$-7I_1 + 2I_2 + 6I_3 = 12 \quad (3)$$

Use the Gaussian elimination method to solve for I_1, I_2 and I_3

Following the above procedure:

(i)
$$2I_1 + 3I_2 - 4I_3 = 26 \quad (1)$$

Equation (2) $-\dfrac{1}{2} \times$ equation (1) gives:

$$0 - 6.5I_2 - I_3 = -100 \quad (2')$$

Equation (3) $-\dfrac{-7}{2}\times$ equation (1) gives:

$$0 + 12.5I_2 - 8I_3 = 103 \qquad (3')$$

(ii)

$$2I_1 + 3I_2 - 4I_3 = 26 \qquad (1)$$

$$0 - 6.5I_2 - I_3 = -100 \qquad (2')$$

Equation (3′) $-\dfrac{12.5}{-6.5}\times$ equation (2′) gives:

$$0 + 0 - 9.923I_3 = -89.308 \qquad (3'')$$

(iii) From equation (3″), $\mathbf{I_3} = \dfrac{-89.308}{-9.923} = \mathbf{9\,mA}$,

from equation (2′), $-6.5I_2 - 9 = -100$,

from which, $\mathbf{I_2} = \dfrac{-100+9}{-6.5} = \mathbf{14\,mA}$

and from equation (1), $2I_1 + 3(14) - 4(9) = 26$,

from which, $\mathbf{I_1} = \dfrac{26-42+36}{2} = \dfrac{20}{2} = \mathbf{10\,mA}$

10 Boolean Algebra and Logic Circuits

10.1 Boolean algebra and switching circuits

Function	Boolean expression	Equivalent electrical circuit	Truth Table
2-input or-function	A + B (i.e. A, or B, or both A and B)	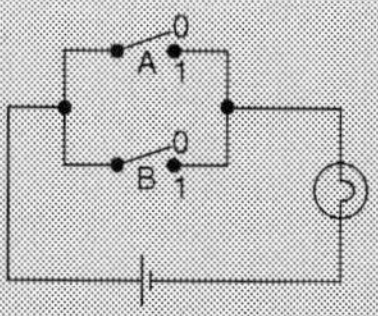 A 0 1 B 0 1	1 2 Input (switches) | 3 Output (lamp) A | B | Z = A + B 0 | 0 | 0 0 | 1 | 1 1 | 0 | 1 1 | 1 | 1
2-input and-function	A . B (i.e. both A and B)	A 0 1 B 0 1	Input (switches) | Output (lamp) A | B | Z = A · B 0 | 0 | 0 0 | 1 | 0 1 | 0 | 0 1 | 1 | 1
Not-function	$\overline{A}$		Input A | Output Z = $\overline{A}$ 0 | 1 1 | 0
3-input or-function	A + B+C	Input A B C Output	Input A B C | Output Z = A + B + C 0 0 0 | 0 0 0 1 | 1 0 1 0 | 1 0 1 1 | 1 1 0 0 | 1 1 0 1 | 1 1 1 0 | 1 1 1 1 | 1

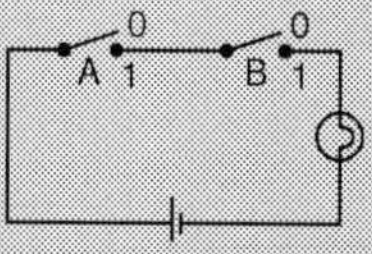

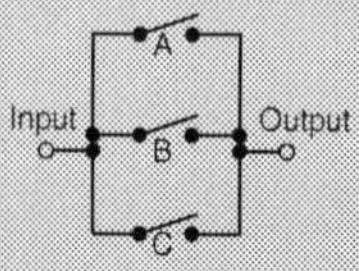

			Input A	B	C	Output $Z = A \cdot B \cdot C$
3-input and-function	A.B.C	Input o—A—B—C—o Output	0	0	0	0
			0	0	1	0
			0	1	0	0
			0	1	1	0
			1	0	0	0
			1	0	1	0
			1	1	0	0
			1	1	1	1

To achieve a given output, it is often necessary to use combinations of switches connected both in series and in parallel. If the output from a switching circuit is given by the Boolean expression: $Z = A.B + \bar{A}.\bar{B}$, the truth table is as shown in Figure 10.5(a). In this table, columns 1 and 2 give all the possible combinations of A and B. Column 3 corresponds to A.B and column 4 to $\bar{A}.\bar{B}$ i.e. a 1 output is obtained when A = 0 and when B = 0. Column 5 is the **or**-function applied to columns 3 and 4 giving an output of $Z = A.B + \bar{A}.\bar{B}$. The corresponding switching circuit is shown in Figure 10.5(b) in which A and B are connected in series to give A.B, $\bar{A}$ and $\bar{B}$ are connected in series to give $\bar{A}.\bar{B}$, and A.B and $\bar{A}.\bar{B}$ are connected in parallel to give $A.B + \bar{A}.\bar{B}$. The circuit symbols used are such that A means the switch is on when A is 1, $\bar{A}$ means the switch is on when A is 0, and so on.

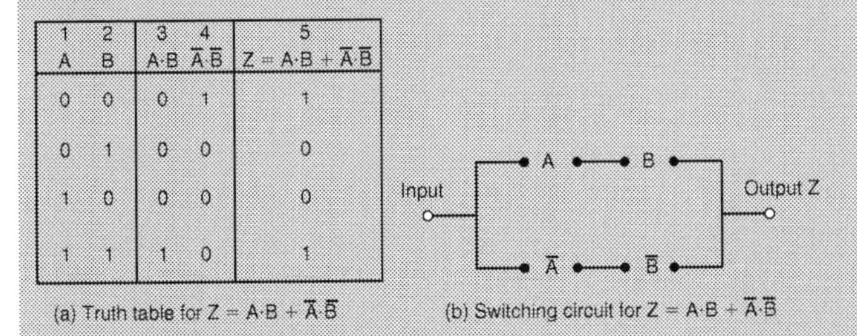

1 A	2 B	3 $A \cdot B$	4 $\bar{A} \cdot \bar{B}$	5 $Z = A \cdot B + \bar{A} \cdot \bar{B}$
0	0	0	1	1
0	1	0	0	0
1	0	0	0	0
1	1	1	0	1

(a) Truth table for $Z = A \cdot B + \bar{A} \cdot \bar{B}$

(b) Switching circuit for $Z = A \cdot B + \bar{A} \cdot \bar{B}$

Figure 10.5

Application: Derive the Boolean expression and construct a truth table for the switching circuit shown in Figure 10.6.

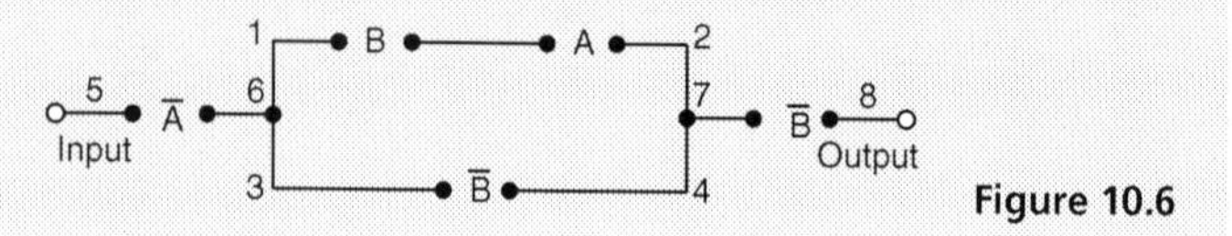

Figure 10.6

The switches between 1 and 2 in Figure 10.6 are in series and have a Boolean expression of $B.A$. The parallel circuit 1 to 2 and 3 to 4 have a Boolean expression of $(B.A + \bar{B})$. The parallel circuit can be treated as a single switching unit, giving the equivalent of switches 5 to 6, 6 to 7 and 7 to 8 in series. Thus the output is given by: $\mathbf{Z = \bar{A}.(B.A + \bar{B}).\bar{B}}$

The truth table is as shown in Table 10.6. Columns 1 and 2 give all the possible combinations of switches A and B. Column 3 is the **and**-function applied to columns 1 and 2, giving $B.A$. Column 4 is $\bar{B}$, i.e. the opposite to column 2. Column 5 is the **or**-function applied to columns 3 and 4. Column 6 is $\bar{A}$, i.e. the opposite to column 1. The output is column 7 and is obtained by applying the **and**-function to columns 4, 5 and 6.

Table 10.6

1 A	2 B	3 $B \cdot A$	4 $\bar{B}$	5 $B \cdot A + \bar{B}$	6 $\bar{A}$	7 $Z = \bar{A} \cdot (B \cdot A + \bar{B}) \cdot \bar{B}$
0	0	0	1	1	1	1
0	1	0	0	0	1	0
1	0	0	1	1	0	0
1	1	1	0	1	0	0

Application: Derive the Boolean expression and construct a truth table for the switching circuit shown in Figure 10.7.

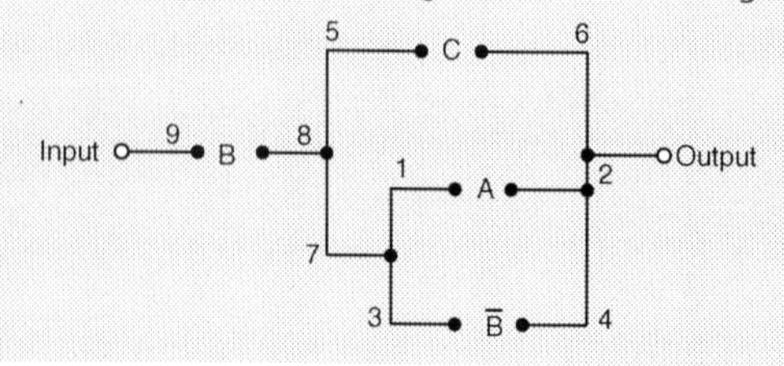

Figure 10.7

The parallel circuit 1 to 2 and 3 to 4 gives $(A + \overline{B})$ and this is equivalent to a single switching unit between 7 and 2. The parallel circuit 5 to 6 and 7 to 2 gives $C + (A + \overline{B})$ and this is equivalent to a single switching unit between 8 and 2. The series circuit 9 to 8 and 8 to 2 gives the output

$$\mathbf{Z = B.[C + (A + \overline{B})]}$$

The truth table is shown in Table 10.7. Columns 1, 2 and 3 give all the possible combinations of A, B and C. Column 4 is $\overline{B}$ and is the opposite to column 2. Column 5 is the **or**-function applied to columns 1 and 4, giving $(A + \overline{B})$. Column 6 is the **or**-function applied to columns 3 and 5 giving: $C + (A + \overline{B})$. The output is given in column 7 and is obtained by applying the **and**-function to columns 2 and 6, giving: $\mathbf{Z = B.[C + (A + \overline{B})]}$

Table 10.7

1 A	2 B	3 C	4 $\overline{B}$	5 $A + \overline{B}$	6 $C + (A + \overline{B})$	7 $Z = B \cdot [C + (A + \overline{B})]$
0	0	0	1	1	1	0
0	0	1	1	1	1	0
0	1	0	0	0	0	0
0	1	1	0	0	1	1
1	0	0	1	1	1	0
1	0	1	1	1	1	0
1	1	0	0	1	1	1
1	1	1	0	1	1	1

Application: Construct a switching circuit to meet the requirements of the Boolean expression:

$$Z = A.\overline{C} + \overline{A}.B + \overline{A}.B.\overline{C}$$

The three terms joined by **or**-functions, (+), indicate three parallel branches, having:

branch 1 A **and** $\overline{C}$ in series
branch 2 $\overline{A}$ **and** B in series
and branch 3 $\overline{A}$ **and** B **and** $\overline{C}$ in series

Hence the required switching circuit is as shown in Figure 10.8.

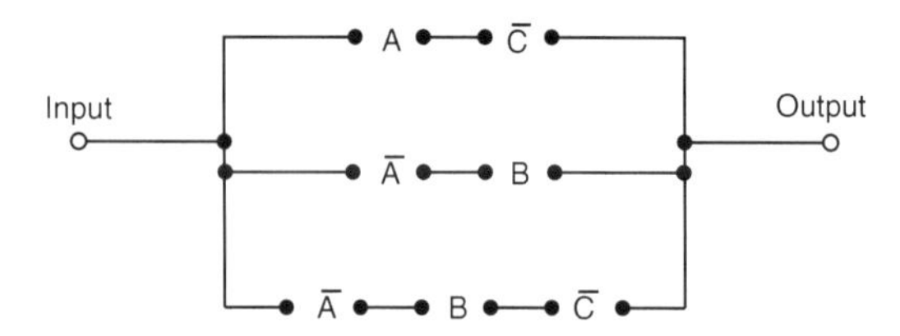

Figure 10.8

The corresponding truth table is shown in Table 10.8.

Column 4 is $\bar{C}$, i.e. the opposite to column 3. Column 5 is $A.\bar{C}$, obtained by applying the **and**-function to columns 1 and 4. Column 6 is $\bar{A}$, the opposite to column 1. Column 7 is $\bar{A}.B$, obtained by applying the **and**-function to columns 2 and 6. Column 8 is $\bar{A}.B.\bar{C}$, obtained by applying the **and**-function to columns 4 and 7. Column 9 is the output, obtained by applying the **or**-function to columns 5, 7 and 8.

Table 10.8

1 A	2 B	3 C	4 $\bar{C}$	5 $A \cdot \bar{C}$	6 $\bar{A}$	7 $\bar{A} \cdot B$	8 $\bar{A} \cdot B \cdot \bar{C}$	9 $Z = A \cdot \bar{C} + \bar{A} \cdot B + \bar{A} \cdot B \cdot \bar{C}$
0	0	0	1	0	1	0	0	0
0	0	1	0	0	1	0	0	0
0	1	0	1	0	1	1	1	1
0	1	1	0	0	1	1	0	1
1	0	0	1	1	0	0	0	1
1	0	1	0	0	0	0	0	0
1	1	0	1	1	0	0	0	1
1	1	1	0	0	0	0	0	0

10.2 Simplifying Boolean expressions

A Boolean expression may be used to describe a complex switching circuit or logic system. If the Boolean expression can be simplified, then the number of switches or logic elements can be reduced resulting in a saving in cost. Three principal ways of simplifying Boolean expressions are:

(a) by using the **laws and rules of Boolean algebra** (see section 10.3),

(b) by applying **de Morgan's laws** (see section 10.4), and

(c) by using **Karnaugh maps** (see section 10.5).

10.3 Laws and rules of Boolean algebra

A summary of the principal laws and rules of Boolean algebra are given in Table 10.9.

Table 10.9

Ref.	Name	Rule or law
1	Commutative laws	$A + B = B + A$
2		$A \cdot B = B \cdot A$
3	Associative laws	$(A + B) + C = A + (B + C)$
4		$(A \cdot B) \cdot C = A \cdot (B \cdot C)$
5	Distributive laws	$A \cdot (B + C) = A \cdot B + A \cdot C$
6		$A + (B \cdot C) = (A + B) \cdot (A + C)$
7	Sum rules	$A + 0 = A$
8		$A + 1 = 1$
9		$A + A = A$
10		$A + \overline{A} = 1$
11	Product rules	$A \cdot 0 = 0$
12		$A \cdot 1 = A$
13		$A \cdot A = A$
14		$A \cdot \overline{A} = 0$
15	Absorption rules	$A + A \cdot B = A$
16		$A \cdot (A + B) = A$
17		$A + \overline{A} \cdot B = A + B$

Application: Simplify the Boolean expression: $\overline{P}.\overline{Q} + \overline{P}.Q + P.\overline{Q}$

With reference to Table 10.9: **Reference**

$$\overline{P}.\overline{Q} + \overline{P}.Q + P.\overline{Q} = \overline{P}.(\overline{Q} + Q) + P.\overline{Q} \qquad 5$$

$$= \overline{P}.1 + P.\overline{Q} \qquad 10$$

$$= \mathbf{\overline{P} + P.\overline{Q}} \qquad 12$$

Application: Simplify $(P + \overline{P}.Q).(Q + \overline{Q}.P)$

With reference to Table 10.9: **Reference**

	Reference
$(P + \overline{P}.Q).(Q + \overline{Q}.P) = P.(Q + \overline{Q}.P) + \overline{P}.Q.(Q + \overline{Q}.P)$	5
$= P.Q + P.\overline{Q}.P + \overline{P}.Q.Q + \overline{P}.Q.\overline{Q}.P$	5
$= P.Q + P.\overline{Q} + \overline{P}.Q + \overline{P}.Q.\overline{Q}.P$	13
$= P.Q + P.\overline{Q} + \overline{P}.Q + 0$	14
$= P.Q + P.\overline{Q} + \overline{P}.Q$	7
$= P.(Q + \overline{Q}) + \overline{P}.Q$	5
$= P.1 + \overline{P}.Q$	10
$= \mathbf{P + \overline{P}.Q}$	12

Application: Simplify $F.G.\overline{H} + F.G.H + \overline{F}.G.H$

With reference to Table 10.9: **Reference**

	Reference
$F.G.\overline{H} + F.G.H + \overline{F}.G.H = F.G.(\overline{H} + H) + \overline{F}.G.H$	5
$= F.G.1 + \overline{F}.G.H$	10
$= F.G + \overline{F}.G.H$	12
$= \mathbf{G.(F + \overline{F}.H)}$	5

Application: Simplify $A.\overline{C} + \overline{A}.(B + C) + A.B.(C + \overline{B})$

With reference to Table 10.9: **Reference**

	Reference
$A.\overline{C} + \overline{A}.(B + C) + A.B.(C + \overline{B})$	
$= A.\overline{C} + \overline{A}.B + \overline{A}.C + A.B.C + A.B.\overline{B}$	5
$= A.\overline{C} + \overline{A}.B + \overline{A}.C + A.B.C + A.0$	14
$= A.\overline{C} + \overline{A}.B + \overline{A}.C + A.B.C$	11
$= A.(\overline{C} + B.C) + \overline{A}.B + \overline{A}.C$	5
$= A.(\overline{C} + B) + \overline{A}.B + \overline{A}.C$	17
$= A.\overline{C} + A.B + \overline{A}.B + \overline{A}.C$	5
$= A.\overline{C} + B.(A + \overline{A}) + \overline{A}.C$	5
$= A.\overline{C} + B.1 + \overline{A}.C$	10
$= \mathbf{A.\overline{C} + B + \overline{A}.C}$	12

10.4 De Morgan's laws

De Morgan's laws state that:

$$\overline{A+B} = \overline{A}\,.\,\overline{B} \qquad \text{and} \qquad \overline{A\,.\,B} = \overline{A} + \overline{B}$$

Application: Simplify the Boolean expression $\overline{\overline{A}\,.\,B} + \overline{\overline{A}+B}$ by using de Morgan's laws and the rules of Boolean algebra

Applying de Morgan's law to the first term gives:

$$\overline{\overline{A}\,.\,B} = \overline{\overline{A}} + \overline{B} = A + \overline{B} \qquad \text{since} \qquad \overline{\overline{A}} = A$$

Applying de Morgan's law to the second term gives:

$$\overline{\overline{A}+B} = \overline{\overline{A}}\,.\,\overline{B} = A\,.\,\overline{B}$$

Thus,

$$\overline{\overline{A}\,.\,B} + \overline{\overline{A}+B} = (A+\overline{B}) + A\,.\,\overline{B}$$

Removing the bracket and reordering gives: $A + A.\overline{B} + \overline{B}$

But, by rule 15, Table 10.9, $A + A\,.\,B = A$. It follows that:

$$A + A\,.\,\overline{B} = A$$

Thus:

$$\mathbf{\overline{\overline{A}\,.\,B} + \overline{\overline{A}+B} = A + \overline{B}}$$

Application: Simplify the Boolean expression $\overline{(A.\overline{B}+C)}.(\overline{A}+\overline{B.\overline{C}})$ by using de Morgan's laws and the rules of Boolean algebra

Applying de Morgan's laws to the first term gives:

$$\overline{(A\,.\,\overline{B}+C)} = \overline{A\,.\,\overline{B}}\,.\,\overline{C} = (\overline{A}+\overline{\overline{B}})\,.\,\overline{C} = (\overline{A}+B)\,.\,\overline{C} = \overline{A}\,.\,\overline{C} + B\,.\,\overline{C}$$

Applying de Morgan's law to the second term gives:

$$(\overline{A} + \overline{B\,.\,\overline{C}}) = \overline{A} + (\overline{B} + \overline{\overline{C}}) = \overline{A} + (\overline{B} + C)$$

Thus

$$\overline{(A\,.\,\overline{B}+C)}\ (\overline{A}+\overline{B\,.\,\overline{C}}) = (\overline{A}\,.\,\overline{C} + B\,.\,\overline{C})\,.\,(\overline{A}+\overline{B}+C)$$
$$= \overline{A}\,.\,\overline{A}\,.\,\overline{C} + \overline{A}\,.\,\overline{B}\,.\,\overline{C} + \overline{A}\,.\,\overline{C}\,.\,C$$
$$+ \overline{A}\,.\,B\,.\,\overline{C} + B\,.\,\overline{B}\,.\,\overline{C} + B\,.\,\overline{C}\,.\,C$$

But from Table 10.9, $\overline{A}.\overline{A} = \overline{A}$ and $\overline{C}.C = B.\overline{B} = 0$

Hence the Boolean expression becomes

$$\overline{A}.\overline{C} + \overline{A}.\overline{B}.\overline{C} + \overline{A}.B.\overline{C} = \overline{A}.\overline{C}(1 + \overline{B} + B)$$

$$= \overline{A}.\overline{C}\,(1 + B) = \overline{A}.\overline{C}(1) = \overline{A}.\overline{C}$$

Thus: $\mathbf{\overline{(A.\overline{B} + C)}.(\overline{A} + \overline{B.\overline{C}}) = \overline{A}.\overline{C}}$

10.5 Karnaugh maps

Summary of procedure when simplifying a Boolean expression using a Karnaugh map

1. Draw a four, eight or sixteen-cell matrix, depending on whether there are two, three or four variables.
2. Mark in the Boolean expression by putting l's in the appropriate cells.
3. Form couples of 8, 4 or 2 cells having common edges, forming the largest groups of cells possible. (Note that a cell containing a 1 may be used more than once when forming a couple. Also note that each cell containing a 1 must be used at least once)
4. The Boolean expression for a couple is given by the variables which are common to all cells in the couple.

(i) Two-variable Karnaugh maps

A truth table for a two-variable expression is shown in Table 10.10(a), the '1' in the third row output showing that $Z = A.\overline{B}$. Each of the four possible Boolean expressions associated with a two-variable function can be depicted as shown in Table 10.10(b) in which one cell is allocated to each row of the truth table. A matrix similar to that shown in Table 10.10(b) can be used to depict $Z = A.\overline{B}$, by putting a 1 in the cell corresponding to $A.\overline{B}$ and 0's in the remaining cells. This method of depicting a Boolean expression is called a two-variable **Karnaugh map**, and is shown in Table 10.10(c).

To simplify a two-variable Boolean expression, the Boolean expression is depicted on a Karnaugh map, as outlined above.

Table 10.10

Inputs		Output	Boolean
A	B	Z	expression
0	0	0	$\overline{A}\cdot\overline{B}$
0	1	0	$\overline{A}\cdot B$
1	0	1	$A\cdot\overline{B}$
1	1	0	$A\cdot B$

(a)

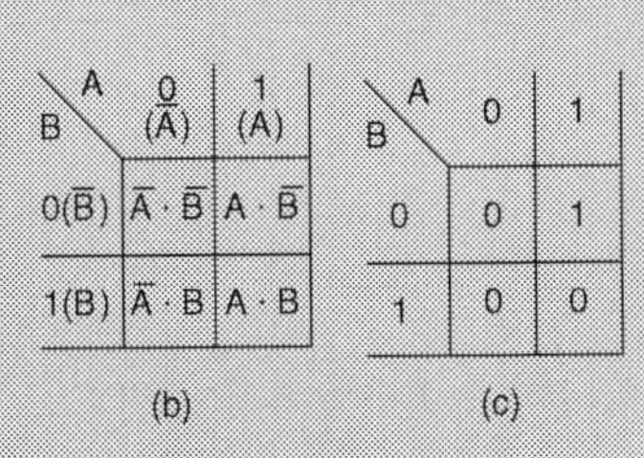

Any cells on the map having either a common vertical side or a common horizontal side are grouped together to form a **couple**. (This is a coupling together of cells, not just combining two together). The simplified Boolean expression for a couple is given by those variables common to all cells in the couple.

(ii) Three-variable Karnaugh maps

A truth table for a three-variable expression is shown in Table 10.11(a), the 1's in the output column showing that: $Z = \overline{A}.\overline{B}.C + \overline{A}.B.C + A.B.\overline{C}$. Each of the eight possible Boolean expressions associated with a three-variable function can be depicted as shown in Table 10.11(b) in which one cell is allocated to each row of the truth table. A matrix similar to that shown in Table 10.11(b) can be used to depict: $Z = \overline{A}.\overline{B}.C + \overline{A}.B.C + A.B.\overline{C}$, by putting 1's in the cells corresponding to the Boolean terms on the right of the Boolean equation and 0's in the remaining cells. This method of depicting a three-variable Boolean expression is called a three-variable Karnaugh map, and is shown in Table 10.11(c).

To simplify a three-variable Boolean expression, the Boolean expression is depicted on a Karnaugh map as outlined above. Any cells on the map having common edges either vertically or horizontally are grouped together to form couples of four cells or two cells. During coupling the horizontal lines at the top and bottom of the cells are taken as a common edge, as are the vertical lines on the left and right of the cells. The simplified Boolean expression for a couple is given by those variables common to all cells in the couple.

Table 10.11

Inputs			Output	Boolean
A	B	C	Z	expression
0	0	0	0	$\bar{A}\cdot\bar{B}\cdot\bar{C}$
0	0	1	1	$\bar{A}\cdot\bar{B}\cdot C$
0	1	0	0	$\bar{A}\cdot B\cdot\bar{C}$
0	1	1	1	$\bar{A}\cdot B\cdot C$
1	0	0	0	$A\cdot\bar{B}\cdot\bar{C}$
1	0	1	0	$A\cdot\bar{B}\cdot C$
1	1	0	1	$A\cdot B\cdot\bar{C}$
1	1	1	0	$A\cdot B\cdot C$

(a)

C \ A·B	00 ($\bar{A}\cdot\bar{B}$)	01 ($\bar{A}\cdot B$)	11 ($A\cdot B$)	10 ($A\cdot\bar{B}$)
0($\bar{C}$)	$\bar{A}\cdot\bar{B}\cdot\bar{C}$	$\bar{A}\cdot B\cdot\bar{C}$	$A\cdot B\cdot\bar{C}$	$A\cdot\bar{B}\cdot\bar{C}$
1(C)	$\bar{A}\cdot\bar{B}\cdot C$	$\bar{A}\cdot B\cdot C$	$A\cdot B\cdot C$	$A\cdot\bar{B}\cdot C$

(b)

C \ A·B	00	01	11	10
0	0	0	1	0
1	1	1	0	0

(c)

(iii) Four-variable Karnaugh maps

A truth table for a four-variable expression is shown in Table 10.12(a), the 1's in the output column showing that:

$$Z = \bar{A}.\bar{B}.C.\bar{D} + \bar{A}.B.C.\bar{D} + A.\bar{B}.C.\bar{D} + A.B.C.\bar{D}$$

Each of the sixteen possible Boolean expressions associated with a four-variable function can be depicted as shown in Table 10.12(b), in which one cell is allocated to each row of the truth table. A matrix similar to that shown in Table 10.12(b) can be used to depict:

$$Z = \bar{A}.\bar{B}.C.\bar{D} + \bar{A}.B.C.\bar{D} + A.\bar{B}.C.\bar{D} + A.B.C.\bar{D}$$

by putting 1's in the cells corresponding to the Boolean terms on the right of the Boolean equation and 0's in the remaining cells. This method of depicting a four-variable expression is called a four-variable Karnaugh map, and is shown in Table 10.12(c).

Table 10.12

Inputs				Output	Boolean
A	B	C	D	Z	expression
0	0	0	0	0	$\overline{A}\cdot\overline{B}\cdot\overline{C}\cdot\overline{D}$
0	0	0	1	0	$\overline{A}\cdot\overline{B}\cdot\overline{C}\cdot D$
0	0	1	0	1	$\overline{A}\cdot\overline{B}\cdot C\cdot\overline{D}$
0	0	1	1	0	$\overline{A}\cdot\overline{B}\cdot C\cdot D$
0	1	0	0	0	$\overline{A}\cdot B\cdot\overline{C}\cdot\overline{D}$
0	1	0	1	0	$\overline{A}\cdot B\cdot\overline{C}\cdot D$
0	1	1	0	1	$\overline{A}\cdot B\cdot C\cdot\overline{D}$
0	1	1	1	0	$\overline{A}\cdot B\cdot C\cdot D$
1	0	0	0	0	$A\cdot\overline{B}\cdot\overline{C}\cdot\overline{D}$
1	0	0	1	0	$A\cdot\overline{B}\cdot\overline{C}\cdot D$
1	0	1	0	1	$A\cdot\overline{B}\cdot C\cdot\overline{D}$
1	0	1	1	0	$A\cdot\overline{B}\cdot C\cdot D$
1	1	0	0	0	$A\cdot B\cdot\overline{C}\cdot\overline{D}$
1	1	0	1	0	$A\cdot B\cdot\overline{C}\cdot D$
1	1	1	0	1	$A\cdot B\cdot C\cdot\overline{D}$
1	1	1	1	0	$A\cdot B\cdot C\cdot D$

(a)

C·D \ A·B	00 ($\overline{A}\cdot\overline{B}$)	01 ($\overline{A}\cdot B$)	11 ($A\cdot B$)	10 ($A\cdot\overline{B}$)
00 ($\overline{C}\cdot\overline{D}$)	$\overline{A}\cdot\overline{B}\cdot\overline{C}\cdot\overline{D}$	$\overline{A}\cdot B\cdot\overline{C}\cdot\overline{D}$	$A\cdot B\cdot\overline{C}\cdot\overline{D}$	$A\cdot\overline{B}\cdot\overline{C}\cdot\overline{D}$
01 ($\overline{C}\cdot D$)	$\overline{A}\cdot\overline{B}\cdot\overline{C}\cdot D$	$\overline{A}\cdot B\cdot\overline{C}\cdot D$	$A\cdot B\cdot\overline{C}\cdot D$	$A\cdot\overline{B}\cdot\overline{C}\cdot D$
11 ($C\cdot D$)	$\overline{A}\cdot\overline{B}\cdot C\cdot D$	$\overline{A}\cdot B\cdot C\cdot D$	$A\cdot B\cdot C\cdot D$	$A\cdot\overline{B}\cdot C\cdot D$
10 ($C\cdot\overline{D}$)	$\overline{A}\cdot\overline{B}\cdot C\cdot\overline{D}$	$\overline{A}\cdot B\cdot C\cdot\overline{D}$	$A\cdot B\cdot C\cdot\overline{D}$	$A\cdot\overline{B}\cdot C\cdot\overline{D}$

(b)

C·D \ A·B	0.0	0.1	1.1	1.0
0.0	0	0	0	0
0.1	0	0	0	0
1.1	0	0	0	0
1.0	1	1	1	1

(c)

To simplify a four-variable Boolean expression, the Boolean expression is depicted on a Karnaugh map as outlined above. Any cells on the map having common edges either vertically or horizontally are grouped together to form couples of eight cells, four cells or two cells. During coupling, the horizontal lines at the top and bottom of the cells may be considered to be common edges, as are the vertical lines on the left and the right of the cells. The simplified Boolean expression for a couple is given by those variables common to all cells in the couple.

Application: Simplify the expression: $\overline{P}.\overline{Q} + \overline{P}.Q$ using Karnaugh map techniques

Using the above procedure:

1. The two-variable matrix is drawn and is shown in Table 10.13.
2. The term $\overline{P}.\overline{Q}$ is marked with a 1 in the top left-hand cell, corresponding to P = 0 and Q = 0; $\overline{P}.Q$ is marked with a 1 in the bottom left-hand cell corresponding to P = 0 and Q = 1.
3. The two cells containing 1's have a common horizontal edge and thus a vertical couple, shown by the broken line, can be formed.
4. The variable common to both cells in the couple is P = 0, i.e. $\overline{P}$ thus

$$\overline{P}.\overline{Q} + \overline{P}.Q = \overline{P}$$

Table 10.13

Q \ P	0	1
0	1	0
1	1	0

Application: Simplify $\overline{X}.Y.\overline{Z} + \overline{X}.\overline{Y}.Z + X.Y.\overline{Z} + X.\overline{Y}.Z$ using Karnaugh map techniques

Table 10.14

Z \ X.Y	0.0	0.1	1.1	1.0
0	0	1	1	0
1	1	0	0	1

Using the above procedure:

1. A three-variable matrix is drawn and is shown in Table 10.14.
2. The 1's on the matrix correspond to the expression given, i.e. for $\overline{X}.Y.\overline{Z}$, X = 0, Y = 1 and Z = 0 and hence corresponds to the cell in the top row and second column, and so on.

3. Two couples can be formed, shown by the broken lines. The couple in the bottom row may be formed since the vertical lines on the left and right of the cells are taken as a common edge.
4. The variables common to the couple in the top row are Y = 1 and Z = 0, that is, $\mathbf{Y.\overline{Z}}$ and the variables common to the couple in the bottom row are Y = 0, Z = 1, that is, $\mathbf{\overline{Y}.Z}$. Hence:

$$\overline{X}.Y.\overline{Z} + \overline{X}.\overline{Y}.Z + X.Y.\overline{Z} + X.\overline{Y}.Z = Y.\overline{Z} + \overline{Y}.Z$$

Application: Simplify $\overline{(P + \overline{Q}.R)} + \overline{(P.Q + \overline{R})}$ using a Karnaugh map technique

The term $(P + \overline{Q}.R)$ corresponds to the cells marked 1 on the matrix in Table 10.15(a), hence $\overline{(P + \overline{Q}.R)}$ corresponds to the cells marked 2. Similarly, $(P.Q + \overline{R})$ corresponds to the cells marked 3 in Table 10.15(a), hence $\overline{(P.Q + \overline{R})}$ corresponds to the cells marked 4. The expression $\overline{(P + \overline{Q}.R)} + \overline{(P.Q + \overline{R})}$ corresponds to cells marked with either a 2 or with a 4 and is shown in Table 10.15(b) by X's. These cells may be coupled as shown by the broken lines. The variables common to the group of four cells is P = 0, i.e. $\mathbf{\overline{P}}$, and those common to the group of two cells are Q = 0, R = 1, i.e. $\mathbf{\overline{Q}.R}$

Thus: $$\overline{(P + \overline{Q}.R)} + \overline{(P.Q + \overline{R})} = \overline{P} + \overline{Q}.R$$

Table 10.15

R \ P.Q	0.0	0.1	1.1	1.0
0	3 2	3 2	3 1	3 1
1	4 1	4 2	3 1	4 1

(a)

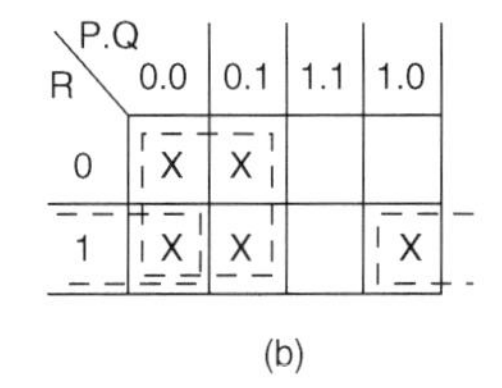

R \ P.Q	0.0	0.1	1.1	1.0
0	X	X		
1	X	X		X

(b)

Application: Simplify the expression: $A.B.\overline{C}.\overline{D} + A.B.C.D + \overline{A}.B.C.D + A.B.C.\overline{D} + \overline{A}.B.C.\overline{D}$ using Karnaugh map techniques

Using the procedure, a four-variable matrix is drawn and is shown in Table 10.16. The 1's marked on the matrix correspond to the expression given. Two couples can be formed and are shown by the broken lines. The four-cell couple has B = 1, C = 1, i.e. **B.C** as the common variables to all four cells and the two-cell couple has $\mathbf{A.B.\overline{D}}$ as the common variables to both cells. Hence, the expression simplifies to: $\mathbf{B.C + A.B.\overline{D}}$ i.e. $\mathbf{B.(C + A.\overline{D})}$

Table 10.16

C.D \ A.B	0.0	0.1	1.1	1.0
0.0			1	
0.1				
1.1		1	1	
1.0		1	1	

10.6 Logic circuits and gates

In practice, logic gates are used to perform the **and**, **or** and **not**-functions introduced earlier. Logic gates can be made from switches, magnetic devices or fluidic devices, but most logic gates in use are electronic devices. Various logic gates are available. For example, the Boolean expression (A.B.C) can be produced using a three-input, **and**-gate and (C + D) by using a two-input **or**-gate. The principal gates in common use are shown in the table below.

Combinational logic networks

In most logic circuits, more than one gate is needed to give the required output. Except for the **invert**-gate, logic gates generally have two, three or four inputs and are confined to one function only. Thus, for example, a two-input, **or**-gate or a four-input **and**-gate can be used when designing a logic circuit.

<table>
<tr><th>Gate type</th><th>Traditional symbol</th><th>IEC Symbol</th><th>Boolean expression</th><th>Truth Table</th></tr>
<tr><td>and-gate</td><td>A B C Z</td><td>A B C & Z</td><td>$Z = A.B.C$</td><td>
<table>
<tr><th colspan="3">INPUTS</th><th>OUTPUT</th></tr>
<tr><th>A</th><th>B</th><th>C</th><th>$Z = A \cdot B \cdot C$</th></tr>
<tr><td>0</td><td>0</td><td>0</td><td>0</td></tr>
<tr><td>0</td><td>0</td><td>1</td><td>0</td></tr>
<tr><td>0</td><td>1</td><td>0</td><td>0</td></tr>
<tr><td>0</td><td>1</td><td>1</td><td>0</td></tr>
<tr><td>1</td><td>0</td><td>0</td><td>0</td></tr>
<tr><td>1</td><td>0</td><td>1</td><td>0</td></tr>
<tr><td>1</td><td>1</td><td>0</td><td>0</td></tr>
<tr><td>1</td><td>1</td><td>1</td><td>1</td></tr>
</table>
</td></tr>
<tr><td>or-gate</td><td>A B C Z</td><td>A B C ≥1 Z</td><td>$Z = A + B + C$</td><td>
<table>
<tr><th colspan="3">INPUTS</th><th>OUTPUT</th></tr>
<tr><th>A</th><th>B</th><th>C</th><th>$Z = A + B + C$</th></tr>
<tr><td>0</td><td>0</td><td>0</td><td>0</td></tr>
<tr><td>0</td><td>0</td><td>1</td><td>1</td></tr>
<tr><td>0</td><td>1</td><td>0</td><td>1</td></tr>
<tr><td>0</td><td>1</td><td>1</td><td>1</td></tr>
<tr><td>1</td><td>0</td><td>0</td><td>1</td></tr>
<tr><td>1</td><td>0</td><td>1</td><td>1</td></tr>
<tr><td>1</td><td>1</td><td>0</td><td>1</td></tr>
<tr><td>1</td><td>1</td><td>1</td><td>1</td></tr>
</table>
</td></tr>
<tr><td>not-gate or invert-gate</td><td>A Z</td><td>A 1 Z</td><td>$Z = \overline{A}$</td><td>
<table>
<tr><th>INPUTS
A</th><th>OUTPUT
$Z = \overline{A}$</th></tr>
<tr><td>0</td><td>1</td></tr>
<tr><td>1</td><td>0</td></tr>
</table>
</td></tr>
<tr><td>nand-gate</td><td>A B C Z</td><td>A B C & Z</td><td>$Z = \overline{A.B.C}$</td><td>
<table>
<tr><th colspan="3">INPUTS</th><th></th><th>OUTPUT</th></tr>
<tr><th>A</th><th>B</th><th>C</th><th>$A \cdot B \cdot C$</th><th>$Z = \overline{A \cdot B \cdot C}$</th></tr>
<tr><td>0</td><td>0</td><td>0</td><td>0</td><td>1</td></tr>
<tr><td>0</td><td>0</td><td>1</td><td>0</td><td>1</td></tr>
<tr><td>0</td><td>1</td><td>0</td><td>0</td><td>1</td></tr>
<tr><td>0</td><td>1</td><td>1</td><td>0</td><td>1</td></tr>
<tr><td>1</td><td>0</td><td>0</td><td>0</td><td>1</td></tr>
<tr><td>1</td><td>0</td><td>1</td><td>0</td><td>1</td></tr>
<tr><td>1</td><td>1</td><td>0</td><td>0</td><td>1</td></tr>
<tr><td>1</td><td>1</td><td>1</td><td>1</td><td>0</td></tr>
</table>
</td></tr>
<tr><td>nor-gate</td><td>A B C Z</td><td>A B C ≥1 Z</td><td>$Z = \overline{A + B + C}$</td><td>
<table>
<tr><th colspan="3">INPUTS</th><th></th><th>OUTPUT</th></tr>
<tr><th>A</th><th>B</th><th>C</th><th>$A + B + C$</th><th>$Z = \overline{A + B + C}$</th></tr>
<tr><td>0</td><td>0</td><td>0</td><td>0</td><td>1</td></tr>
<tr><td>0</td><td>0</td><td>1</td><td>1</td><td>0</td></tr>
<tr><td>0</td><td>1</td><td>0</td><td>1</td><td>0</td></tr>
<tr><td>0</td><td>1</td><td>1</td><td>1</td><td>0</td></tr>
<tr><td>1</td><td>0</td><td>0</td><td>1</td><td>0</td></tr>
<tr><td>1</td><td>0</td><td>1</td><td>1</td><td>0</td></tr>
<tr><td>1</td><td>1</td><td>0</td><td>1</td><td>0</td></tr>
<tr><td>1</td><td>1</td><td>1</td><td>1</td><td>0</td></tr>
</table>
</td></tr>
</table>

Gate type	Traditional symbol	IEC Symbol	Boolean expression	Truth Table
xor-gate		=1	$Z = A \oplus B$	Inputs: A B / Output: Z = A XOR B 0 0 → 0 0 1 → 1 1 0 → 1 1 1 → 0
xnor-gate		=1	$Z = \overline{A \oplus B}$	Inputs: A B / Output: Z = A XNOR B 0 0 → 1 0 1 → 0 1 0 → 0 1 1 → 1

Application: Devise a logic system to meet the requirements of: $Z = A.\overline{B} + C$

With reference to Figure 10.23 an **invert**-gate, shown as (1), gives $\overline{B}$. The **and**-gate, shown as (2), has inputs of A and $\overline{B}$, giving $A.\overline{B}$. The **or**-gate, shown as (3), has inputs of $A.\overline{B}$ and C, giving:

$$\mathbf{Z = A.\overline{B} + C}$$

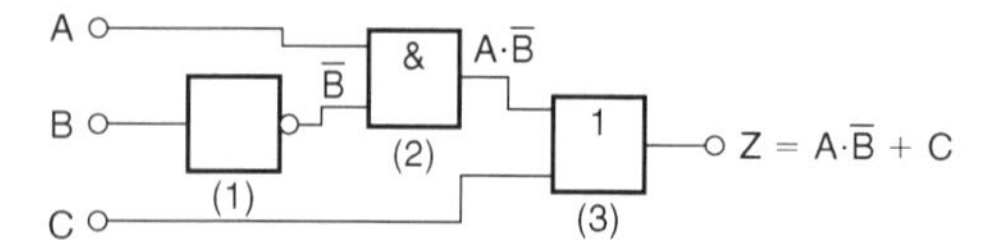

Figure 10.23

Application: Devise a logic system to meet the requirements of $(P + \overline{Q}).(\overline{R} + S)$

The logic system is shown in Figure 10.24. The given expression shows that two **invert**-functions are needed to give $\overline{Q}$ and $\overline{R}$ and these are shown as gates (1) and (2). Two **or**-gates, shown as (3) and (4), give $(P + \overline{Q})$ and $(\overline{R} + S)$ respectively. Finally, an **and**-gate, shown as (5), gives the required output, $\mathbf{Z = (P + \overline{Q}).(\overline{R} + S)}$

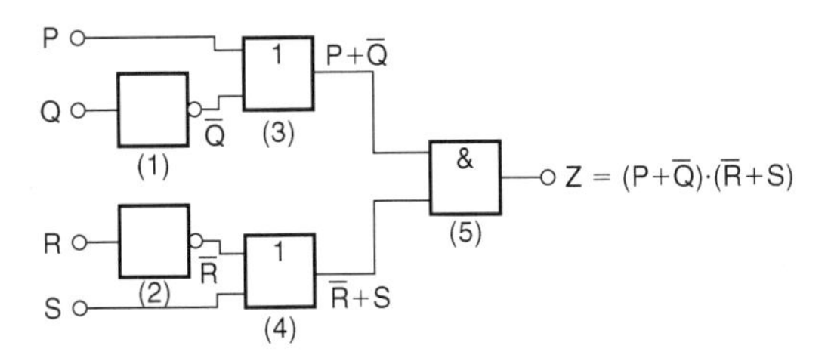

Figure 10.24

Application: Devise a logic circuit to meet the requirements of the output given in Table 10.24, using as few gates as possible

Table 10.24

Inputs			Output
A	B	C	Z
0	0	0	0
0	0	1	0
0	1	0	0
0	1	1	0
1	0	0	0
1	0	1	1
1	1	0	1
1	1	1	1

The '1' outputs in rows 6, 7 and 8 of Table 10.24 show that the Boolean expression is: $Z = A.\overline{B}.C + A.B.\overline{C} + A.B.C$

The logic circuit for this expression can be built using three, 3-input **and**-gates and one, 3-input **or**-gate, together with two **invert**-gates. However, the number of gates required can be reduced by using the techniques introduced earlier, resulting in the cost of the circuit being reduced. Any of the techniques can be used, and in this case, the rules of Boolean algebra (see Table 10.9) are used.

$$Z = A.\overline{B}.C + A.B.\overline{C} + A.B.C = A.[\overline{B}.C + B.\overline{C} + B.C]$$
$$= A.[\overline{B}.C + B(\overline{C} + C)] = A.[\overline{B}.C + B]$$
$$= A.[B + \overline{B}.C] = \mathbf{A.[B + C]}$$

The logic circuit to give this simplified expression is shown in Figure 10.25.

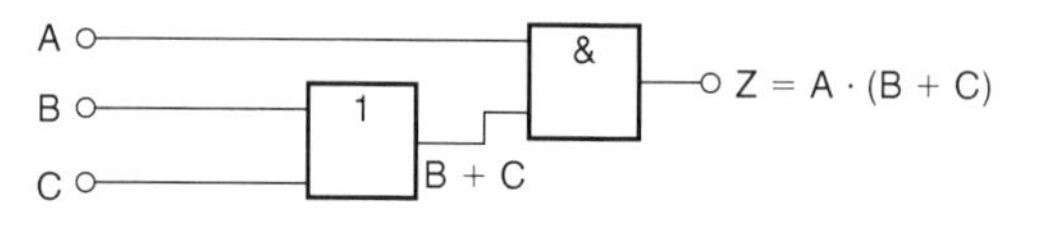

Figure 10.25

Application: Simplify the expression:

$$Z = \overline{P}.\overline{Q}.\overline{R}.\overline{S} + \overline{P}.\overline{Q}.\overline{R}.S + \overline{P}.Q.\overline{R}.\overline{S} + \overline{P}.Q.\overline{R}.S + P.\overline{Q}.\overline{R}.\overline{S}$$

and devise a logic circuit to give this output

The given expression is simplified using the Karnaugh map techniques introduced earlier. Two couples are formed as shown in Figure 10.26(a) and the simplified expression becomes:

$$Z = \overline{Q}.\overline{R}.\overline{S} + \overline{P}.\overline{R} \quad \text{i.e.} \quad \mathbf{Z = \overline{R}.(\overline{P} + \overline{Q}.\overline{S})}$$

The logic circuit to produce this expression is shown in Figure 10.26(b).

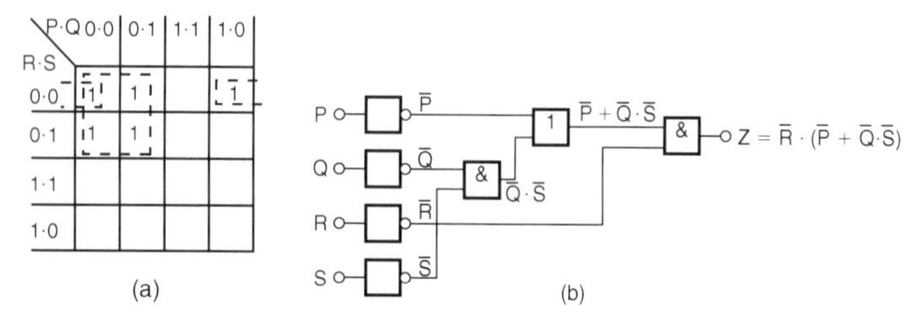

Figure 10.26

10.7 Universal logic gates

The function of any of the five logic gates in common use can be obtained by using either **nand**-gates or **nor**-gates and when used in this manner, the gate selected is called a **universal gate**.

Application: Show how **invert**, **and**, **or** and **nor**-functions can be produced using nand-gates only

A single input to a **nand**-gate gives the **invert**-function, as shown in Figure 10.27(a). When two **nand**-gates are connected, as shown in Figure 10.27(b), the output from the first gate is $\overline{A.B.C}$ and this is inverted by the second gate, giving $Z = \overline{\overline{A.B.C}} = A.B.C$ i.e. the **and**-function is produced. When $\overline{A}$, $\overline{B}$ and $\overline{C}$ are the inputs to a **nand**-gate, the output is $\overline{\overline{A}.\overline{B}.\overline{C}}$

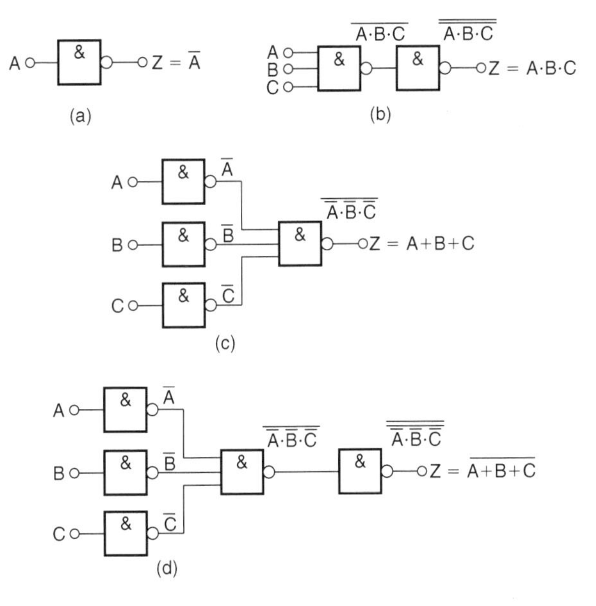

Figure 10.27

By de Morgan's law, $\overline{\overline{A}.\overline{B}.\overline{C}} = \overline{\overline{A}} + \overline{\overline{B}} + \overline{\overline{C}} = A + B + C$, i.e. a **nand**-gate is used to produce the **or**-function. The logic circuit is shown in Figure 10.27(c). If the output from the logic circuit in Figure 10.27(c) is inverted by adding an additional **nand**-gate, the output becomes the invert of an **or**-function, i.e. the **nor**-function, as shown in Figure 10.27(d).

Application: Show how **invert**, **or**, **and** and **nand**-functions can be produced by using **nor**-gates only

A single input to a **nor**-gate gives the **invert**-function, as shown in Figure 10.28(a). When two **nor**-gates are connected, as shown in Figure 10.28(b), the output from the first gate is $\overline{A + B + C}$ and this is inverted by the second gate, giving $Z = \overline{\overline{A + B + C}} = A + B + C$, i.e. the **or**-function is produced. Inputs of $\overline{A}$, $\overline{B}$ and $\overline{C}$ to a **nor**-gate give an output of $\overline{\overline{A} + \overline{B} + \overline{C}}$

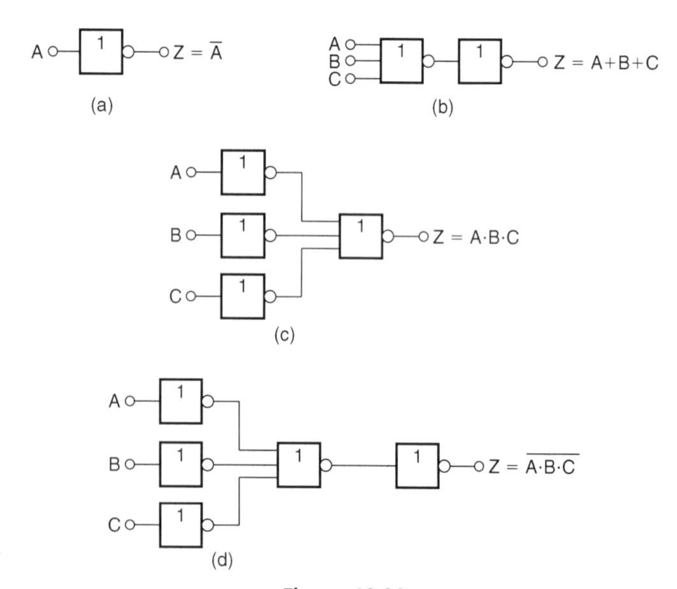

Figure 10.28

By de Morgan's law, $\overline{\overline{A}+\overline{B}+\overline{C}} = \overline{\overline{A}}.\overline{\overline{B}}.\overline{\overline{C}} = A.B.C$, i.e. the **nor**-gate can be used to produce the **and**-function. The logic circuit is shown in Figure 10.28(c). When the output of the logic circuit, shown in Figure 10.28(c), is inverted by adding an additional **nor**-gate, the output then becomes the invert of an **or**-function, i.e. the **nor**-function as shown in Figure 10.28(d).

Application: Design a logic circuit, using **nand**-gates having not more than three inputs, to meet the requirements of the Boolean expression: $Z = \overline{A} + \overline{B} + C + \overline{D}$

When designing logic circuits, it is often easier to start at the output of the circuit. The given expression shows there are four variables joined by **or**-functions. From the principles introduced above, if a four-input **nand**-gate is used to give the expression given, the inputs are $\overline{\overline{A}}$, $\overline{\overline{B}}$, $\overline{C}$ and $\overline{\overline{D}}$ that is A, B, $\overline{C}$ and D. However, the problem states that three-inputs are not to be exceeded so two of the variables are joined, i.e. the inputs to the three-input **nand**-gate, shown as gate (1) in Figure 10.29, is A.B, $\overline{C}$ and D. From above, the **and**-function is generated by using two **nand**-gates connected in series, as shown by gates (2) and (3) in Figure 10.29. The logic circuit required to produce the given expression is as shown in Figure 10.29.

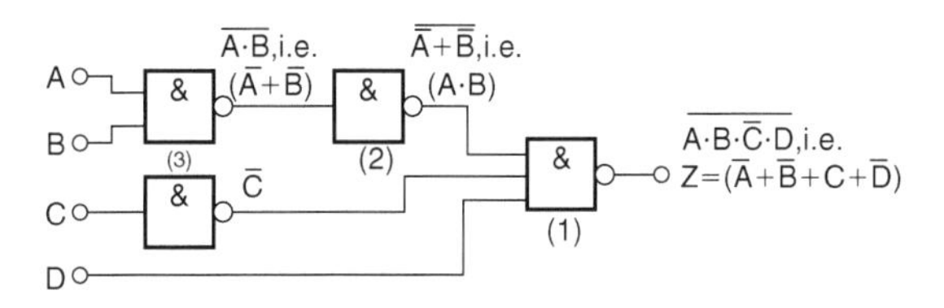

Figure 10.29

Application: Using **nor**-gates only, design a logic circuit to meet the requirements of the expression: $Z = \overline{D}(\overline{A} + B + \overline{C})$

It is usual in logic circuit design to start the design at the output. From earlier, the **and**-function between $\overline{D}$ and the terms in the bracket can be produced by using inputs of $\overline{\overline{D}}$ and $\overline{\overline{A} + B + \overline{C}}$ to a **nor**-gate, i.e. by de Morgan's law, inputs of D and $A.\overline{B}.C$. Inputs of $\overline{A}.B$ and $\overline{C}$ to a **nor**-gate give an output of $\overline{\overline{A} + B + \overline{C}}$, which by de Morgan's law is $A.\overline{B}.C$. The logic circuit to produce the required expression is as shown in Figure 10.30.

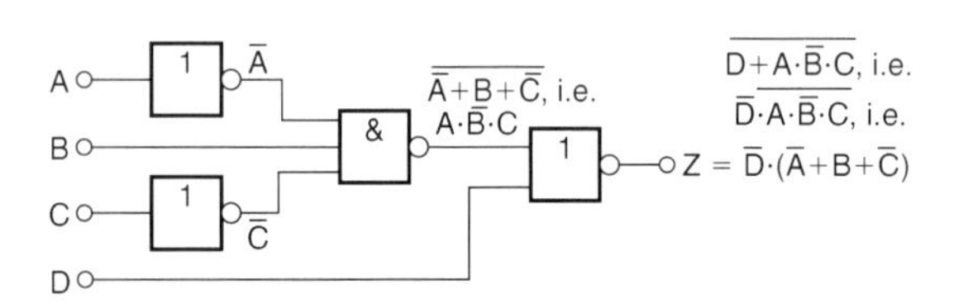

Figure 10.30

Application: An alarm indicator in a grinding mill complex should be activated if (a) the power supply to all mills is off and (b) the hopper feeding the mills is less than 10% full, and (c) if less than two of the three grinding mills are in action. Devise a logic system to meet these requirements

Let variable A represent the power supply on to all the mills, then $\overline{A}$ represents the power supply off. Let B represent the hopper feeding the mills being more than 10% full, then $\overline{B}$ represents the hopper being less than 10% full. Let C, D and E represent the three mills respectively being in action, then $\overline{C}$, $\overline{D}$ and $\overline{E}$ represent the three mills respectively not being in action. The required expression to activate the alarm is: $Z = \overline{A}.\overline{B}.(\overline{C} + \overline{D} + \overline{E})$.

There are three variables joined by **and**-functions in the output, indicating that a three-input **and**-gate is required, having inputs of $\overline{A}$, $\overline{B}$ and $(\overline{C} + \overline{D} + \overline{E})$. The term $(\overline{C} + \overline{D} + \overline{E})$ is produced by a three-input

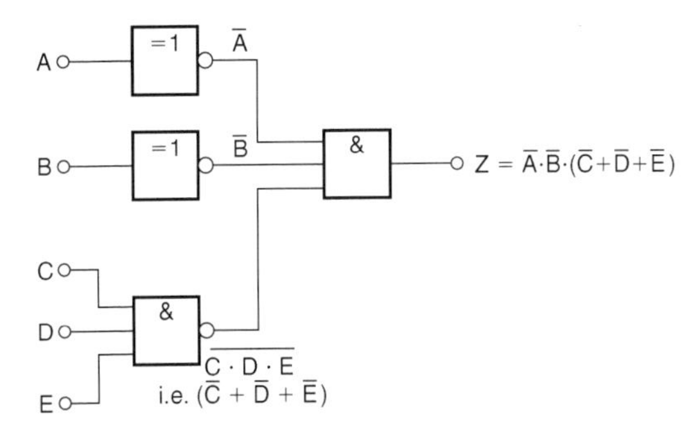

Figure 10.31

nand-gate. When variables C, D and E are the inputs to a **nand**-gate, the output is $\overline{C . D . E}$, which, by De Morgan's law is $(\overline{C} + \overline{D} + \overline{E})$. Hence the required logic circuit is as shown in Figure 10.31.

---

11 Differential Calculus and its Applications

11.1 Common standard derivatives

y or f(x)	$\frac{dy}{dx}$ or f′(x)
ax^n	anx^{n-1}
sin ax	a cos ax
cos ax	−a sin ax
tan ax	$a\sec^2 ax$
sec ax	a sec ax tan ax
cosec ax	−a cosec ax cot ax
cot ax	$-a\,\mathrm{cosec}^2\, ax$
e^{ax}	ae^{ax}
ln ax	$\frac{1}{x}$

Application: Differentiate $y = 5x^4 + 4x - \frac{1}{2x^2} + \frac{1}{\sqrt{x}} - 3$ with respect to x

$y = 5x^4 + 4x - \frac{1}{2x^2} + \frac{1}{\sqrt{x}} - 3$ is rewritten as:

$$y = 5x^4 + 4x - \frac{1}{2}x^{-2} + x^{-\frac{1}{2}} - 3$$

Thus $\frac{dy}{dx} = (5)(4)x^{4-1} + (4)(1)x^{1-1} - \frac{1}{2}(-2)x^{-2-1} + (1)\left(-\frac{1}{2}\right)x^{-\frac{1}{2}-1} - 0$

$$= 20x^3 + 4 + x^{-3} - \frac{1}{2}x^{-\frac{3}{2}}$$

i.e. $\mathbf{\frac{dy}{dx} = 20x^3 + 4 + \frac{1}{x^3} - \frac{1}{2\sqrt{x^3}}}$

Application: Find the differential coefficient of $y = 3\sin 4x - 2\cos 3x$

When $y = 3\sin 4x - 2\cos 3x$

then $\dfrac{dy}{dx} = (3)(4\cos 4x) - (2)(-3\sin 3x) = \mathbf{12\cos 4x + 6\sin 3x}$

Application: Determine the derivative of $f(\theta) = \dfrac{2}{e^{3\theta}} + 6\ln 2\theta$

$$f(\theta) = \frac{2}{e^{3\theta}} + 6\ln 2\theta = 2e^{-3\theta} + 6\ln 2\theta$$

Hence, $f'(\theta) = (2)(-3)e^{-3\theta} + 6\left(\dfrac{1}{\theta}\right) = -6e^{-3\theta} + \dfrac{6}{\theta} = \mathbf{\dfrac{-6}{e^{3\theta}} + \dfrac{6}{\theta}}$

11.2 Products and quotients

When $y = uv$, and u and v are both functions of x, then:

$$\frac{dy}{dx} = u\frac{dv}{dx} + v\frac{du}{dx}$$

When $y = \dfrac{u}{v}$, and u and v are both functions of x then:

$$\frac{dy}{dx} = \frac{v\dfrac{du}{dx} - u\dfrac{dv}{dx}}{v^2}$$

Application: Find the differential coefficient of $y = 3x^2 \sin 2x$

$3x^2 \sin 2x$ is a product of two terms $3x^2$ and $\sin 2x$

Let $u = 3x^2$ and $v = \sin 2x$

Using the product rule: $\dfrac{dy}{dx} = u \quad \dfrac{dv}{dx} + v \quad \dfrac{du}{dx}$

$\qquad\qquad\qquad\qquad\quad \downarrow \quad \downarrow \qquad \downarrow \quad \downarrow$

gives: $\dfrac{dy}{dx} = (3x^2)(2\cos 2x) + (\sin 2x)(6x)$

i.e. $$\frac{dy}{dx} = 6x^2 \cos 2x + 6x \sin 2x$$
$$= \mathbf{6x\,(x \cos 2x + \sin 2x)}$$

Application: Find the differential coefficient of $y = \dfrac{4 \sin 5x}{5x^4}$

$\dfrac{4 \sin 5x}{5x^4}$ is a quotient. Let $u = 4 \sin 5x$ and $v = 5x^4$

$$\frac{dy}{dx} = \frac{v\dfrac{du}{dx} - u\dfrac{dv}{dx}}{v^2} = \frac{(5x^4)(20 \cos 5x) - (4 \sin 5x)(20x^3)}{(5x^4)^2}$$
$$= \frac{100x^4 \cos 5x - 80x^3 \sin 5x}{25x^8}$$
$$= \frac{20x^3[5x \cos 5x - 4 \sin 5x]}{25x^8}$$

i.e. $$\mathbf{\frac{dy}{dx} = \frac{4}{5x^5}(5x \cos 5x - 4 \sin 5x)}$$

Application: Determine the differential coefficient of $y = \tan ax$

$y = \tan ax = \dfrac{\sin ax}{\cos ax}$. Differentiation of tan ax is thus treated as a quotient with $u = \sin ax$ and $v = \cos ax$

$$\frac{dy}{dx} = \frac{v\dfrac{du}{dx} - u\dfrac{dv}{dx}}{v^2} = \frac{(\cos ax)(a \cos ax) - (\sin ax)(-a \sin ax)}{(\cos ax)^2}$$
$$= \frac{a \cos^2 ax + a \sin^2 ax}{(\cos ax)^2} = \frac{a(\cos^2 ax + \sin^2 ax)}{\cos^2 ax}$$
$$= \frac{a}{\cos^2 ax} \quad \text{since } \cos^2 ax + \sin^2 ax = 1$$

Hence, $\mathbf{\dfrac{dy}{dx} = a \sec^2 ax}$ since $\sec^2 ax = \dfrac{1}{\cos^2 ax}$

11.3 Function of a function

It is often easier to make a substitution before differentiating.

If y is a function of x then: $\dfrac{dy}{dx} = \dfrac{dy}{du} \times \dfrac{du}{dx}$

This is known as the **'function of a function'** rule (or sometimes the **chain rule**).

Application: Differentiate $y = (3x - 1)^9$

If $y = (3x - 1)^9$ then, by making the substitution $u = (3x - 1)$, $y = u^9$, which is of the 'standard' form.

Hence, $\dfrac{dy}{du} = 9u^8$ and $\dfrac{du}{dx} = 3$

Then $\dfrac{dy}{dx} = \dfrac{dy}{du} \times \dfrac{du}{dx} = (9u^8)(3) = 27u^8$

Rewriting u as (3x − 1) gives: $\mathbf{\dfrac{dy}{dx} = 27(3x - 1)^8}$

Application: Determine the differential coefficient of $y = \sqrt{3x^2 + 4x - 1}$

$$y = \sqrt{3x^2 + 4x - 1} = (3x^2 + 4x - 1)^{1/2}$$

Let $u = 3x^2 + 4x - 1$ then $y = u^{1/2}$

Hence $\dfrac{du}{dx} = 6x + 4$ and $\dfrac{dy}{du} = \dfrac{1}{2}u^{-1/2} = \dfrac{1}{2\sqrt{u}}$

Using the function of a function rule,

$$\frac{dy}{dx} = \frac{dy}{du} \times \frac{du}{dx} = \left(\frac{1}{2\sqrt{u}}\right)(6x + 4) = \frac{3x + 2}{\sqrt{u}}$$

i.e. $$\mathbf{\frac{dy}{dx} = \frac{3x + 2}{\sqrt{(3x^2 + 4x - 1)}}}$$

11.4 Successive differentiation

When a function $y = f(x)$ is differentiated with respect to x the differential coefficient is written as $\frac{dy}{dx}$ or $f'(x)$. If the expression is differentiated again, the second differential coefficient is obtained and is written as $\frac{d^2y}{dx^2}$ or $f''(x)$. By successive differentiation further higher derivatives such as $\frac{d^3y}{dx^3}$ and $\frac{d^4y}{dx^4}$ may be obtained.

Thus, if $y = 3x^4$, $\frac{dy}{dx} = 12x^3$, $\frac{d^2y}{dx^2} = 36x^2$, $\frac{d^3y}{dx^3} = 72x$,

$\frac{d^4y}{dx^4} = 72$ and $\frac{d^5y}{dx^5} = 0$

Application: If $f(x) = 2x^5 - 4x^3 + 3x - 5$ determine $f''(x)$

If $f(x) = 2x^5 - 4x^3 + 3x - 5$

then $\quad f'(x) = 10x^4 - 12x^2 + 3$

and $\quad \mathbf{f''(x)} = 40x^3 - 24x = \mathbf{4x(10x^2 - 6)}$

Application: Evaluate $\frac{d^2y}{d\theta^2}$ when $\theta = 0$ given $y = 4\sec 2\theta$

Since $y = 4\sec 2\theta$, then $\frac{dy}{d\theta} = (4)(2)\sec 2\theta \tan 2\theta$

$= 8\sec 2\theta \tan 2\theta$ (i.e. a product)

$$\frac{d^2y}{d\theta^2} = (8\sec 2\theta)(2\sec^2 2\theta) + (\tan 2\theta)[(8)(2)\sec 2\theta \tan 2\theta]$$

$$= 16\sec^3 2\theta + 16\sec 2\theta \tan^2 2\theta$$

When $\theta = 0$, $\frac{d^2y}{d\theta^2} = 16\sec^3 0 + 16\sec 0 \tan^2 0$

$= 16(1) + 16(1)(0) = \mathbf{16}$

11.5 Differentiation of hyperbolic functions

y or f(x)	$\frac{dy}{dx}$ or f′(x)
sinh ax	a cosh ax
cosh ax	a sinh ax
tanh ax	a sech²ax
sech ax	−a sech ax tanh ax
cosech ax	−a cosech ax coth ax
coth ax	−a cosech²ax

Application: Differentiate the following with respect to x:
(a) $y = 4\text{sh } 2x - \frac{3}{7}\text{ch } 3x$ (b) $y = 5\text{th } \frac{x}{2} - 2\coth 4x$

(a) $\frac{dy}{dx} = 4(2\cosh 2x) - \frac{3}{7}(3\sinh 3x) = \mathbf{8\cosh 2x - \frac{9}{7}\sinh 3x}$

(b) $\frac{dy}{dx} = 5\left(\frac{1}{2}\text{sec h}^2\,\frac{x}{2}\right) - 2(-4\,\text{cosech}^2 4x)$

$= \mathbf{\frac{5}{2}\text{sec h}^2\,\frac{x}{2} + 8\,\text{cosech}^2 4x}$

Application: Differentiate the following with respect to the variable:
(a) $y = 4\sin 3t\,\text{ch } 4t$ (b) $y = \ln(\text{sh } 3\theta) - 4\,\text{ch}^2 3\theta$

(a) $y = 4\sin 3t\,\text{ch } 4t$ (i.e. a product)

$\frac{dy}{dx} = (4\sin 3t)(4\,\text{sh } 4t) + (\text{ch } 4t)(4)(3\cos 3t)$

$= 16\sin 3t\,\text{sh } 4t + 12\,\text{ch } 4t\cos 3t$

$= \mathbf{4(4\sin 3t\,\text{sh } 4t + 3\cos 3t\,\text{ch } 4t)}$

(b) $y = \ln(\text{sh } 3\theta) - 4 \text{ ch}^2 3\theta$ (i.e. a function of a function)

$$\frac{dy}{d\theta} = \left(\frac{1}{\text{sh } 3\theta}\right)(3 \text{ ch } 3\theta) - (4)(2 \text{ ch } 3\theta)(3 \text{ sh } 3\theta)$$

$$= 3 \coth 3\theta - 24 \text{ ch } 3\theta \text{ sh } 3\theta = \mathbf{3(\coth 3\theta - 8 \text{ ch } 3\theta \text{ sh } 3\theta)}$$

11.6 Rates of change using differentiation

If a quantity y depends on and varies with a quantity x then the rate of change of y with respect to x is $\frac{dy}{dx}$. Thus, for example, the rate of change of pressure p with height h is $\frac{dp}{dh}$.

A rate of change with respect to time is usually just called 'the rate of change', the 'with respect to time' being assumed. Thus, for example, a rate of change of current, i, is $\frac{di}{dt}$ and a rate of change of temperature, θ, is $\frac{d\theta}{dt}$, and so on.

Application: Newtons law of cooling is given by $\theta = \theta_0 e^{-kt}$, where the excess of temperature at zero time is θ_0°C and at time t seconds is θ°C. Determine the rate of change of temperature after 40 s, given that $\theta_0 = 16$°C and $k = -0.03$

The rate of change of temperature is $\frac{d\theta}{dt}$

Since $\theta = \theta_0 e^{-kt}$ then $\frac{d\theta}{dt} = (\theta_0)(-k)e^{-kt} = -k\theta_0 e^{-kt}$

When $\theta_0 = 16$, $k = -0.03$ and $t = 40$

then $\frac{d\theta}{dt} = -(-0.03)(16)e^{-(-0.03)(40)}$

$= 0.48\, e^{1.2} = \mathbf{1.594°C/s}$

Application: The luminous intensity I candelas of a lamp at varying voltage V is given by: $I = 4 \times 10^{-4} V^2$. Determine the voltage at which the light is increasing at a rate of 0.6 candelas per volt

The rate of change of light with respect to voltage is given by $\frac{dI}{dV}$

Since $I = 4 \times 10^{-4}V^2$, $\frac{dI}{dV} = (4 \times 10^{-4})(2)V = 8 \times 10^{-4} V$

When the light is increasing at 0.6 candelas per volt then

$$+0.6 = 8 \times 10^{-4}V, \text{ from which,}$$

$$\text{voltage } V = \frac{0.6}{8 \times 10^{-4}} = 0.075 \times 10^{+4} = \mathbf{750 \text{ volts}}$$

11.7 Velocity and acceleration

If a body moves a distance x metres in a time t seconds then:

(i) distance, $x = f(t)$

(ii) velocity, $v = f'(t)$ or $\frac{dx}{dt}$, which is the gradient of the distance/time graph

(iii) acceleration, $a = \frac{dv}{dt} = f''(x)$ or $\frac{d^2x}{dt^2}$, which is the gradient of the velocity/time graph.

Application: The distance x metres travelled by a vehicle in time t seconds after the brakes are applied is given by $x = 20t - \frac{5}{3}t^2$. Determine (a) the speed of the vehicle (in km/h) at the instant the brakes are applied, and (b) the distance the car travels before it stops

(a) Distance, $x = 20t - \frac{5}{3}t^2$. Hence velocity, $v = \frac{dx}{dt} = 20 - \frac{10}{3}t$

At the instant the brakes are applied, time = 0

Hence **velocity, v** $= 20\,\text{m/s} = \frac{20 \times 60 \times 60}{1000}\,\text{km/h} = \mathbf{72\,km/h}$

(Note: changing from m/s to km/h merely involves multiplying by 3.6.)

(b) When the car finally stops, the velocity is zero,

i.e. $v = 20 - \dfrac{10}{3}t = 0$, from which, $20 = \dfrac{10}{3}t$, giving $t = 6\,s$.

Hence the distance travelled before the car stops is given by:

$$x = 20t - \frac{5}{3}t^2 = 20(6) - \frac{5}{3}(6)^2 = 120 - 60 = \mathbf{60\,m}$$

Application: The angular displacement θ radians of a flywheel varies with time t seconds and follows the equation $\theta = 9t^2 - 2t^3$. Determine (a) the angular velocity and acceleration of the flywheel when time, $t = 1\,s$, and (b) the time when the angular acceleration is zero.

(a) Angular displacement $\theta = 9t^2 - 2t^3$ rad

Angular velocity, $\omega = \dfrac{d\theta}{dt} = 18t - 6t^2$ rad/s

When time $t = 1\,s$, $\omega = 18(1) - 6(1)^2 = \mathbf{12\ rad/s}$

Angular acceleration, $\alpha = \dfrac{d^2\theta}{dt^2} = 18 - 12t$ rad/s

When time $t = 1\,s$, $\alpha = 18 - 12(1) = \mathbf{6\ rad/s^2}$

(b) When the angular acceleration is zero, $18 - 12t = 0$, from which, $18 = 12t$, giving time, $\mathbf{t = 1.5\ s}$

Application: The displacement x cm of the slide valve of an engine is given by: $x = 2.2\cos 5\pi t + 3.6\sin 5\pi t$. Evaluate the velocity (in m/s) when time $t = 30\,ms$.

Displacement $x = 2.2\cos 5\pi t + 3.6\sin 5\pi t$

$$\text{Velocity } v = \frac{dx}{dt} = (2.2)(-5\pi)\sin 5\pi t + (3.6)(5\pi)\cos 5\pi t$$
$$= -11\pi\sin 5\pi t + 18\pi\cos 5\pi t \text{ cm/s}$$

When time $t = 30\,ms$,

$$\begin{aligned}\text{velocity} &= -11\pi\sin(5\pi \times 30 \times 10^{-3}) + 18\pi\cos(5\pi \times 30 \times 10^{-3})\\ &= -11\pi\sin 0.4712 + 18\pi\cos 0.4712\\ &= -11\pi\sin 27^\circ + 18\pi\cos 27^\circ\\ &= -15.69 + 50.39\\ &= 34.7\text{ cm/s} = \mathbf{0.347\ m/s}\end{aligned}$$

11.8 Turning points

Procedure for finding and distinguishing between stationary points

(i) Given $y = f(x)$, determine $\frac{dy}{dx}$ (i.e. $f'(x)$)

(ii) Let $\frac{dy}{dx} = 0$ and solve for the values of x

(iii) Substitute the values of x into the original equation, $y = f(x)$, to find the corresponding y-ordinate values. This establishes the co-ordinates of the stationary points.

To determine the nature of the stationary points:

Either

(iv) Find $\frac{d^2y}{dx^2}$ and substitute into it the values of x found in (ii).

If the result is: (a) positive – the point is a minimum one,
(b) negative – the point is a maximum one,
(c) zero – the point is a point of inflexion

or

(v) Determine the sign of the gradient of the curve just before and just after the stationary points. If the sign change for the gradient of the curve is:

(a) positive to negative – the point is a maximum one
(b) negative to positive – the point is a minimum one
(c) positive to positive or negative to negative – the point is a point of inflexion

Application: Find the maximum and minimum values of the curve $y = x^3 - 3x + 5$

Since $y = x^3 - 3x + 5$ then $\frac{dy}{dx} = 3x^2 - 3$

For a maximum or minimum value $\frac{dy}{dx} = 0$

Hence, $3x^2 - 3 = 0$, from which, $3x^2 = 3$ and $x = \pm 1$

When $x = 1$, $y = (1)^3 - 3(1) + 5 = 3$

When $x = -1$, $y = (-1)^3 - 3(-1) + 5 = 7$

Hence, **(1, 3) and (−1, 7) are the co-ordinates of the turning points**.

Considering the point (1, 3):

If x is slightly less than 1, say 0.9, then $\frac{dy}{dx} = 3(0.9)^2 - 3$, which is negative.

If x is slightly more than 1, say 1.1, then $\frac{dy}{dx} = 3(1.1)^2 - 3$, which is positive.

Since the gradient changes from negative to positive, **the point (1, 3) is a minimum point**.

Considering the point (−1, 7):

If x is slightly less than −1, say −1.1, then $\frac{dy}{dx} = 3(-1.1)^2 - 3$, which is positive.

If x is slightly more than −1, say −0.9, then $\frac{dy}{dx} = 3(-0.9)^2 - 3$, which is negative.

Since the gradient changes from positive to negative, **the point (−1, 7) is a maximum point**.

Since $\frac{dy}{dx} = 3x^2 - 3$, then $\frac{d^2y}{dx^2} = 6x$

When $x = 1$, $\frac{d^2y}{dx^2}$ is positive, hence (1, 3) is a **minimum value**.

When $x = -1$, $\frac{d^2y}{dx^2}$ is negative, hence (−1, 7) is a **maximum value**.

Thus the maximum value is 7 and the minimum value is 3.

It can be seen that the second differential method of determining the nature of the turning points is, in this case, quicker than investigating the gradient.

Application: Determine the area of the largest piece of rectangular ground that can be enclosed by 100 m of fencing, if part of an existing straight wall is used as one side

Let the dimensions of the rectangle be x and y as shown in Figure 11.1, where PQ represents the straight wall.

From Figure 11.1, $x + 2y = 100$ (1)

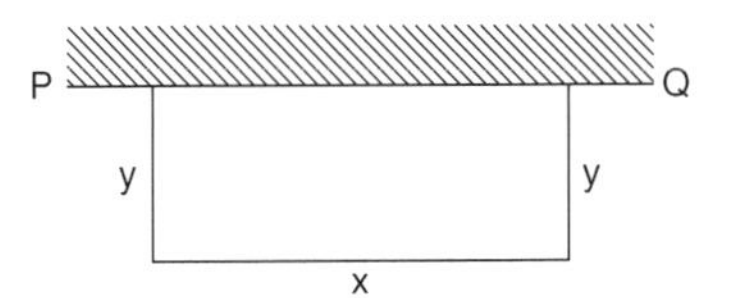

Figure 11.1

Area of rectangle, $A = xy$ (2)

Since the maximum area is required, a formula for area A is needed in terms of one variable only. From equation (1), $x = 100 - 2y$

Hence, area, $A = xy = (100 - 2y)y = 100y - 2y^2$

$$\frac{dA}{dy} = 100 - 4y = 0, \text{ for a turning point, from which, } y = 25\,\text{m}.$$

$$\frac{d^2A}{dy^2} = -4, \text{ which is negative, giving a maximum value.}$$

When $y = 25$ m, $x = 50$ m from equation (1).

Hence, the **maximum possible area** $= xy = (50)(25) =$ **$1250\,\text{m}^2$**

Application: An open rectangular box with square ends is fitted with an overlapping lid which covers the top and the front face. Determine the maximum volume of the box if 6 m^2 of metal are used in its construction

A rectangular box having square ends of side x and length y is shown in Figure 11.2.

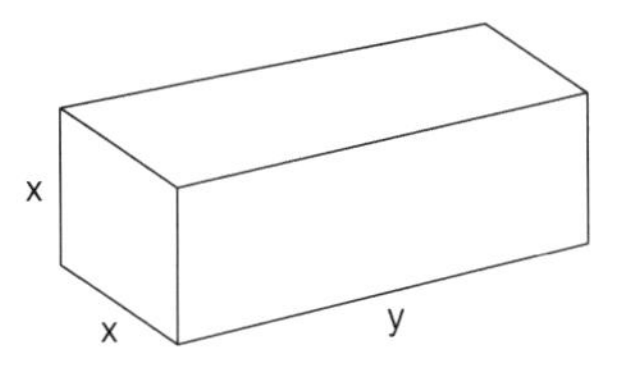

Figure 11.2

Surface area of box, A, consists of two ends and five faces (since the lid also covers the front face).

Hence, $A = 2x^2 + 5xy = 6$ (1)

Since it is the maximum volume required, a formula for the volume in terms of one variable only is needed. Volume of box, $V = x^2y$

From equation (1), $y = \frac{6 - 2x^2}{5x} = \frac{6}{5x} - \frac{2x}{5}$ (2)

Hence, volume $V = x^2y = x^2\left(\frac{6}{5x} - \frac{2x}{5}\right) = \frac{6x}{5} - \frac{2x^3}{5}$

$\frac{dV}{dx} = \frac{6}{5} - \frac{6x^2}{5} = 0$ for a maximum or minimum value.

Hence, $6 = 6x^2$, giving $x = 1$ m ($x = -1$ is not possible, and is thus neglected).

$\frac{d^2V}{dx^2} = \frac{-12x}{5}$. When $x = 1$, $\frac{d^2V}{dx^2}$ is negative, giving a maximum value.

From equation (2), when $x = 1$, $y = \frac{6}{5} - \frac{2}{5} = \frac{4}{5}$

Hence, **the maximum volume of the box,**

$$\mathbf{V} = x^2y = (1)^2\left(\frac{4}{5}\right) = \mathbf{\frac{4}{5}\ m^3}$$

11.9 Tangents and normals

The **equation of the tangent** to a curve $y = f(x)$ at the point (x_1, y_1) is given by:

$$\mathbf{y - y_1 = m(x - x_1)}$$

where $m = \frac{dy}{dx}$ = gradient of the curve at (x_1, y_1).

The **equation of the normal** to a curve at the point (x_1, y_1) is given by:

$$\mathbf{y - y_1 = -\frac{1}{m}(x - x_1)}$$

Application: Find the equation of the tangent to the curve $y = x^2 - x - 2$ at the point $(1, -2)$

Gradient, $m = \frac{dy}{dx} = 2x - 1$

At the point $(1, -2)$, $x = 1$ and $m = 2(1) - 1 = 1$

Hence the equation of the tangent is: $y - y_1 = m(x - x_1)$

i.e. $y - -2 = 1(x - 1)$

i.e. $y + 2 = x - 1$

or $\mathbf{y = x - 3}$

The graph of $y = x^2 - x - 2$ is shown in Figure 11.3. The line AB is the tangent to the curve at the point C, i.e. $(1, -2)$, and the equation of this line is $y = x - 3$.

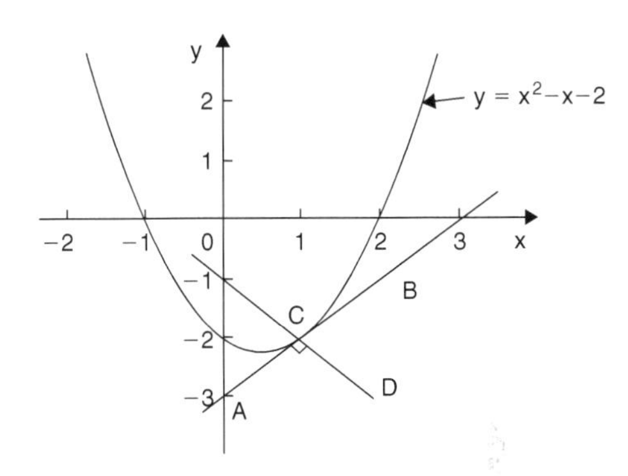

Figure 11.3

Application: Find the equation of the normal to the curve $y = x^2 - x - 2$ at the point $(1, -2)$

$m = 1$ from above, hence the equation of the normal is:

$$y - -2 = -\frac{1}{1}(x - 1)$$

i.e. $y + 2 = -x + 1$ or $\mathbf{y = -x - 1}$

Thus the line CD in Figure 11.3 has the equation $y = -x - 1$

11.10 Small changes using differentiation

If y is a function of x, i.e. $y = f(x)$, and the approximate change in y corresponding to a small change δx in x is required, then:

$$\frac{\delta y}{\delta x} \approx \frac{dy}{dx}$$

and $$\boldsymbol{\delta y \approx \frac{dy}{dx} \cdot \delta x} \quad \text{or} \quad \boldsymbol{\delta y \approx f'(x) \cdot \delta x}$$

Application: The time of swing T of a pendulum is given by $T = k\sqrt{\ell}$, where ℓ is a constant. Determine the percentage change in the time of swing if the length of the pendulum ℓ changes from 32.1 cm to 32.0 cm

If $T = k\sqrt{\ell} = k\ell^{\frac{1}{2}}$, then $\dfrac{dT}{d\ell} = k\left(\dfrac{1}{2}\ell^{-\frac{1}{2}}\right) = \dfrac{k}{2\sqrt{\ell}}$

Approximate change in T, $\delta t \approx \dfrac{dT}{d\ell}\delta\ell \approx \left(\dfrac{k}{2\sqrt{\ell}}\right)\delta\ell$

$$\approx \left(\frac{k}{2\sqrt{\ell}}\right)(-0.1) \quad \text{(negative since } \ell \text{ decreases)}$$

$$\text{Percentage error} = \left(\frac{\text{approximate change in T}}{\text{original value of T}}\right)100\%$$

$$= \frac{\left(\frac{k}{2\sqrt{\ell}}\right)(-0.1)}{k\sqrt{\ell}} \times 100\% = \left(\frac{-0.1}{2\ell}\right)100\%$$

$$= \left(\frac{-0.1}{2(32.1)}\right)100\% = \mathbf{-0.156\%}$$

Hence, the change in the time of swing is a decrease of 0.156%

11.11 Parametric equations

The following are some of the more **common parametric equations**, and Figure 11.4 shows typical shapes of these curves.

(a) Ellipse $x = a\cos\theta,\ y = b\sin\theta$

(b) Parabola $x = at^2,\ y = 2at$

(c) Hyperbola $x = a\sec\theta,\ y = b\tan\theta$

(d) Rectangular hyperbola $x = ct,\ y = \dfrac{c}{t}$

(e) Cardioid $x = a(2\cos\theta - \cos 2\theta),$
$y = a(2\sin\theta - \sin 2\theta)$

(f) Astroid $x = a\cos^3\theta,\ y = a\sin^3\theta$

(g) Cycloid $x = a(\theta - \sin\theta),\ y = a(1 - \cos\theta)$

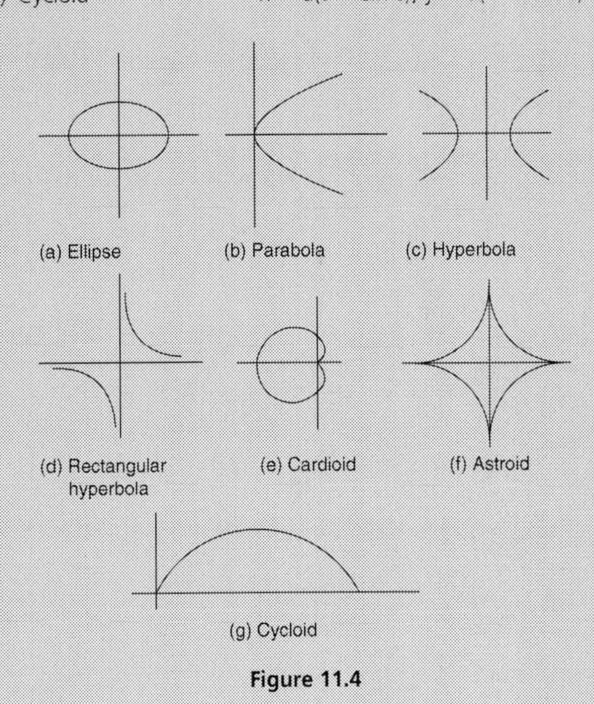

Figure 11.4

Differentiation in parameters

When x and y are both functions of θ, then:

$$\frac{dy}{dx} = \frac{\frac{dy}{d\theta}}{\frac{dx}{d\theta}} \qquad (1)$$

and

$$\frac{d^2y}{dx^2} = \frac{\frac{d}{d\theta}\left(\frac{dy}{dx}\right)}{\frac{dx}{d\theta}} \qquad (2)$$

Application: Given $x = 5\theta - 1$ and $y = 2\theta(\theta - 1)$, determine $\frac{dy}{dx}$ in terms of θ.

$x = 5\theta - 1$, hence $\frac{dx}{d\theta} = 5$

$y = 2\theta(\theta - 1) = 2\theta^2 - 2\theta$, hence $\frac{dy}{d\theta} = 4\theta - 2 = 2(2\theta - 1)$

From equation (1), $\mathbf{\frac{dy}{dx}} = \frac{\frac{dy}{d\theta}}{\frac{dx}{d\theta}} = \frac{2(2\theta - 1)}{5}$ or $\mathbf{\frac{2}{5}(2\theta - 1)}$

Application: The parametric equations of a cycloid are $x = 4(\theta - \sin\theta)$, $y = 4(1 - \cos\theta)$. Determine

(a) $\frac{dy}{dx}$ (b) $\frac{d^2y}{dx^2}$

(a) $x = 4(\theta - \sin\theta)$, hence $\frac{dx}{d\theta} = 4 - 4\cos\theta = 4(1 - \cos\theta)$

$y = 4(1 - \cos\theta)$, hence $\frac{dy}{d\theta} = 4\sin\theta$

From equation (1), $\mathbf{\frac{dy}{dx}} = \frac{\frac{dy}{d\theta}}{\frac{dx}{d\theta}} = \frac{4\sin\theta}{4(1 - \cos\theta)} = \mathbf{\frac{\sin\theta}{(1 - \cos\theta)}}$

(b) From equation (2),

$$\frac{\mathbf{d^2y}}{\mathbf{dx^2}} = \frac{\dfrac{d}{d\theta}\left(\dfrac{\sin\theta}{1-\cos\theta}\right)}{4(1-\cos\theta)} = \frac{\dfrac{(1-\cos\theta)(\cos\theta)-(\sin\theta)(\sin\theta)}{(1-\cos\theta)^2}}{4(1-\cos\theta)}$$

$$= \frac{\cos\theta-\cos^2\theta-\sin^2\theta}{4(1-\cos\theta)^3} = \frac{\cos\theta-(\cos^2\theta+\sin^2\theta)}{4(1-\cos\theta)^3}$$

$$= \frac{\cos\theta-1}{4(1-\cos\theta)^3} = \frac{-(1-\cos\theta)}{4(1-\cos\theta)^3} = \mathbf{\frac{-1}{4(1-\cos\theta)^2}}$$

Application: When determining the surface tension of a liquid, the radius of curvature ρ, of part of the surface is given by: $\rho = \dfrac{\sqrt{\left[1+\left(\dfrac{dy}{dx}\right)^2\right]^3}}{\dfrac{d^2y}{dx^2}}$

Find the radius of curvature of the part of the surface having the parametric equations

$$x = 3t^2, \; y = 6t \text{ at the point } t = 2$$

$x = 3t^2$, hence $\dfrac{dx}{dt} = 6t$ and $y = 6t$, hence $\dfrac{dy}{dt} = 6$

From equation (1),

$$\frac{dy}{dx} = \frac{\dfrac{dy}{dt}}{\dfrac{dx}{dt}} = \frac{6}{6t} = \frac{1}{t}$$

From equation (2),

$$\frac{d^2y}{dt^2} = \frac{\dfrac{d}{dt}\left(\dfrac{dy}{dx}\right)}{\dfrac{dx}{dt}} = \frac{\dfrac{d}{dt}\left(\dfrac{1}{t}\right)}{6t} = \frac{\dfrac{d}{dt}(t^{-1})}{6t} = \frac{-t^{-2}}{6t} = \frac{-\dfrac{1}{t^2}}{6t} = \frac{-1}{6t^3}$$

Hence radius of curvature, $\rho = \dfrac{\sqrt{\left[1+\left(\dfrac{dy}{dx}\right)^2\right]^3}}{\dfrac{d^2y}{dx^2}} = \dfrac{\sqrt{\left[1+\left(\dfrac{1}{t}\right)^2\right]^3}}{\dfrac{-1}{6t^3}}$

When t = 2,

$$\rho = \frac{\sqrt{\left[1+\left(\frac{1}{2}\right)^2\right]^3}}{-\frac{1}{6(2)^3}} = \frac{\sqrt{(1.25)^3}}{-\frac{1}{48}} = -48\sqrt{1.25^3} = \mathbf{-67.08}$$

11.12 Differentiating implicit functions

$$\frac{d}{dx}[f(y)] = \frac{d}{dy}[f(y)] \times \frac{dy}{dx} \qquad (3)$$

Sometimes with equations involving, say, y and x, it is impossible to make y the subject of the formula. The equation is then called an **implicit function** and examples of such functions include $y^3 + 2x^2 = y^2 - x$ and $\sin y = x^2 + 2xy$

Application: Differentiate u = sin 3t with respect to x

$$\frac{du}{dx} = \frac{du}{dt} \times \frac{dt}{dx} = \frac{d}{dt}(\sin 3t) \times \frac{dt}{dx} = \mathbf{3\cos 3t\,\frac{dt}{dx}}$$

Application: Differentiate u = 4 ln 5y with respect to t

$$\frac{du}{dt} = \frac{du}{dy} \times \frac{dy}{dt} = \frac{d}{dy}(4\ln 5y) \times \frac{dy}{dt} = \mathbf{\left(\frac{4}{y}\right)\frac{dy}{dt}}$$

Application: Determine $\dfrac{d}{dx}(x^2y)$

$$\frac{d}{dx}(x^2y) = (x^2)\frac{d}{dx}(y) + (y)\frac{d}{dx}(x^2), \quad \text{by the product rule}$$

$$= (x^2)\left(1\frac{dy}{dx}\right) + y(2x), \quad \text{by using equation (3)}$$

$$= \mathbf{x^2\frac{dy}{dx} + 2xy}$$

Application: Determine $\frac{d}{dx}\left(\frac{3y}{2x}\right)$

$$\frac{d}{dx}\left(\frac{3y}{2x}\right) = \frac{(2x)\frac{d}{dx}(3y) - (3y)\frac{d}{dx}(2x)}{(2x)^2} = \frac{(2x)\left(3\frac{dy}{dx}\right) - (3y)(2)}{4x^2}$$

$$= \frac{6x\frac{dy}{dx} - 6y}{4x^2} = \mathbf{\frac{3}{2x^2}\left(x\frac{dy}{dx} - y\right)}$$

Application: Given $3x^2 + y^2 - 5x + y = 2$ determine $\frac{dy}{dx}$

Differentiating term by term with respect to x gives:

$$\frac{d}{dx}(3x^2) + \frac{d}{dx}(y^2) - \frac{d}{dx}(5x) + \frac{d}{dx}(y) = \frac{d}{dx}(2)$$

i.e. $6x + 2y\frac{dy}{dx} - 5 + 1\frac{dy}{dx} = 0$ using equation (3) and standard derivatives.

Rearranging gives: $(2y + 1)\frac{dy}{dx} = 5 - 6x$

from which, $\mathbf{\frac{dy}{dx} = \frac{5 - 6x}{2y + 1}}$

Application: Determine the values of $\frac{dy}{dx}$ when $x = 4$ given that $x^2 + y^2 = 25$

Differentiating each term in turn with respect to x gives:

$$\frac{d}{dx}(x^2) + \frac{d}{dx}(y^2) = \frac{d}{dx}(25) \quad \text{i.e.} \quad 2x + 2y\frac{dy}{dx} = 0$$

Hence
$$\frac{dy}{dx} = -\frac{2x}{2y} = -\frac{x}{y}$$

Since $x^2 + y^2 = 25$, when $x = 4$, $y = \sqrt{(25 - 4^2)} = \pm 3$

Thus, when $x = 4$ and $y = \pm 3$, $\mathbf{\frac{dy}{dx}} = -\frac{4}{\pm 3} = \mathbf{\pm\frac{4}{3}}$

$x^2 + y^2 = 25$ is the equation of a circle, centre at the origin and radius 5, as shown in Figure 11.5. At $x = 4$, the two gradients are shown.

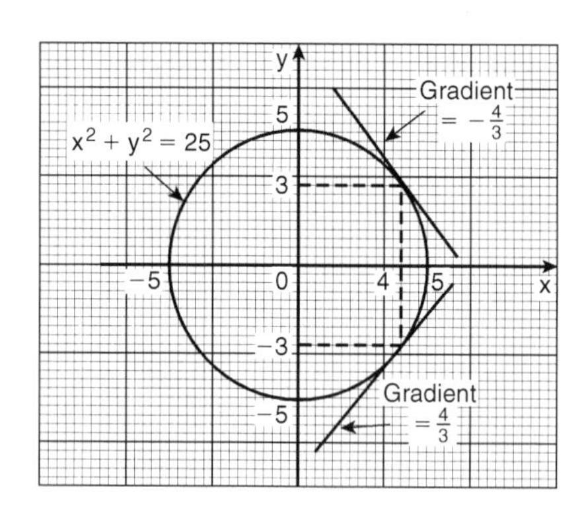

Figure 11.5

Above, $x^2 + y^2 = 25$ was differentiated implicitly; actually, the equation could be transposed to $y = \sqrt{(25 - x^2)}$ and differentiated using the function of a function rule. This gives

$$\frac{dy}{dx} = \frac{1}{2}(25 - x^2)^{-\frac{1}{2}}(-2x) = -\frac{x}{\sqrt{(25 - x^2)}}$$

and when $x = 4$, $\frac{dy}{dx} = -\frac{4}{\sqrt{(25 - 4^2)}} = \pm\frac{4}{3}$ as obtained above.

11.13 Differentiation of logarithmic functions

Logarithmic differentiation is achieved with knowledge of (i) the laws of logarithms, (ii) the differential coefficients of logarithmic functions, and (iii) the differentiation of implicit functions.

(i) **The laws of logarithms** are:

1. $\log(A \times B) = \log A + \log B$
2. $\log\left(\frac{A}{B}\right) = \log A - \log B$
3. $\log A^n = n \log A$

(ii) The differential coefficient of the logarithmic function ln x is given by:

$$\frac{d}{dx}(\ln x) = \frac{1}{x}$$

More generally, it may be shown that:

$$\frac{d}{dx}[\ln f(x)] = \frac{f'(x)}{f(x)} \qquad (4)$$

(iii) Differentiation of implicit functions is obtained using:

$$\frac{d}{dx}[f(y)] = \frac{d}{dy}[f(y)] \times \frac{dy}{dx} \qquad (5)$$

Application: If $y = \ln(3x^2 + 2x - 1)$ determine $\frac{dy}{dx}$

If $y = \ln(3x^2 + 2x - 1)$ then $\frac{dy}{dx} = \frac{6x + 2}{3x^2 + 2x - 1}$

Application: If $y = \ln(\sin 3x)$ determine $\frac{dy}{dx}$

If $y = \ln(\sin 3x)$ then $\frac{dy}{dx} = \frac{3 \cos 3x}{\sin 3x} = 3 \cot 3x$

Application: Differentiate $y = \frac{(1+x)^2\sqrt{(x-1)}}{x\sqrt{(x+2)}}$ with respect to x

(i) Taking Napierian logarithms of both sides of the equation gives:

$$\ln y = \ln\left\{\frac{(1+x)^2\sqrt{(x-1)}}{x\sqrt{(x+2)}}\right\} = \ln\left\{\frac{(1+x)^2(x-1)^{1/2}}{x(x+2)^{1/2}}\right\}$$

(ii) Applying the laws of logarithms gives:

$$\ln y = \ln(1+x)^2 + \ln(x-1)^{1/2} - \ln x - \ln(x+2)^{1/2}$$

by laws 1 and 2

i.e. $$\ln y = 2\ln(1+x) + \frac{1}{2}\ln(x-1) - \ln x - \frac{1}{2}\ln(x+2)$$

by law 3

(iii) Differentiating each term in turn with respect to x using equations (4) and (5) gives:

$$\frac{1}{y}\frac{dy}{dx} = \frac{2}{(1+x)} + \frac{\frac{1}{2}}{(x-1)} - \frac{1}{x} - \frac{\frac{1}{2}}{(x+2)}$$

(iv) Rearranging the equation to make $\frac{dy}{dx}$ the subject gives:

$$\frac{dy}{dx} = y\left\{\frac{2}{(1+x)} + \frac{1}{2(x-1)} - \frac{1}{x} - \frac{1}{2(x+2)}\right\}$$

(v) Substituting for y in terms of x gives:

$$\mathbf{\frac{dy}{dx} = \frac{(1+x)^2\sqrt{(x-1)}}{x\sqrt{(x+2)}}\left\{\frac{2}{(1+x)} + \frac{1}{2(x-1)} - \frac{1}{x} - \frac{1}{2(x+2)}\right\}}$$

Application: Determine $\frac{dy}{dx}$ given $y = x^x$

Taking Napierian logarithms of both sides of $y = x^x$ gives:

$$\ln y = \ln x^x = x \ln x \qquad \text{by law 3}$$

Differentiating both sides with respect to x gives:

$$\frac{1}{y}\frac{dy}{dx} = (x)\left(\frac{1}{x}\right) + (\ln x)(1), \quad \text{using the product rule}$$

i.e. $\frac{1}{y}\frac{dy}{dx} = 1 + \ln x$ from which, $\frac{dy}{dx} = y(1 + \ln x)$

i.e. $$\mathbf{\frac{dy}{dx} = x^x(1 + \ln x)}$$

Application: Determine the differential coefficient of $y = \sqrt[x]{(x-1)}$ and evaluate $\frac{dy}{dx}$ when $x = 2$.

$y = \sqrt[x]{(x-1)} = (x-1)^{1/x}$ since by the laws of indices: $\sqrt[n]{a^m} = a^{\frac{m}{n}}$

Taking Napierian logarithms of both sides gives:

$$\ln y = \ln(x-1)^{1/x} = \frac{1}{x}\ln(x-1) \qquad \text{by law 3}$$

Differentiating each side with respect to *x* gives:

$$\frac{1}{y}\frac{dy}{dx} = \left(\frac{1}{x}\right)\left(\frac{1}{x-1}\right) + [\ln(x-1)]\left(\frac{-1}{x^2}\right) \qquad \text{by the product rule}$$

Hence $$\frac{dy}{dx} = y\left\{\frac{1}{x(x-1)} - \frac{\ln(x-1)}{x^2}\right\}$$

i.e. $$\mathbf{\frac{dy}{dx} = \sqrt[x]{(x-1)}\left\{\frac{1}{x(x-1)} - \frac{\ln(x-1)}{x^2}\right\}}$$

When $x = 2$, $$\frac{dy}{dx} = \sqrt[2]{(1)}\left\{\frac{1}{2(1)} - \frac{\ln(1)}{4}\right\} = \pm 1\left\{\frac{1}{2} - 0\right\} = \pm\frac{1}{2}$$

11.14 Differentiation of inverse trigonometric functions

If $y = 3x - 2$, then by transposition, $x = \frac{y+2}{3}$. The function $x = \frac{y+2}{3}$ is called the **inverse function** of $y = 3x - 2$.

Inverse trigonometric functions are denoted by prefixing the function with $^{-1}$ or 'arc'. For example, if $y = \sin x$, then $x = \sin^{-1} y$ or $x = \arcsin y$. Similarly, if $y = \cos x$, then $y = \cos^{-1} y$ or $x = \arccos y$, and so on. A sketch of each of the inverse trigonometric functions is shown in Figure 11.6.

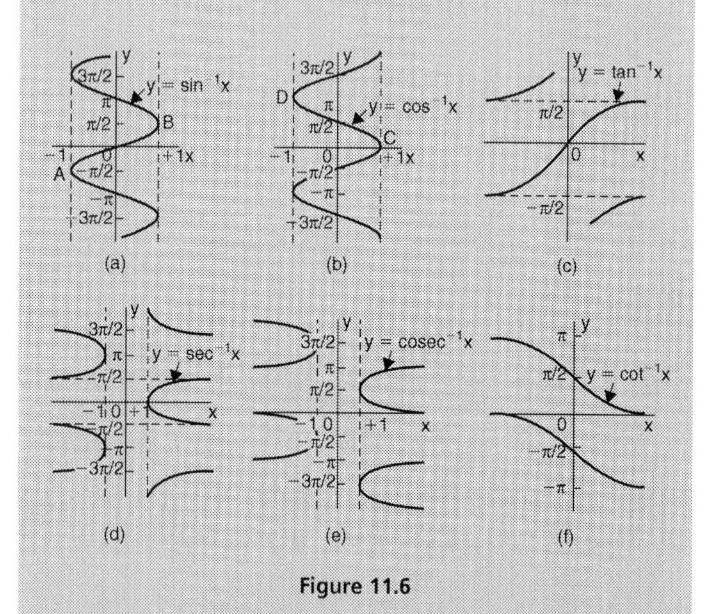

Figure 11.6

Table 11.1 Differential coefficients of inverse trigonometric functions

y or f(x)	$\frac{dy}{dx}$ or f′(x)
(i) $\sin^{-1}\frac{x}{a}$	$\frac{1}{\sqrt{a^2 - x^2}}$
$\sin^{-1} f(x)$	$\frac{f'(x)}{\sqrt{1 - [f(x)]^2}}$

(ii) $\cos^{-1}\frac{x}{a}$	$\frac{-1}{\sqrt{a^2-x^2}}$
$\cos^{-1} f(x)$	$\frac{-f'(x)}{\sqrt{1-[f(x)]^2}}$
(iii) $\tan^{-1}\frac{x}{a}$	$\frac{a}{a^2+x^2}$
$\tan^{-1} f(x)$	$\frac{f'(x)}{1+[f(x)]^2}$
(iv) $\sec^{-1}\frac{x}{a}$	$\frac{a}{x\sqrt{x^2-a^2}}$
$\sec^{-1} f(x)$	$\frac{f'(x)}{f(x)\sqrt{[f(x)]^2-1}}$
(v) $\text{cosec}^{-1}\frac{x}{a}$	$\frac{-a}{x\sqrt{x^2-a^2}}$
$\text{cosec}^{-1} f(x)$	$\frac{-f'(x)}{f(x)\sqrt{[f(x)]^2-1}}$
(vi) $\cot^{-1}\frac{x}{a}$	$\frac{-a}{a^2+x^2}$
$\cot^{-1} f(x)$	$\frac{-f'(x)}{1+[f(x)]^2}$

Application: Find $\frac{dy}{dx}$ given $y = \sin^{-1}5x^2$

From Table 11.1(i), if $y = \sin^{-1}f(x)$ then $\frac{dy}{dx} = \frac{f'(x)}{\sqrt{1-[f(x)]^2}}$

Hence, if $y = \sin^{-1}5x^2$ then $f(x) = 5x^2$ and $f'(x) = 10x$

Thus, $\frac{dy}{dx} = \frac{10x}{\sqrt{1-(5x^2)^2}} = \mathbf{\frac{10x}{\sqrt{1-25x^4}}}$

Application: Find the differential coefficient of $y = \ln(\cos^{-1}3x)$

Let $u = \cos^{-1}3x$ then $y = \ln u$

By the function of a function rule,

$$\frac{dy}{dx} = \frac{dy}{du}\cdot\frac{du}{dx} = \frac{1}{u}\times\frac{d}{dx}(\cos^{-1}3x)$$

$$= \frac{1}{\cos^{-1}3x}\left\{\frac{-3}{\sqrt{1-(3x)^2}}\right\}$$

i.e. $$\mathbf{\frac{d}{dx}[\ln(\cos^{-1}3x)] = \frac{-3}{\sqrt{1-9x^2}\,\cos^{-1}3x}}$$

Application: Find $\frac{dy}{dt}$ given $y = \tan^{-1}\frac{3}{t^2}$

Using the general form from Table 11.1(iii),

$$f(t) = \frac{3}{t^2} = 3t^{-2} \quad \text{from which,} \quad f'(t) = \frac{-6}{t^3}$$

Hence, $$\frac{d}{dt}\left(\tan^{-1}\frac{3}{t^2}\right) = \frac{f'(t)}{1+[f(t)]^2} = \frac{-\frac{6}{t^3}}{\left\{1+\left(\frac{3}{t^2}\right)^2\right\}} = \frac{-\frac{6}{t^3}}{\frac{t^4+9}{t^4}}$$

$$= \left(-\frac{6}{t^3}\right)\left(\frac{t^4}{t^4+9}\right) = -\mathbf{\frac{6t}{t^4+9}}$$

11.15 Differentiation of inverse hyperbolic functions

Inverse hyperbolic functions are denoted by prefixing the function with $^{-1}$ or 'ar'. For example, if $y = \sinh x$, then $x = \sinh^{-1}y$ or $x = \text{arsinh } y$. Similarly, if $y = \text{sech } x$, then $x = \text{sech}^{-1}y$ or $x = \text{arsech } y$, and so on. A sketch of each of the inverse hyperbolic functions is shown in Figure 11.7.

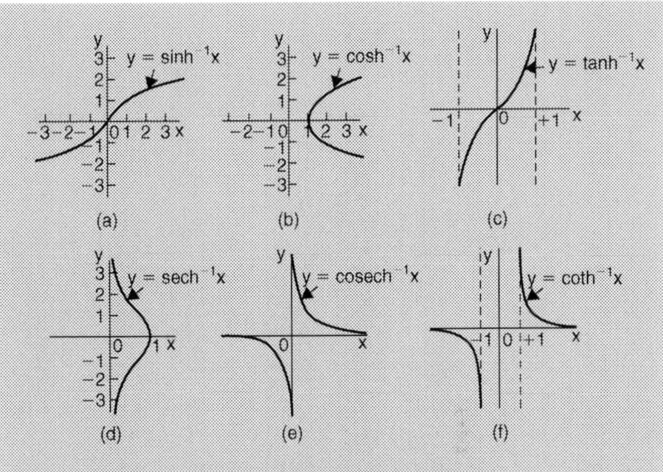

Figure 11.7

Table 11.2 Differential coefficients of inverse trigonometric functions

y or f(x)	$\frac{dy}{dx}$ or f′(x)
(i) $\sinh^{-1}\frac{x}{a}$	$\frac{1}{\sqrt{x^2 + a^2}}$
$\sinh^{-1} f(x)$	$\frac{f'(x)}{\sqrt{[f(x)]^2 + 1}}$
(ii) $\cosh^{-1}\frac{x}{a}$	$\frac{1}{\sqrt{x^2 - a^1}}$
$\cosh^{-1} f(x)$	$\frac{f'(x)}{\sqrt{[f(x)]^2 - 1}}$
(iii) $\tanh^{-1}\frac{x}{a}$	$\frac{a}{a^2 - x^2}$
$\tanh^{-1} f(x)$	$\frac{f'(x)}{1 - [f(x)]^2}$

(iv) $\text{sech}^{-1}\frac{x}{a}$	$\frac{-a}{x\sqrt{a^2 - x^2}}$
$\text{sech}^{-1} f(x)$	$\frac{-f'(x)}{f(x)\sqrt{1 - [f(x)]^2}}$
(v) $\text{cosech}^{-1}\frac{x}{a}$	$\frac{-a}{x\sqrt{x^2 + a^2}}$
$\text{cosech}^{-1} f(x)$	$\frac{-f'(x)}{f(x)\sqrt{[f(x)]^2 + 1}}$
(vi) $\coth^{-1}\frac{x}{a}$	$\frac{a}{a^2 - x^2}$
$\coth^{-1} f(x)$	$\frac{f'(x)}{1 - [f(x)]^2}$

Application: Find the differential coefficient of $y = \sinh^{-1}2x$

From Table 11.2(i), $\frac{d}{dx}[\sinh^{-1}f(x)] = \frac{f'(x)}{\sqrt{[f(x)]^2 + 1}}$

Hence $\frac{d}{dx}(\sinh^{-1}2x) = \frac{2}{\sqrt{[(2x)^2 + 1]}} = \mathbf{\frac{2}{\sqrt{[4x^2 + 1]}}}$

Application: Determine $\frac{d}{dx}[\cosh^{-1}\sqrt{(x^2 + 1)}]$

If $y = \cosh^{-1}f(x)$, $\frac{dy}{dx} = \frac{f'(x)}{\sqrt{\{[f(x)]^2 - 1\}}}$

If $y = \cosh^{-1}\sqrt{(x^2 + 1)}$, then $f(x) = \sqrt{(x^2 + 1)}$ and

$f'(x) = \frac{1}{2}(x + 1)^{-1/2}(2x) = \frac{x}{\sqrt{(x^2 + 1)}}$

Hence $\frac{d}{dx}\left[\cosh^{-1}\sqrt{(x^2+1)}\right] = \frac{\frac{x}{\sqrt{(x^2+1)}}}{\sqrt{\left\{[\sqrt{(x^2+1)}]^2 - 1\right\}}} = \frac{\frac{x}{\sqrt{(x^2+1)}}}{\sqrt{(x^2+1-1)}}$

$$= \frac{\frac{x}{\sqrt{(x^2+1)}}}{x} = \mathbf{\frac{1}{\sqrt{(x^2+1)}}}$$

Application: Find the differential coefficient of $y = \text{sech}^{-1}(2x - 1)$

From Table 11.2(iv), $\frac{d}{dx}[\text{sech}^{-1} f(x)] = \frac{-f'(x)}{f(x)\sqrt{1-[f(x)]^2}}$

Hence

$$\frac{d}{dx}[\text{sech}^{-1}(2x-1)] = \frac{-2}{(2x-1)\sqrt{[1-(2x-1)^2]}}$$

$$= \frac{-2}{(2x-1)\sqrt{[1-(4x^2-4x+1)]}}$$

$$= \frac{-2}{(2x-1)\sqrt{(4x-4x^2)}} = \frac{-2}{(2x-1)\sqrt{[4x(1-x)]}}$$

$$= \frac{-2}{(2x-1)\,2\sqrt{[x(1-x)]}} = \mathbf{\frac{-1}{(2x-1)\sqrt{[x(1-x)]}}}$$

Logarithmic forms of the inverse hyperbolic functions

Inverse hyperbolic functions may be evaluated most conveniently when expressed in a **logarithmic form**.

$$\sinh^{-1}\frac{x}{a} = \ln\left\{\frac{x+\sqrt{a^2+x^2}}{a}\right\} \qquad (6)$$

$$\cosh^{-1}\frac{x}{a} = \ln\left\{\frac{x+\sqrt{x^2-a^2}}{a}\right\} \qquad (7)$$

and $$\tanh^{-1}\frac{x}{a} = \frac{1}{2}\ln\left(\frac{a+x}{a-x}\right) \quad (8)$$

A calculator with inverse hyperbolic functions may also be used to evaluate such functions.

Application: Evaluate $\sinh^{-1}\frac{3}{4}$

To evaluate $\sinh^{-1}\frac{3}{4}$ let x = 3 and a = 4 in equation (6).

Then, $\sinh^{-1}\frac{3}{4} = \ln\left\{\frac{3+\sqrt{4^2+3^2}}{4}\right\} = \ln\left(\frac{3+5}{4}\right) = \ln 2 = \mathbf{0.6931}$

Application: Evaluate, correct to 4 decimal places, $\sinh^{-1}2$

From equation (6), with x = 2 and a = 1,

$$\sinh^{-1}2 = \ln\left\{\frac{2+\sqrt{1^2+2^2}}{1}\right\} = \ln\left(2+\sqrt{5}\right) = \ln 4.2361$$

$$= \mathbf{1.4436,\ correct\ to\ 4\ decimal\ places}$$

Application: Evaluate $\cosh^{-1}1.4$, correct to 3 decimal places

From equation (7), $$\cosh^{-1}\frac{x}{a} = \ln\left\{\frac{x \pm \sqrt{x^2-a^2}}{a}\right\}$$

and $\cosh^{-1}1.4 = \cosh^{-1}\frac{14}{10} = \cosh^{-1}\frac{7}{5}$ hence, x = 7 and a = 5

Then, $$\cosh^{-1}\frac{7}{5} = \ln\left\{\frac{7+\sqrt{7^2-5^2}}{5}\right\} = \ln 2.3798$$

$$= \mathbf{0.867},\ \text{correct to 3 decimal places}$$

Application: Evaluate $\tanh^{-1}\frac{3}{5}$, correct to 4 decimal places

From equation (8),

$$\tanh^{-1}\frac{x}{a} = \frac{1}{2}\ln\left(\frac{a+x}{a-x}\right); \quad \text{substituting } x = 3 \text{ and } a = 5 \text{ gives:}$$

$$\tanh^{-1}\frac{3}{5} = \frac{1}{2}\ln\left(\frac{5+3}{5-3}\right) = \frac{1}{2}\ln 4$$

$$= \mathbf{0.6931}, \text{ correct to 4 decimal places}$$

11.16 Partial differentiation

When differentiating a function having two variables, one variable is kept constant and the differential coefficient of the other variable is found with respect to that variable. The differential coefficient obtained is called a **partial derivative** of the function.

First order partial derivatives

If $V = \pi r^2 h$ then $\frac{\partial V}{\partial r}$ means 'the partial derivative of V with respect to r, with h remaining constant'

Thus $\quad \frac{\partial V}{\partial r} = (\pi h)\frac{d}{dr}(r^2) = (\pi h)(2r) = 2\pi rh$

Similarly, $\frac{\partial V}{\partial h}$ means 'the partial derivative of V with respect to h, with r remaining constant'

Thus $\quad \frac{\partial V}{\partial h} = (\pi r^2)\frac{d}{dh}(h) = (\pi r^2)(1) = \pi r^2$

Second order partial derivatives

(i) Differentiating $\frac{\partial V}{\partial r}$ with respect to r, keeping h constant, gives

$$\frac{\partial}{\partial r}\left(\frac{\partial V}{\partial r}\right), \text{ which is written as } \frac{\partial^2 V}{\partial r^2}$$

Thus $\quad$ if $V = \pi r^2 h$ $\quad$ then $\quad \frac{\partial^2 V}{\partial r^2} = \frac{\partial}{\partial r}(2\pi rh) = \mathbf{2\pi h}$

(ii) Differentiating $\frac{\partial V}{\partial h}$ with respect to h, keeping r constant, gives

$$\frac{\partial}{\partial h}\left(\frac{\partial V}{\partial h}\right), \text{ which is written as } \frac{\partial^2 V}{\partial h^2}$$

Thus $\frac{\partial^2 V}{\partial h^2} = \frac{\partial}{\partial h}(\pi r^2) = \mathbf{0}$

(iii) Differentiating $\frac{\partial V}{\partial h}$ with respect to r, keeping h constant, gives

$$\frac{\partial}{\partial r}\left(\frac{\partial V}{\partial h}\right), \text{ which is written as } \frac{\partial^2 V}{\partial r \partial h}$$

Thus $\frac{\partial^2 V}{\partial r \partial h} = \frac{\partial}{\partial r}\left(\frac{\partial V}{\partial h}\right) = \frac{\partial}{\partial r}(\pi r^2) = \mathbf{2\pi r}$

(iv) Differentiating $\frac{\partial V}{\partial r}$ with respect to h, keeping r constant, gives

$$\frac{\partial}{\partial h}\left(\frac{\partial V}{\partial r}\right), \text{ which is written as } \frac{\partial^2 V}{\partial h \partial r}$$

Thus $\frac{\partial^2 V}{\partial h \partial r} = \frac{\partial}{\partial h}\left(\frac{\partial V}{\partial r}\right) = \frac{\partial}{\partial h}(2\pi rh) = \mathbf{2\pi r}$

$\frac{\partial^2 V}{\partial r^2}, \frac{\partial^2 V}{\partial h^2}, \frac{\partial^2 V}{\partial r \partial h}$ and $\frac{\partial^2 V}{\partial h \partial r}$ are examples of **second order partial derivatives**. It is seen from (iii) and (iv) that $\frac{\partial^2 V}{\partial r \partial h} = \frac{\partial^2 V}{\partial h \partial r}$ and such a result is always true for continuous functions.

Application: If $Z = 5x^4 + 2x^3y^2 - 3y$ determine $\frac{\partial Z}{\partial x}$ and $\frac{\partial Z}{\partial y}$

If $Z = 5x^4 + 2x^3y^2 - 3y$

then $\frac{\partial Z}{\partial x} = \frac{d}{dx}(5x^4) + (2y^2)\frac{d}{dx}(x^3) - (3y)\frac{d}{dx}(1)$

$$= 20x^3 + (2y^2)(3x^2) - (3y)(0) = \mathbf{20x^3 + 6x^2y^2}$$

and $$\frac{\partial Z}{\partial y} = (5x^4)\frac{d}{dy}(1) + (2x^3)\frac{d}{dy}(y^2) - 3\frac{d}{dy}(y)$$
$$= 0 + (2x^3)(2y) - 3 = \mathbf{4x^3y - 3}$$

Application: The time of oscillation, t, of a pendulum is given by: $t = 2\pi\sqrt{\frac{l}{g}}$ where l is the length of the pendulum and g the free fall acceleration due to gravity. Find $\frac{\partial t}{\partial l}$ and $\frac{\partial t}{\partial g}$

To find $\frac{\partial t}{\partial 1}$, g is kept constant.

$$t = 2\pi\sqrt{\frac{l}{g}} = \left(\frac{2\pi}{\sqrt{g}}\right)\sqrt{l} = \left(\frac{2\pi}{\sqrt{g}}\right)l^{1/2}$$

Hence, $$\frac{\partial t}{\partial l} = \left(\frac{2\pi}{\sqrt{g}}\right)\frac{d}{dl}(l^{1/2}) = \left(\frac{2\pi}{\sqrt{g}}\right)\left(\frac{1}{2}l^{-1/2}\right) = \left(\frac{2\pi}{\sqrt{g}}\right)\left(\frac{1}{2\sqrt{l}}\right) = \mathbf{\frac{\pi}{\sqrt{lg}}}$$

To find $\frac{\partial t}{\partial g}$, l is kept constant.

$$t = 2\pi\sqrt{\frac{l}{g}} = (2\pi\sqrt{l})\left(\sqrt{\frac{l}{g}}\right) = (2\pi\sqrt{l})\,g^{-1/2}$$

Hence

$$\frac{\partial t}{\partial g} = (2\pi\sqrt{l})\left(-\frac{1}{2}g^{-3/2}\right) = (2\pi\sqrt{l})\left(\frac{-l}{2\sqrt{g^3}}\right) = \frac{-\pi\sqrt{l}}{\sqrt{g^3}} = \mathbf{-\pi\sqrt{\frac{l}{g^3}}}$$

Application: Given $Z = 4x^2y^3 - 2x^3 + 7y^2$ find

(a) $\frac{\partial^2 Z}{\partial x^2}$ (b) $\frac{\partial^2 Z}{\partial y^2}$ (c) $\frac{\partial^2 Z}{\partial x \partial y}$ (d) $\frac{\partial^2 Z}{\partial y \partial x}$

(a) $$\frac{\partial Z}{\partial x} = 8xy^3 - 6x^2$$

$$\frac{\partial^2 Z}{\partial x^2} = \frac{\partial}{\partial x}\left(\frac{\partial Z}{\partial x}\right) = \frac{\partial}{\partial x}(8xy^3 - 6x^2) = \mathbf{8y^3 - 12x}$$

(b) $\frac{\partial Z}{\partial y} = 12x^2y^2 + 14y$

$$\frac{\partial^2 Z}{\partial y^2} = \frac{\partial}{\partial y}\left(\frac{\partial Z}{\partial y}\right) = \frac{\partial}{\partial y}(12x^2y^2 + 14y) = \mathbf{24x^2y + 14}$$

(c) $\frac{\partial^2 Z}{\partial x \partial y} = \frac{\partial}{\partial x}\left(\frac{\partial Z}{\partial y}\right) = \frac{\partial}{\partial x}(12x^2y^2 + 14y) = \mathbf{24xy^2}$

(d) $\frac{\partial^2 Z}{\partial y \partial x} = \frac{\partial}{\partial y}\left(\frac{\partial Z}{\partial x}\right) = \frac{\partial}{\partial y}(8xy^3 - 6x^2) = \mathbf{24xy^2}$

11.17 Total differential

If $Z = f(u, v, w, \ldots)$, then the **total differential, dZ**, is given by:

$$\mathbf{dZ = \frac{\partial Z}{\partial u}du + \frac{\partial Z}{\partial v}dv + \frac{\partial Z}{\partial w}dw + \ldots\ldots} \qquad (9)$$

Application: If $Z = f(u, v, w)$ and $Z = 3u^2 - 2v + 4w^3v^2$ determine the total differential dZ

Total differential, $dZ = \frac{\partial Z}{\partial u}du + \frac{\partial Z}{\partial v}dv + \frac{\partial Z}{\partial w}dw$

$\frac{\partial Z}{\partial u} = 6u$ (i.e. v and w are kept constant)

$\frac{\partial Z}{\partial v} = -2 + 8w^3v$ (i.e. u and w are kept constant)

$\frac{\partial Z}{\partial w} = 12w^2v^2$ (i.e. u and v are kept constant)

Hence, $\mathbf{dZ = 6u\,du + (8vw^3 - 2)dv + (12v^2w^2)dw}$

11.18 Rates of change using partial differentiation

If $Z = f(u, v, w, \ldots)$ and $\frac{du}{dt}, \frac{dv}{dt}, \frac{dw}{dt}, \ldots\ldots$ denote the rate of change of u, v, w, ...respectively, then the rate of change of Z, $\frac{dZ}{dt}$, is given by:

$$\frac{dZ}{dt} = \frac{\partial Z}{\partial u}\frac{du}{dt} + \frac{\partial Z}{\partial v}\frac{dv}{dt} + \frac{\partial Z}{\partial w}\frac{dw}{dt} + \ldots \quad (9)$$

Application: If the height of a right circular cone is increasing at 3 mm/s and its radius is decreasing at 2 mm/s, find the rate at which the volume is changing (in cm^3/s) when the height is 3.2 cm and the radius is 1.5 cm.

Volume of a right circular cone, $V = \frac{1}{3}\pi r^2 h$

Using equation (9), the rate of change of volume,

$$\frac{dV}{dt} = \frac{\partial V}{\partial r}\frac{dr}{dt} + \frac{\partial V}{\partial h}\frac{dh}{dt}$$

$\frac{\partial V}{\partial r} = \frac{2}{3}\pi rh$ and $\frac{\partial V}{\partial h} = \frac{1}{3}\pi r^2$

Since the height is increasing at 3 mm/s, i.e. 0.3 cm/s, then $\frac{dh}{dt} = +0.3$ and since the radius is decreasing at 2 mm/s, i.e. 0.2 cm/s, then $\frac{dr}{dt} = -0.2$

Hence, $\frac{dV}{dt} = \left(\frac{2}{3}\pi rh\right)(-0.2) + \left(\frac{1}{3}\pi r^2\right)(+0.3) = \frac{-0.4}{3}\pi rh + 0.1\pi r^2$

However, $h = 3.2$ cm and $r = 1.5$ cm.

$$\text{Hence } \frac{dV}{dt} = \frac{-0.4}{3}\pi(1.5)(3.2) + (0.1)\pi(1.5)^2$$
$$= -2.011 + 0.707 = -1.304 \text{ cm}^3/\text{s}$$

Thus, the rate of change of volume is 1.30 cm^3/s decreasing

11.19 Small changes using partial differentiation

If $Z = f(u, v, w,...)$ and $\delta u, \delta v, \delta w,...$ denote **small changes** in u, v, w,... respectively, then the corresponding approximate change δZ in Z is given by:

$$\delta Z \approx \frac{\partial Z}{\partial u}\delta u + \frac{\partial Z}{\partial v}\delta v + \frac{\partial Z}{\partial w}\delta w + ... \qquad (10)$$

Application: If the modulus of rigidity $G = (R^4\theta)/L$, where R is the radius, θ the angle of twist and L the length, find the approximate percentage error in G when R is increased by 2%, θ is reduced by 5% and L is increased by 4%

From equation (10), $\delta G \approx \frac{\partial G}{\partial R}\delta R + \frac{\partial G}{\partial \theta}\delta\theta + \frac{\partial G}{\partial L}\delta L$

Since $G = \frac{R^4\theta}{L}$, $\frac{\partial G}{\partial R} = \frac{4R^3\theta}{L}$, $\frac{\partial G}{\partial \theta} = \frac{R^4}{L}$ and $\frac{\partial G}{\partial L} = \frac{-R^4\theta}{L^2}$

Since R is increased by 2%, $\delta R = \frac{2}{100}R = 0.02\,R$

Similarly, $\delta\theta = -0.05\theta$ and $\delta L = 0.04\,L$

Hence $\delta G \approx \left(\frac{4R^3\theta}{L}\right)(0.02\,R) + \left(\frac{R^4}{L}\right)(-0.05\,\theta) + \left(-\frac{R^4\theta}{L^2}\right)(0.04\,L)$

$$\approx \frac{R^4\theta}{L}[0.08 - 0.05 - 0.04] \approx -0.01\frac{R^4\theta}{L}$$

i.e. $\delta G \approx -\frac{1}{100}G$

Hence the approximate percentage error in G is a 1% decrease.

Application: If the second moment of area I of a rectangle is given by $I = \frac{bl^3}{3}$, find the approximate error in the calculated value of I, if b and l are measured as 40 mm and 90 mm respectively and the measurement errors are −5 mm in b and +8 mm in l.

Using equation (10), the approximate error in I, $\delta I \approx \frac{\partial I}{\partial b}\delta b + \frac{\partial I}{\partial l}\delta l$

$$\frac{\partial I}{\partial b} = \frac{l^3}{3} \text{ and } \frac{\partial I}{\partial l} = \frac{3bl^2}{3} = bl^2$$

$\delta b = -5\,\text{mm}$ and $\delta l = +8\,\text{mm}$

Hence $\delta I \approx \left(\frac{l^3}{3}\right)(-5) + (bl^2)(+8)$

Since $b = 40\,\text{mm}$ and $l = 90\,\text{mm}$ then

$$\begin{aligned}\delta I &\approx \left(\frac{90^3}{3}\right)(-5) + 40(90)^2(8)\\ &\approx -1{,}215{,}000 + 2{,}592{,}000\\ &\approx 1{,}377{,}000\,\text{mm}^4 \approx 137.7\,\text{cm}^4\end{aligned}$$

Hence, the approximate error in the calculated value of I is a 137.7 cm^4 increase.

11.20 Maxima, minima and saddle points of functions of two variables

Procedure to determine maxima, minima and saddle points for functions of two variables

Given $z = f(x, y)$:

(i) determine $\frac{\partial z}{\partial x}$ and $\frac{\partial z}{\partial y}$

(ii) for stationary points, $\frac{\partial z}{\partial x} = 0$ and $\frac{\partial z}{\partial y} = 0$,

(iii) solve the simultaneous equations $\frac{\partial z}{\partial x} = 0$ and $\frac{\partial z}{\partial y} = 0$ for x and y, which gives the co-ordinates of the stationary points,

(iv) determine $\frac{\partial^2 z}{\partial x^2}$, $\frac{\partial^2 z}{\partial y^2}$ and $\frac{\partial^2 z}{\partial x \partial y}$

(v) for each of the co-ordinates of the stationary points, substitute values of x and y into $\frac{\partial^2 z}{\partial x^2}$, $\frac{\partial^2 z}{\partial y^2}$ and $\frac{\partial^2 z}{\partial x \partial y}$ and evaluate each,

(vi) evaluate $\left(\frac{\partial^2 z}{\partial x \partial y}\right)^2$ for each stationary point,

(vii) substitute the values of $\frac{\partial^2 z}{\partial x^2}$, $\frac{\partial^2 z}{\partial y^2}$ and $\frac{\partial^2 z}{\partial x \partial y}$ into the equation

$$\Delta = \left(\frac{\partial^2 z}{\partial x \partial y}\right)^2 - \left(\frac{\partial^2 z}{\partial x^2}\right)\left(\frac{\partial^2 z}{\partial y^2}\right) \text{ and evaluate,}$$

(viii) (a) if $\boldsymbol{\Delta > 0}$ then the stationary point is a **saddle point**

(b) if $\boldsymbol{\Delta < 0}$ and $\boldsymbol{\frac{\partial^2 z}{\partial x^2} < 0}$, then the stationary point is a **maximum point**, and

(c) if $\boldsymbol{\Delta < 0}$ and $\boldsymbol{\frac{\partial^2 z}{\partial x^2} < 0}$, then the stationary point is a **minimum point**

Application: Determine the co-ordinates of the stationary point and its nature for the function $z = (x - 1)^2 + (y - 2)^2$

Following the above procedure:

(i) $\frac{\partial z}{\partial x} = 2(x - 1)$ and $\frac{\partial z}{\partial y} = 2(y - 2)$

(ii) $2(x - 1) = 0$ (1)

$2(y - 2) = 0$ (2)

(iii) From equations (1) and (2), x = 1 and y = 2, thus the only stationary point exists at (1, 2)

(iv) Since $\frac{\partial z}{\partial x} = 2(x - 1) = 2x - 2$, $\frac{\partial^2 z}{\partial x^2} = 2$

and since $\frac{\partial z}{\partial y} = 2(y - 2) = 2y - 4$, $\frac{\partial^2 z}{\partial y^2} = 2$

and $\dfrac{\partial^2 z}{\partial x \partial y} = \dfrac{\partial}{\partial x}\left(\dfrac{\partial z}{\partial y}\right) = \dfrac{\partial}{\partial x}(2y - 4) = 0$

(v) $\dfrac{\partial^2 z}{\partial x^2} = \dfrac{\partial^2 z}{\partial y^2} = 2$ and $\dfrac{\partial^2 z}{\partial x \partial y} = 0$

(vi) $\left(\dfrac{\partial^2 z}{\partial x \partial y}\right)^2 = 0$

(vii) $\Delta = (0)^2 - (2)(2) = -4$

(viii) Since $\Delta < 0$ and $\dfrac{\partial^2 z}{\partial x^2} > 0$, **the stationary point (1, 2) is a minimum**.

The surface $z = (x - 1)^2 + (y - 2)^2$ is shown in three dimensions in Figure 11.8. Looking down towards the x–y plane from above, it is possible to produce a **contour map**. A contour is a line on a map that gives places having the same vertical height above a datum line (usually the mean sea-level on a geographical map). A contour map for $z = (x - 1)^2 + (y - 2)^2$ is shown in Figure 11.9. The values of z are shown on the map and these give an indication of the rise and fall to a stationary point.

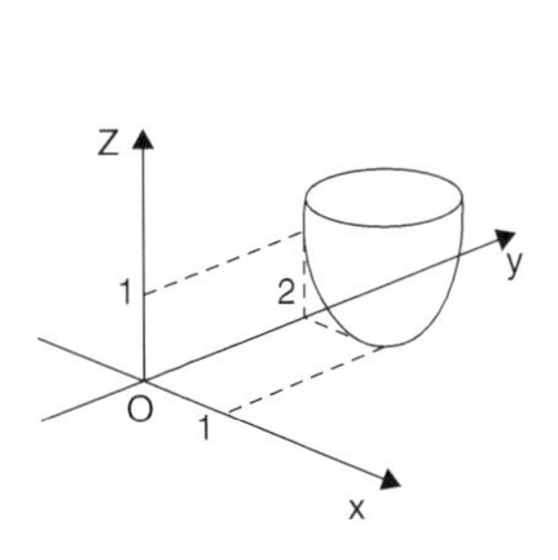

Figure 11.8

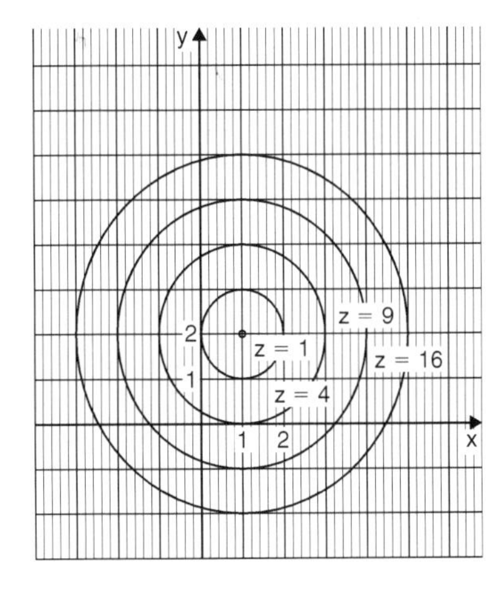

Figure 11.9

Application: Find the co-ordinates of the stationary points on the surface $z = (x^2 + y^2)^2 - 8(x^2 - y^2)$, and distinguish between them

Following the procedure:

(i) $\dfrac{\partial z}{\partial x} = 2(x^2 + y^2)2x - 16x$ and $\dfrac{\partial z}{\partial y} = 2(x^2 + y^2)2y + 16y$

(ii) for stationary points,

$$2(x^2 + y^2)2x - 16x = 0$$

i.e. $$4x^3 + 4xy^2 - 16x = 0 \qquad (1)$$

and $$2(x^2 + y^2)2y + 16y = 0$$

i.e. $$4y(x^2 + y^2 + 4) = 0 \qquad (2)$$

(iii) From equation (1), $y^2 = \dfrac{16x - 4x^3}{4x} = 4 - x^2$

Substituting $y^2 = 4 - x^2$ in equation (2) gives

$$4y(x^2 + 4 - x^2 + 4) = 0$$

i.e. $$32y = 0 \text{ and } y = 0$$

When $y = 0$ in equation (1), $4x^3 - 16x = 0$

i.e. $$4x(x^2 - 4) = 0$$

from which, $$x = 0 \text{ or } x = \pm 2$$

The co-ordinates of the stationary points are (0, 0), (2, 0) and (−2, 0)

(iv) $\dfrac{\partial^2 z}{\partial x^2} = 12x^2 + 4y^2 - 16$, $\dfrac{\partial^2 z}{\partial y^2} = 4x^2 + 12y^2 + 16$

and $\dfrac{\partial^2 z}{\partial x \partial y} = 8xy$

(v) For the point (0, 0), $\dfrac{\partial^2 z}{\partial x^2} = -16$, $\dfrac{\partial^2 z}{\partial y^2} = 16$ and $\dfrac{\partial^2 z}{\partial x \partial y} = 0$

For the point (2, 0), $\dfrac{\partial^2 z}{\partial x^2} = 32$, $\dfrac{\partial^2 z}{\partial y^2} = 32$ and $\dfrac{\partial^2 z}{\partial x \partial y} = 0$

For the point (−2, 0), $\dfrac{\partial^2 z}{\partial x^2} = 32$, $\dfrac{\partial^2 z}{\partial y^2} = 32$ and $\dfrac{\partial^2 z}{\partial x \partial y} = 0$

(vi) $\left(\dfrac{\partial^2 z}{\partial x \partial y}\right)^2 = 0$ for each stationary point

(vii) $\Delta_{(0,0)} = (0)^2 - (-16)(16) = 256$
$\Delta_{(2,0)} = (0)^2 - (32)(32) = -1024$
$\Delta_{(-2,0)} = (0)^2 - (32)(32) = -1024$

(viii) Since $\Delta_{(0,0)} > 0$, **the point (0, 0) is a saddle point**

Since $\Delta_{(2,0)} < 0$ and $\left(\dfrac{\partial^2 z}{\partial x^2}\right)_{(2,0)} > 0$, **the point (2, 0) is a minimum point**

Since $\Delta_{(-2,0)} < 0$ and $\left(\dfrac{\partial^2 z}{\partial x^2}\right)_{(-2,0)} > 0$, **the point (−2, 0) is a minimum point**

Looking down towards the x–y plane from above, an approximate contour map can be constructed to represent the value of z. Such a map is shown in Figure 11.10. To produce a contour map requires a

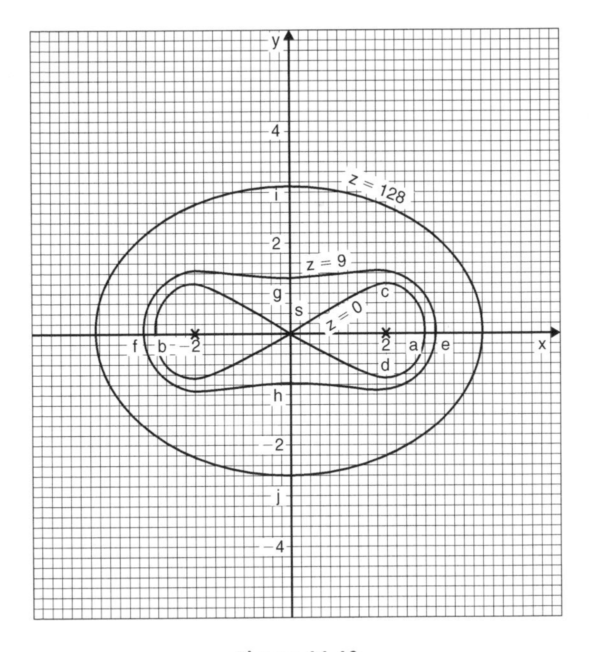

Figure 11.10

large number of x–y co-ordinates to be chosen and the values of z at each co-ordinate calculated. Here are a few examples of points used to construct the contour map.

When z = 0, $0 = (x^2 + y^2)^2 - 8(x^2 - y^2)$

In addition, when, say, $y = 0$ (i.e. on the x-axis)

$$0 = x^4 - 8x^2 \text{ i.e. } x^2(x^2 - 8) = 0$$

from which, $x = 0$ or $x = \pm\sqrt{8}$

Hence the contour $z = 0$ crosses the x-axis at 0 and $\pm\sqrt{8}$, i.e. at co-ordinates (0, 0), (2.83, 0) and (−2.83, 0) shown as points S, a and b respectively.

When z = 0 and x = 2 then $0 = (4 + y^2)^2 - 8(4 - y^2)$

i.e. $0 = 16 + 8y^2 + y^4 - 32 + 8y^2$

i.e. $0 = y^4 + 16y^2 - 16$

Let $y^2 = p$, then $p^2 + 16p - 16 = 0$

and $$p = \frac{-16 \pm \sqrt{16^2 - 4(1)(-16)}}{2} = \frac{-16 \pm 17.89}{2}$$

$$= 0.945 \text{ or } -16.945$$

Hence $y = \sqrt{p} = \sqrt{(0.945)}$ or $\sqrt{(-16.945)} = \pm 0.97$ or complex roots

Hence the $z = 0$ contour passes through the co-ordinates (2, 0.97) and (2, −0.97) shown as c and d in Figure 11.10.

Similarly, for the **z = 9** contour, when $y = 0$,

$$9 = (x^2 + 0^2)^2 - 8(x^2 - 0^2)$$

i.e. $$9 = x^4 - 8x^2$$

i.e. $$x^4 - 8x^2 - 9 = 0$$

Hence $(x^2 - 9)(x^2 + 1) = 0$ from which, $x = \pm 3$ or complex roots

Thus the $z = 9$ contour passes through (3, 0) and (−3, 0), shown as e and f in Figure 11.10.

If $z = 9$ and $x = 0$, $9 = y^4 + 8y^2$

i.e. $$y^4 + 8y^2 - 9 = 0$$

i.e. $$(y^2 + 9)(y^2 - 1) = 0$$

from which, $y = \pm 1$ or complex roots

Thus the $z = 9$ contour also passes through (0, 1) and (0, −1), shown as g and h in Figure 11.10.

When, say, $x = 4$ and $y = 0$, $z = (4^2)^2 - 8(4^2) = 128$

When $z = 128$ and $x = 0$, $128 = y^4 + 8y^2$

i.e. $y^4 + 8y^2 - 128 = 0$

i.e. $(y^2 + 16)(y^2 - 8) = 0$

from which, $y = \pm\sqrt{8}$ or complex roots

Thus the $z = 128$ contour passes through (0, 2.83) and (0, −2.83), shown as i and j in Figure 11.10.

In a similar manner many other points may be calculated with the resulting approximate contour map shown in Figure 11.10. It is seen that two 'hollows' occur at the minimum points, and a 'cross-over' occurs at the saddle point S, which is typical of such contour maps.

Application: An open rectangular container is to have a volume of $62.5\,m^3$. Find the least surface area of material required

Let the dimensions of the container be x, y and z as shown in Figure 11.11.

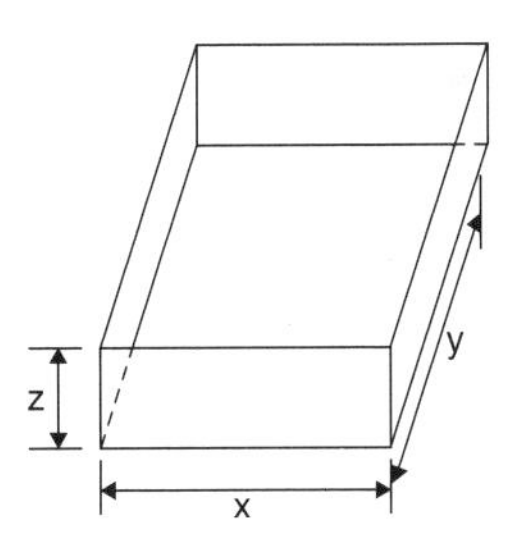

Figure 11.11

$$\text{Volume } V = xyz = 62.5 \qquad (1)$$

$$\text{Surface area, } S = xy + 2yz + 2xz \qquad (2)$$

From equation (1), $z = \dfrac{62.5}{xy}$

Substituting in equation (2) gives:

$$S = xy + 2y\left(\frac{62.5}{xy}\right) + 2x\left(\frac{62.5}{xy}\right)$$

i.e. $S = xy + \dfrac{125}{x} + \dfrac{125}{y}$ which is a function of two variables

$$\frac{\partial S}{\partial x} = y - \frac{125}{x^2} = 0 \text{ for a stationary point, } \quad \text{hence } x^2y = 125 \quad (3)$$

$$\frac{\partial S}{\partial y} = x - \frac{125}{y^2} = 0 \text{ for a stationary point, } \quad \text{hence } xy^2 = 125 \quad (4)$$

Dividing equation (3) by (4) gives: $\dfrac{x^2y}{xy^2} = 1$ i.e. $\dfrac{x}{y} = 1$ i.e. $x = y$

Substituting $y = x$ in equation (3) gives $x^3 = 125$, from which, $x = 5\,\text{m}$.

Hence $y = 5\,\text{m}$ also.

From equation (1), $(5)(5)(z) = 62.5$ from which, $z = \dfrac{62.5}{25} = 2.5\,\text{m}$

$$\frac{\partial^2 S}{\partial x^2} = \frac{250}{x^3}, \quad \frac{\partial^2 S}{\partial y^2} = \frac{250}{y^3} \quad \text{and} \quad \frac{\partial^2 S}{\partial x \partial y} = 1$$

When $x = y = 5$, $\dfrac{\partial^2 S}{\partial x^2} = 2$, $\dfrac{\partial^2 S}{\partial y^2} = 2$ and $\dfrac{\partial^2 S}{\partial x \partial y} = 1$

$\Delta = (1)^2 - (2)(2) = -3$

Since $\Delta < 0$ and $\dfrac{\partial^2 S}{\partial x^2} > 0$, then the surface area S is a **minimum**

Hence the minimum dimensions of the container to have a volume of $62.5\,\text{m}^3$ are **5 m by 5 m by 2.5 m**

From equation (2), **minimum surface area,**

$$\mathbf{S} = (5)(5) + 2(5)(2.5) + 2(5)(2.5) = \mathbf{75\,m^2}$$

12 Integral Calculus and its Applications

12.1 Standard integrals

Table 12.1

$\int ax^n\, dx$	$= \dfrac{ax^{n+1}}{n+1} + c$	(except when $n = -1$)
$\int \cos ax\, dx$	$= \dfrac{1}{a}\sin ax + c$	
$\int \sin ax\, dx$	$= -\dfrac{1}{a}\cos ax + c$	
$\int \sec^2 ax\, dx$	$= \dfrac{1}{a}\tan ax + c$	
$\int \operatorname{cosec}^2 ax\, dx$	$= -\dfrac{1}{a}\cot ax + c$	
$\int \operatorname{cosec} ax \cot ax\, dx$	$= -\dfrac{1}{a}\operatorname{cosec} ax + c$	
$\int \sec ax \tan ax\, dx$	$= \dfrac{1}{a}\sec ax + c$	
$\int e^{ax}\, dx$	$= \dfrac{1}{a}e^{ax} + c$	
$\int \dfrac{1}{x}\, dx$	$= \ln x + c$	

Application: Find $\int 3x^4\, dx$

$$\int 3x^4\, dx = \frac{3x^{4+1}}{4+1} + c = \mathbf{\frac{3}{5}x^5 + c}$$

Application: Find $\int \frac{2}{x^2}\,dx$

$$\int \frac{2}{x^2}\,dx = \int 2x^{-2}\,dx = \frac{2x^{-2+1}}{-2+1} + c = \frac{2x^{-1}}{-1} + c = \mathbf{\frac{-2}{x} + c}$$

Application: Find $\int \sqrt{x}\,dx$

$$\int \sqrt{x}\,dx = \int x^{\frac{1}{2}}\,dx = \frac{x^{\frac{1}{2}+1}}{\frac{1}{2}+1} + c = \frac{x^{\frac{3}{2}}}{\frac{3}{2}} + c = \mathbf{\frac{2}{3}\sqrt{x^3} + c}$$

Application: Find $\int \frac{-5}{9\sqrt[4]{t^3}}\,dt$

$$\int \frac{-5}{9\sqrt[4]{t^3}}\,dt = \int \frac{-5}{9t^{\frac{3}{4}}}\,dt = -\frac{5}{9}\int t^{-\frac{3}{4}}\,dt$$

$$= -\frac{5}{9}\left(\frac{t^{-\frac{3}{4}+1}}{-\frac{3}{4}+1}\right) + c = -\frac{5}{9}\left(\frac{t^{\frac{1}{4}}}{\frac{1}{4}}\right) + c$$

$$= -\left(\frac{5}{9}\right)\left(\frac{4}{1}\right)t^{\frac{1}{4}} + c = \mathbf{-\frac{20}{9}\sqrt[4]{t} + c}$$

Application: Find $\int \frac{(1+\theta)^2}{\sqrt{\theta}}\,d\theta$

$$\int \frac{(1+\theta)^2}{\sqrt{\theta}}\,d\theta = \int \frac{1+2\theta+\theta^2}{\sqrt{\theta}}\,d\theta$$

$$= \int \left(\frac{1}{\theta^{\frac{1}{2}}} + \frac{2\theta}{\theta^{\frac{1}{2}}} + \frac{\theta^2}{\theta^{\frac{1}{2}}}\right)d\theta = \int \left(\theta^{-\frac{1}{2}} + 2\theta^{1-\frac{1}{2}} + \theta^{2-\frac{1}{2}}\right)d\theta$$

$$= \int \left(\theta^{-\frac{1}{2}} + 2\theta^{\frac{1}{2}} + \theta^{\frac{3}{2}}\right)d\theta$$

$$= \frac{\theta^{-\frac{1}{2}+1}}{-\frac{1}{2}+1} + \frac{2\theta^{\frac{1}{2}+1}}{\frac{1}{2}+1} + \frac{\theta^{\frac{3}{2}+1}}{\frac{3}{2}+1} + c = \frac{\theta^{\frac{1}{2}}}{\frac{1}{2}} + \frac{2\theta^{\frac{3}{2}}}{\frac{3}{2}} + \frac{\theta^{\frac{5}{2}}}{\frac{5}{2}} + c$$

$$= 2\theta^{\frac{1}{2}} + \frac{4}{3}\theta^{\frac{3}{2}} + \frac{2}{5}\theta^{\frac{5}{2}} + c = \mathbf{2\sqrt{\theta} + \frac{4}{3}\sqrt{\theta^3} + \frac{2}{5}\sqrt{\theta^5} + c}$$

Application: Find $\int (4\cos 3x - 5\sin 2x)dx$

$$\int (4\cos 3x - 5\sin 2x)\,dx = (4)\left(\frac{1}{3}\right)\sin 3x - (5)\left(-\frac{1}{2}\right)\cos 2x$$

$$= \mathbf{\frac{4}{3}\sin 3x + \frac{5}{2}\cos 2x + c}$$

Application: Find $\int (7\sec^2 4t + 3\,\text{cosec}^2\, 2t)dt$

$$\int (7\sec^2 4t + 3\,\text{cosec}^2\, 2t)dt = (7)\left(\frac{1}{4}\right)\tan 4t + (3)\left(-\frac{1}{2}\right)\cot 2t + c$$

$$= \mathbf{\frac{7}{4}\tan 4t - \frac{3}{2}\cot 2t + c}$$

Application: Find $\int \frac{2}{3e^{4t}}dt$

$$\int \frac{2}{3e^{4t}}dt = \frac{2}{3}\int e^{-4t}\,dt = \left(\frac{2}{3}\right)\left(-\frac{1}{4}\right)e^{-4t} + c$$

$$= -\frac{1}{6}e^{-4t} + c = \mathbf{-\frac{1}{6e^{4t}} + c}$$

Application: Find $\int \frac{3}{5x}dx$

$$\int \frac{3}{5x}dx = \frac{3}{5}\int \frac{1}{x}dx = \mathbf{\frac{3}{5}\ln x + c}$$

Definite Integrals

Application: Evaluate $\int_{-2}^{3}(4-x^2)dx$

$$\int_{-2}^{3}(4-x^2)\,dx = \left[4x - \frac{x^3}{3}\right]_{-2}^{3} = \left(4(3) - \frac{3^3}{3}\right) - \left(4(-2) - \frac{(-2)^3}{3}\right)$$

$$= (12-9) - \left(-8 - \frac{-8}{3}\right) = (3) - \left(-5\frac{1}{3}\right) = \mathbf{8\frac{1}{3}}$$

Application: Evaluate $\int_{0}^{\pi/2} 3\sin 2x\,dx$

$$\int_{0}^{\pi/2} 3\sin 2x\,dx = \left[(3)\left(-\frac{1}{2}\right)\cos 2x\right]_{0}^{\pi/2} = \left[-\frac{3}{2}\cos 2x\right]_{0}^{\pi/2}$$

$$= \left\{-\frac{3}{2}\cos 2\left(\frac{\pi}{2}\right)\right\} - \left\{-\frac{3}{2}\cos 2(0)\right\}$$

$$= \left\{-\frac{3}{2}(-1)\right\} - \left\{-\frac{3}{2}(1)\right\} = \frac{3}{2} + \frac{3}{2} = \mathbf{3}$$

Application: Evaluate $\int_{1}^{2} 4\cos 3t\,dt$

$$\int_{1}^{2} 4\cos 3t\,dt = \left[(4)\left(\frac{1}{3}\right)\sin 3t\right]_{1}^{2} = \left[\frac{4}{3}\sin 3t\right]_{1}^{2} = \left\{\frac{4}{3}\sin 6\right\} - \left\{\frac{4}{3}\sin 3\right\}$$

(Note that limits of trigonometric functions are always expressed in ***radians***, thus, for example, sin 6 means the sine of 6 radians = −0.279415..)

Hence, $$\int_{1}^{2} 4\cos 3t\,dt = \left\{\frac{4}{3}(-0.279415..)\right\} - \left\{\frac{4}{3}(0.141120..)\right\}$$

$$= (-0.37255) - (0.18816) = \mathbf{-0.5607}$$

Application: Evaluate $\int_1^2 4e^{2x}\,dx$

$$\int_1^2 4e^{2x}\,dx = \left[\frac{4}{2}e^{2x}\right]_1^2 = 2[e^{2x}]_1^2 = 2[e^4 - e^2]$$

$$= 2[54.5982 - 7.3891] = \mathbf{94.42}$$

Application: Evaluate $\int_1^4 \frac{3}{4u}\,du$

$$\int_1^4 \frac{3}{4u}\,du = \left[\frac{3}{4}\ln u\right]_1^4 = \frac{3}{4}[\ln 4 - \ln 1] = \frac{3}{4}[1.3863 - 0] = \mathbf{1.040}$$

12.2 Non-standard integrals

Functions that require integrating are not always in the 'standard form' shown above. However, it is often possible to change a function into a form that can be integrated by using either:

1. an algebraic substitution,
2. trigonometric and hyperbolic substitutions,
3. partial fractions,
4. $t = \tan\frac{\theta}{2}$ substitution,
5. integration by parts, or
6. reduction formulae.

12.3 Integration using algebraic substitutions

Application: Determine $\int \cos(3x + 7)\,dx$

$\int \cos(3x + 7)\,dx$ is not a standard integral of the form shown in Table 12.1, page 303, thus an algebraic substitution is made.

Let $u = 3x + 7$ then $\frac{du}{dx} = 3$ and rearranging gives: $dx = \frac{du}{3}$

$$\text{Hence} \int \cos(3x + 7)dx = \int (\cos u)\frac{du}{3}$$

$$= \int \frac{1}{3}\cos u\, du, \text{ which is a standard integral}$$

$$= \frac{1}{3}\sin u + c$$

Rewriting u as (3x + 7) gives: $\int \cos(3x + 7)dx = \mathbf{\frac{1}{3}\sin(3x + 7) + c}$, which may be checked by differentiating it.

Application: Find $\int (2x - 5)^7\, dx$

Let $u = (2x - 5)$ then $\frac{du}{dx} = 2$ and $dx = \frac{du}{2}$

Hence,

$$\int (2x - 5)^7\, dx = \int u^7 \frac{du}{2} = \frac{1}{2}\int u^7 du = \frac{1}{2}\left(\frac{u^8}{8}\right) + c = \frac{1}{16}u^8 + c$$

Rewriting u as (2x − 5) gives: $\int \mathbf{(2x - 5)^7\, dx = \frac{1}{16}(2x - 5)^8 + c}$

Application: Evaluate $\int_0^{\pi/6} 24\sin^5\theta\cos\theta\, d\theta$

Let $u = \sin\theta$ then $\frac{du}{d\theta} = \cos\theta$ and $d\theta = \frac{du}{\cos\theta}$

$$\text{Hence,} \int 24\sin^5\theta\cos\theta\, d\theta = \int 24u^5\cos\theta\frac{du}{\cos\theta}$$

$$= 24\int u^5\, du, \text{ by cancelling}$$

$$= 24\frac{u^6}{6} + c = 4u^6 + c = 4(\sin\theta)^6 + c$$

$$= 4\sin^6\theta + c$$

$$\text{Thus,} \quad \int_0^{\pi/6} 24 \sin^5 \theta \cos \theta \, d\theta = \left[4 \sin^6 \theta\right]_0^{\pi/6}$$

$$= 4\left[\left(\sin \frac{\pi}{6}\right)^6 - (\sin 0)^6\right]$$

$$= 4\left[\left(\frac{1}{2}\right)^6 - 0\right] = \mathbf{\frac{1}{16}} \text{ or } \mathbf{0.0625}$$

Application: Determine $\int \frac{2x}{\sqrt{(4x^2 - 1)}} dx$

Let $u = 4x^2 - 1$ then $\frac{du}{dx} = 8x$ and $dx = \frac{du}{8x}$

Hence $\int \frac{2x}{\sqrt{(4x^2 - 1)}} dx = \int \frac{2x}{\sqrt{u}} \frac{du}{8x} = \frac{1}{4} \int \frac{1}{\sqrt{u}} du$, by cancelling

$$= \frac{1}{4} \int u^{-\frac{1}{2}} = \frac{1}{4}\left[\frac{u^{-\frac{1}{2}+1}}{-\frac{1}{2}+1}\right] + c = \frac{1}{4}\left[\frac{u^{\frac{1}{2}}}{\frac{1}{2}}\right] + c$$

$$= \frac{1}{2}\sqrt{u} + c = \mathbf{\frac{1}{2}\sqrt{4x^2 - 1} + c}$$

Change of limits

When evaluating definite integrals involving substitutions it is sometimes more convenient to **change the limits** of the integral.

Application: Evaluate $\int_1^3 5x\sqrt{2x^2 + 7}dx$, taking positive values of square roots only:

Let $u = 2x^2 + 7$, then $\frac{du}{dx} = 4x$ and $dx = \frac{du}{4x}$

When $x = 3$, $u = 2(3)^2 + 7 = 25$ and when $x = 1$, $u = 2(1)^2 + 7 = 9$

Hence, $\int_{x=1}^{x=3} 5x\sqrt{2x^2+7} = \int_{u=9}^{u=25} 5x\sqrt{u}\,\frac{du}{4x}$

$$= \frac{5}{4}\int_9^{25} \sqrt{u}\,du = \frac{5}{4}\int_9^{25} u^{\frac{1}{2}}du$$

Thus the limits have been changed, and it is unnecessary to change the integral back in terms of x.

Thus, $\int_{x=1}^{x=3} 5x\sqrt{2x^2+7dx} = \frac{5}{4}\left[\frac{u^{3/2}}{3/2}\right]_9^{25} = \frac{5}{6}\left[\sqrt{u^3}\right]_9^{25}$

$$= \frac{5}{6}\left[\sqrt{25^3} - \sqrt{9^3}\right] = \frac{5}{6}(125 - 27) = \mathbf{81\frac{2}{3}}$$

12.4 Integration using trigonometric and hyperbolic substitutions

Table 12.2 Integrals using trigonometric substitutions

f(x)	$\int$ **f(x) dx**	**Method**
1. $\cos^2 x$	$\frac{1}{2}\left(x + \frac{\sin 2x}{2}\right) + c$	Use $\cos 2x = 2\cos^2 x - 1$
2. $\sin^2 x$	$\frac{1}{2}\left(x - \frac{\sin 2x}{2}\right) + c$	Use $\cos 2x = 1 - 2\sin^2 x$
3. $\tan^2 x$	$\tan x - x + c$	Use $1 + \tan^2 x = \sec^2 x$
4. $\cot^2 x$	$-\cot x - x + c$	Use $\cot^2 x + 1 = \mathrm{cosec}^2 x$
5. $\cos^m x \sin^n x$	(a) If either m or n is odd (but not both), use $\cos^2 x + \sin^2 x = 1$ (b) If both m and n are even, use either $\cos 2x = 2\cos^2 x - 1$ or $\cos 2x = 1 - 2\sin^2 x$	
6. sin A cos B		Use $\frac{1}{2}$ [sin(A + B) + sin(A − B)]

Table 12.2 Continued

$f(x)$	$\int f(x)\,dx$	Method
7. $\cos A \sin B$		Use $\frac{1}{2}[\sin(A+B) - \sin(A-B)]$
8. $\cos A \cos B$		Use $\frac{1}{2}[\cos(A+B) + \cos(A-B)]$
9. $\sin A \sin B$		Use $-\frac{1}{2}[\cos(A+B) - \cos(A-B)]$
10. $\frac{1}{\sqrt{(a^2 - x^2)}}$	$\sin^{-1}\frac{x}{a} + c$	Use $x = a\sin\theta$ substitution
11. $\sqrt{a^2 - x^2}$	$\frac{a^2}{2}\sin^{-1}\frac{x}{a} + \frac{x}{2}\sqrt{a^2 - x^2} + c$	Use $x = a\sin\theta$ substitution
12. $\frac{1}{a^2 + x^2}$	$\frac{1}{a}\tan^{-1}\frac{x}{a} + c$	Use $x = a\tan\theta$ substitution
13. $\frac{1}{\sqrt{(x^2 + a^2)}}$	$\sinh^{-1}\frac{x}{a} + c$ or $\ln\left\{\frac{x + \sqrt{(x^2 + a^2)}}{a}\right\} + c$	Use $x = a\sinh\theta$ substitution
14. $\sqrt{(x^2 + a^2)}$	$\frac{a^2}{2}\sinh^{-1}\frac{x}{a} + \frac{x}{2}\sqrt{(x^2 + a^2)} + c$	
15. $\frac{1}{\sqrt{(x^2 - a^2)}}$	$\cosh^{-1}\frac{x}{a} + c$ or $\ln\left\{\frac{x + \sqrt{(x^2 - a^2)}}{a}\right\} + c$	Use $x = a\cosh\theta$ substitution
16. $\sqrt{(x^2 - a^2)}$	$\frac{x}{2}\sqrt{(x^2 - a^2)} - \frac{a^2}{2}\cosh^{-1}\frac{x}{a} + c$	

Application: Evaluate $\int_0^{\pi/4} 2\cos^2 4t\,dt$

Since $\cos 2t = 2\cos^2 t - 1$ (from Chapter 5),

then $\cos^2 t = \frac{1}{2}(1 + \cos 2t)$ and $\cos^2 4t = \frac{1}{2}(1 + \cos 8t)$

Hence $\int_0^{\pi/4} 2\cos^2 4t\,dt = 2\int_0^{\pi/4} \frac{1}{2}(1 + \cos 8t)\,dt = \left[t + \frac{\sin 8t}{8}\right]_0^{\pi/4}$

$$= \left[\frac{\pi}{4} + \frac{\sin 8\left(\frac{\pi}{4}\right)}{8}\right] - \left[0 + \frac{\sin 0}{8}\right]$$

$$= \frac{\pi}{4} \text{ or } \mathbf{0.7854}$$

Application: Find $3\int \tan^2 4x\,dx$

Since $1 + \tan^2 x = \sec^2 x$, then $\tan^2 x = \sec^2 x - 1$ and $\tan^2 4x = \sec^2 4x - 1$

Hence, $3\int \tan^2 4x\,dx = 3\int (\sec^2 4x - 1)\,dx = \mathbf{3\left(\frac{\tan 4x}{4} - x\right) + c}$

Application: Determine $\int \sin^5\theta\,d\theta$

Since $\cos^2\theta + \sin^2\theta = 1$ then $\sin^2\theta = (1 - \cos^2\theta)$

Hence, $\int \sin^5\theta\,d\theta = \int \sin\theta(\sin^2\theta)^2\,d\theta = \int \sin\theta(1 - \cos^2\theta)^2\,d\theta$

$$= \int \sin\theta(1 - 2\cos^2\theta + \cos^4\theta)\,d\theta$$

$$= \int (\sin\theta - 2\sin\theta\cos^2\theta + \sin\theta\cos^4\theta)\,d\theta$$

$$= \mathbf{-\cos\theta + \frac{2\cos^3\theta}{3} - \frac{\cos^5\theta}{5} + c}$$

[Whenever a power of a cosine is multiplied by a sine of power 1, or vice-versa, the integral may be determined by inspection as follows.

In general, $\int \cos^n\theta \sin\theta \, d\theta = \frac{-\cos^{n+1}\theta}{(n+1)} + c$

and $\int \sin^n\theta \cos\theta \, d\theta = \frac{\sin^{n+1}\theta}{(n+1)} + c$

Alternatively, an algebraic substitution may be used.]

Application: Evaluate $\int_0^{\pi/2} \sin^2 x \cos^3 x \, dx$

$$\int_0^{\pi/2} \sin^2 x \cos^3 x \, dx = \int_0^{\pi/2} \sin^2 x \cos^2 x \cos x \, dx$$

$$= \int_0^{\pi/2} \sin^2 x(1 - \sin^2 x)\cos x \, dx$$

$$= \int_0^{\pi/2} (\sin^2 x \cos x - \sin^4 \cos x)dx$$

$$= \left[\frac{\sin^3 x}{3} - \frac{\sin^5 x}{5}\right]_0^{\pi/2}$$

$$= \left[\frac{\left(\sin\frac{\pi}{2}\right)^3}{3} - \frac{\left(\sin\frac{\pi}{2}\right)^5}{5}\right] - [0 - 0]$$

$$= \frac{1}{3} - \frac{1}{5} = \mathbf{\frac{2}{15}} \text{ or } \mathbf{0.1333}$$

Application: Find $\int \sin^2 t \cos^4 t \, dt$

$$\int \sin^2 t \cos^4 t \, dt = \int \sin^2 t(\cos^2 t)^2 \, dt = \int \left(\frac{1 - \cos 2t}{2}\right)\left(\frac{1 + \cos 2t}{2}\right)^2 dt$$

$$= \frac{1}{8}\int (1 - \cos 2t)(1 + 2\cos 2t + \cos^2 2t) \, dt$$

$$= \frac{1}{8}\int (1 + 2\cos 2t + \cos^2 2t - \cos 2t - 2\cos^2 2t - \cos^3 2t)\,dt$$

$$= \frac{1}{8}\int (1 + \cos 2t - \cos^2 2t - \cos^3 2t)\,dt$$

$$= \frac{1}{8}\int \left[1 + \cos 2t - \left(\frac{1 + \cos 4t}{2}\right) - \cos 2t(1 - \sin^2 2t)\right] dt$$

$$= \frac{1}{8}\int \left(\frac{1}{2} - \frac{\cos 4t}{2} + \cos 2t \sin^2 2t\right) dt$$

$$= \mathbf{\frac{1}{8}\left(\frac{t}{2} - \frac{\sin 4t}{8} + \frac{\sin^3 2t}{6}\right) + c}$$

Application: Determine $\int \sin 3t \cos 2t\,dt$

$$\int \sin 3t \cos 2t\,dt = \int \frac{1}{2}[\sin(3t + 2t) + \sin(3t - 2t)]\,dt,$$

from 6 of Table 12.2,

$$= \frac{1}{2}\int (\sin 5t + \sin t)dt = \mathbf{\frac{1}{2}\left(\frac{-\cos 5t}{5} - \cos t\right) + c}$$

Application: Evaluate $\int_0^1 2\cos 6\theta \cos\theta\,d\theta$, correct to 4 decimal places

$$\int_0^1 2\cos 6\theta \cos\theta\,d\theta = 2\int_0^1 \frac{1}{2}[\cos(6\theta + \theta) + \cos(6\theta - \theta)]\,d\theta,$$

from 8 of Table 12.2

$$= \int_0^1 (\cos 7\theta + \cos 5\theta)\,d\theta$$

$$= \left[\frac{\sin 7\theta}{7} + \frac{\sin 5\theta}{5}\right]_0^1$$

$$= \left(\frac{\sin 7}{7} + \frac{\sin 5}{5}\right) - \left(\frac{\sin 0}{7} + \frac{\sin 0}{5}\right)$$

'sin 7' means 'the sine of 7 radians' ($\equiv 401.07°$) and $\sin 5 \equiv 286.48°$

Hence, $\int_0^1 2\cos 6\theta \cos\theta \, d\theta = (0.09386 + -0.19178) - (0)$

$= \mathbf{-0.0979}$, correct to 4 decimal places

Application: Evaluate $\int_0^4 \sqrt{16 - x^2}\, dx$

From 11 of Table 12.2,

$$\int_0^4 \sqrt{16 - x^2}\, dx = \left[\frac{16}{2}\sin^{-1}\frac{x}{4} + \frac{x}{2}\sqrt{(16 - x^2)}\right]_0^4$$

$$= \left[8\sin^{-1}1 + 2\sqrt{0}\right] - [8\sin^{-1}0 + 0]$$

$$= 8\sin^{-1}1 = 8\left(\frac{\pi}{2}\right) = \mathbf{4\pi} \text{ or } \mathbf{12.57}$$

Application: Evaluate $\int_0^2 \frac{1}{(4 + x^2)}\, dx$

From 12 of Table 12.2, $\int_0^2 \frac{1}{(4 + x^2)}\, dx = \frac{1}{2}\left[\tan^{-1}\frac{x}{2}\right]_0^2$ since a = 2

$$= \frac{1}{2}(\tan^{-1}1 - \tan^{-1}0)$$

$$= \frac{1}{2}\left(\frac{\pi}{4} - 0\right)$$

$$= \frac{\boldsymbol{\pi}}{\mathbf{8}} \text{ or } \mathbf{0.3927}$$

Application: Evaluate $\int_0^2 \frac{1}{\sqrt{(x^2 + 4)}}\, dx$, correct to 4 decimal places

$$\int_0^2 \frac{1}{\sqrt{(x^2 + 4)}}\, dx = \left[\sinh^{-1}\frac{x}{2}\right]_0^2 \quad \text{or} \quad \left[\ln\left\{\frac{x + \sqrt{(x^2 + 4)}}{2}\right\}\right]_0^2$$

from 13 of Table 12.2, where a = 2

Using the logarithmic form,

$$\int_0^2 \frac{1}{\sqrt{(x^2+4)}}\,dx = \left[\ln\left(\frac{2+\sqrt{8}}{2}\right) - \ln\left(\frac{0+\sqrt{4}}{2}\right)\right]$$

$$= \ln 2.4142 - \ln 1$$

$$= \mathbf{0.8814}, \text{ correct to 4 decimal places}$$

Application: Determine $\int \frac{2x-3}{\sqrt{(x^2-9)}}\,dx$

$$\int \frac{2x-3}{\sqrt{(x^2+9)}}\,dx = \int \frac{2x}{\sqrt{(x^2-9)}}\,dx - \int \frac{3}{\sqrt{(x^2-9)}}\,dx$$

The first integral is determined using the algebraic substitution $u = (x^2 - 9)$, and the second integral is of the form $\int \frac{1}{\sqrt{(x^2-a^2)}}\,dx$ (see 15 of Table 12.2)

Hence,

$$\int \frac{2x}{\sqrt{(x^2-9)}}\,dx - \int \frac{3}{\sqrt{(x^2-9)}}\,dx = \mathbf{2\sqrt{(x^2-9)} - 3\cosh^{-1}\frac{x}{3} + c}$$

Application: Evaluate $\int_2^3 \sqrt{(x^2-4)}\,dx$

$$\int_2^3 \sqrt{(x^2-4)}\,dx = \left[\frac{x}{2}\sqrt{(x^2-4)} - \frac{4}{2}\cosh^{-1}\frac{x}{2}\right]_2^3$$

from 16 of Table 12.2, when a = 2,

$$= \left(\frac{3}{2}\sqrt{5} - 2\cosh^{-1}\frac{3}{2}\right) - \left(0 - 2\cosh^{-1}1\right)$$

$$= \mathbf{1.429}, \text{ by calculator}$$

or since $\cosh^{-1}\frac{x}{a} = \ln\left\{\frac{x+\sqrt{(x^2-a^2)}}{a}\right\}$

then $\cosh^{-1}\frac{3}{2} = \ln\left\{\frac{3+\sqrt{(3^2-2^2)}}{2}\right\}$

$$= \ln 2.6180 = 0.9624$$

Similarly, $\cosh^{-1} 1 = 0$

Hence, $$\int_2^3 \sqrt{(x^2-4)}\, dx = \left[\frac{3}{2}\sqrt{5} - 2(0.9624)\right] - [0]$$

$$= \mathbf{1.429}, \text{ correct to 4 significant figures}$$

12.5 Integration using partial fractions

1. Linear factors

Application: Determine $\displaystyle\int \frac{11-3x}{x^2+2x-3}\, dx$

As shown on page 42: $$\frac{11-3x}{x^2+2x-3} \equiv \frac{2}{(x-1)} - \frac{5}{(x+3)}$$

Hence $$\int \frac{11-3x}{x^2+2x-3}\, dx = \int \left\{\frac{2}{(x-1)} - \frac{5}{(x+3)}\right\} dx$$

$$= \mathbf{2\ln(x-1) - 5\ln(x+3) + c}$$

(by algebraic substitutions (see section 12.3))

or $\mathbf{\ln\left\{\dfrac{(x-1)^2}{(x+3)^5}\right\} + c}$ by the laws of logarithms

Application: Evaluate $\displaystyle\int_2^3 \frac{x^3-2x^2-4x-4}{x^2+x-2}\, dx$, correct to 4 significant figures

By dividing out and resolving into partial fractions, it was shown on page 43:

$$\frac{x^3-2x^2-4x-4}{x^2+x-2} \equiv x-3+\frac{4}{(x+2)} - \frac{3}{(x-1)}$$

Hence,

$$\int_2^3 \frac{x^3-2x^2-4x-4}{x^2+x-2}\, dx \equiv \int_2^3 \left\{x-3+\frac{4}{(x+2)} - \frac{3}{(x-1)}\right\} dx$$

$$= \left[\frac{x^2}{2} - 3x + 4\ln(x+2) - 3\ln(x-1)\right]_2^3$$

$$= \left(\frac{9}{2} - 9 + 4\ln 5 - 3\ln 2\right) - (2 - 6 + 4\ln 4 - 3\ln 1)$$

$$= \mathbf{-1.687}, \text{ correct to 4 significant figures}$$

---

2. Repeated linear factors

Application: Find $\int \frac{5x^2 - 2x - 19}{(x+3)(x-1)^2}\,dx$

It was shown on page 44:

$$\frac{5x^2 - 2x - 19}{(x+3)(x-1)^2} \equiv \frac{2}{(x+3)} + \frac{2}{(x-1)} - \frac{4}{(x-1)^2}$$

Hence, $\int \frac{5x^2 - 2x - 19}{(x+3)(x-1)^2}\,dx \equiv \int \left\{\frac{2}{(x+3)} + \frac{3}{(x-1)} - \frac{4}{(x-1)^2}\right\} dx$

$$= \mathbf{2\ln(x+3) + 3\ln(x-1) + \frac{4}{(x-1)} + c}$$

$$\text{or } \mathbf{\ln\{(x+3)^2 (x-1)^3\} + \frac{4}{(x-1)} + c}$$

---

3. Quadratic factors

Application: Find $\int \frac{3 + 6x + 4x^2 - 2x^3}{x^2(x^2+3)}\,dx$

It was shown on page 45: $\frac{3 + 6x + 4x^2 - 2x^2}{x^2(x^2+3)} \equiv \frac{2}{x} + \frac{1}{x^2} + \frac{3 - 4x}{(x^2+3)}$

Thus,

$$\int \frac{3+6x+4x^2-2x^3}{x^2(x^2+3)}\,dx \equiv \int \left(\frac{2}{x}+\frac{1}{x^2}+\frac{3-4x}{(x^2+3)}\right)dx$$

$$= \int \left\{\frac{2}{x}+\frac{1}{x^2}+\frac{3}{(x^2+3)}-\frac{4x}{(x^2+3)}\right\}dx$$

$$\int \frac{3}{(x^2+3)}\,dx = 3\int \frac{1}{x^2+\left(\sqrt{3}\right)^2}\,dx = \frac{3}{\sqrt{3}}\tan^{-1}\frac{x}{\sqrt{3}},$$

from 12, Table 12.2, page 311.

$\int \frac{4x}{x^2+3}\,dx$ is determined using the algebraic substitution $u = (x^2+3)$

Hence, $$\int \left\{\frac{2}{x}+\frac{1}{x^2}+\frac{3}{(x^2+3)}-\frac{4x}{(x^2+3)}\right\}dx$$

$$= \mathbf{2\ln x - \frac{1}{x} + \frac{3}{\sqrt{3}}\tan^{-1}\frac{x}{\sqrt{3}} - 2\ln(x^2+3) + c}$$

$$\text{or } \mathbf{\ln\left(\frac{x}{x^2+3}\right)^2 - \frac{1}{x} + \sqrt{3}\tan^{-1}\frac{x}{\sqrt{3}} + c}$$

12.6 The $t = \tan\frac{\theta}{2}$ substitution

To determine $\int \frac{1}{a\cos\theta + b\sin\theta + c}\,d\theta$, where a, b and c are constants, if $t = \tan\frac{\theta}{2}$ then:

$$\mathbf{\sin\theta = \frac{2t}{(1+t^2)}} \qquad (1)$$

$$\mathbf{\cos\theta = \frac{1-t^2}{1+t^2}} \qquad (2)$$

$$\mathbf{d\theta = \frac{2dt}{1+t^2}} \qquad (3)$$

Application: Determine $\int \frac{d\theta}{\sin\theta}$

If $t = \tan\frac{\theta}{2}$ then $\sin\theta = \frac{2t}{1+t^2}$ and $d\theta = \frac{2dt}{1+t^2}$ from equations (1) and (3).

Thus, $$\int \frac{d\theta}{\sin\theta} = \int \frac{1}{\frac{2t}{1+t^2}}\left(\frac{2dt}{1+t^2}\right) = \int \frac{1}{t}\,dt = \ln t + c$$

Hence, $$\boldsymbol{\int \frac{d\theta}{\sin\theta} = \ln\left(\tan\frac{\theta}{2}\right) + c}$$

Application: Determine $\int \frac{dx}{\cos x}$

If $\tan\frac{x}{2}$ then $\cos x = \frac{1-t^2}{1+t^2}$ and $dx = \frac{2dt}{1+t^2}$ from equations (2) and (3).

Thus $$\int \frac{dx}{\cos x} = \int \frac{1}{\frac{1-t^2}{1+t^2}}\left(\frac{2dt}{1+t^2}\right) = \int \frac{2}{1-t^2}\,dt$$

$\frac{2}{1-t^2}$ may be resolved into partial fractions (see section 2.10)

Let $$\frac{2}{1-t^2} = \frac{2}{(1-t)(1+t)} = \frac{A}{(1-t)} + \frac{B}{(1+t)} = \frac{A(1+t)+B(1-t)}{(1-t)(1+t)}$$

Hence $2 = A(1+t) + B(1-t)$

When $t = 1$, $2 = 2A$, from which, $A = 1$

When $t = -1$, $2 = 2B$, from which, $B = 1$

Hence, $$\int \frac{2}{1-t^2} = \int \left(\frac{1}{1-t} + \frac{1}{1+t}\right) dt = -\ln(1-t) + \ln(1+t) + c$$
$$= \ln\left\{\frac{(1+t)}{(1-t)}\right\} + c$$

Thus, $$\int \frac{dx}{\cos x} = \ln\left\{\frac{1+\tan\frac{x}{2}}{1-\tan\frac{x}{2}}\right\} + c$$

Note that since $\tan\frac{\pi}{4} = 1$, the above result may be written as:

$$\int \frac{dx}{\cos x} = \ln\left\{\frac{\tan\frac{\pi}{4}+\tan\frac{x}{2}}{\tan\frac{\pi}{4}-\tan\frac{x}{2}}\right\} + c = \ln\left\{\tan\left(\frac{\pi}{4}+\frac{x}{2}\right)\right\} + c$$

from compound angles, chapter 5.

Application: Determine $\int \frac{d\theta}{5+4\cos\theta}$

If $t = \tan\frac{\theta}{2}$ then $\cos\theta = \frac{1-t^2}{1+t^2}$ and $d\theta = \frac{2\,dt}{1+t^2}$ from equations (2) and (3).

Thus, $$\int \frac{d\theta}{5+4\cos\theta} = \int \frac{1}{5+4\left(\frac{1-t^2}{1+t^2}\right)}\left(\frac{2\,dt}{1+t^2}\right)$$

$$= \int \frac{1}{\frac{5(1+t^2)+4(1-t^2)}{1+t^2}}\left(\frac{2\,dt}{(1+t^2)}\right)$$

$$= 2\int \frac{dt}{t^2+9} = 2\int \frac{dt}{t^2+3^2} = 2\left(\frac{1}{3}\tan^{-1}\frac{t}{3}\right) + c$$

Hence, $$\int \frac{d\theta}{5+4\cos\theta} = \frac{2}{3}\tan^{-1}\left(\frac{1}{3}\tan\frac{\theta}{2}\right) + c$$

Application: Determine $\int \frac{dx}{\sin x + \cos x}$

If $t = \tan\frac{x}{2}$ then $\sin x = \frac{2t}{1+t^2}$, $\cos x = \frac{1-t^2}{1+t^2}$ and $dx = \frac{2\,dt}{1+t^2}$ from equations (1), (2) and (3).

Thus, $$\int \frac{dx}{\sin x + \cos x} = \int \frac{\frac{2\,dt}{1+t^2}}{\left(\frac{2t}{1+t^2}\right) + \left(\frac{1-t^2}{1+t^2}\right)} = \int \frac{\frac{2\,dt}{1+t^2}}{\frac{2t+1-t^2}{1+t^2}}$$

$$= \int \frac{2\,dt}{1+2t-t^2} = \int \frac{-2\,dt}{t^2-2t-1}$$

$$= \int \frac{-2\,dt}{(t-1)^2 - 2} = \int \frac{2\,dt}{\left(\sqrt{2}\right)^2 - (t-1)^2}$$

$$= 2\left[\frac{1}{2\sqrt{2}} \ln\left\{\frac{\sqrt{2}+(t-1)}{\sqrt{2}-(t-1)}\right\}\right] + c$$

by using partial fractions $\int \frac{1}{a^2 - x^2}\,dx = \frac{1}{2a}\ln\left(\frac{a+x}{a-x}\right)$

i.e. $$\int \frac{dx}{\sin x + \cos x} = \mathbf{\frac{1}{\sqrt{2}} \ln\left\{\frac{\sqrt{2} - 1 + \tan\frac{x}{2}}{\sqrt{2} + 1 - \tan\frac{x}{2}}\right\} + c}$$

Application: Determine $\int \frac{dx}{7 - 3\sin x + 6\cos x}$

From equations (1) to (3),

$$\int \frac{dx}{7 - 3\sin x + 6\cos x} = \int \frac{\frac{2\,dt}{1+t^2}}{7 - 3\left(\frac{2t}{1+t^2}\right) + 6\left(\frac{1-t^2}{1+t^2}\right)}$$

$$= \int \frac{\frac{2\,dt}{1+t^2}}{\frac{7(1+t^2) - 3(2t) + 6(1-t^2)}{1+t^2}}$$

$$= \int \frac{2\,dt}{7 + 7t^2 - 6t + 6 - 6t^2}$$

$$= \int \frac{2\,dt}{t^2 - 6t + 13} = \int \frac{2\,dt}{(t-3)^2 + 2^2}$$

$$= 2\left[\frac{1}{2}\tan^{-1}\left(\frac{t-3}{2}\right)\right] + c$$

from 12 of Table 12.2, page 311.

Hence, $$\int \frac{dx}{7 - 3\sin x + 6\cos x} = \mathbf{\tan^{-1}\left(\frac{\tan\frac{x}{2} - 3}{2}\right) + c}$$

12.7 Integration by parts

If u and v are both functions of x, then:

$$\int u\frac{dv}{dx}\,dx = \int uv - \int v\frac{du}{dx}\,dx$$

or

$$\int \mathbf{u\,dv = uv - \int v\,du}$$

This is known as the **integration by parts formula**.

Application: Determine $\int x\cos x\,dx$

From the integration by parts formula, $\int u dv = uv - \int v du$

Let $u = x$, from which $\frac{du}{dx} = 1$, i.e. $du = dx$

and let $dv = \cos x\,dx$, from which $v = \int \cos x\,dx = \sin x$

Expressions for u, du, v and dv are now substituted into the 'by parts' formula as shown below.

$\int$ u	dv	=	u	v	$- \int$ v	du
$\int$ x	cos x dx	=	(x)	(sin x)	$- \int$ (sin x)	(dx)

i.e. $\int x\cos x\,dx = x\sin x - (-\cos x) + c = \mathbf{x\sin x + \cos x + c}$

[This result may be checked by differentiating the right-hand side, i.e.

$$\frac{d}{dx}(x \sin x + \cos x + c) = [(x)(\cos x) + (\sin x)(1)] - \sin x + 0$$

using the product rule

$$= x \cos x,$$

which is the function being integrated]

Application: Find $\int 3t\, e^{2t}\, dt$

Let $u = 3t$, from which, $\frac{du}{dt} = 3$, i.e. $du = 3dt$

and let $dv = e^{2t}\, dt$, from which, $v = \int e^{2t} dt = \frac{1}{2} e^{2t}$

Substituting into $\int u dv = uv - \int v du$ gives:

$$\int 3t\, e^{2t} dt = (3t)\left(\frac{1}{2}e^{2t}\right) - \int \left(\frac{1}{2}e^{2t}\right)(3\, dt) = \frac{3}{2} t\, e^{2t} - \frac{3}{2}\int e^{2t} dt$$

$$= \frac{3}{2} t\, e^{2t} - \frac{3}{2}\left(\frac{e^{2t}}{2}\right) + c$$

Hence, $\boldsymbol{\int 3t\, e^{2t} dt = \frac{3}{2} e^{2t}\left(t - \frac{1}{2}\right) + c}$, which may be checked by differentiating

Application: Evaluate $\int_0^{\pi/2} 2\theta \sin\theta\, d\theta$

Let $u = 2\theta$, from which, $\frac{du}{d\theta} = 2$, i.e. $du = 2d\theta$

and let $dv = \sin\theta\, d\theta$, from which, $v = \int \sin\theta\, d\theta = -\cos\theta$

Substituting into $\int u\, dv = uv - \int v\, du$ gives:

$$\int 2\theta \sin\theta\, d\theta = (2\theta)(-\cos\theta) - \int (-\cos\theta)(2\, d\theta)$$

$$= -2\theta\cos\theta + 2\int \cos\theta\, d\theta = -2\theta\cos\theta + 2\sin\theta + c$$

Hence, $\int_0^{\pi/2} 2\theta \sin\theta \, d\theta = \left[-2\theta\cos\theta + 2\sin\theta\right]_0^{\pi/2}$

$$= \left[-2\left(\frac{\pi}{2}\right)\cos\frac{\pi}{2} + 2\sin\frac{\pi}{2}\right] - [0 + 2\sin 0]$$

$$= (-0 + 2) - (0 + 0) = \mathbf{2}$$

since $\cos\frac{\pi}{2} = 0$ and $\sin\frac{\pi}{2} = 1$

Application: Determine $\int x^2 \sin x \, dx$

Let $u = x^2$, from which, $\frac{du}{dx} = 2x$, i.e. $du = 2x\,dx$,

and let $dv = \sin x\,dx$, from which, $v = \int \sin x\,dx = -\cos x$

Substituting into $\int u\,dv = uv - \int v\,du$ gives:

$$\int x^2 \sin x\,dx = (x^2)(-\cos x) - \int (-\cos x)(2x\,dx)$$

$$= -x^2\cos x + 2\left[\int x\cos x\,dx\right]$$

The integral, $\int x\cos x\,dx$, is not a 'standard integral' and it can only be determined by using the integration by parts formula again.

From the first application, page 323, $\int x\cos x\,dx\,x = x\sin x + \cos x$

Hence, $\int x^2\sin x\,dx = -x^2\cos x + 2\{x\sin x + \cos x\} + c$

$$= -x^2\cos x + 2x\sin x + 2\cos x + c$$

$$= \mathbf{(2 - x^2)\cos x + 2x\sin x + c}$$

In general, if the algebraic term of a product is of power n, then the integration by parts formula is applied n times.

Application: Find $\int x \ln x\,dx$

The logarithmic function is chosen as the 'u part'

Thus, when $u = \ln x$, then $\frac{du}{dx} = \frac{1}{x}$, i.e. $du = \frac{dx}{x}$

Letting $dv = x\,dx$ gives $v = \int x\,dx = \frac{x^2}{2}$

Substituting into $\int u\,dv = uv - \int v\,du$ gives:

$$\int x \ln x\,dx = (\ln x)\left(\frac{x^2}{2}\right) - \int\left(\frac{x^2}{2}\right)\frac{dx}{x}$$

$$= \frac{x^2}{2}\ln x - \frac{1}{2}\int x\,dx = \frac{x^2}{2}\ln x - \frac{1}{2}\left(\frac{x^2}{2}\right) + c$$

Hence, $\int x \ln x\,dx = \mathbf{\frac{x^2}{2}\left(\ln x - \frac{1}{2}\right) + c}$ or $\mathbf{\frac{x^2}{4}(2\ln x - 1) + c}$

12.8 Reduction formulae

$$\int x^n e^x\,dx = I_n = x^n e^x - n\,I_{n-1} \tag{4}$$

$$\int x^n \cos x\,dx = I_n = x^n \sin x + nx^{n-1}\cos x - n(n-1)I_{n-2} \tag{5}$$

$$\int_0^{\pi} x^n \cos x\,dx = I_n = -n\pi^{n-1} - n(n-1)I_{n-2} \tag{6}$$

$$\int x^n \sin x\,dx = I_n = -x^n \cos x + nx^{n-1}\sin x - n(n-1)I_{n-2} \tag{7}$$

$$\int \sin^n x\,dx = I_n = -\frac{1}{n}\sin^{n-1} x \cos x + \frac{n-1}{n}I_{n-2} \tag{8}$$

$$\int \cos^n x\,dx = I_n = \frac{1}{n}\cos^{n-1} \sin x + \frac{n-1}{n}I_{n-2} \tag{9}$$

$$\int_0^{\pi/2} \sin^n x\,dx = \int_0^{\pi/2} \cos^n x\,dx = I_n = \frac{n-1}{n}I_{n-2} \tag{10}$$

$$\int \tan^n x\,dx = I_n = \frac{\tan^{n-1} x}{n-1} - I_{n-2} \tag{11}$$

$$\int (\ln x)^n\,dx = I_n = x(\ln x)^n - n\,I_{n-1} \tag{12}$$

When using integration by parts, an integral such as $\int x^2e^x\,dx$ requires integration by parts twice. Similarly, $\int x^3e^x\,dx$ requires integration by parts three times. Thus, integrals such as $\int x^5e^x\,dx$, $\int x^6\cos x\,dx$ and $\int x^8\sin 2x\,dx$ for example, would take a long time to determine using integration by parts. **Reduction formulae** provide a quicker method for determining such integrals.

Integrals of the form $\int x^ne^x\,dx$

Application: Determine $\int x^3e^x\,dx$ using a reduction formula

From equation (4), $I_n = x^ne^x - nI_{n-1}$

Hence $\int x^3e^x\,dx = I_3 = x^3e^x - 3I_2$

$$I_2 = x^2e^x - 2I_1$$

$$I_1 = x^1e^x - 1I_0$$

and $I_0 = \int x^0e^x\,dx = \int e^x\,dx = e^x$

Thus
$$\int x^3e^x\,dx = x^3e^x - 3[x^2e^x - 2I_1]$$
$$= x^3e^x - 3[x^2e^x - 2(xe^x - I_0)]$$
$$= x^3e^x - 3[x^2e^x - 2(xe^x - e^x)]$$
$$= x^3e^x - 3x^2e^x + 6(xe^x - e^x)$$
$$= x^3e^x - 3x^2e^x + 6xe^x - 6e^x$$

i.e. $\boldsymbol{\int x^3e^x\,dx = e^x(x^3 - 3x^2 + 6x - 6) + c}$

Integrals of the form $\int x^n\cos x\,dx$

Application: Determine $\int x^2\cos x\,dx$ using a reduction formula

Using the reduction formula of equation (5):

$$\int x^2 \cos x\, dx = I_2 = x^2 \sin x + 2x^1 \cos x - 2(1)I_0$$

and $$I_0 = \int x^0 \cos x\, dx = \int \cos x\, dx = \sin x$$

Hence $$\int \mathbf{x^2 \cos x\, dx = x^2 \sin x + 2x \cos x - 2 \sin x + c}$$

Application: Evaluate $\int_1^2 4t^3 \cos t\, dt$, correct to 4 significant figures

From equation (5),

$$\int t^3 \cos t\, dt = I_3 = t^3 \sin t + 3t^2 \cos t - 3(2)I_1$$

and $$I_1 = t^1 \sin t + 1t^0 \cos t - 1(0)I_{n-2}$$
$$= t \sin t + \cos t$$

Hence $$\int t^3 \cos t\, dt = t^3 \sin t + 3t^2 \cos t - 3(2)\,[t \sin t + \cos t]$$
$$= t^3 \sin t + 3t^2 \cos t - 6t \sin t - 6 \cos t$$

Thus, $$\int_1^2 4t^3 \cos t\, dt = \left[4\,(t^3 \sin t + 3t^2 \cos t - 6t \sin t - 6 \cos t)\right]_1^2$$
$$= [4\,(8 \sin 2 + 12 \cos 2 - 12 \sin 2 - 6 \cos 2)]$$
$$- [4\,(\sin 1 + 3 \cos 1 - 6 \sin 1 - 6 \cos 1)]$$
$$= (-24.53628) - (-23.31305)$$
$$= \mathbf{-1.223}$$

Integrals of the form $\int x^n \sin x\, dx$

Application: Determine $\int x^3 \sin x\, dx$ using a reduction formula

Using equation (7),

$$\int x^3 \sin x\, dx = I_3 = -x^3 \cos x + 3x^2 \sin x - 3(2)I_1$$

and $$I_1 = -x^1 \cos x + 1x^0 \sin x = -x \cos x + \sin x$$

Hence,

$$\int x^3 \sin x\,dx = -x^3\cos x + 3x^2 \sin x - 6[-x\cos x + \sin x]$$

$$= \mathbf{-x^3\cos x + 3x^2\sin x + 6x\cos x - 6\sin x + c}$$

Integrals of the form $\int \sin^n x\,dx$

Application: Determine $\int \sin^4 x\,dx$ using a reduction formula

Using equation (8), $\int \sin^4 x\,dx = I_4 = -\frac{1}{4}\sin^3 x\cos x + \frac{3}{4}I_2$

$I_2 = -\frac{1}{2}\sin^1 x\cos x + \frac{1}{2}I_0$ and $I_0 = \int \sin^0 x\,dx = \int 1\,dx = x$

Hence

$$\int \sin^4 x\,dx = I_4 = -\frac{1}{4}\sin^3 x\cos x + \frac{3}{4}\left[-\frac{1}{2}\sin x\cos x + \frac{1}{2}(x)\right]$$

$$= \mathbf{-\frac{1}{4}\sin^3 x\cos x - \frac{3}{8}\sin x\cos x + \frac{3}{8}x + c}$$

Integrals of the form $\int \cos^n x\,dx$

Application: Determine $\int \cos^4 x\,dx$ using a reduction formula

Using equation (9), $\int \cos^4 x\,dx = I_4 = \frac{1}{4}\cos^3 x\sin x + \frac{3}{4}I_2$

and $I_2 = \frac{1}{2}\cos x\sin x + \frac{1}{2}I_0$ and $I_0 = \int \cos^0 x\,dx = \int 1\,dx = x$

Hence, $\int \cos^4 x\,dx = \frac{1}{4}\cos^3 x\sin x + \frac{3}{4}\left(\frac{1}{2}\cos x\sin x + \frac{1}{2}x\right)$

$$= \mathbf{\frac{1}{4}\cos^3 x\sin x + \frac{3}{8}\cos x\sin x + \frac{3}{8}x + c}$$

Application: Evaluate $\int_0^{\pi/2} \cos^5 x\, dx$

From equation (10), $\int_0^{\pi/2} \cos^n x\, dx = I_n = \frac{n-1}{n} I_{n-2}$ (This is usually known as **Wallis's formula**)

Thus, $\int_0^{\pi/2} \cos^5 x\, dx = \frac{4}{5} I_3$

$I_3 = \frac{2}{3} I_1$ and $I_1 = \int_0^{\pi/2} \cos^1 x\, dx = [\sin x]_0^{\pi/2} = (1 - 0) = 1$

Hence $\int_0^{\pi/2} \cos^5 x\, dx = \frac{4}{5} I_3 = \frac{4}{5}\left[\frac{2}{3} I_1\right] = \frac{4}{5}\left[\frac{2}{3}(1)\right] = \mathbf{\frac{8}{15}}$

Further reduction formulae

Application: Determine $\int \tan^7 x\, dx$

From equation (11), $I_n = \frac{\tan^{n-1} x}{n-1} - I_{n-2}$

When n = 7, $I_7 = \int \tan^7 x\, dx = \frac{\tan^6 x}{6} - I_5$

$I_5 = \frac{\tan^4 x}{4} - I_3$ and $I_3 = \frac{\tan^2 x}{2} - I_1$

$I_1 = \int \tan x\, dx = \ln(\sec x)$ using $\tan x = \frac{\sin x}{\cos x}$ and letting $u = \cos x$

Thus $\int \tan^7 x\, dx = \frac{\tan^6 x}{6} - \left[\frac{\tan^4 x}{4} - \left(\frac{\tan^2 x}{2} - (\ln(\sec x))\right)\right]$

Hence,

$$\int \tan^7 x\, dx = \frac{1}{6}\tan^6 x - \frac{1}{4}\tan^4 x + \frac{1}{2}\tan^2 x - \ln(\sec x) + c$$

Application: Evaluate $\int_0^{\pi/2} \sin^2 t \cos^6 t\,dt$ using a reduction formula

$$\int_0^{\pi/2} \sin^2 t \cos^6 t\,dt = \int_0^{\pi/2} (1-\cos^2 t)\cos^6 t\,dt$$
$$= \int_0^{\pi/2} \cos^6 t\,dt - \int_0^{\pi/2} \cos^8 t\,dt$$

If $I_n = \int_0^{\pi/2} \cos^n t\,dt$ then $\int_0^{\pi/2} \sin^2 t \cos^6 t\,dt = I_6 - I_8$

and from equation (10), $I_6 = \frac{5}{6} I_4 = \frac{5}{6}\left[\frac{3}{4} I_2\right] = \frac{5}{6}\left[\frac{3}{4}\left(\frac{1}{2} I_0\right)\right]$

and $\quad I_0 = \int_0^{\pi/2} \cos^0 t\,dt = \int_0^{\pi/2} 1\,dt = [x]_0^{\pi/2} = \frac{\pi}{2}$

Hence $\quad I_6 = \frac{5}{6} \cdot \frac{3}{4} \cdot \frac{1}{2} \cdot \frac{\pi}{2} = \frac{15\pi}{96}$ or $\frac{5\pi}{32}$

Similarly, $\quad I_8 = \frac{7}{8} I_6 = \frac{7}{8} \cdot \frac{5\pi}{32}$

Thus $\int_0^{\pi/2} \sin^2 t \cos^6 t\,dt = I_6 - I_8 = \frac{5\pi}{32} - \frac{7}{8} \cdot \frac{5\pi}{32} = \frac{1}{8} \cdot \frac{5\pi}{32} = \boldsymbol{\frac{5\pi}{256}}$

12.9 Numerical integration

The trapezoidal rule states:

$$\int_a^b y\,dx \approx \left(\begin{array}{c}\text{width of}\\ \text{interval}\end{array}\right)\left\{\frac{1}{2}\left(\begin{array}{c}\text{first + last}\\ \text{ordinate}\end{array}\right) + \left(\begin{array}{c}\text{sum of remaining}\\ \text{ordinates}\end{array}\right)\right\} \quad (13)$$

The mid-ordinate rule states:

$$\int_a^b y\,dx \approx \text{(width of interval)(sum of mid-ordinates)} \quad (14)$$

Simpson's rule states:

$$\int_a^b y\,dx \approx \frac{1}{3}\left(\begin{array}{c}\text{width of}\\ \text{interval}\end{array}\right)\left\{\left[\left(\begin{array}{c}\text{first + last}\\ \text{ordinate}\end{array}\right)+4\left(\begin{array}{c}\text{sum of even}\\ \text{ordinates}\end{array}\right)\right.\right.$$
$$\left.\left.+2\left(\begin{array}{c}\text{sum of remaining}\\ \text{odd ordinates}\end{array}\right)\right]\right\} \qquad (15)$$

Application: Using the trapezoidal rule with 8 intervals, evaluate $\int_1^3 \frac{2}{\sqrt{x}}\,dx$, correct to 3 decimal places

With 8 intervals, the width of each is $\frac{3-1}{8}$ i.e. 0.25 giving ordinates at 1.00, 1.25, 1.50, 1.75, 2.00, 2.25, 2.50, 2.75 and 3.00. Corresponding values of $\frac{2}{\sqrt{x}}$ are shown in the table below.

x	1.00	1.25	1.50	1.75	2.00	2.25	2.50	2.75	3.00
$\frac{2}{\sqrt{x}}$	2.0000	1.7889	1.6330	1.5119	1.4142	1.3333	1.2649	1.2060	1.1547

From equation (13):

$$\int_1^3 \frac{2}{\sqrt{x}}\,dx \approx (0.25)\left\{\begin{array}{l}\frac{1}{2}(2.000+1.1547)+1.7889\\ +1.6330+1.5119+1.4142\\ +1.3333+1.2649+1.2060\end{array}\right\}$$

$$= \mathbf{2.932}, \text{ correct to 3 decimal places}$$

The greater the number of intervals chosen (i.e. the smaller the interval width) the more accurate will be the value of the definite integral. The exact value is found when the number of intervals is infinite, which is, of course, what the process of integration is based upon. Using integration:

$$\int_1^3 \frac{2}{\sqrt{x}}\,dx = \int_1^3 2x^{-\frac{1}{2}}\,dx$$

$$= \left[\frac{2x^{(-1/2)+1}}{-\frac{1}{2}+1}\right]_1^3 = \left[4x^{1/2}\right]_1^3$$

$$= 4\left[\sqrt{x}\right]_1^3 = 4\left[\sqrt{3}-\sqrt{1}\right]$$

$= \mathbf{2.928}$, correct to 3 decimal places

Application: Using the trapezoidal rule, evaluate $\int_0^{\pi/2} \frac{1}{1+\sin x}\,dx$ using 6 intervals

With 6 intervals, each will have a width of $\frac{\frac{\pi}{2}-0}{6}$, i.e. $\frac{\pi}{12}$ rad (or 15°) and the ordinates occur at 0, $\frac{\pi}{12}, \frac{\pi}{6}, \frac{\pi}{4}, \frac{\pi}{3}, \frac{5\pi}{12}$ and $\frac{\pi}{2}$

Corresponding values of $\frac{1}{1+\sin x}$ are shown in the table below.

x	0	$\frac{\pi}{12}$ (or 15°)	$\frac{\pi}{6}$ (or 30°)	$\frac{\pi}{4}$ (or 45°)
$\frac{1}{1+\sin x}$	1.0000	0.79440	0.66667	0.58579
x	$\frac{\pi}{3}$ (or 60°)	$\frac{5\pi}{12}$ (or 75°)	$\frac{\pi}{2}$ (or 90°)	
$\frac{1}{1+\sin x}$	0.53590	0.50867	0.50000	

From equation (13):

$$\int_0^{\pi/2} \frac{1}{1+\sin x}\,dx \approx \left(\frac{\pi}{12}\right)\left\{\begin{array}{l}\frac{1}{2}(1.00000+0.50000)+0.79440 \\ \quad +0.66667+0.58579 \\ \quad +0.53590+0.50867\end{array}\right\}$$

$= \mathbf{1.006}$, correct to 4 significant figures

Application: Using the mid-ordinate rule with 8 intervals, evaluate $\int_1^3 \frac{2}{\sqrt{x}}\,dx$, correct to 3 decimal places

With 8 intervals, each will have a width of 0.25 and the ordinates will occur at 1.00, 1.25, 1.50, 1.75, ... and thus mid-ordinates at 1.125, 1.375, 1.625,1.875......

Corresponding values of $\frac{2}{\sqrt{x}}$ are shown in the following table.

x	1.125	1.375	1.625	1.875	2.125	2.375	2.625	2.875
$\frac{2}{\sqrt{x}}$	1.8856	1.7056	1.5689	1.4606	1.3720	1.2978	1.2344	1.1795

From equation (14):

$$\int_1^3 \frac{2}{\sqrt{x}}\,dx \approx (0.25)[1.8856+1.7056+1.5689+1.4606+1.3720 +1.2978+1.2344+1.1795]$$

$= \mathbf{2.926}$, correct to 3 decimal places

As previously, the greater the number of intervals the nearer the result is to the true value (of 2.928, correct to 3 decimal places).

Application: Using Simpson's rule with 8 intervals, evaluate $\int_1^3 \frac{2}{\sqrt{x}}\,dx$, correct to 3 decimal places:

With 8 intervals, each will have a width of $\frac{3-1}{8}$, i.e. 0.25 and the ordinates occur at 1.00, 1.25, 1.50, 1.75, ..., 3.0. The values of the ordinates are as shown in the table above

Thus, from equation (15):

$$\int_1^3 \frac{2}{\sqrt{x}}\,dx \approx \frac{1}{3}(0.25)[(2.0000 + 1.1547) + 4(1.7889 + 1.5119 + 1.3333 + 1.2060) + 2(1.6330 + 1.4142 + 1.2649)]$$

$$= \frac{1}{3}(0.25)[3.1547 + 23.3604 + 8.6242]$$

$$= \mathbf{2.928}, \text{ correct to 3 decimal places}$$

It is noted that the latter answer is exactly the same as that obtained by integration. In general, Simpson's rule is regarded as the most accurate of the three approximate methods used in numerical integration.

Application: An alternating current i has the following values at equal intervals of 2.0 milliseconds.

Time (ms)	0	2.0	4.0	6.0	8.0	10.0	12.0
Current i (A)	0	3.5	8.2	10.0	7.3	2.0	0

Charge, q, in millicoulombs, is given by $q = \int_0^{12.0} i\,dt$. Use Simpson's rule to determine the approximate charge in the 12 millisecond period

From equation (15):

$$\text{Charge, } q = \int_0^{12.0} i\,dt \approx \frac{1}{3}(2.0)[(0 + 0) + 4(3.5 + 10.0 + 2.0) + 2(8.2 + 7.3)]$$

$$= \mathbf{62\,mC}$$

12.10 Area under and between curves

The area shown shaded in Figure 12.1 is given by:

total shaded area $= \int_a^b f(x)\,dx - \int_b^c f(x)\,dx + \int_c^d f(x)\,dx$

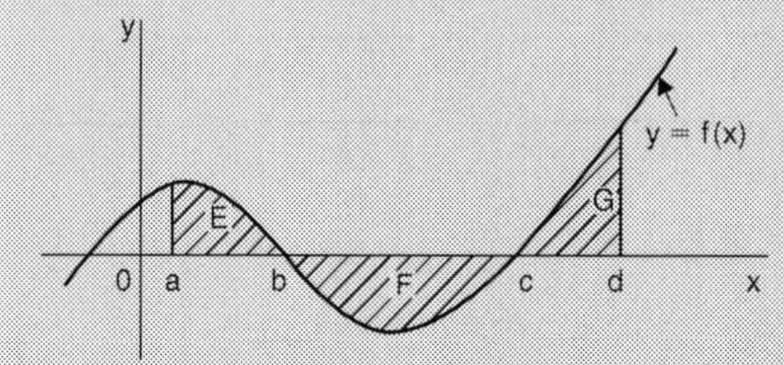

Figure 12.1

The area shown shaded in Figure 12.2, is given by:

shaded area $= \int_a^b [f_2(x) - f_1(x)]dx$

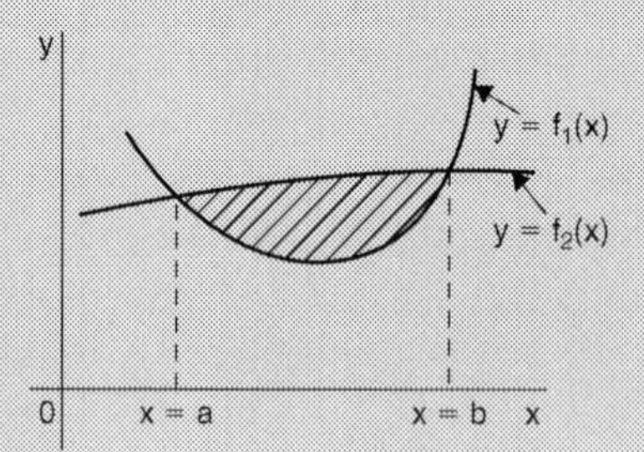

Figure 12.2

Application: The velocity v of a body t seconds after a certain instant is $(2t^2 + 5)$m/s. Find by integration how far it moves in the interval from t = 0 to t = 4 s

Since $2t^2 + 5$ is a quadratic expression, the curve $v = 2t^2 + 5$ is a parabola cutting the v-axis at v = 5, as shown in Figure 12.3.

The distance travelled is given by the area under the v/t curve, shown shaded in Figure 12.3.

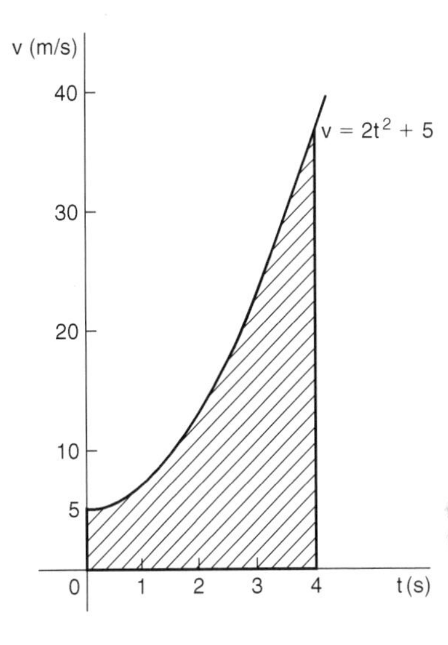

Figure 12.3

By integration,

$$\text{shaded area} = \int_0^4 v\,dt = \int_0^4 (2t^2 + 5)\,dt = \left[\frac{2t^3}{3} + 5t\right]_0^4$$

i.e. **distance travelled = 62.67 m**

Application: Determine the area enclosed by the curve $y = x^3 + 2x^2 - 5x - 6$ and the x-axis between $x = -3$ and $x = 2$

A table of values is produced and the graph sketched as shown in Figure 12.4 where the area enclosed by the curve and the x-axis is shown shaded.

x	−3	−2	−1	0	1	2
x^3	−27	−8	−1	0	1	8
$2x^2$	18	8	2	0	2	8
$-5x$	15	10	5	0	−5	−10
-6	−6	−6	−6	−6	−6	−6
y	0	4	0	−6	−8	0

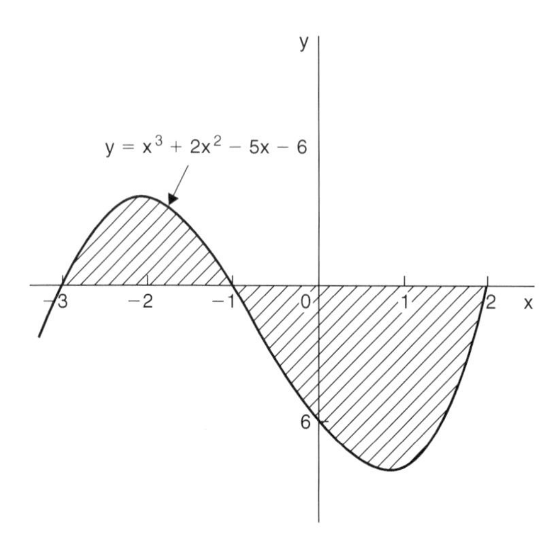

Figure 12.4

Shaded area $= \int_{-3}^{-1} y\,dx - \int_{-1}^{2} y\,dx$, the minus sign before the second integral being necessary since the enclosed area is below the x-axis.

Hence, shaded area

$$= \int_{-3}^{-1} (x^3 + 2x^2 - 5x - 6)\,dx - \int_{-1}^{2} (x^3 + 2x^2 - 5x - 6)\,dx$$

$$= \left[\frac{x^4}{4} + \frac{2x^3}{3} - \frac{5x^2}{2} - 6x\right]_{-3}^{-1} - \left[\frac{x^4}{4} + \frac{2x^3}{3} - \frac{5x^2}{2} - 6x\right]_{-1}^{2}$$

$$= \left[\left\{\frac{1}{4} - \frac{2}{3} - \frac{5}{2} + 6\right\} - \left\{\frac{81}{4} - 18 - \frac{45}{2} + 18\right\}\right]$$

$$- \left[\left\{4 + \frac{16}{3} - 10 - 12\right\} - \left\{\frac{1}{4} - \frac{2}{3} - \frac{5}{2} + 6\right\}\right]$$

$$= \left[\left\{3\frac{1}{12}\right\} - \left\{-2\frac{1}{4}\right\}\right] - \left[\left\{-12\frac{2}{3}\right\} - \left\{3\frac{1}{12}\right\}\right] = \left[5\frac{1}{3}\right] - \left[-15\frac{3}{4}\right]$$

$$= \mathbf{21\frac{1}{12}} \text{ or } \mathbf{21.083 \text{ square units}}$$

Application: Find the area enclosed by the curve $y = \sin 2x$, the x-axis and the ordinates $x = 0$ and $x = \frac{\pi}{3}$

A sketch of $y = \sin 2x$ is shown in Figure 12.5.
(Note that $y = \sin 2x$ has a period of $\frac{2\pi}{2}$, i.e. π radians)

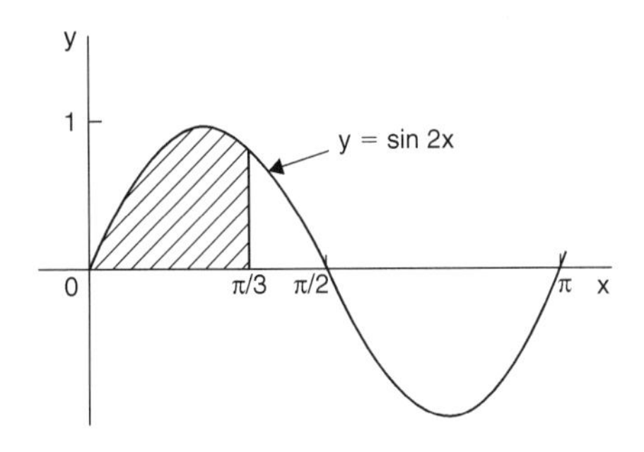

Figure 12.5

$$\text{Shaded area} = \int_0^{\pi/3} y\,dx = \int_0^{\pi/3} \sin 2x\,dx$$

$$= \left[-\frac{1}{2}\cos 2x\right]_0^{\pi/3} = \left\{-\frac{1}{2}\cos\frac{2\pi}{3}\right\} - \left\{-\frac{1}{2}\cos 0\right\}$$

$$= \left\{-\frac{1}{2}\left(-\frac{1}{2}\right)\right\} - \left\{-\frac{1}{2}(1)\right\} = \frac{1}{4} + \frac{1}{2} = \frac{3}{4} \text{ square units}$$

Application: Determine the area between the curve $y = x^3 - 2x^2 - 8x$ and the x-axis

$y = x^3 - 2x^2 - 8x = x(x^2 - 2x - 8) = x(x + 2)(x - 4)$
When $y = 0$, $x = 0$ or $(x + 2) = 0$ or $(x - 4) = 0$, i.e. when $y = 0$, $x = 0$ or -2 or 4, which means that the curve crosses the x-axis at 0, -2, and 4. Since the curve is a continuous function, only one other co-ordinate value needs to be calculated before a sketch of the curve can be produced. When $x = 1$, $y = -9$, showing that the part of the curve between $x = 0$ and $x = 4$ is negative. A sketch of $y = x^3 - 2x^2 - 8x$ is shown in Figure 12.6. (Another method of sketching Figure 12.6 would have been to draw up a table of values.)

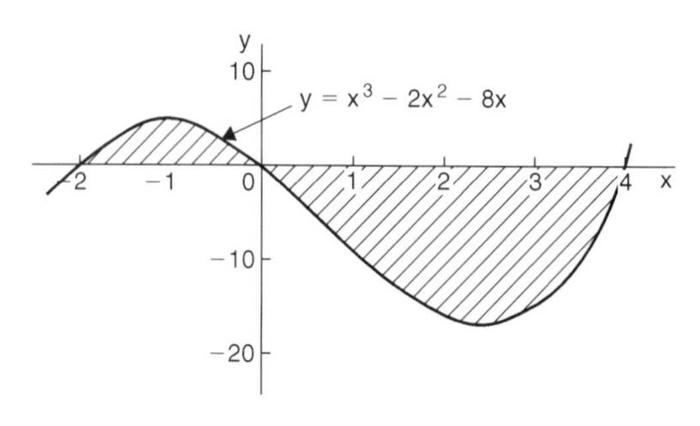

Figure 12.6

$$\text{Shaded area} = \int_{-2}^{0} (x^3 - 2x^2 - 8x)\,dx - \int_{0}^{4} (x^3 - 2x^2 - 8x)\,dx$$

$$= \left[\frac{x^4}{4} - \frac{2x^3}{3} - \frac{8x^2}{2}\right]_{-2}^{0} - \left[\frac{x^4}{4} - \frac{2x^3}{3} - \frac{8x^2}{2}\right]_{0}^{4}$$

$$= \left(6\frac{2}{3}\right) - \left(-42\frac{2}{3}\right)$$

$$= \mathbf{49\frac{1}{3}\ \textbf{square units}}$$

Application: Determine the area enclosed between the curves $y = x^2 + 1$ and $y = 7 - x$

At the points of intersection the curves are equal. Thus, equating the y values of each curve gives:

$$x^2 + 1 = 7 - x$$

from which, $\quad x^2 + x - 6 = 0$

Factorising gives: $\quad (x - 2)(x + 3) = 0$

from which $\quad x = 2$ and $x = -3$

By firstly determining the points of intersection the range of x-values has been found. Tables of values are produced as shown below.

x	−3	−2	−1	0	1	2
$y = x^2 + 1$	10	5	2	1	2	5

x	−3	0	2
$y = 7 - x$	10	7	5

A sketch of the two curves is shown in Figure 12.7.

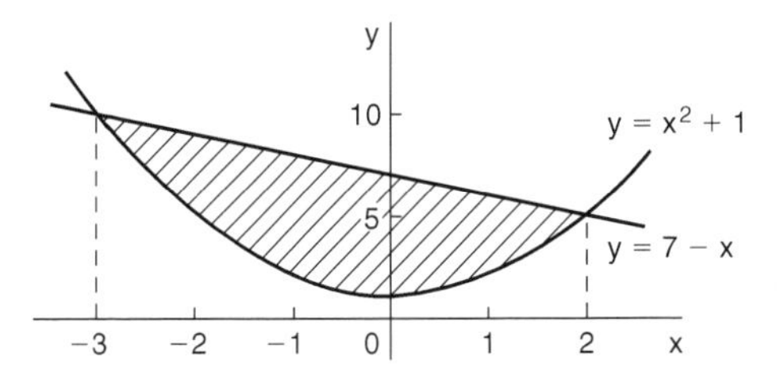

Figure 12.7

Shaded area

$$= \int_{-3}^{2} (7 - x)\,dx - \int_{-3}^{2} (x^2 + 1)\,dx = \int_{-3}^{2} [(7 - x) - (x^2 + 1)]\,dx$$

$$= \int_{-3}^{2} (6 - x - x^2)\,dx = \left[6x - \frac{x^2}{2} - \frac{x^3}{3}\right]_{-3}^{2}$$

$$= \left(12 - 2 - \frac{8}{3}\right) - \left(-18 - \frac{9}{2} + 9\right) = \left(7\frac{1}{3}\right) - \left(-13\frac{1}{2}\right)$$

$$= \mathbf{20\frac{5}{6}\ sq.\ units}$$

Application: Calculate the area enclosed by the curves $y = x^2$ and $y^2 = 8x$

At the points of intersection the co-ordinates of the curves are equal.

When $y = x^2$ then $y^2 = x^4$

Hence, at the points of intersection $x^4 = 8x$, by equating the y^2 values.

Thus $x^4 - 8x = 0$, from which $x(x^3 - 8) = 0$, i.e. $x = 0$ or $(x^3 - 8) = 0$

Hence at the points of intersection $x = 0$ or $x = 2$.

When $x = 0$, $y = 0$ and when $x = 2$, $y = 2^2 = 4$

Hence the points of intersection of the curves $y = x^2$ and $y^2 = 8x$ are (0, 0) and (2, 4).

A sketch of $y = x^2$ and $y^2 = 8x$ is shown in Figure 12.8.

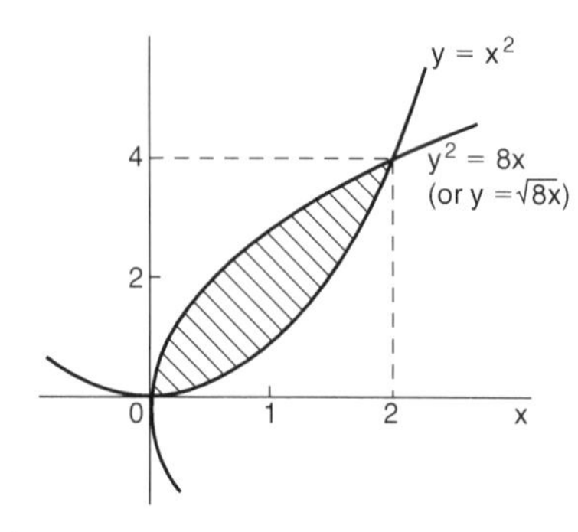

Figure 12.8

Shaded area

$$= \int_0^2 \left(\sqrt{8x} - x^2\right) dx = \int_0^2 \left(\left(\sqrt{8}\right) x^{1/2} - x^2\right) dx = \left[\left(\sqrt{8}\right)\frac{x^{3/2}}{3/2} - \frac{x^3}{3}\right]_0^2$$

$$= \left\{\frac{\sqrt{8}\sqrt{8}}{3/2} - \frac{8}{3}\right\} - \{0\} = \frac{16}{3} - \frac{8}{3} = \frac{8}{3} = \mathbf{2\frac{2}{3}\ \textbf{sq. units}}$$

Application: Determine by integration the area bounded by the three straight lines $y = 4 - x$, $y = 3x$ and $3y = x$

Each of the straight lines are shown sketched in Figure 12.9.

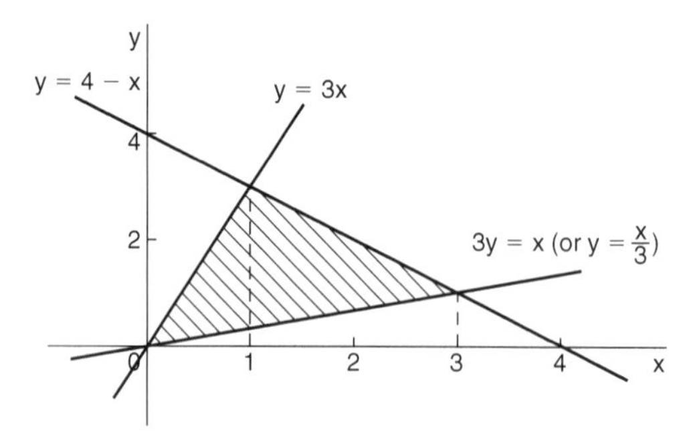

Figure 12.9

$$\text{Shaded area} = \int_0^1 \left(3x - \frac{x}{3}\right) dx + \int_1^3 \left[(4 - x) - \frac{x}{3}\right] dx$$

$$= \left[\frac{3x^2}{2} - \frac{x^2}{6}\right]_0^1 + \left[4x - \frac{x^2}{2} - \frac{x^2}{6}\right]_1^3$$

$$= \left[\left(\frac{3}{2} - \frac{1}{6}\right) - (0)\right] + \left[\left(12 - \frac{9}{2} - \frac{9}{6}\right) - \left(4 - \frac{1}{2} - \frac{1}{6}\right)\right]$$

$$= \left(1\frac{1}{3}\right) + \left(6 - 3\frac{1}{3}\right)$$

$= \mathbf{4}$ **square units**

12.11 Mean or average values

The mean or average value of the curve shown in Figure 12.10, between $x = a$ and $x = b$, is given by:

mean or average value, $\bar{y} = \dfrac{1}{b - a}\displaystyle\int_a^b f(x)dx$

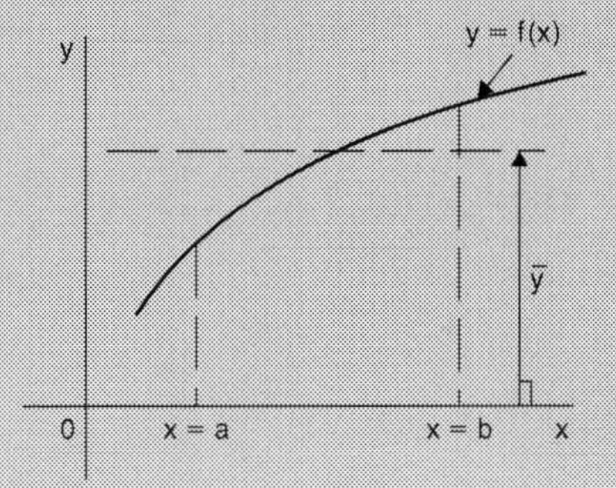

Figure 12.10

Application: Determine the mean value of $y = 5x^2$ between $x = 1$ and $x = 4$

Mean value, $\bar{y} = \dfrac{1}{4 - 1}\displaystyle\int_1^4 y\,dx = \frac{1}{3}\int_1^4 5x^2\,dx$

$$= \frac{1}{3}\left[\frac{5x^3}{3}\right]_1^4 = \frac{5}{9}\left[x^3\right]_1^4 = \frac{5}{9}(64 - 1)$$

$= \mathbf{35}$

Application: A sinusoidal voltage is given by $v = 100 \sin \omega t$ volts. Determine the mean value of the voltage over half a cycle using integration

Half a cycle means the limits are 0 to π radians.

Mean value, $$\bar{v} = \frac{1}{\pi - 0}\int_0^{\pi} v\, d(\omega t) = \frac{1}{\pi}\int_0^{\pi} 100 \sin \omega t\, d(\omega t)$$

$$= \frac{100}{\pi}[-\cos \omega t]_0^{\pi} = \frac{100}{\pi}[(-\cos \pi) - (-\cos 0)]$$

$$= \frac{100}{\pi}[(+1) - (-1)] = \frac{200}{\pi} = \mathbf{63.66\ volts}$$

[Note that for a sine wave, **mean value** $= \frac{2}{\pi} \times$ **maximum value**

In this case, mean value $= \frac{2}{\pi} \times 100 = 63.66\,V$]

Application: The number of atoms, N, remaining in a mass of material during radioactive decay after time t seconds is given by $N = N_0 e^{-\lambda t}$, where N_0 and λ are constants. Determine the mean number of atoms in the mass of material for the time period $t = 0$ and $t = \frac{1}{\lambda}$

Mean number of atoms

$$= \frac{1}{\frac{1}{\lambda} - 0}\int_0^{1/\lambda} N\, dt = \frac{1}{\frac{1}{\lambda}}\int_0^{1/\lambda} N_0 e^{-\lambda t}\, dt = \lambda N_0 \int_0^{1/\lambda} e^{-\lambda t}\, dt$$

$$= \lambda N_0 \left[\frac{e^{-\lambda t}}{-\lambda}\right]_0^{1/\lambda} = -N_0\,[e^{-\lambda(1/\lambda)} - e^0] = -N_0\,[e^{-1} - e^0]$$

$$= +N_0\left[e^0 - e^{-1}\right] = N_0\left[1 - e^{-1}\right] = \mathbf{0.632\ N_0}$$

12.12 Root mean square values

With reference to Figure 12.10, the r.m.s. value of $y = f(x)$ over the range $x = a$ to $x = b$ is given by:

$$\textbf{r.m.s. value} = \sqrt{\left\{\frac{1}{b-a}\int_a^b y^2\,dx\right\}}$$

The r.m.s. value of an alternating current is defined as 'that current which will give the same heating effect as the equivalent direct current'.

Application: Determine the r.m.s. value of $y = 2x^2$ between $x = 1$ and $x = 4$

$$\text{R.m.s. value} = \sqrt{\left\{\frac{1}{4-1}\int_1^4 y^2\,dx\right\}} = \sqrt{\left\{\frac{1}{3}\int_1^4 (2x^2)^2\,dx\right\}}$$

$$= \sqrt{\left\{\frac{1}{3}\int_1^4 4x^4\,dx\right\}}$$

$$= \sqrt{\left\{\frac{4}{3}\left[\frac{x^5}{5}\right]_1^4\right\}} = \sqrt{\left\{\frac{4}{15}(1024-1)\right\}} = \sqrt{272.8} = \mathbf{16.5}$$

Application: A sinusoidal voltage has a maximum value of 100 V. Calculate its r.m.s. value

A sinusoidal voltage v having a maximum value of 10 V may be written as $v = 10\sin\theta$. Over the range $\theta = 0$ to $\theta = \pi$,

$$\text{r.m.s. value} = \sqrt{\left\{\frac{1}{\pi-0}\int_0^\pi v^2\,d\theta\right\}} = \sqrt{\left\{\frac{1}{\pi}\int_0^\pi (100\sin\theta)^2\,d\theta\right\}}$$

$$= \sqrt{\left\{\frac{10000}{\pi}\int_0^\pi \sin^2\theta\,d\theta\right\}}$$ which is not a 'standard' integral

It is shown in chapter 5 that $\cos 2A = 1 - 2\sin^2 A$

Rearranging $\cos 2A = 1 - 2\sin^2 A$ gives $\sin^2 A = \frac{1}{2}(1 - \cos 2A)$

Hence,

$$\sqrt{\left\{\frac{10000}{\pi}\int_0^{\pi}\sin^2\theta\,d\theta\right\}} = \sqrt{\left\{\frac{10000}{\pi}\int_0^{\pi}\frac{1}{2}(1-\cos 2\theta)\,d\theta\right\}}$$

$$= \sqrt{\left\{\frac{10000}{\pi}\frac{1}{2}\left[\theta - \frac{\sin 2\theta}{2}\right]_0^{\pi}\right\}}$$

$$= \sqrt{\left\{\frac{10000}{\pi}\frac{1}{2}\left[\left(\pi - \frac{\sin 2\pi}{2}\right) - \left(0 - \frac{\sin 0}{2}\right)\right]\right\}}$$

$$= \sqrt{\left\{\frac{10000}{\pi}\frac{1}{2}[\pi]\right\}} = \sqrt{\left\{\frac{10000}{2}\right\}} = \frac{100}{\sqrt{2}}$$

$$= \mathbf{70.71\ volts}$$

[Note that for a sine wave, **r.m.s. value =** $\mathbf{\frac{1}{\sqrt{2}}}$ **× maximum value**.

In this case, r.m.s. value $= \frac{1}{\sqrt{2}} \times 100 = 70.71\,V$]

Application: In a frequency distribution the average distance from the mean, y, is related to the variable, x, by the equation $y = 2x^2 - 1$. Determine, correct to 3 significant figures, the r.m.s. deviation from the mean for values of x from −1 to +4

R.m.s. deviation

$$= \sqrt{\left\{\frac{1}{4 - -1}\int_{-1}^{4} y^2\,dx\right\}} = \sqrt{\left\{\frac{1}{5}\int_{-1}^{4}(2x^2 - 1)^2\,dx\right\}}$$

$$= \sqrt{\left\{\frac{1}{5}\int_{-1}^{4}(4x^4 - 4x^2 + 1)\,dx\right\}}$$

$$= \sqrt{\left\{\frac{1}{5}\left[\frac{4x^5}{5} - \frac{4x^3}{3} + x\right]_{-1}^{4}\right\}}$$

$$= \sqrt{\left\{\frac{1}{5}\left[\left(\frac{4}{5}(4)^5 - \frac{4}{3}(4)^3 + 4\right) - \left(\frac{4}{5}(-1)^5 - \frac{4}{3}(-1)^3 + (-1)\right)\right]\right\}}$$

$$= \sqrt{\left\{\frac{1}{5}[(737.87) - (-0.467)]\right\}} = \sqrt{\left\{\frac{1}{5}[738.34]\right\}}$$

$= \sqrt{147.67} = 12.152 =$ **12.2, correct to 3 significant figures.**

12.13 Volumes of solids of revolution

With reference to Figure 12.11, the volume of solid of revolution V obtained by rotating the shaded area through one revolution is given by:

$$\mathbf{V = \int_a^b \pi y^2\, dx} \quad \textbf{about the x-axis}$$

$$\mathbf{V = \int_c^d \pi x^2\, dy} \quad \textbf{about the y-axis}$$

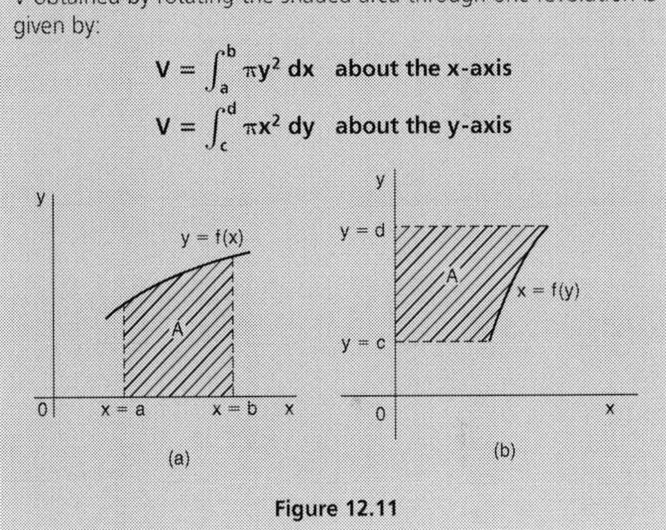

Figure 12.11

Application: The curve $y = x^2 + 4$ is rotated one revolution about (a) the x-axis, and (b) the y-axis, between the limits $x = 1$ and $x = 4$. Determine the volume of the solid of revolution produced in each case

(a) Revolving the shaded area shown in Figure 12.12 about the x-axis 360° produces a solid of revolution given by:

$$\text{Volume} = \int_1^4 \pi y^2\, dx = \int_1^4 \pi (x^2 + 4)^2\, dx$$

$$= \int_1^4 \pi(x^4 + 8x^2 + 16)\,dx = \pi\left[\frac{x^5}{5} + \frac{8x^3}{3} + 16x\right]_1^4$$

$$= \pi[(204.8 + 170.67 + 64) - (0.2 + 2.67 + 16)]$$

$$= \mathbf{420.6\pi \text{ cubic units}}$$

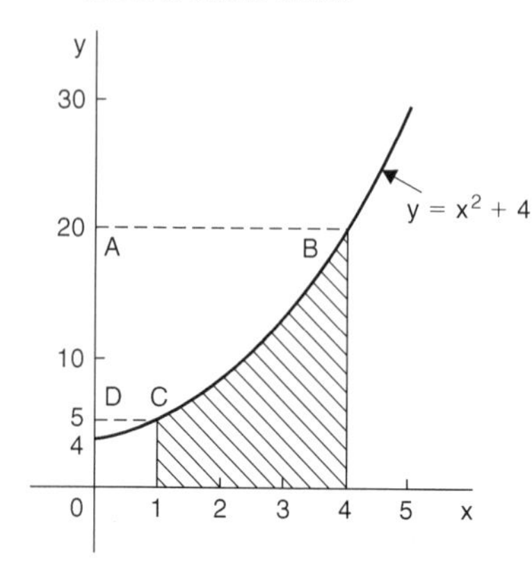

Figure 12.12

(b) The volume produced when the curve $y = x^2 + 4$ is rotated about the y-axis between $y = 5$ (when $x = 1$) and $y = 20$ (when $x = 4$), i.e. rotating area ABCD of Figure 12.12 about the y-axis is given by:

$$\text{volume} = \int_5^{20} \pi x^2\,dy$$

Since $y = x^2 + 4$, then $x^2 = y - 4$

Hence,

$$\text{volume} = \int_5^{20} \pi(y - 4)\,dy = \pi\left[\frac{y^2}{2} - 4y\right]_5^{20} = \pi[(120) - (-7.5)]$$

$$= \mathbf{127.5\pi \text{ cubic units}}$$

Application: The area enclosed by the curve $y = 3e^{\frac{x}{3}}$, the x-axis and ordinates $x = -1$ and $x = 3$ is rotated 360° about the x-axis. Determine the volume generated.

A sketch of $y = 3e^{\frac{x}{3}}$ is shown in Figure 12.13.

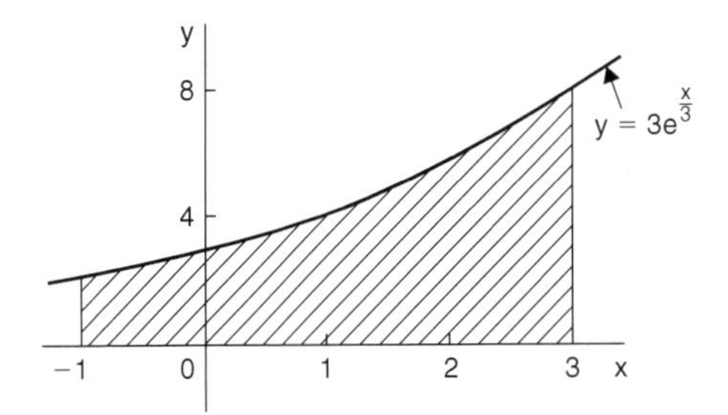

Figure 12.13

When the shaded area is rotated 360° about the x-axis then:

$$\text{volume generated} = \int_{-1}^{3} \pi y^2 \, dx = \int_{-1}^{3} \pi \left(3e^{\frac{x}{3}}\right)^2 dx = 9\pi \int_{-1}^{3} e^{\frac{2x}{3}} \, dx$$

$$= 9\pi \left[\frac{e^{\frac{2x}{3}}}{\frac{2}{3}} \right]_{-1}^{3} = \frac{27\pi}{2}\left(e^2 - e^{-\frac{2}{3}}\right)$$

$$= \mathbf{92.82\pi \text{ cubic units}}$$

Application: Calculate the volume of a frustum of a sphere of radius 4 cm that lies between two parallel planes at 1 cm and 3 cm from the centre and on the same side of it

The volume of a frustum of a sphere may be determined by integration by rotating the curve $x^2 + y^2 = 4^2$ (i.e. a circle, centre 0, radius 4) one revolution about the x-axis, between the limits $x = 1$ and $x = 3$ (i.e. rotating the shaded area of Figure 12.14).

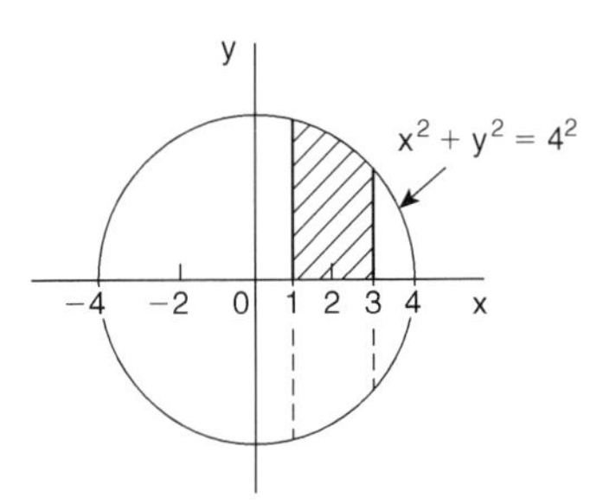

Figure 12.14

$$\text{Volume of frustum} = \int_1^3 \pi y^2\,dx = \int_1^3 \pi(4^2 - x^2)\,dx$$

$$= \pi\left[16x - \frac{x^3}{3}\right]_1^3 = \pi\left[(39) - \left(15\frac{2}{3}\right)\right]$$

$$= \mathbf{23\frac{1}{3}\pi \text{ cubic units}}$$

12.14 Centroids

Centroid of area between a curve and the x-axis

If $\bar{x}$ and $\bar{y}$ denote the co-ordinates of the centroid C of area A in Figure 12.15 then:

$$\bar{x} = \frac{\int_a^b xy\,dx}{\int_a^b y\,dx} \quad \text{and} \quad \bar{y} = \frac{\frac{1}{2}\int_a^b y^2\,dx}{\int_a^b y\,dx}$$

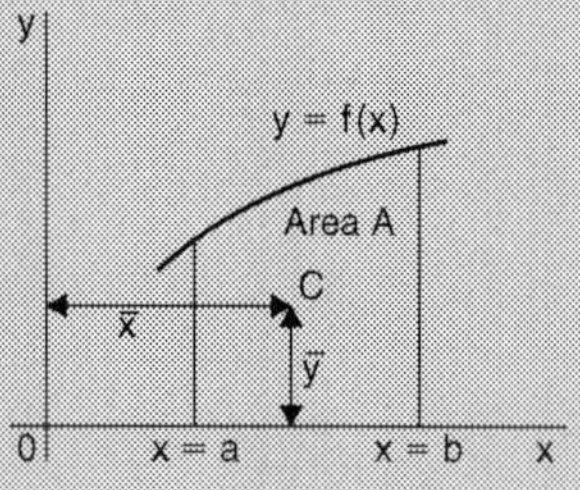

Figure 12.15

Centroid of area between a curve and the y-axis

If $\bar{x}$ and $\bar{y}$ denote the co-ordinates of the centroid C of area A in Figure 12.16 then:

$$\bar{x} = \frac{\frac{1}{2}\int_c^d x^2\,dy}{\int_c^d x\,dy} \quad \text{and} \quad \bar{y} = \frac{\int_c^d xy\,dy}{\int_c^d x\,dy}$$

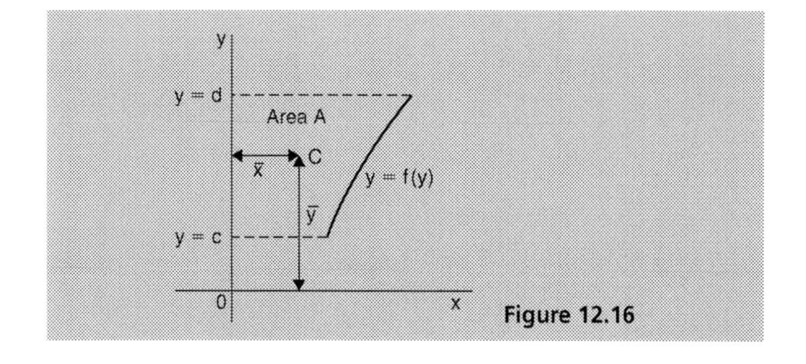

Figure 12.16

Application: Find the position of the centroid of the area bounded by the curve $y = 3x^2$, the x-axis and the ordinates $x = 0$ and $x = 2$

If $(\bar{x}, \bar{y})$ are the co-ordinates of the centroid of the given area then:

$$\bar{x} = \frac{\int_0^2 xy\,dx}{\int_0^2 y\,dx} = \frac{\int_0^2 x(3x^2)\,dx}{\int_0^2 3x^2\,dx} = \frac{\int_0^2 3x^3\,dx}{\int_0^2 3x^2\,dx}$$

$$= \frac{\left[\frac{3x^4}{4}\right]_0^2}{\left[x^3\right]_0^2} = \frac{12}{8} = \mathbf{1.5}$$

$$\bar{y} = \frac{\frac{1}{2}\int_0^2 y^2\,dx}{\int_0^2 y\,dx} = \frac{\frac{1}{2}\int_0^2 (3x^2)^2\,dx}{8} = \frac{\frac{1}{2}\int_0^2 9x^4\,dx}{8}$$

$$= \frac{\frac{9}{2}\left[\frac{x^5}{5}\right]_0^2}{8} = \frac{\frac{9}{2}\left(\frac{32}{5}\right)}{8} = \frac{18}{5} = \mathbf{3.6}$$

Hence the centroid lies at (1.5, 3.6)

Application: Locate the position of the centroid enclosed by the curves $y = x^2$ and $y^2 = 8x$

Figure 12.17 shows the two curves intersecting at (0, 0) and (2, 4). These are the same curves as used in the application on page 341, where the shaded area was calculated as $2\frac{2}{3}$ square units. Let the co-ordinates of centroid C be $\bar{x}$ and $\bar{y}$

By integration, $$\bar{x} = \frac{\int_0^2 xy\,dx}{\int_0^2 y\,dx}$$

The value of y is given by the height of the typical strip shown in Figure 12.17, i.e. $y = \sqrt{8x} - x^2$

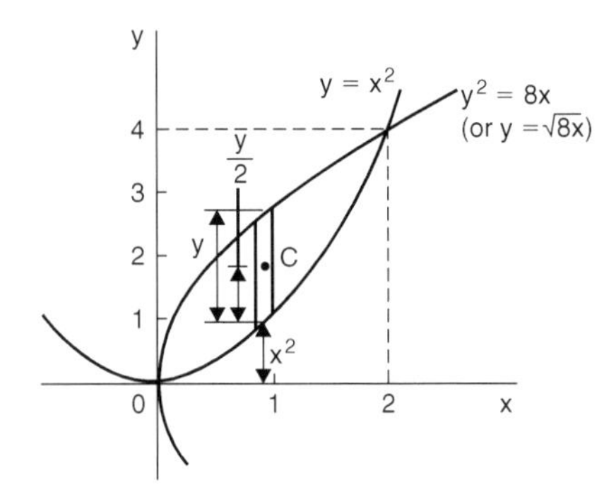

Figure 12.17

Hence,

$$\bar{x} = \frac{\int_0^2 x\left(\sqrt{8x} - x^2\right)dx}{2\frac{2}{3}} = \frac{\int_0^2 (\sqrt{8}\,x^{3/2} - x^3)}{2\frac{2}{3}} = \frac{\left[\sqrt{8}\,\frac{x^{5/2}}{\frac{5}{2}} - \frac{x^4}{4}\right]_0^2}{2\frac{2}{3}}$$

$$= \left(\frac{\sqrt{8}\,\frac{\sqrt{2^5}}{\frac{5}{2}} - 4}{2\frac{2}{3}}\right) = \frac{2\frac{2}{5}}{2\frac{2}{3}} = 0.9$$

Care needs to be taken when finding $\bar{y}$ in such examples as this.

From Figure 12.17, $y = \sqrt{8x} - x^2$ and $\dfrac{y}{2} = \dfrac{1}{2}\left(\sqrt{8x} - x^2\right)$

The perpendicular distance from centroid C of the strip to OX is $\dfrac{1}{2}\left(\sqrt{8x} - x^2\right) + x^2$

Taking moments about 0x gives:

$$(\text{total area})(\bar{y}) = \sum_{x=0}^{x=2} (\text{area of strip})(\text{perpendicular distance of centroid of strip to } 0x)$$

Hence,
$$(\text{area})(\bar{y}) = \int_0^2 \left[\sqrt{8x} - x^2\right]\left[\frac{1}{2}\left(\sqrt{8x} - x^2\right) + x^2\right] dx$$

i.e.
$$\left(2\frac{2}{3}\right)(\bar{y}) = \int_0^2 \left[\sqrt{8x} - x^2\right]\left(\frac{\sqrt{8x}}{2} + \frac{x^2}{2}\right) dx$$

$$= \int_0^2 \left(\frac{8x}{2} - \frac{x^4}{2}\right) dx = \left[\frac{8x^2}{4} - \frac{x^5}{10}\right]_0^2$$

$$= \left(8 - 3\frac{1}{5}\right) - (0) = 4\frac{4}{5}$$

Hence
$$\bar{y} = \frac{4\frac{4}{5}}{2\frac{2}{3}} = 1.8$$

Thus the position of the centroid of the enclosed area in Figure 12.17 is at (0.9, 1.8)

Application: Locate the centroid of the area enclosed by the curve $y = 2x^2$, the y-axis and ordinates $y = 1$ and $y = 4$, correct to 3 decimal places

$$\bar{x} = \frac{\frac{1}{2}\int_1^4 x^2\,dy}{\int_1^4 x\,dy} = \frac{\frac{1}{2}\int_1^4 \frac{y}{2}\,dy}{\int_1^4 \sqrt{\frac{y}{2}}\,dy} = \frac{\frac{1}{2}\left[\frac{y^2}{4}\right]_1^4}{\left[\frac{2y^{3/2}}{3\sqrt{2}}\right]_1^4} = \frac{\frac{15}{8}}{\frac{14}{3\sqrt{2}}} = 0.568$$

$$\bar{y} = \frac{\int_1^4 xy\,dy}{\int_1^4 x\,dy} = \frac{\int_1^4 \sqrt{\frac{y}{2}}(y)\,dy}{\frac{14}{3\sqrt{2}}} = \frac{\int_1^4 \frac{y^{3/2}}{\sqrt{2}}\,dy}{\frac{14}{3\sqrt{2}}}$$

$$= \frac{\frac{1}{\sqrt{2}}\left[\frac{y^{5/2}}{\frac{5}{2}}\right]_1^4}{\frac{14}{3\sqrt{2}}} = \frac{\frac{2}{5\sqrt{2}}(31)}{\frac{14}{3\sqrt{2}}} = 2.657$$

Hence the position of the centroid is at (0.568, 2.657)

12.15 Theorem of Pappus

A theorem of Pappus states:

'If a plane area is rotated about an axis in its own plane but not intersecting it, the volume of the solid formed is given by the product of the area and the distance moved by the centroid of the area'.

With reference to Figure 12.18, when the curve $y = f(x)$ is rotated one revolution about the x-axis between the limits $x = a$ and $x = b$, the volume V generated is given by:

$$\text{volume } V = (A)(2\pi\bar{y}), \text{ from which, } \bar{\mathbf{y}} = \frac{\mathbf{V}}{\mathbf{2\pi A}}$$

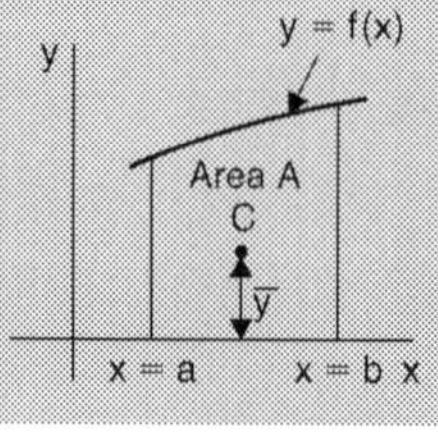

Figure 12.18

Application: Determine the position of the centroid of a semicircle of radius r by using the theorem of Pappus

A semicircle is shown in Figure 12.19 with its diameter lying on the x-axis and its centre at the origin.

Area of semicircle $= \dfrac{\pi r^2}{2}$. When the area is rotated about the x-axis one revolution a sphere is generated of volume $\dfrac{4}{3}\pi r^3$

Let centroid C be at a distance $\overline{y}$ from the origin as shown in Figure 12.19.

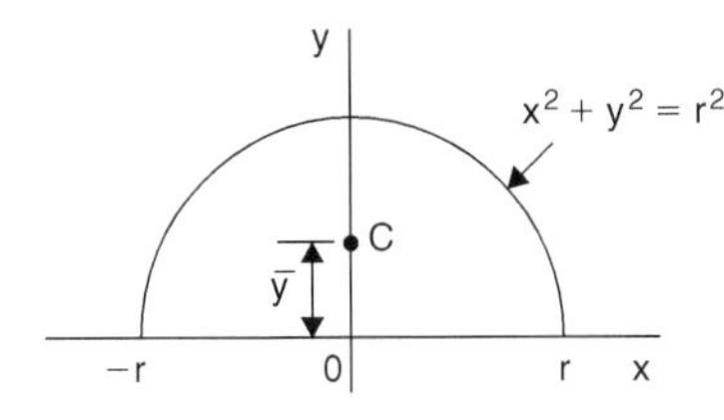

Figure 12.19

From the theorem of Pappus, volume generated = area × distance moved through by centroid

i.e. $$\frac{4}{3}\pi r^3 = \left(\frac{\pi r^2}{2}\right)(2\pi\overline{y})$$

Hence $$\overline{y} = \frac{\frac{4}{3}\pi r^3}{\pi^2 r^2} = \frac{4r}{3\pi}$$

[By integration,

$$\overline{y} = \frac{\frac{1}{2}\int_{-r}^{r} y^2\,dx}{\text{area}} = \frac{\frac{1}{2}\int_{-r}^{r} (r^2 - x^2)\,dx}{\frac{\pi r^2}{2}} = \frac{\frac{1}{2}\left[r^2 x - \frac{x^3}{3}\right]_{-r}^{r}}{\frac{\pi r^2}{2}}$$

$$= \frac{\frac{1}{2}\left[\left(r^3 - \frac{r^3}{3}\right) - \left(-r^3 + \frac{r^3}{3}\right)\right]}{\frac{\pi r^2}{2}} = \frac{4r}{3\pi}]$$

Hence the centroid of a semicircle lies on the axis of symmetry, distance $\boldsymbol{\dfrac{4r}{3\pi}}$ (or 0.424 r) from its diameter.

Application: (a) Calculate the area bounded by the curve $y = 2x^2$, the x-axis and ordinates $x = 0$ and $x = 3$ (b) If the area in part (a) is revolved (i) about the x-axis and (ii) about the y-axis, find the volumes of the solids produced, and (c) locate the position of the centroid using (i) integration, and (ii) the theorem of Pappus

(a) The required area is shown shaded in Figure 12.20.

$$\text{Area} = \int_0^3 y\,dx = \int_0^3 2x^2\,dx = \left[\frac{2x^3}{3}\right]_0^3 = \textbf{18 square units}$$

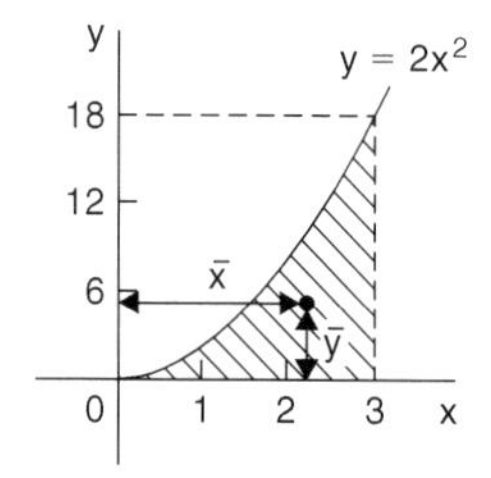

Figure 12.20

(b) (i) When the shaded area of Figure 12.20 is revolved 360° about the x-axis, the volume generated

$$= \int_0^3 \pi y^2\,dx = \int_0^3 \pi(2x^2)^2\,dx = \int_0^3 4\pi x^4\,dx$$

$$= 4\pi\left[\frac{x^5}{5}\right]_0^3 = 4\pi\left(\frac{243}{5}\right)$$

$$= \mathbf{194.4\pi \text{ cubic units}}$$

(ii) When the shaded area of Figure 12.20 is revolved 360° about the y-axis, the volume generated = (volume generated by $x = 3$) − (volume generated by $y = 2x^2$)

$$= \int_0^{18} \pi(3)^2\,dy - \int_0^{18} \pi\left(\frac{y}{2}\right)dy = \pi\int_0^{18}\left(9 - \frac{y}{2}\right)dy$$

$$= \pi\left[9y - \frac{y^2}{4}\right]_0^{18}$$

$$= \mathbf{81\pi \text{ cubic units}}$$

(c) If the co-ordinates of the centroid of the shaded area in Figure 12.20 are $(\overline{x}, \overline{y})$ then:

(i) by integration,

$$\overline{x} = \frac{\int_0^3 xy\,dx}{\int_0^3 y\,dx} = \frac{\int_0^3 x(2x^2)\,dx}{18} = \frac{\int_0^3 2x^3\,dx}{18} = \frac{\left[\frac{2x^4}{4}\right]_0^3}{18}$$

$$= \frac{81}{36} = \mathbf{2.25}$$

$$\overline{y} = \frac{\frac{1}{2}\int_0^3 y^2\,dx}{\int_0^3 y\,dx} = \frac{\frac{1}{2}\int_0^3 (2x^2)^2\,dx}{18} = \frac{\frac{1}{2}\int_0^3 4x^4\,dx}{18}$$

$$= \frac{\frac{1}{2}\left[\frac{4x^5}{5}\right]_0^3}{18} = \mathbf{5.4}$$

(ii) using the theorem of Pappus:

Volume generated when shaded area is revolved about $Oy = (\text{area})(2\pi\overline{x})$

i.e. $81\pi = (18)(2\pi\overline{x})$, from which, $\overline{x} = \frac{81\pi}{36\pi} = \mathbf{2.25}$

Volume generated when shaded area is revolved about $Ox = (\text{area})(2\pi\overline{y})$

i.e. $194.4\pi = (18)(2\pi\overline{y})$, from which, $\overline{y} = \frac{194.4\pi}{36\pi} = \mathbf{5.4}$

Hence, the centroid of the shaded area in Figure 12.20 is at (2.25, 5.4)

Application: A metal disc has a radius of 5.0 cm and is of thickness 2.0 cm. A semicircular groove of diameter 2.0 cm is machined centrally around the rim to form a pulley. Using Pappus' theorem, determine the volume and mass of metal removed and the volume and mass of the pulley if the density of the metal is 8000 kg/m^3

A side view of the rim of the disc is shown in Figure 12.21.

When area PQRS is rotated about axis XX the volume generated is that of the pulley.

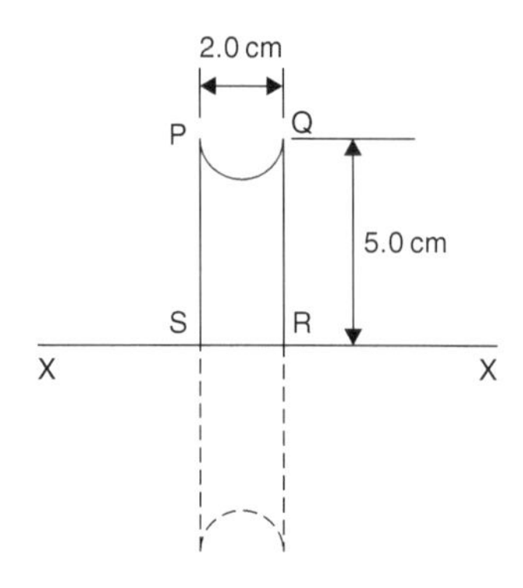

Figure 12.21

The centroid of the semicircular area removed is at a distance of $\frac{4r}{3\pi}$ from its diameter (see earlier example), i.e. $\frac{4(1.0)}{3\pi}$, i.e. 0.424 cm from PQ. Thus the distance of the centroid from XX is (5.0 – 0.424), i.e. 4.576 cm.

The distance moved through in one revolution by the centroid is $2\pi(4.576)$ cm.

Area of semicircle $= \frac{\pi r^2}{2} = \frac{\pi(1.0)^2}{2} = \frac{\pi}{2}\ \text{cm}^2$.

By the theorem of Pappus,

volume generated = area × distance moved by centroid

$$= \left(\frac{\pi}{2}\right)(2\pi)(4.576)$$

i.e. **volume of metal removed = 45.16 cm³**

Mass of metal removed = density × volume

$$= 8000\ \text{kg/m}^3 \times \frac{45.16}{10^6}\ \text{m}^3$$

$$= \mathbf{0.361\ kg}\ \text{or}\ \mathbf{361\ g}$$

Volume of pulley = volume of cylindrical disc − volume of metal removed

$$= \pi(5.0)^2(2.0) - 45.16 = \mathbf{111.9\ cm^3}$$

Mass of pulley = density × volume

$$= 8000\ \text{kg/m}^3 \times \frac{111.9}{10^6}\ \text{m}^3 = \mathbf{0.895\ kg\ or\ 895\ g}$$

12.16 Second moments of area

Table 12.3 Summary of standard results of the second moments of areas of regular sections

Shape	Position of axis	Second moment of area, I	Radius of gyration, k
Rectangle	(1) Coinciding with b	$\frac{bl^3}{3}$	$\frac{l}{\sqrt{3}}$
length l	(2) Coinciding with l	$\frac{lb^3}{3}$	$\frac{b}{\sqrt{3}}$
breadth b	(3) Through centroid, parallel to b	$\frac{bl^3}{12}$	$\frac{l}{\sqrt{12}}$
	(4) Through centroid, parallel to l	$\frac{lb^3}{12}$	$\frac{b}{\sqrt{12}}$
Triangle	(1) Coinciding with b	$\frac{bh^3}{12}$	$\frac{h}{\sqrt{6}}$
Perpendicular height h	(2) Through centroid, parallel to base	$\frac{bh^3}{36}$	$\frac{h}{\sqrt{18}}$
base b	(3) Through vertex, parallel to base	$\frac{bh^3}{4}$	$\frac{h}{\sqrt{2}}$
Circle radius r	(1) Through centre, perpendicular to plane (i.e. polar axis)	$\frac{\pi r^4}{2}$	$\frac{r}{\sqrt{2}}$
	(2) Coinciding with diameter	$\frac{\pi r^4}{4}$	$\frac{r}{2}$
	(3) About a tangent	$\frac{5\pi r^4}{4}$	$\frac{\sqrt{5}}{2}r$
Semicircle radius r	Coinciding with diameter	$\frac{\pi r^4}{8}$	$\frac{r}{2}$

Parallel axis theorem

If C is the centroid of area A in Figure 12.22, then:

$$I_{DD} = I_{GG} + Ad^2$$

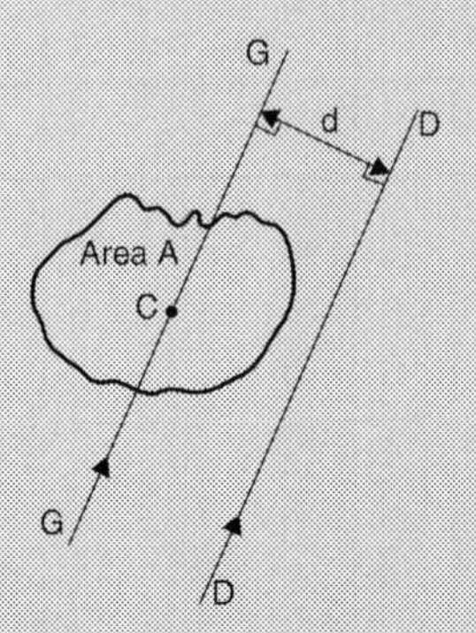

Figure 12.22

Perpendicular axis theorem

If OX and OY lie in the plane of area A in Figure 12.23, then:

$$I_{OZ} = I_{OX} + I_{OY}$$

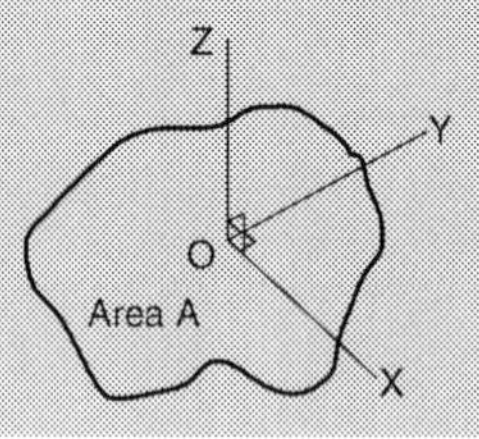

Figure 12.23

Application: Determine the second moment of area and the radius of gyration about axes AA, BB and CC for the rectangle shown in Figure 12.24

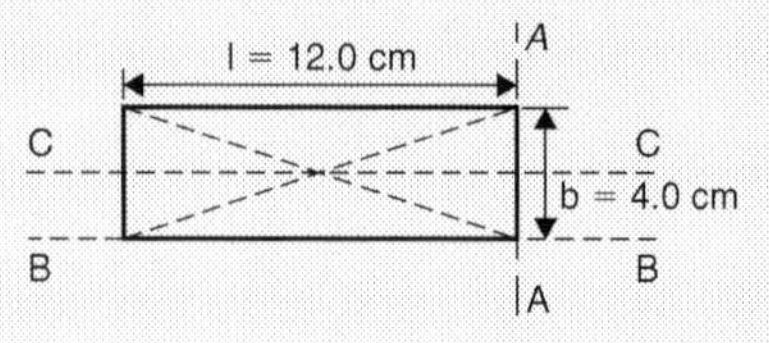

Figure 12.24

From Table 12.3, the second moment of area about axis AA,

$$\mathbf{I_{AA}} = \frac{bl^3}{3} = \frac{(4.0)(12.0)^3}{3} = \mathbf{2304\ cm^4}$$

Radius of gyration, $$\mathbf{k_{AA}} = \frac{l}{\sqrt{3}} = \frac{12.0}{\sqrt{3}} = \mathbf{6.93\ cm}$$

Similarly, $$\mathbf{I_{BB}} = \frac{lb^3}{3} = \frac{(12.0)(4.0)^3}{3} = \mathbf{256\ cm^4}$$

and $$\mathbf{k_{BB}} = \frac{b}{\sqrt{3}} = \frac{4.0}{\sqrt{3}} = \mathbf{2.31\,cm}$$

The second moment of area about the centroid of a rectangle is $\frac{bl^3}{12}$ when the axis through the centroid is parallel with the breadth b. In this case, the axis CC is parallel with the length l

Hence $$\mathbf{I_{CC}} = \frac{lb^3}{12} = \frac{(12.0)(4.0)^3}{12} = \mathbf{64\ cm^4}$$

and $$\mathbf{k_{CC}} = \frac{b}{\sqrt{12}} = \frac{4.0}{\sqrt{12}} = \mathbf{1.15\ cm}$$

Application: Find the second moment of area and the radius of gyration about axis PP for the rectangle shown in Figure 12.25

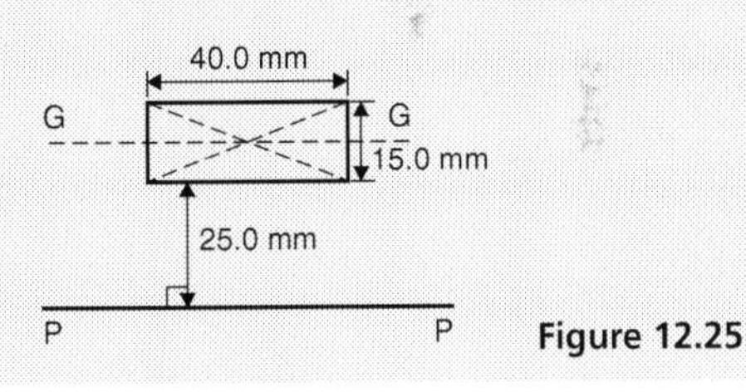

Figure 12.25

$$I_{GG} = \frac{lb^3}{12} \text{ where } l = 40.0\,\text{mm and } b = 15.0\,\text{mm}$$

Hence $$I_{GG} = \frac{(40.0)(15.0)^3}{12} = 11250\,\text{mm}^4$$

From the parallel axis theorem, $I_{PP} = I_{GG} + Ad^2$, where $A = 40.0 \times 15.0 = 600\,\text{mm}^2$ and $d = 25.0 + 7.5 = 32.5\,\text{mm}$, the perpendicular distance between GG and PP.

Hence, $\mathbf{I_{PP}} = 11250 + (600)(32.5)^2 = \mathbf{645000\,mm^4}$

$I_{PP} = Ak_{PP}^2$, from which, $\mathbf{k_{PP}} = \sqrt{\frac{I_{PP}}{\text{area}}} = \sqrt{\left(\frac{645000}{600}\right)} = \mathbf{32.79\,mm}$

Application: Determine the second moment of area and radius of gyration about axis QQ of the triangle BCD shown in Figure 12.26

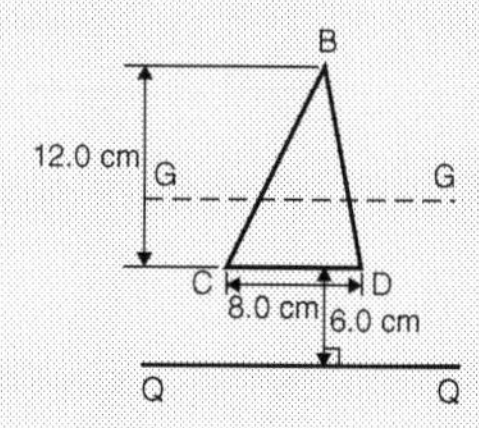

Figure 12.26

Using the parallel axis theorem: $I_{QQ} = I_{GG} + Ad^2$, where I_{GG} is the second moment of area about the centroid of the triangle, i.e. $\frac{bh^3}{36} = \frac{(8.0)(12.0)^3}{36} = 384\,\text{cm}^4$,

A is the area of the triangle $= \frac{1}{2}bh = \frac{1}{2}(8.0)(12.0) = 48\,\text{cm}^2$ and d is the distance between axes GG and QQ $= 6.0 + \frac{1}{3}(12.0) = 10\,\text{cm}$

Hence the second moment of area about axis QQ,

$\mathbf{I_{QQ}} = 384 + (48)(10)^2 = \mathbf{5184\,cm^4}$

Radius of gyration, $\mathbf{K_{QQ}} = \sqrt{\frac{I_{QQ}}{\text{area}}} = \sqrt{\left(\frac{5184}{48}\right)} = \mathbf{10.4\,cm}$

Application: Determine the second moment of area and radius of gyration of the circle shown in Figure 12.27 about axis YY

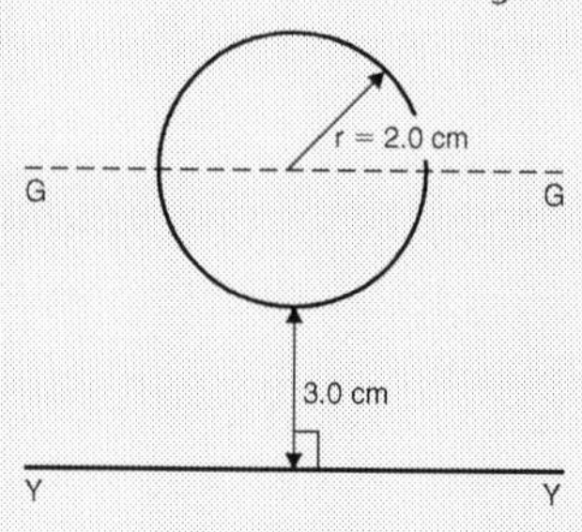

Figure 12.27

In Figure 12.27, $I_{GG} = \frac{\pi r^4}{4} = \frac{\pi}{4}(2.0)^4 = 4\pi \text{ cm}^4$

Using the parallel axis theorem, $I_{YY} = I_{GG} + Ad^2$, where d = 3.0 + 2.0 = 5.0 cm.

Hence $\mathbf{I_{YY}} = 4\pi + [\pi(2.0)^2](5.0)^2 = 4\pi + 100\pi = 104\pi = \mathbf{327\ cm^4}$

Radius of gyration, $\mathbf{k_{YY}} = \sqrt{\frac{I_{YY}}{\text{area}}} = \sqrt{\left(\frac{104\pi}{\pi(2.0)^2}\right)} = \sqrt{26} = \mathbf{5.10\ cm}$

Application: Determine the second moment of area and radius of gyration for the semicircle shown in Figure 12.28 about axis XX

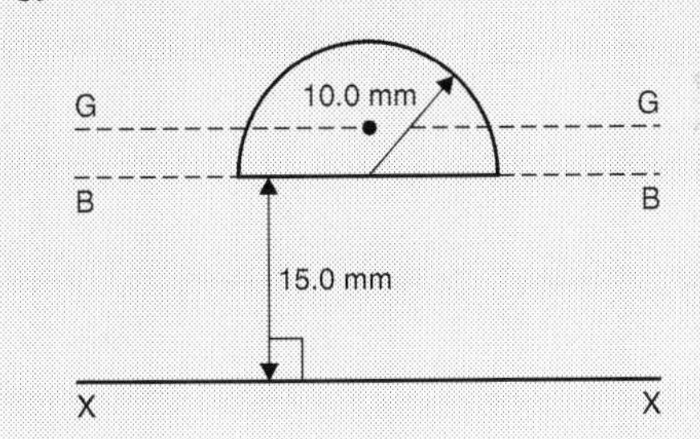

Figure 12.28

The centroid of a semicircle lies at $\frac{4r}{3\pi}$ from its diameter

Using the parallel axis theorem: $I_{BB} = I_{GG} + Ad^2$,

where $I_{BB} = \frac{\pi r^4}{8}$ (from Table 12.3) $= \frac{\pi(10.0)^4}{8} = 3927 \text{ mm}^4$,

$A = \frac{\pi r^2}{2} = \frac{\pi(10.0)^2}{2} = 157.1 \text{mm}^2$

and $d = \frac{4r}{3\pi} = \frac{4(10.0)}{3\pi} = 4.244 \text{ mm}$

Hence, $3927 = I_{GG} + (157.1)(4.244)^2$

i.e. $3927 = I_{GG} + 2830$, from which, $I_{GG} = 3927 - 2830 = 1097 \text{ mm}^4$

Using the parallel axis theorem again: $I_{XX} = I_{GG} + A(15.0 + 4.244)^2$

i.e. $\mathbf{I_{XX}} = 1097 + (157.1)(19.244)^2 = 1097 + 58179 = 59276 \text{ mm}^4$ or $\mathbf{59280\ mm^4}$, correct to 4 significant figures.

Radius of gyration, $\mathbf{k_{XX}} = \sqrt{\frac{I_{XX}}{\text{area}}} = \sqrt{\left(\frac{59276}{157.1}\right)} = \mathbf{19.42\ mm}$

Application: Determine the polar second moment of area of the propeller shaft cross-section shown in Figure 12.29

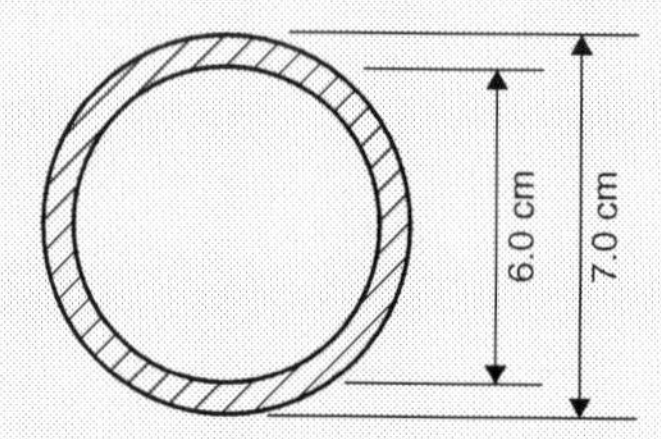

Figure 12.29

The polar second moment of area of a circle $= \dfrac{\pi r^4}{2}$

The polar second moment of area of the shaded area is given by the polar second moment of area of the 7.0 cm diameter circle minus the polar second moment of area of the 6.0 cm diameter circle.

Hence the polar second moment of area of the cross-section shown

$$= \frac{\pi}{2}\left(\frac{7.0}{2}\right)^4 - \frac{\pi}{2}\left(\frac{6.0}{2}\right)^4 = 235.7 - 127.2 = \mathbf{108.5\ cm^4}$$

Application: Determine the second moment of area and radius of gyration of a rectangular lamina of length 40 mm and width 15 mm about an axis through one corner, perpendicular to the plane of the lamina

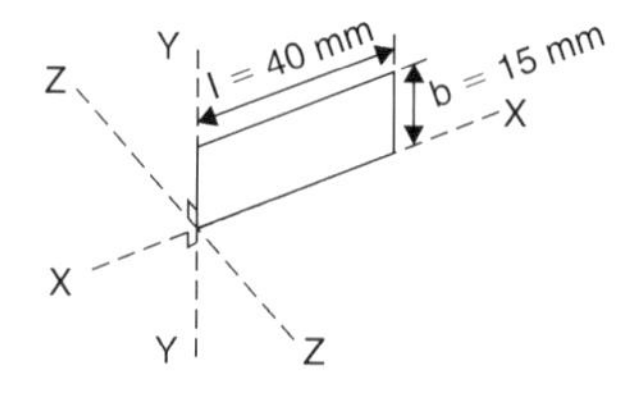

Figure 12.30

The lamina is shown in Figure 12.30.

From the perpendicular axis theorem: $I_{ZZ} = I_{XX} + I_{YY}$

$$I_{XX} = \frac{lb^3}{3} = \frac{(40)(15)^3}{3} = 45000\ \text{mm}^4$$

and $$I_{YY} = \frac{bl^3}{3} = \frac{(15)(40)^3}{3} = 320000\ \text{mm}^4$$

Hence $I_{ZZ} = 45000 + 320000 = \mathbf{365000\,mm^4}$ or $\mathbf{36.5\,cm^4}$

$$\text{Radius of gyration, } \mathbf{k_{ZZ}} = \sqrt{\frac{I_{ZZ}}{\text{area}}} = \sqrt{\left(\frac{365000}{(40)(15)}\right)}$$

$$= \mathbf{24.7\,mm} \text{ or } \mathbf{2.47\,cm}$$

Application: Determine correct to 3 significant figures, the second moment of area about axis XX for the composite area shown in Figure 12.31.

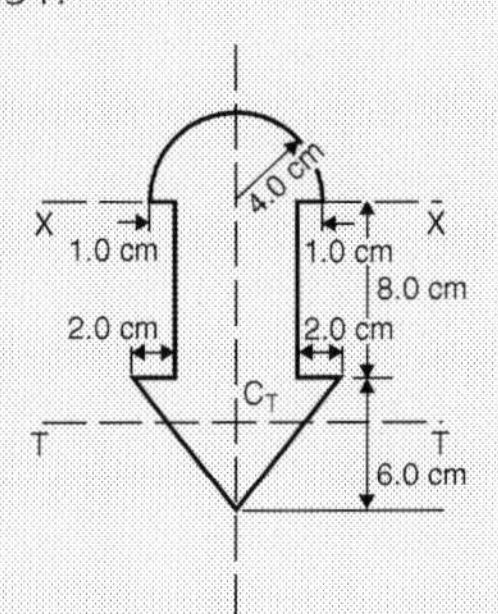

Figure 12.31

$$\text{For the semicircle, } I_{XX} = \frac{\pi r^4}{8} = \frac{\pi(4.0)^4}{8} = 100.5\,\text{cm}^4$$

$$\text{For the rectangle } I_{XX} = \frac{bl^3}{3} = \frac{(6.0)(8.0)^3}{3} = 1024\,\text{cm}^4$$

For the triangle, about axis TT through centroid C_T,

$$I_{TT} = \frac{bh^3}{36} = \frac{(10)(6.0)^3}{36} = 60\,\text{cm}^4$$

By the parallel axis theorem, the second moment of area of the triangle about axis XX

$$= 60 + \left[\frac{1}{2}(10)(6.0)\right]\left[8.0 + \frac{1}{3}(6.0)\right]^2 = 3060\,\text{cm}^4$$

Total second moment of area about XX $= 100.5 + 1024 + 3060$

$= 4184.5 = \mathbf{4180\,cm^4}$, correct to 3 significant figures

13 Differential Equations

13.1 The solution of equations of the form $\frac{dy}{dx} = f(x)$

A differential equation of the form $\frac{dy}{dx} = f(x)$ is solved by direct integration, i.e.

$$y = \int f(x)dx$$

Application: Find the particular solution of the differential equation $5\frac{dy}{dx} + 2x = 3$, given the boundary conditions $y = 1\frac{2}{5}$ when $x = 2$

Since $5\frac{dy}{dx} + 2x = 3$ then $\frac{dy}{dx} = \frac{3 - 2x}{5} = \frac{3}{5} - \frac{2x}{5}$

Hence, $y = \int \left(\frac{3}{5} - \frac{2x}{5}\right) dx$

i.e. $y = \frac{3x}{5} - \frac{x^2}{5} + c$, which is the general solution.

Substituting the boundary conditions $y = 1\frac{2}{5}$ and $x = 2$ to evaluate c gives:

$1\frac{2}{5} = \frac{6}{5} - \frac{4}{5} + c$, from which, $c = 1$.

Hence the particular solution is $\mathbf{y = \frac{3x}{5} - \frac{x^2}{5} + 1}$

13.2 The solution of equations of the form $\frac{dy}{dx} = f(y)$

A differential equation of the form $\frac{dy}{dx} = f(y)$ is initially rearranged to give $dx = \frac{dy}{f(y)}$ and then the solution is obtained by direct integration, i.e.

$$\int \mathbf{dx} = \int \frac{\mathbf{dy}}{\mathbf{f(y)}}$$

Application:

(a) The variation of resistance, R ohms, of an aluminium conductor with temperature θ°C is given by $\frac{dR}{d\theta} = \alpha R$, where α is the temperature coefficient of resistance of aluminium. If $R = R_0$ when $\theta = 0$°C, solve the equation for R.

(b) If $\alpha = 38 \times 10^{-4}$/°C, determine the resistance of an aluminium conductor at 50°C, correct to 3 significant figures, when its resistance at 0°C is 24.0 Ω

(a) $\frac{dR}{d\theta} = \alpha R$ is of the form $\frac{dy}{dx} = f(y)$

Rearranging gives: $d\theta = \frac{dR}{\alpha R}$

Integrating both sides gives: $\int d\theta = \int \frac{dR}{\alpha R}$

i.e. $\theta = \frac{1}{\alpha}\ln R + c$, which is the general solution

Substituting the boundary conditions $R = R_0$ when $\theta = 0$ gives:

$$0 = \frac{1}{\alpha}\ln R_0 + c \text{ from which } c = -\frac{1}{\alpha}\ln R_0$$

Hence the particular solution is

$$\theta = \frac{1}{\alpha}\ln R - \frac{1}{\alpha}\ln R_0 = \frac{1}{\alpha}(\ln R - \ln R_0)$$

i.e. $\theta = \frac{1}{\alpha}\ln\left(\frac{R}{R_0}\right)$ or $\alpha\theta = \ln\left(\frac{R}{R_0}\right)$

Hence $e^{\alpha\theta} = \frac{R}{R_0}$ from which, $\mathbf{R = R_0\, e^{\alpha\theta}}$

(b) Substituting $\alpha = 38 \times 10^{-4}$, $R_0 = 24.0$ and $\theta = 50$ into $R = R_0\, e^{\alpha\theta}$ gives the resistance at 50°C, i.e.

$\mathbf{R_{50}} = 24.0\ e^{(38\times 10^{-4}\times 50)} = \mathbf{29.0\ ohms}$

13.3 The solution of equations of the form $\frac{dy}{dx} = f(x) \,.\, f(y)$

A differential equation of the form $\frac{dy}{dx} = f(x).f(y)$, where f(x) is a function of x only and f(y) is a function of y only, may be rearranged as $\frac{dy}{f(y)} = f(x)dx$, and then the solution is obtained by direct integration, i.e.

$$\int \frac{dy}{f(y)} = \int f(x)dx$$

Application: Solve the equation $4xy\frac{dy}{dx} = y^2 - 1$

Separating the variables gives: $\left(\frac{4y}{y^2 - 1}\right)dy = \frac{1}{x}dx$

Integrating both sides gives: $\int \left(\frac{4y}{y^2 - 1}\right) dy = \int \left(\frac{1}{x}\right) dx$

Using the substitution $u = y^2 - 1$, the general solution is:

$$\mathbf{2 \ln (y^2 - 1) = \ln x + c} \qquad (1)$$

or $\ln (y^2 - 1)^2 - \ln x = c$

from which, $\ln \left\{\frac{(y^2 - 1)^2}{x}\right\} = c$

and $$\mathbf{\frac{(y^2 - 1)^2}{x} = e^c} \qquad (2)$$

If in equation (1), $c = \ln A$, where A is a different constant,

then $\ln (y^2 - 1)^2 = \ln x + \ln A$

i.e. $\ln (y^2 - 1)^2 = \ln Ax$

i.e. $$\mathbf{(y^2 - 1)^2 = Ax} \qquad (3)$$

Equations (1) to (3) are thus three valid solutions of the differential equations $4xy \frac{dy}{dx} = y^2 - 1$

Application: The current i in an electric circuit containing resistance R and inductance L in series with a constant voltage source E is given by the differential equation $E - L\left(\frac{di}{dt}\right) = Ri$.
Solve the equation to find i in terms of time t, given that when $t = 0, i = 0$

In the R–L series circuit shown in Figure 13.1, the supply p.d., E, is given by

$$E = V_R + V_l$$

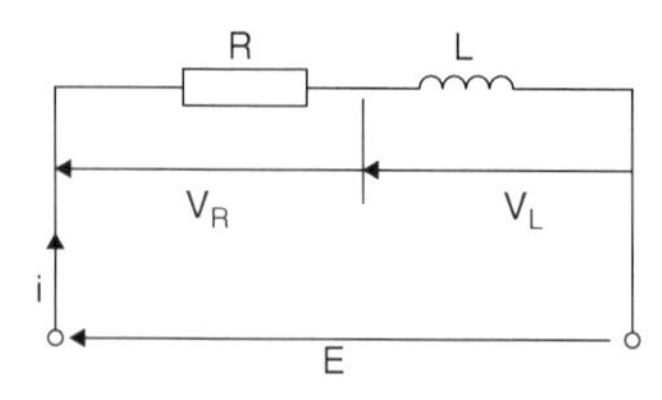

Figure 13.1

$V_R = iR$ and $V_L = L\dfrac{di}{dt}$

Hence $E = iR + L\dfrac{di}{dt}$ from which $E - L\dfrac{di}{dt} = Ri$

Most electrical circuits can be reduced to a differential equation.

Rearranging $E - L\dfrac{di}{dt} = Ri$ gives: $\dfrac{di}{dt} = \dfrac{E - Ri}{L}$

and separating the variables gives: $\dfrac{di}{E - Ri} = \dfrac{dt}{L}$

Integrating both sides gives: $\displaystyle\int \frac{di}{E - Ri} = \int \frac{dt}{L}$

Hence the general solution is: $-\dfrac{1}{R}\ln(E - Ri) = \dfrac{t}{L} + c$

(by making a substitution $u = E - Ri$, see chapter 12)

When $t = 0$, $i = 0$, thus $-\dfrac{1}{R}\ln E = c$

Thus the particular solution is: $-\dfrac{1}{R}\ln(E - Ri) = \dfrac{t}{L} - \dfrac{1}{R}\ln E$

Transposing gives: $-\dfrac{1}{R}\ln(E - Ri) + \dfrac{1}{R}\ln E = \dfrac{t}{L}$

$$\frac{1}{R}[\ln E - \ln(E - Ri)] = \frac{t}{L}$$

$$\ln\left(\frac{E}{E - Ri}\right) = \frac{Rt}{L} \text{ from which } \frac{E}{E - Ri} = e^{\frac{Rt}{L}}$$

Hence $\frac{E - Ri}{E} = e^{-\frac{Rt}{L}}$ and $E - Ri = E\,e^{-\frac{Rt}{L}}$ and $Ri = E - E\,e^{-\frac{Rt}{L}}$

Hence current, $\mathbf{i = \frac{E}{R}\left(1 - e^{-\frac{Rt}{L}}\right)}$ which represents the law of growth of current in an inductive circuit as shown in Figure 13.2.

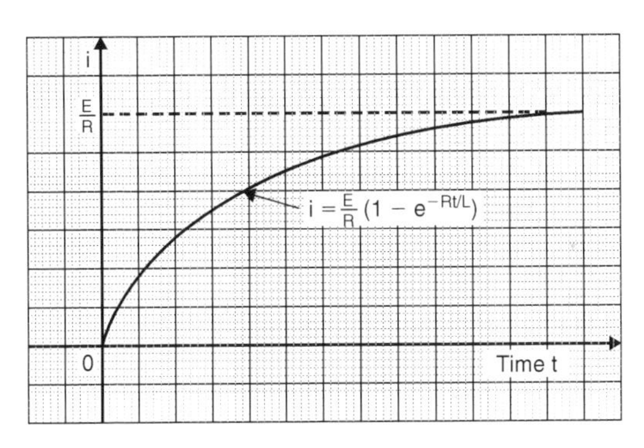

Figure 13.2

13.4 Homogeneous first order differential equations

Procedure to solve differential equations of the form $P\frac{dy}{dx} = Q$

1. Rearrange $P\frac{dy}{dx} = Q$ into the form $\frac{dy}{dx} = \frac{P}{Q}$.
2. Make the substitution $y = vx$ (where v is a function of x), from which, $\frac{dy}{dx} = v(1) + x\frac{dv}{dx}$ by the product rule.
3. Substitute for both y and $\frac{dy}{dx}$ in the equation $\frac{dy}{dx} = \frac{P}{Q}$.

   Simplify, by cancelling, and an equation results in which the variables are separable.

4. Separate the variables and solve.
5. Substitute $v = \frac{y}{x}$ to solve in terms of the original variables.

Application: Determine the particular solution of the equation $x\frac{dy}{dx} = \frac{x^2 + y^2}{y}$, given the boundary conditions that $x = 1$ when $y = 4$

Using the above procedure:

1. Rearranging $x\frac{dy}{dx} = \frac{x^2 + y^2}{y}$ gives $\frac{dy}{dx} = \frac{x^2 + y^2}{xy}$ which is homogeneous in x and y since each of the three terms on the right hand side are of the same degree (i.e. degree 2).
2. Let $y = vx$ then $\frac{dy}{dx} = v(1) + x\frac{dv}{dx}$
3. Substituting for y and $\frac{dy}{dx}$ in the equation $\frac{dy}{dx} = \frac{x^2 + y^2}{xy}$ gives:

$$v + x\frac{dv}{dx} = \frac{x^2 + (vx)^2}{x(vx)} = \frac{x^2 + v^2x^2}{vx^2} = \frac{1 + v^2}{v}$$

4. Separating the variables give:

$$x\frac{dv}{dx} = \frac{1 + v^2}{v} - v = \frac{1 + v^2 - v^2}{v} = \frac{1}{v}$$

Hence, $v\,dv = \frac{1}{x}dx$

Integrating both sides gives:

$$\int v\,dv = \int \frac{1}{x}dx \text{ i.e. } \frac{v^2}{2} = \ln x + c$$

5. Replacing v by $\frac{y}{x}$ gives: $\frac{y^2}{2x^2} = \ln x + c$, which is the general solution.

   When x = 1, y = 4, thus: $\frac{16}{2} = \ln 1 + c$, from which, c = 8

   Hence, **the particular solution is:** $\mathbf{\frac{y^2}{2x^2} = \ln x + 8}$ or

   $$\mathbf{y^2 = 2x^2 (\ln x + 8)}$$

13.5 Linear first order differential equations

Procedure to solve differential equations of the form $\frac{dy}{dx} + Py = Q$

1. Rearrange the differential equation into the form $\frac{dy}{dx} + Py = Q$, where P and Q are functions of x
2. Determine $\int P\,dx$
3. Determine the integrating factor $e^{\int P\,dx}$
4. Substitute $e^{\int P\,dx}$ into the equation:

   $$y\,e^{\int P\,dx} = \int e^{\int P\,dx}\,Q\,dx \qquad (1)$$

5. Integrate the right hand side of equation (1) to give the general solution of the differential equation. Given boundary conditions, the particular solution may be determined.

Application: Solve the differential equation $\frac{1}{x}\frac{dy}{dx} + 4y = 2$, given the boundary conditions x = 0 when y = 4

Using the above procedure:

1. Rearranging gives $\frac{dy}{dx} + 4xy = 2x$, which is of the form

   $\frac{dy}{dx} + Py = Q$ where P = 4x and Q = 2x

2. $\int P\,dx = \int 4x\,dx = 2x^2$

3. Integrating factor, $e^{\int P\,dx} = e^{2x^2}$

4. Substituting into equation (1) gives:

$$y\,e^{2x^2} = \int e^{2x^2}(2x)\,dx$$

5. Hence the general solution is: $y\,e^{2x^2} = \frac{1}{2}e^{2x^2} + c$, by using the substitution $u = 2x^2$

   When x = 0, y = 4, thus $4e^0 = \frac{1}{2}e^0 + c$, from which, $c = \frac{7}{2}$

   Hence the particular solution is:

$$y\,e^{2x^2} = \frac{1}{2}e^{2x^2} + \frac{7}{2}$$

i.e. $\mathbf{y = \frac{1}{2} + \frac{7}{2}e^{-2x^2}}$ or $\mathbf{y = \frac{1}{2}(1 + 7e^{-2x^2})}$

13.6 Second order differential equations of the form $a\frac{d^2y}{dx^2}+b\frac{dy}{dx}+cy=0$

Procedure to solve differential equations of the form

$$a\frac{d^2y}{dx^2}+b\frac{dy}{dx}+cy=0$$

1. Rewrite the differential equation $a\frac{d^2y}{dx^2}+b\frac{dy}{dx}+cy=0$ as $(aD^2+bD+c)y=0$
2. Substitute m for D and solve the auxiliary equation $am^2+bm+c=0$ for m
3. If the roots of the auxiliary equation are:
   (a) **real and different,** say $m=\alpha$ and $m=\beta$, then the general solution is: $\mathbf{y=Ae^{\alpha x}+Be^{\beta x}}$
   (b) **real and equal**, say $m=\alpha$ twice, then the general solution is: $\mathbf{y=(Ax+B)e^{\alpha x}}$
   (c) **complex**, say $m=\alpha\pm j\beta$, then the general solution is: $\mathbf{y=e^{\alpha x}\{A\cos\beta x+B\sin\beta x\}}$
4. Given boundary conditions, constants A and B may be determined and the **particular solution** of the differential equation obtained. The particular solution obtained with differential equations may be verified by substituting expressions for y, $\frac{dy}{dx}$ and $\frac{d^2y}{dx^2}$ into the original equation.

Application: The oscillations of a heavily damped pendulum satisfy the differential equation $\frac{d^2x}{dt^2}+6\frac{dx}{dt}+8x=0$, where x cm is the displacement of the bob at time t seconds.

The initial displacement is equal to + 4 cm and the initial velocity $\left(\text{i.e. }\frac{dx}{dt}\right)$ is 8 cm/s. Solve the equation for x.

Using the above procedure:

1. $\frac{d^2x}{dt^2} + 6\frac{dx}{dt} + 8x = 0$ in D-operator form is $(D^2 + 6D + 8)x = 0$, where $D \equiv \frac{d}{dt}$

2. The auxiliary equation is $m^2 + 6m + 8 = 0$
Factorising gives: $(m + 2)(m + 4) = 0$, from which, $m = -2$ or $m = -4$

3. Since the roots are real and different, **the general solution is:**

   $\mathbf{x = Ae^{-2t} + Be^{-4t}}$

4. Initial displacement means that time $t = 0$. At this instant, $x = 4$

   Thus $$4 = A + B \quad (1)$$

   Velocity, $\frac{dx}{dt} = -2Ae^{-2t} - 4Be^{-4t}$

   $\frac{dx}{dt} = 8$ cm/s when $t = 0$, thus $8 = -2A - 4B \quad (2)$

   From equations (1) and (2), $A = 12$ and $B = -8$

   Hence the particular solution is: $x = 12e^{-2t} - 8e^{-4t}$

   i.e. **displacement,** $\mathbf{x = 4(3e^{-2t} - 2e^{-4t})}$ **cm**

Application: The equation $\frac{d^2i}{dt^2} + \frac{R}{L}\frac{di}{dt} + \frac{1}{LC}i = 0$ represents a current i flowing in an electrical circuit containing resistance R, inductance L and capacitance C connected in series. If $R = 200$ ohms, $L = 0.20$ henry and $C = 20 \times 10^{-6}$ farads, solve the equation for i given the boundary conditions that when $t = 0$, $i = 0$ and $\frac{di}{dt} = 100$

Using the procedure:

1. $\frac{d^2i}{dt^2} + \frac{R}{L}\frac{di}{dt} + \frac{1}{LC}i = 0$ in D-operator form is

   $\left(D^2 + \frac{R}{L}D + \frac{1}{LC}\right)i = 0$ where $D \equiv \frac{d}{dt}$

2. The auxiliary equation is $m^2 + \frac{R}{L}m + \frac{1}{LC} = 0$

   Hence, $m = \dfrac{-\frac{R}{L} \pm \sqrt{\left[\left(\frac{R}{L}\right)^2 - 4(1)\left(\frac{1}{LC}\right)\right]}}{2}$

   When $R = 200$, $L = 0.20$ and $C = 20 \times 10^{-6}$,

   then $m = \dfrac{-\frac{200}{0.20} \pm \sqrt{\left[\left(\frac{200}{0.20}\right)^2 - \frac{4}{(0.20)(20 \times 10^{-6})}\right]}}{2}$

   $= \dfrac{-1000 \pm \sqrt{0}}{2} = -500$

3. Since the two roots are real and equal (i.e. -500 twice, since for a second order differential equation there must be two solutions), **the general solution is: $i = (At + B)e^{-500t}$**

4. When $t = 0$, $i = 0$, hence $B = 0$

   $\frac{di}{dt} = (At + B)(-500e^{-500t}) + (e^{-500t})(A)$ by the product rule

   When $t = 0$, $\frac{di}{dt} = 100$, thus $100 = -500B + A$

   i.e. $A = 100$, since $B = 0$

   Hence the particular solution is: $i = 100\,te^{-500t}$

Application: The equation of motion of a body oscillating on the end of a spring is $\frac{d^2x}{dt^2} + 100x = 0$, where x is the displacement in metres of the body from its equilibrium position after time t seconds. Determine x in terms of t given that at time $t = 0$, $x = 2\,m$ and $\frac{dx}{dt} = 0$

An equation of the form $\frac{d^2x}{dt^2} + m^2x = 0$ is a differential equation representing simple harmonic motion (S.H.M.). Using the procedure:

1. $\frac{d^2x}{dt^2} + 100x = 0$ in D-operator form is $(D^2 + 100)x = 0$

2. The auxiliary equation is $m^2 + 100 = 0$, i.e. $m^2 = -100$ and
   $m = \sqrt{-100}$
   i.e. $m = \pm j10$

3. Since the roots are complex, the general solution is:
   $x = e^0(A\cos 10t + B\sin 10t)$,

   i.e. $\mathbf{x = (A\cos 10t + B\sin 10t)}$ **metres**

4. When $t = 0$, $x = 2$, thus $2 = A$

   $\frac{dx}{dt} = -10A\ \sin 10t + 10B\cos 10t$

   When $t = 0$, $\frac{dx}{dt} = 0$ thus $0 = -10A\sin 0 + 10B\cos 0$ i.e. $B = 0$

   Hence the particular solution is: $\mathbf{x = 2\cos 10t}$ **metres**

13.7 Second order differential equations of the form $a\frac{d^2y}{dx^2} + b\frac{dy}{dx} + cy = f(x)$

Procedure to solve differential equations of the form $a\frac{d^2y}{dx^2} + b\frac{dy}{dx} + cy = f(x)$

1. Rewrite the given differential equation as $(aD^2 + bD + c)y = f(x)$
2. Substitute m for D, and solve the auxiliary equation $am^2 + bm + c = 0$ for m
3. Obtain the **complementary function**, u, which is achieved using the same procedure as on page 375
4. To determine the **particular integral**, v , firstly assume a particular integral which is suggested by f(x), but which contains undetermined coefficients. Table 13.1 gives some suggested substitutions for different functions f(x).
5. Substitute the suggested P.I. into the differential equation $(aD^2 + bD + c)v = f(x)$ and equate relevant coefficients to find the constants introduced.
6. The general solution is given by y = C.F. + P.I. i.e. $y = u + v$
7. Given boundary conditions, arbitrary constants in the C.F. may be determined and the particular solution of the differential equation obtained.

Table 13.1 Form of particular integral for different functions

Type	Straightforward cases Try as particular integral:	'Snag' cases Try as particular integral:
(a) f(x) = a constant	$v = k$	$v = kx$ (used when C.F. contains a constant)

Table 13.1 Continued

Type	Straightforward cases Try as particular integral:	'Snag' cases Try as particular integral:
(b) $f(x)$ = polynomial (i.e. $f(x) = L + Mx + Nx^2 + ..$ where any of the coefficients may be zero)	$v = a + bx + cx^2 + ..$	
(c) $f(x)$ = an exponential function (i.e. $f(x) = Ae^{ax}$)	$v = ke^{ax}$	(i) $v = kxe^{ax}$ (used when e^{ax} appears in the C.F.) (ii) $v = kx^2e^{ax}$ (used when e^{ax} **and** xe^{ax} both appear in the C.F.)
(d) $f(x)$ = a sine or cosine function (i.e. $f(x) = a\sin px + b\cos px$ where a or b may be zero)	$v = A\sin px + B\cos px$	$v = x(A\sin px + B\cos px)$ (used when sin px and/or cos px appears in the C.F.)
(e) $f(x)$ = a sum e.g. (i) $f(x) = 4x^2 - 3\sin 2x$ (ii) $f(x) = 2 - x + e^{3x}$	(i) $v = ax^2 + bx + c + d\sin 2x + e\cos 2x$ (ii) $v = ax + b + ce^{3x}$	
(f) $f(x)$ = a product e.g. $f(x) = 2e^x\cos 2x$	$v = e^x(A\sin 2x + B\cos 2x)$	

Application: In a galvanometer the deflection θ satisfies the differential equation $\frac{d^2\theta}{dt^2} + 4\frac{d\theta}{dt} + 4\theta = 8$. Solve the equation for θ given that when $t = 0$, $\theta = \frac{d\theta}{dt} = 2$

1. $\frac{d^2\theta}{dt^2} + 4\frac{d\theta}{dt} + 4\theta = 8$ in D-operator form is: $(D^2 + 4D + 4)\theta = 8$

2. Auxiliary equation is: $m^2 + 4m + 4 = 0$
   i.e. $(m + 2)(m + 2) = 0$
   from which, $m = -2$ twice

3. Hence, C.F., $u = (At + B)e^{-2t}$

4. Let the particular integral, P.I., $v = k$

5. Substituting $v = k$ gives: $(D^2 + 4D + 4)k = 8$

   $D(k) = 0$ and $D^2(k) = D(0) = 0$

   Hence, $4k = 8$ from which, $k = 2$

   Hence, P.I., $v = 2$

6. The general solution, $\theta = u + v = (At + B)e^{-2t} + 2$

7. $t = 0$ and $\theta = 2$, hence, $2 = B + 2$ from which, $B = 0$

   $$\frac{d\theta}{dt} = (At + B)(-2e^{-2t}) + (e^{-2t})(A)$$

   $x = 0$ and $\frac{d\theta}{dt} = 2$, hence, $2 = -2B + A$ from which, $A = 2$

   Hence, $\theta = 2te^{-2t} + 2$

   i.e. $\boldsymbol{\theta = 2(te^{-2t} + 1)}$

Application: Solve $2\frac{d^2y}{dx^2} - 11\frac{dy}{dx} + 12y = 3x - 2$

1. $2\frac{d^2y}{dx^2} - 11\frac{dy}{dx} + 12y = 3x - 2$ in D-operator form is

   $(2D^2 - 11D + 12)y = 3x - 2$

2. Substituting m for D gives the auxiliary equation
   $2m^2 - 11m + 12 = 0$

   Factorising gives: $(2m - 3)(m - 4) = 0$, from which, $m = \frac{3}{2}$ or $m = 4$

3. Since the roots are real and different, the C.F.,

   $\boldsymbol{u = Ae^{\frac{3}{2}x} + Be^{4x}}$

4. Since $f(x) = 3x - 2$ is a polynomial, let the P.I., $v = ax + b$ (see Table 13.1(b))

5. Substituting $v = ax + b$ into $(2D^2 - 11D + 12)v = 3x - 2$ gives:

$$(2D^2 - 11D + 12)(ax + b) = 3x - 2,$$

i.e. $$2D^2(ax + b) - 11D(ax + b) + 12(ax + b) = 3x - 2$$

i.e. $$0 - 11a + 12ax + 12b = 3x - 2$$

Equating the coefficients of x gives: $12a = 3$, from which, $a = \frac{1}{4}$

Equating the constant terms gives: $-11a + 12b = -2$

i.e. $-11\left(\frac{1}{4}\right) + 12b = -2$ from which, $12b = -2 + \frac{11}{4} = \frac{3}{4}$

i.e. $b = \frac{1}{16}$

Hence the P.I., $\mathbf{v} = ax + b = \mathbf{\frac{1}{4}x + \frac{1}{16}}$

6. The general solution is given by $y = u + v$

i.e. $\mathbf{y = Ae^{\frac{3}{2}x} + Be^{4x} + \frac{1}{4}x + \frac{1}{16}}$

Application: Solve $\frac{d^2y}{dx^2} - 2\frac{dy}{dx} + y = 3e^{4x}$ given that when $x = 0$, $y = -\frac{2}{3}$ and $\frac{dy}{dx} = 4\frac{1}{3}$

1. $\frac{d^2y}{dx^2} - 2\frac{dy}{dx} + y = 3e^{4x}$ in D-operator form is

$(D^2 - 2D + 1)y = 3e^{4x}$

2. Substituting m for D gives the auxiliary equation $m^2 - 2m + 1 = 0$

Factorising gives: $(m - 1)(m - 1) = 0$, from which, $m = 1$ twice

3. Since the roots are real and equal the C.F., $\mathbf{u = (Ax + B)e^x}$

4. Let the particular integral, $v = ke^{4x}$ (see Table 13.1(c))

5. Substituting $v = ke^{4x}$ into $(D^2 - 2D + 1)v = 3e^{4x}$ gives:

$$(D^2 - 2D + 1)ke^{4x} = 3e^{4x}$$

i.e. $D^2(ke^{4x}) - 2D(ke^{4x}) + 1(ke^{4x}) = 3e^{4x}$

i.e. $16ke^{4x} - 8ke^{4x} + ke^{4x} = 3e^{4x}$

Hence $9ke^{4x} = 3e^{4x}$, from which, $k = \frac{1}{3}$

Hence the P.I., $\mathbf{v = ke^{4x} = \frac{1}{3}e^{4x}}$

6. The general solution is given by $y = u + v$, i.e.

$$\mathbf{y = (Ax + B)e^x + \frac{1}{3}e^{4x}}$$

7. When $x = 0$, $y = -\frac{2}{3}$ thus $-\frac{2}{3} = (0 + B)e^0 + \frac{1}{3}e^0$, from which, $B = -1$

$$\frac{dy}{dx} = (Ax + B)e^x + e^x(A) + \frac{4}{3}e^{4x}$$

When $x = 0$, $\frac{dy}{dx} = 4\frac{1}{3}$, thus $\frac{13}{3} = B + A + \frac{4}{3}$ from which, $A = 4$, since $B = -1$

Hence the particular solution is: $\mathbf{y = (4x - 1)e^x + \frac{1}{3}e^{4x}}$

Application: $L\frac{d^2q}{dt^2} + R\frac{dq}{dt} + \frac{1}{C}q = V_0 \sin wt$ represents the variation of capacitor charge in an electric circuit. Determine an expression for q at time t seconds given that $R = 40\,\Omega$, $L = 0.02\,H$, $C = 50 \times 10^{-6}F$, $V_0 = 540.8\,V$ and $\omega = 200\,rad/s$ and given the boundary conditions that when $t = 0$, $q = 0$ and $\frac{dq}{dt} = 4.8$

$L\frac{d^2q}{dt^2} + R\frac{dq}{dt} + \frac{1}{C}q = V_0 \sin\omega t$ in D-operator form is:

$$\left(LD^2 + RD + \frac{1}{C}\right)q = V_0 \sin\omega t$$

The auxiliary equation is: $Lm^2 + Rm + \frac{1}{C} = 0$
and

$$m = \frac{-R \pm \sqrt{R^2 - \frac{4L}{C}}}{2L} = \frac{-40 \pm \sqrt{40^2 - \frac{4(0.02)}{50 \times 10^{-6}}}}{2(0.02)} = \frac{-40 \pm \sqrt{0}}{0.04} = -1000$$

Hence, C.F., $u = (At + B)e^{-1000t}$

Let P.I., $v = A\sin\omega t + B\cos\omega t$

$$\left(LD^2 + RD + \frac{1}{C}\right)[A\sin\omega t + B\cos\omega t] = V_0 \sin\omega t$$

$D(v) = A\omega\cos\omega t - B\omega\sin\omega t$ and $D^2(v) = -A\omega^2\sin\omega t - B\omega^2\cos\omega t$

Thus,

$$\left(LD^2 + RD + \frac{1}{C}\right)v = 0.02(-A\omega^2\sin\omega t - B\omega^2\cos\omega t) + 40(A\omega\cos\omega t - B\omega\sin\omega t) + \frac{1}{50 \times 10^{-6}}(A\sin\omega t + B\cos\omega t) = V_0\sin\omega t$$

i.e. $-800A\sin 200t - 800B\cos 200t + 8000A\cos 200t - 8000B\sin 200t + 20000A\sin 200t + 20000B\cos 200t = 540.8\sin 200t$

Hence, $-800A - 8000B + 20000A = 540.8$

and $-800B + 8000A + 20000B = 0$

i.e. $19200A - 8000B = 540.8$ (1)

and $8000A + 19200B = 0$ (2)

$8 \times (1)$ gives: $153600A - 64000B = 4326.4$ (3)

$19.2 \times (2)$ gives: $153600A + 368640B = 0$ (4)

(3)–(4) gives: $-432640B = 4326.4$

from which, $B = \dfrac{4326.4}{432640} = -0.01$

Substituting in (1) gives: $19200A - 8000(-0.01) = 540.8$

i.e. $19200A + 80 = 540.8$

and $A = \dfrac{540.8 - 80}{19200} = \dfrac{460.8}{19200} = 0.024$

Hence, P.I., $v = 0.024 \sin 200t - 0.01 \cos 200t$

Thus, $q = u + v = (At + B)e^{-1000t} + 0.024 \sin 200t - 0.01 \cos 200t$

When $t = 0$, $q = 0$, hence, $0 = B - 0.01$ from which, $B = 0.01$

$$\frac{dq}{dt} = (At + B)\left(-1000e^{-1000t}\right) + Ae^{-1000t} + (0.024)(200)\cos 200t + (0.01)(200)\sin 200t$$

When $t = 0$, $\dfrac{dq}{dt} = 4.8$, hence, $4.8 = -1000B + A + 4.8$

i.e. $A = 1000B = 1000(0.01) = 10$

Thus, $\mathbf{q = (10t + 0.01)e^{-1000t} + 0.024 \sin 200t - 0.010 \cos 200t}$

13.8 Numerical methods for first order differential equations

Euler's method

$$\mathbf{y_1 = y_0 + h(y')_0} \qquad (1)$$

Application: Obtain a numerical solution of the differential equation $\dfrac{dy}{dx} = 3(1 + x) - y$ given the initial conditions that $x = 1$ when $y = 4$, for the range $x = 1.0$ to $x = 2.0$ with intervals of 0.2.

$$\frac{dy}{dx} = y' = 3(1 + x) - y$$

With $x_0 = 1$ and $y_0 = 4$, $\mathbf{(y')_0} = 3(1 + 1) - 4 = \mathbf{2}$

By Euler's method: $y_1 = y_0 + h(y')_0$, from equation (1)

Hence $\mathbf{y_1} = 4 + (0.2)(2) = \mathbf{4.4}$, since $h = 0.2$

At point Q in Figure 13.3, $x_1 = 1.2$, $y_1 = 4.4$

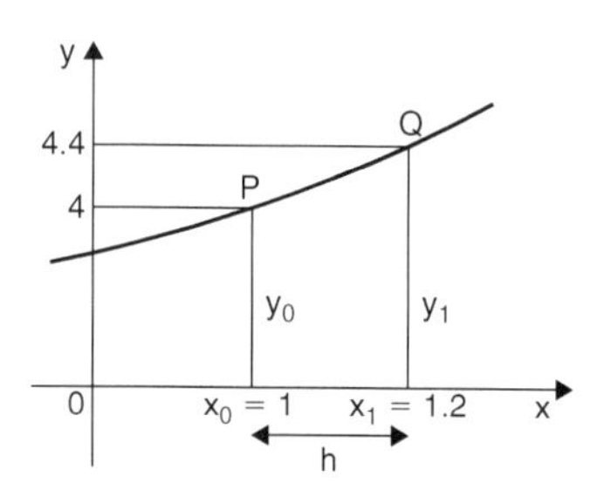

Figure 13.3

and $(y')_1 = 3(1 + x_1) - y_1$

i.e. $\mathbf{(y')_1} = 3(1 + 1.2) - 4.4 = \mathbf{2.2}$

If the values of x, y and y′ found for point Q are regarded as new starting values of x_0, y_0 and $(y')_0$, the above process can be repeated and values found for the point R shown in Figure 13.4.

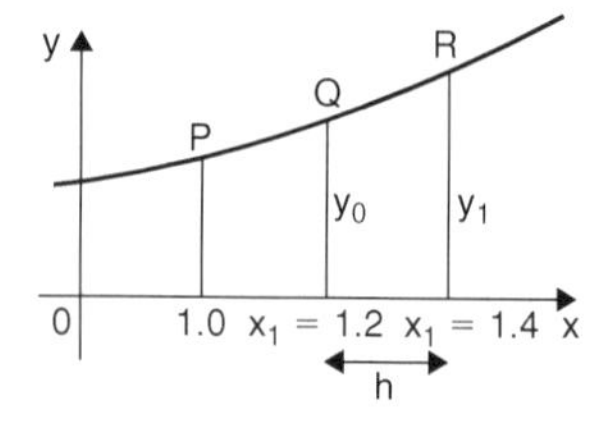

Figure 13.4

Thus at point R, $\mathbf{y_1} = y_0 + h(y')_0$ from equation (1)

$$= 4.4 + (0.2)(2.2) = \mathbf{4.84}$$

When $x_1 = 1.4$ and $y_1 = 4.84$, $\mathbf{(y')_1} = 3(1 + 1.4) - 4.84 = \mathbf{2.36}$

This step by step Euler's method can be continued and it is easiest to list the results in a table, as shown in Table 13.2. The results for lines 1 to 3 have been produced above.

Table 13.2

	x_0	y_0	$(y')_0$
1.	1	4	2
2.	1.2	4.4	2.2
3.	1.4	4.84	2.36
4.	1.6	5.312	2.488
5.	1.8	5.8096	2.5904
6.	2.0	6.32768	

For line 4, where $x_0 = 1.6$: $\mathbf{y_1} = y_0 + h(y')_0$

$$= 4.84 + (0.2)(2.36) = \mathbf{5.312}$$

and $\mathbf{(y')_0} = 3(1 + 1.6) - 5.312 = \mathbf{2.488}$

For line 5, where $x_0 = 1.8$: $\mathbf{y_1} = y_0 + h(y')_0$

$$= 5.312 + (0.2)(2.488) = \mathbf{5.8096}$$

and $\mathbf{(y')_0} = 3(1 + 1.8) - 5.8096 = \mathbf{2.5904}$

For line 6, where $x_0 = 2.0$: $\mathbf{y_1} = y_0 + h\,(y')_0$

$$= 5.8096 + (0.2)(2.5904) = \mathbf{6.32768}$$

(As the range is 1.0 to 2.0 there is no need to calculate $(y')_0$ in line 6)

The particular solution is given by the value of y against x.

A graph of the solution of $\frac{dy}{dx} = 3(1 + x) - y$ with initial conditions $x = 1$ and $y = 4$ is shown in Figure 13.5.

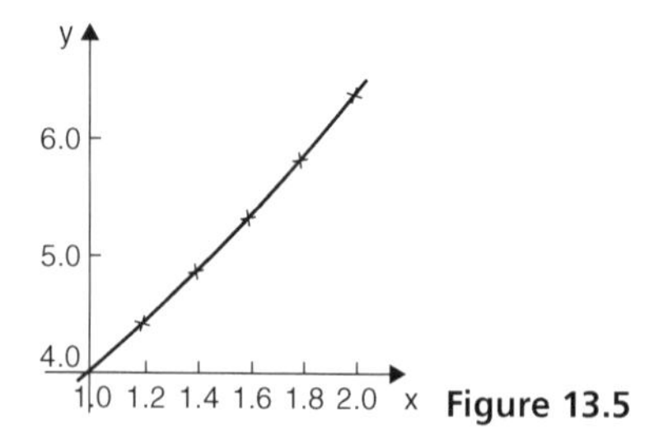

Figure 13.5

In practice it is probably best to plot the graph as each calculation is made, which checks that there is a smooth progression and that no calculation errors have occurred.

Euler-Cauchy method

$$\mathbf{y_{P_1} = y_0 + h(y')_0} \qquad (2)$$

$$\mathbf{y_{C_1} = y_0 + \frac{1}{2}h[(y')_0 + f(x_1, y_{P_1})]} \qquad (3)$$

Application: Applying the Euler-Cauchy method, solve the differential equation $\frac{dy}{dx} = y - x$ in the range 0(0.1)0.5, given the initial conditions that at $x = 0$, $y = 2$

$\frac{dy}{dx} = y' = y - x$

Since the initial conditions are $x_0 = 0$ and $y_0 = 2$ then $(y')_0 = 2 - 0 = 2$

Interval $h = 0.1$, hence $x_1 = x_0 + h = 0 + 0.1 = 0.1$

From equation (2), $y_{P_1} = y_0 + h(y')_0 = 2 + (0.1)(2) = 2.2$

From equation (3), $y_{C_1} = y_0 + \frac{1}{2}h[(y')_0 + f(x_1, y_{P_1})]$

$$= y_0 + \frac{1}{2}h[(y')_0 + (y_{P_1} - x_1)] \text{ in this case}$$

$$= 2 + \frac{1}{2}(0.1)[2 + (2.2 - 0.1)] = \mathbf{2.205}$$

$$(y')_1 = y_{C_1} - x_1 = 2.205 - 0.1 = 2.105$$

If a table of values is produced, as in Euler's method, lines 1 and 2 has so far been determined for Table 13.3.

Table 13.3

	x	y	y'
1.	0	2	2
2.	0.1	2.205	2.105
3.	0.2	2.421025	2.221025
4.	0.3	2.649232625	2.349232625
5.	0.4	2.89090205	2.49090205
6.	0.5	3.147446765	

The results in line 2 are now taken as x_0, y_0 and $(y')_0$ for the next interval and the process is repeated.

For line 3, $x_1 = 0.2$

$$y_{P_1} = y_0 + h(y')_0 = 2.205 + (0.1)(2.105) = 2.4155$$

$$y_{C_1} = y_0 + \frac{1}{2}h[(y')_0 + f(x_1, y_{P_1})]$$

$$= 2.205 + \frac{1}{2}(0.1)\,[2.105 + (2.4155 - 0.2)] = \mathbf{2.421025}$$

$$(y')_0 = y_{C_1} - x_1 = 2.421025 - 0.2 = 2.221025$$

For line 4, $x_1 = 0.3$

$$y_{P_1} = y_0 + h(y')_0 = 2.421025 + (0.1)(2.221025) = 2.6431275$$

$$y_{C_1} = y_0 + \frac{1}{2}h[(y')_0 + f(x_1, y_{P_1})]$$

$$= 2.421025 + \frac{1}{2}(0.1)\,[2.221025 + (2.6431275 - 0.3)]$$

$$= \mathbf{2.649232625}$$

$$(y')_0 = y_{C_1} - x_1 = 2.649232625 - 0.3 = 2.349232625$$

For line 5, $x_1 = 0.4$

$$y_{P_1} = y_0 + h(y')_0 = 2.649232625 + (0.1)(2.349232625)$$
$$= 2.884155887$$

$$y_{C_1} = y_0 + \frac{1}{2}h[(y')_0 + f(x_1, y_{P_1})]$$
$$= 2.649232625 + \frac{1}{2}(0.1)\,[2.349232625 + (2.884155887 - 0.4)] = \mathbf{2.89090205}$$

$$(y')_0 = y_{C_1} - x_1 = 2.89090205 - 0.4 = 2.49090205$$

For line 6, $x_1 = 0.5$

$$y_{P_1} = y_0 + h(y')_0 = 2.89090205 + (0.1)(2.49090205)$$
$$= 3.139992255$$

$$y_{C_1} = y_0 + \frac{1}{2}h[(y')_0 + f(x_1, y_{P_1})]$$
$$= 2.89090205 + \frac{1}{2}(0.1)\,[2.49090205 + (3.139992255 - 0.5)] = \mathbf{3.147446765}$$

Runge-Kutta method

To solve the differential equation $\frac{dy}{dx} = f(x, y)$ given the initial condition $y = y_0$ at $x = x_0$ for a range of values of $x = x_0(h)x_n$:

1. Identify x_0, y_0 and h, and values of $x_1, x_2, x_3, \ldots$
2. Evaluate $k_1 = f(x_n, y_n)$ starting with $n = 0$
3. Evaluate $k_2 = f\left(x_n + \frac{h}{2}, y_n + \frac{h}{2}k_1\right)$
4. Evaluate $k_3 = f\left(x_n + \frac{h}{2}, y_n + \frac{h}{2}k_2\right)$

5. Evaluate $k_4 = f(x_n + h, y_n + hk_3)$
6. Use the values determined from steps 2 to 5 to evaluate:

$$y_{n+1} = y_n + \frac{h}{6}\{k_1 + 2k_2 + 2k_3 + k_4\}$$

7. Repeat steps 2 to 6 for n = 1, 2, 3, ...

Application: Use the Runge-Kutta method to solve the differential equation: $\frac{dy}{dx} = y - x$ in the range 0(0.1)0.5, given the initial conditions that at x = 0, y = 2

Using the above procedure:

1. $x_0 = 0$, $y_0 = 2$ and since h = 0.1, and the range is from x = 0 to x = 0.5, then
   $x_1 = 0.1$, $x_2 = 0.2$, $x_3 = 0.3$, $x_4 = 0.4$, and $x_5 = 0.5$

Let n = 0 to determine y_1:

2. $k_1 = f(x_0, y_0) = f(0, 2)$; since $\frac{dy}{dx} = y - x$, $f(0, 2) = 2 - 0 = \mathbf{2}$

3. $k_2 = f\left(x_0 + \frac{h}{2}, y_0 + \frac{h}{2}k_1\right) = f\left(0 + \frac{0.1}{2}, 2 + \frac{0.1}{2}(2)\right) = f(0.05, 2.1)$

$$= 2.1 - 0.05 = \mathbf{2.05}$$

4. $k_3 = f\left(x_0 + \frac{h}{2}, y_0 + \frac{h}{2}k_2\right) = f\left(0 + \frac{0.1}{2}, 2 + \frac{0.1}{2}(2.05)\right)$

$$= f(0.05, 2.1025)$$

$$= 2.1025 - 0.05 = \mathbf{2.0525}$$

5. $k_4 = f(x_0 + h, y_0 + hk_3) = f(0 + 0.1, 2 + 0.1(2.0525))$

$$= f(0.1, 2.20525)$$

$$= 2.20525 - 0.1 = \mathbf{2.10525}$$

6. $y_{n+1} = y_n + \frac{h}{6}\{k_1 + 2k_2 + 2k_3 + k_4\}$ and when n = 0:

$$\mathbf{y_1} = y_0 + \frac{h}{6}\{k_1 + 2k_2 + 2k_3 + k_4\}$$
$$= 2 + \frac{0.1}{6}\{2 + 2(2.05) + 2(2.0525) + 2.10525\}$$
$$= 2 + \frac{0.1}{6}\{12.31025\} = \mathbf{2.205171}$$

A table of values may be constructed as shown in Table 13.4. The working has been shown for the first two rows.

Table 13.4

n	x_n	k_1	k_2	k_3	k_4	y_n
0	**0**					**2**
1	**0.1**	2.0	2.05	2.0525	2.10525	**2.205171**
2	**0.2**	2.105171	2.160430	2.163193	2.221490	**2.421403**
3	**0.3**	2.221403	2.282473	2.285527	2.349956	**2.649859**
4	**0.4**	2.349859	2.417339	2.420726	2.491932	**2.891824**
5	**0.5**	2.491824	2.566415	2.570145	2.648838	**3.148720**

Let n = 1 to determine y_2:

2. $k_1 = f(x_1, y_1) = f(0.1, 2.205171)$; since $\frac{dy}{dx} = y - x$,

$$f(0.1, 2.205171) = 2.205171 - 0.1 = \mathbf{2.105171}$$

3. $$k_2 = f\left(x_1 + \frac{h}{2}, y_1 + \frac{h}{2}k_1\right)$$
$$= f\left(0.1 + \frac{0.1}{2}, 2.205171 + \frac{0.1}{2}(2.105171)\right)$$
$$= f(0.15, 2.31042955)$$
$$= 2.31042955 - 0.15 = \mathbf{2.160430}$$

4. $k_3 = f\left(x_1 + \frac{h}{2}, y_1 + \frac{h}{2}k_2\right)$

$$= f\left(0.1 + \frac{0.1}{2},\ 2.205171 + \frac{0.1}{2}(2.160430)\right)$$

$$= f(0.15, 2.3131925) = 2.3131925 - 0.15 = \mathbf{2.163193}$$

5. $k_4 = f(x_1 + h, y_1 + hk_3) = f(0.1 + 0.1, 2.205171 + 0.1(2.163193))$

$$= f(0.2, 2.421490) = 2.421490 - 0.2$$

$$= \mathbf{2.221490}$$

6. $y_{n+1} = y_n + \frac{h}{6}\{k_1 + 2k_2 + 2k_3 + k_4\}$ and when $n = 1$:

$$\mathbf{y_2} = y_1 + \frac{h}{6}\{k_1 + 2k_2 + 2k_3 + k_4\}$$

$$= 2.205171 + \frac{0.1}{6}\{2.105171 + 2(2.160430) + 2(2.163193) + 2.221490\}$$

$$= 2.205171 + \frac{0.1}{6}\{12.973907\} = \mathbf{2.421403}$$

This completes the third row of Table 13.4. In a similar manner y_3, y_4 and y_5 can be calculated and the results are as shown in Table 13.4.

This problem is the same as the application on page 388 which used the Euler-Cauchy method, and a comparison of results can be made.

The differential equation $\frac{dy}{dx} = y - x$ may be solved analytically using the integrating factor method shown on page 373, with the solution: $\mathbf{y = x + 1 + e^x}$

Substituting values of x of 0, 0.1, 0.2, …., 0.5 will give the exact values. A comparison of the results obtained by Euler's method (which is left to the reader to produce), the Euler-Cauchy method and the Runga-Kutta method, together with the exact values is shown in Table 13.5.

Table 13.5

x	Euler's method y	Euler-Cauchy method Y	Runge-Kutta method y	Exact value $y = x + 1 + e^x$
0	2	2	2	2
0.1	2.2	2.205	2.205171	2.205170918
0.2	2.41	2.421025	2.421403	2.421402758
0.3	2.631	2.649232625	2.649859	2.649858808
0.4	2.8641	2.89090205	2.891824	2.891824698
0.5	3.11051	3.147446765	3.148720	3.148721271

It is seen from Table 13.5 that **the Runge-Kutta method is exact, correct to 5 decimal places**.

13.9 Power series methods of solving ordinary differential equations

Leibniz's theorem

To find the n'th derivative of a product $y = uv$:

$$y^{(n)} = (uv)^{(n)} = u^{(n)}v + nu^{(n-1)}v^{(1)} + \frac{n(n-1)}{2!}u^{(n-2)}v^{(2)} + \frac{n(n-1)(n-2)}{3!}u^{(n-3)}v^{(3)} + \cdots \quad (4)$$

Application: Find the 5'th derivative of $y = x^4 \sin x$

If $y = x^4 \sin x$, then using Leibniz's equation with $u = \sin x$ and $v = x^4$ gives:

$$y^{(n)} = \left[\sin\left(x + \frac{n\pi}{2}\right)x^4\right] + n\left[\sin\left(x + \frac{(n-1)\pi}{2}\right)4x^3\right]$$

$$+ \frac{n(n-1)}{2!}\left[\sin\left(x + \frac{(n-2)\pi}{2}\right)12x^2\right]$$

$$+ \frac{n(n-1)(n-2)}{3!}\left[\sin\left(x + \frac{(n-3)\pi}{2}\right)24x\right]$$

$$+ \frac{n(n-1)(n-2)(n-3)}{4!}\left[\sin\left(x + \frac{(n-4)\pi}{2}\right)24\right]$$

and

$$y^{(5)} = x^4 \sin\left(x + \frac{5\pi}{2}\right) + 20x^3 \sin(x + 2\pi) + \frac{(5)(4)}{2}(12x^2)\sin\left(x + \frac{3\pi}{2}\right)$$

$$+ \frac{(5)(4)(3)}{(3)(2)}(24x)\sin(x + \pi) + \frac{(5)(4)(3)(2)}{(4)(3)(2)}(24)\sin\left(x + \frac{\pi}{2}\right)$$

Since $\sin\left(x + \frac{5\pi}{2}\right) \equiv \sin\left(x + \frac{\pi}{2}\right) \equiv \cos x$, $\sin(x + 2\pi) \equiv \sin x$,

$$\sin\left(x + \frac{3\pi}{2}\right) \equiv -\cos x,$$

and $\sin(x + \pi) \equiv -\sin x$,

then $y^{(5)} = x^4 \cos x + 20x^3 \sin x + 120x^2(-\cos x)$

$$+ 240x(-\sin x) + 120 \cos x$$

i.e. $\mathbf{y^{(5)} = (x^4 - 120x^2 + 120)\cos x + (20x^3 - 240x)\sin x}$

Leibniz-Maclaurin method

(i) Differentiate the given equation n times, using the Leibniz theorem of equation (4),

(ii) rearrange the result to obtain the recurrence relation at $x = 0$,

(iii) determine the values of the derivatives at $x = 0$, i.e. find $(y)_0$ and $(y')_0$,

(iv) substitute in the Maclaurin expansion for $y = f(x)$ (see page 54, equation (5)),

(v) simplify the result where possible and apply boundary condition (if given).

Application: Determine the power series solution of the differential equation: $\frac{d^2y}{dx^2} + x\frac{dy}{dx} + 2y = 0$ using Leibniz-Maclaurin's method, given the boundary conditions $x = 0$, $y = 1$ and $\frac{dy}{dx} = 2$

Following the above procedure:

(i) The differential equation is rewritten as: $y'' + xy' + 2y = 0$ and from the Leibniz theorem of equation (4), each term is differentiated n times, which gives:

$$y^{(n+2)} + \left\{y^{(n+1)}(x) + n\,y^{(n)}(1) + 0\right\} + 2\,y^{(n)} = 0$$

i.e. $$y^{(n+2)} + x\,y^{(n+1)} + (n+2)\,y^{(n)} = 0 \qquad (5)$$

(ii) At $x = 0$, equation (5) becomes:

$$y^{(n+2)} + (n+2)\,y^{(n)} = 0$$

from which, $$y^{(n+2)} = -(n+2)\,y^{(n)}$$

This equation is called a **recurrence relation** or **recurrence formula**, because each recurring term depends on a previous term.

(iii) Substituting $n = 0, 1, 2, 3, \ldots$ will produce a set of relationships between the various coefficients. For

$n = 0,\quad (y'')_0 = -2\,(y)_0$

$n = 1,\quad (y''')_0 = -3\,(y')_0$

$n = 2,\ (y^{(4)})_0 = -4(y'')_0 = -4\left\{-2(y)_0\right\} = 2 \times 4(y)_0$

$n = 3,\ (y^{(5)})_0 = -5(y''')_0 = -5\left\{-3(y')_0\right\} = 3 \times 5(y')_0$

$n = 4,\ (y^{(6)})_0 = -6(y^{(4)})_0 = -6\left\{2 \times 4(y)_0\right\} = -2 \times 4 \times 6(y)_0$

$n = 5,\ (y^{(7)})_0 = -7(y^{(5)})_0 = -7\left\{3 \times 5(y')_0\right\} = -3 \times 5 \times 7(y')_0$

$$n = 6,\ (y^{(8)})_0 = -8(y^{(6)})_0 = -8\left\{-2 \times 4 \times 6\,(y)_0\right\}$$
$$= 2 \times 4 \times 6 \times 8(y)_0$$

(iv) Maclaurin's theorem from page 54 may be written as:

$$y = (y)_0 + x\,(y')_0 + \frac{x^2}{2!}(y'')_0 + \frac{x^3}{3!}(y''')_0 + \frac{x^4}{4!}(y^{(4)})_0 + \ldots$$

Substituting the above values into Maclaurin's theorem gives:

$$y = (y)_0 + x\,(y')_0 + \frac{x^2}{2!}\left\{-2(y)_0\right\} + \frac{x^3}{3!}\left\{-3\,(y')_0\right\}$$
$$+ \frac{x^4}{4!}\left\{2 \times 4\,(y)_0\right\} + \frac{x^5}{5!}\left\{3 \times 5\,(y')_0\right\} + \frac{x^6}{6!}\left\{-2 \times 4 \times 6\,(y)_0\right\}$$
$$+ \frac{x^7}{7!}\left\{-3 \times 5 \times 7(y')_0\right\} + \frac{x^8}{8!}\left\{2 \times 4 \times 6 \times 8\,(y)_0\right\}$$

(v) Collecting similar terms together gives:

$$y = (y)_0\left\{1 - \frac{2x^2}{2!} + \frac{2 \times 4x^4}{4!} - \frac{2 \times 4 \times 6x^6}{6!} + \frac{2 \times 4 \times 6 \times 8x^8}{8!} - \ldots\right\}$$
$$+ (y')_0\left\{x - \frac{3x^3}{3!} + \frac{3 \times 5x^5}{5!} - \frac{3 \times 5 \times 7x^7}{7!} + \ldots\right\}$$

i.e.
$$y = (y)_0\left\{1 - \frac{x^2}{1} + \frac{x^4}{1 \times 3} - \frac{x^6}{3 \times 5} + \frac{x^8}{3 \times 5 \times 7} - \ldots\right\}$$
$$+ (y')_0\left\{\frac{x}{1} - \frac{x^3}{1 \times 2} + \frac{x^5}{2 \times 4} - \frac{x^7}{2 \times 4 \times 6} + \ldots\right\}$$

The boundary conditions are that at $x = 0$, $y = 1$ and $\frac{dy}{dx} = 2$, i.e. $(y)_0 = 1$ and $(y')_0 = 2$.

Hence, the power series solution of the differential equation: $\frac{d^2y}{dx^2} + x\frac{dy}{dx} + 2y = 0$ is:

$$\mathbf{y = \left\{1 - \frac{x^2}{1} + \frac{x^4}{1 \times 3} - \frac{x^6}{3 \times 5} + \frac{x^8}{3 \times 5 \times 7} - \ldots\right\}}$$
$$\mathbf{+ 2\left\{\frac{x}{1} - \frac{x^3}{1 \times 2} + \frac{x^5}{2 \times 4} - \frac{x^7}{2 \times 4 \times 6} + \ldots\right\}}$$

Frobenius method

A differential equation of the form $y'' + Py' + Qy = 0$, where P and Q are both functions of x, can be represented by a power series as follows:

(i) Assume a trial solution of the form

$$\mathbf{y = x^c\{a_0 + a_1x + a_2x^2 + a_3x^3 + \cdots + a_rx^r + \cdots\}}$$

(ii) differentiate the trial series,

(iii) substitute the results in the given differential equation,

(iv) equate coefficients of corresponding powers of the variable on each side of the equation; this enables index c and coefficients $a_1, a_2, a_3, \ldots$ from the trial solution, to be determined.

Application: Determine, using the Frobenius method, the general power series solution of the differential equation:

$$3x\frac{d^2y}{dx^2} + \frac{dy}{dx} - y = 0$$

The differential equation may be rewritten as: $3xy'' + y' - y = 0$

(i) Let a trial solution be of the form

$$y = x^c\{a_0 + a_1x + a_2x^2 + a_3x^3 + \cdots + a_rx^r + \cdots\} \quad (6)$$

where $a_0 \neq 0$,

$$\text{i.e. } y = a_0x^c + a_1x^{c+1} + a_2x^{c+2} + a_3x^{c+3} + \cdots + a_rx^{c+r} + \cdots \quad (7)$$

(ii) Differentiating equation (7) gives:

$$y' = a_0cx^{c-1} + a_1(c+1)x^c + a_2(c+2)x^{c+1} + \cdots + a_r(c+r)x^{c+r-1} + \cdots$$

and

$$y'' = a_0c(c-1)x^{c-2} + a_1c(c+1)x^{c-1} + a_2(c+1)(c+2)x^c + \cdots + a_r(c+r-1)(c+r)x^{c+r-2} + \cdots$$

(iii) Substituting y, y′ and y″ into each term of the given equation $3xy'' + y' - y = 0$ gives:

$$3xy'' = 3a_0c(c-1)x^{c-1} + 3a_1c(c+1)x^c + 3a_2(c+1)(c+2)x^{c+1} + \ldots + 3a_r(c+r-1)(c+r)x^{c+r-1} + \cdots \quad \text{(a)}$$

$$y' = a_0cx^{c-1} + a_1(c+1)x^c + a_2(c+2)x^{c+1} + \cdots + a_r(c+r)x^{c+r-1} + \cdots \quad \text{(b)}$$

$$-y = -a_0x^c - a_1x^{c+1} - a_2x^{c+2} - a_3x^{c+3} - \cdots - a_rx^{c+r} - \cdots \quad \text{(c)}$$

(iv) The sum of these three terms forms the left-hand side of the equation. Since the right-hand side is zero, the coefficients of each power of x can be equated to zero.

For example, the coefficient of x^{c-1} is equated to zero giving:

$$3a_0c(c-1) + a_0c = 0$$

or $$a_0c[3c - 3 + 1] = \mathbf{a_0\, c(3c-2) = 0} \quad (8)$$

The coefficient of x^c is equated to zero giving:

$$3a_1c(c+1) + a_1(c+1) - a_0 = 0$$

i.e. $$a_1(3c^2 + 3c + c + 1) - a_0 = a_1(3c^2 + 4c + 1) - a_0 = 0$$

or $$\mathbf{a_1(3c+1)(c+1) - a_0 = 0} \quad (9)$$

In each of series (a), (b) and (c) an x^c term is involved, after which, a general relationship can be obtained for x^{c+r}, where $r \geq 0$.

In series (a) and (b), terms in x^{c+r-1} are present; replacing r by $(r+1)$ will give the corresponding terms in x^{c+r}, which occurs in all three equations, i.e.

in series (a), $3a_{r+1}(c+r)(c+r+1)x^{c+r}$

in series (b), $a_{r+1}(c+r+1)x^{c+r}$

in series (c), $-a_rx^{c+r}$

Equating the total coefficients of x^{c+r} to zero gives:

$$3a_{r+1}(c+r)(c+r+1) + a_{r+1}(c+r+1) - a_r = 0$$

which simplifies to:

$$\mathbf{a_{r+1}\{(c+r+1)(3c+3r+1)\} - a_r = 0} \quad (10)$$

Equation (8), which was formed from the coefficients of the lowest power of x, i.e. x^{c-1}, is called the **indicial equation**, from which the value of c is obtained. From equation (8), since $a_0 \neq 0$, then

$$\mathbf{c = 0} \text{ or } \mathbf{c = \frac{2}{3}}$$

(a) When c = 0:

From equation (9), if $c = 0$, $a_1(1 \times 1) - a_0 = 0$, i.e. $\mathbf{a_1 = a_0}$

From equation (10), if $c = 0$, $a_{r+1}(r + 1)(3r + 1) - a_r = 0$,

$$\text{i.e. } \mathbf{a_{r+1} = \frac{a_r}{(r+1)(3r+1)}} \qquad r \geq 0$$

Thus,

$$\text{when } r = 1,\ a_2 = \frac{a_1}{(2 \times 4)} = \frac{a_0}{(2 \times 4)} \text{ since } a_1 = a_0$$

$$\text{when } r = 2,\ a_3 = \frac{a_2}{(3 \times 7)} = \frac{a_0}{(2 \times 4)(3 \times 7)} \text{ or } \frac{a_0}{(2 \times 3)(4 \times 7)}$$

$$\text{when } r = 3,\ a_4 = \frac{a_3}{(4 \times 10)} = \frac{a_0}{(2 \times 3 \times 4)(4 \times 7 \times 10)}$$

and so on.

From equation (6), the trial solution was:

$$y = x^c\{a_0 + a_1x + a_2x^2 + a_3x^3 + \cdots + a_rx^r + \cdots\}$$

Substituting $c = 0$ and the above values of $a_1, a_2, a_3, \ldots$ into the trial solution gives:

$$y = x^0 \left\{ a_0 + a_0x + \left(\frac{a_0}{(2 \times 4)}\right)x^2 + \left(\frac{a_0}{(2 \times 3)(4 \times 7)}\right)x^3 + \left(\frac{a_0}{(2 \times 3 \times 4)(4 \times 7 \times 10)}\right)x^4 + \cdots \right\}$$

$$\text{i.e. } y = a_0 \left\{ 1 + x + \frac{x^2}{(2 \times 4)} + \frac{x^3}{(2 \times 3)(4 \times 7)} + \frac{x^4}{(2 \times 3 \times 4)(4 \times 7 \times 10)} + \cdots \right\} \qquad (11)$$

(b) When $c = \frac{2}{3}$:

From equation (9), if $c = \frac{2}{3}$, $a_1(3)\left(\frac{5}{3}\right) - a_0 = 0$, i.e. $\mathbf{a_1 = \frac{a_0}{5}}$

From equation (10),

$$\text{if } c = \frac{2}{3},\ a_{r+1}\left(\frac{2}{3} + r + 1\right)(2 + 3r + 1) - a_r = 0,$$

i.e. $a_{r+1}\left(r + \frac{5}{3}\right)(3r + 3) - a_r = a_{r+1}(3r^2 + 8r + 5) - a_r = 0,$

i.e. $\mathbf{a_{r+1} = \frac{a_r}{(r+1)(3r+5)}}$ $\qquad r \geq 0$

Thus, when $r = 1$, $a_2 = \frac{a_1}{(2 \times 8)} = \frac{a_0}{(2 \times 5 \times 8)}$ since $a_1 = \frac{a_0}{5}$

when $r = 2$, $a_3 = \frac{a_2}{(3 \times 11)} = \frac{a_0}{(2 \times 3)(5 \times 8 \times 11)}$

when $r = 3$, $a_4 = \frac{a_3}{(4 \times 14)} = \frac{a_0}{(2 \times 3 \times 4)(5 \times 8 \times 11 \times 14)}$

and so on.

From equation (6), the trial solution was:

$$y = x^c\{a_0 + a_1x + a_2x^2 + a_3x^3 + \cdots + a_rx^r + \cdots\}$$

Substituting $c = \frac{2}{3}$ and the above values of a_1, a_2, a_3, ... into the trial solution gives:

$$y = x^{\frac{2}{3}}\left\{a_0 + \left(\frac{a_0}{5}\right)x + \left(\frac{a_0}{2 \times 5 \times 8}\right)x^2 + \left(\frac{a_0}{(2 \times 3)(5 \times 8 \times 11)}\right)x^3 + \left(\frac{a_0}{(2 \times 3 \times 4)(5 \times 8 \times 11 \times 14)}\right)x^4 + \cdots\right\}$$

i.e.

$$y = a_0x^{\frac{2}{3}}\left\{1 + \frac{x}{5} + \frac{x^2}{(2 \times 5 \times 8)} + \frac{x^3}{(2 \times 3)(5 \times 8 \times 11)} + \frac{x^4}{(2 \times 3 \times 4)(5 \times 8 \times 11 \times 14)} + \cdots\right\} \qquad (12)$$

Since a_0 is an arbitrary (non-zero) constant in each solution, its value could well be different.

Let $a_0 = A$ in equation (11), and $a_0 = B$ in equation (12). Also, if the first solution is denoted by u(x) and the second by v(x), then the general solution of the given differential equation is $y = u(x) + v(x)$. Hence,

$$\mathbf{y = A\left\{1 + x + \frac{x^2}{(2 \times 4)} + \frac{x^3}{(2 \times 3)(4 \times 7)} + \frac{x^4}{(2 \times 3 \times 4)(4 \times 7 \times 10)} + \cdots\right\}}$$

$$\mathbf{+ Bx^{\frac{2}{3}}\left\{1 + \frac{x}{5} + \frac{x^2}{(2 \times 5 \times 8)} + \frac{x^3}{(2 \times 3)(5 \times 8 \times 11)} + \frac{x^4}{(2 \times 3 \times 4)(5 \times 8 \times 11 \times 14)} + \cdots\right\}}$$

Bessel's equation

The solution of $x^2 \dfrac{d^2y}{dx^2} + x\dfrac{dy}{dx} + (x^2 - v^2)y = 0$

is: $$y = Ax^v\left\{1 - \frac{x^2}{2^2(v+1)} + \frac{x^4}{2^4 \times 2!(v+1)(v+2)} - \frac{x^6}{2^6 \times 3!(v+1)(v+2)(v+3)} + \cdots\right\}$$

$$+ Bx^{-v}\left\{1 + \frac{x^2}{2^2(v-1)} + \frac{x^4}{2^4 \times 2!(v-1)(v-2)} + \frac{x^6}{2^6 \times 3!(v-1)(v-2)(v-3)} + \cdots\right\}$$

or, in terms of **Bessel functions** and **gamma functions**:

$y = A\,J_v(x) + BJ_{-v}(x)$

$$= A\left(\frac{x}{2}\right)^{v}\left\{\frac{1}{\Gamma(v+1)} - \frac{x^2}{2^2(1!)\Gamma(v+2)} + \frac{x^4}{2^4(2!)\Gamma(v+4)} - \cdots\right\}$$

$$+ B\left(\frac{x}{2}\right)^{-v}\left\{\frac{1}{\Gamma(1-v)} - \frac{x^2}{2^2(1!)\Gamma(2-v)} + \frac{x^4}{2^4(2!)\Gamma(3-v)} - \cdots\right\}$$

In general terms: $J_v(x) = \left(\frac{x}{2}\right)^{v} \sum_{k=0}^{\infty} \frac{(-1)^k x^{2k}}{2^{2k}(k!)\Gamma(v+k+1)}$

and $J_{-v}(x) = \left(\frac{x}{2}\right)^{-v} \sum_{k=0}^{\infty} \frac{(-1)^k x^{2k}}{2^{2k}(k!)\Gamma(k-v+1)}$

and in particular:

$$J_n(x) = \left(\frac{x}{2}\right)^{n}\left\{\frac{1}{n!} - \frac{1}{(n+1)!}\left(\frac{x}{2}\right)^2 + \frac{1}{(2!)(n+2)!}\left(\frac{x}{2}\right)^4 - \cdots\right\}$$

$$J_0(x) = 1 - \frac{x^2}{2^2(1!)^2} + \frac{x^4}{2^4(2!)^2} - \frac{x^6}{2^6(3!)^2} + \cdots$$

and $J_1(x) = \frac{x}{2} - \frac{x^3}{2^3(1!)(2!)} + \frac{x^5}{2^5(2!)(3!)} - \frac{x^7}{2^7(3!)(4!)} + \cdots$

Legendre's equation

The solution of $(1-x^2)\frac{d^2y}{dx^2} - 2x\frac{dy}{dx} + k(k+1)y = 0$

is: $y = a_0\left\{1 - \frac{k(k+1)}{2!}x^2 + \frac{k(k+1)(k-2)(k+3)}{4!}x^4 - \cdots\right\}$

$$+ a_1\left\{x - \frac{(k-1)(k+2)}{3!}x^3 + \frac{(k-1)(k-3)(k+2)(k+4)}{5!}x^5 - \cdots\right\} \quad (13)$$

Legendre's polynomials

Application: Determine the Legendre polynomial $P_3(x)$

Since in $P_3(x)$, $n = k = 3$, then from the second part of equation (13), i.e. the odd powers of x:

$$y = a_1\left\{x - \frac{(k-1)(k+2)}{3!}x^3 + \frac{(k-1)(k-3)(k+2)(k+4)}{5!}x^5 - \ldots\right\}$$

i.e. $$y = a_1\left\{x - \frac{(2)(5)}{3!}x^3 + \frac{(2)(0)(5)(7)}{5!}x^5\right\} = a_1\left\{x - \frac{5}{3}x^3 + 0\right\}$$

a_1 is chosen to make $y = 1$ when $x = 1$.

i.e. $$1 = a_1\left\{1 - \frac{5}{3}\right\} = a_1\left(-\frac{2}{3}\right) \text{ from which, } a_1 = -\frac{3}{2}$$

Hence, $P_3(x) = -\frac{3}{2}\left(x - \frac{5}{3}x^3\right)$ or $\mathbf{P_3(x) = \frac{1}{2}(5x^3 - 3x)}$

Rodrigue's formula

$$\mathbf{P_n(x) = \frac{1}{2^n n!}\frac{d^n(x^2-1)^n}{dx^n}}$$

Application: Determine the Legendre polynomial $P_3(x)$ using Rodrigue's formula

In Rodrigue's formula, $P_n(x) = \frac{1}{2^n\,n!}\frac{d^n(x^2-1)^n}{dx^n}$ and when $n = 3$,

$$P_3(x) = \frac{1}{2^3\,3!}\frac{d^3(x^2-1)^3}{dx^3} = \frac{1}{2^3(6)}\frac{d^3(x^2-1)(x^4-2x^2+1)}{dx^3}$$

$$= \frac{1}{(8)(6)}\frac{d^3(x^6-3x^4+3x^2-1)}{dx^3}$$

$$\frac{d(x^6-3x^4+3x^2-1)}{dx} = 6x^5 - 12x^3 + 6x$$

$$\frac{d(6x^5 - 12x^3 + 6x)}{dx} = 30x^4 - 36x^2 + 6$$

and $$\frac{d(30x^4 - 36x^2 + 6)}{dx} = 120x^3 - 72x$$

Hence, $$P_3(x) = \frac{1}{(8)(6)} \frac{d^3(x^6 - 3x^4 + 3x^2 - 1)}{dx^3}.$$

$$= \frac{1}{(8)(6)}(120x^3 - 72x) = \frac{1}{8}(20x^3 - 12x)$$

i.e. $\mathbf{P_3(x) = \frac{1}{2}(5x^3 - 3x)}$ the same as in the previous application.

13.10 Solution of partial differential equations

By direct partial integration

Application: Solve the differential equation $\frac{\partial^2 u}{\partial x^2} = 6x^2\,(2y - 1)$ given the boundary conditions that at $x = 0$, $\frac{\partial u}{\partial x} = \sin 2y$ and $u = \cos y$

Since $\frac{\partial^2 u}{\partial x^2} = 6x^2\,(2y - 1)$ then integrating partially with respect to x gives:

$$\frac{\partial u}{\partial x} = \int 6x^2\,(2y - 1)\,dx = (2y - 1)\int 6x^2\,dx = (2y - 1)\frac{6x^3}{3} + f(y)$$

$$= 2x^3\,(2y - 1) + f(y)$$

where f(y) is an arbitrary function.

From the boundary conditions, when $x = 0$, $\frac{\partial u}{\partial x} = \sin 2y$

Hence, $\sin 2y = 2(0)^3\,(2y - 1) + f(y)$ from which, $f(y) = \sin 2y$

Now $\frac{\partial u}{\partial x} = 2x^3(2y - 1) + \sin 2y$

Integrating partially with respect to x gives:

$$u = \int [2x^3(2y-1) + \sin 2y]\,dx = \frac{2x^4}{4}(2y-1) + x(\sin 2y) + F(y)$$

From the boundary conditions, when $x = 0$, $u = \cos y$, hence

$$\cos y = \frac{(0)^4}{2}(2y-1) + (0)\sin 2y + F(y)$$

from which, $F(y) = \cos y$

Hence, the solution of $\frac{\partial^2 u}{\partial x^2} = 6x^2(2y-1)$ for the given boundary conditions is:

$$\mathbf{u = \frac{x^4}{2}(2y-1) + x\sin y + \cos y}$$

The wave equation

The **wave equation** is given by: $\frac{\partial^2 u}{\partial x^2} = \frac{1}{c^2}\frac{\partial^2 u}{\partial t^2}$

where $c^2 = \frac{T}{\rho}$, with T being the tension in a string and ρ being the mass /unit length of the string.

Summary of solution of the wave equation

1. Identify clearly the initial and boundary conditions.
2. Assume a solution of the form $u = XT$ and express the equations in terms of X and T and their derivatives.
3. Separate the variables by transposing the equation and equate each side to a constant, say, μ; two separate equations are obtained, one in x and the other in t.
4. Let $\mu = -p^2$ to give an oscillatory solution.

5. The two solutions are of the form: $X = A\cos px + B\sin px$ and $T = C\cos cpt + D\sin cpt$
   Then $u(x, t) = \{A\cos px + B\sin px\}\{C\cos cpt + D\sin cpt\}$
6. Apply the boundary conditions to determine constants A and B.
7. Determine the general solution as an infinite sum.
8. Apply the remaining initial and boundary conditions and determine the coefficients A_n and B_n from equations (14) and (15) below:

$$\mathbf{A_n = \frac{2}{L}\int_0^L f(x)\sin\frac{n\pi x}{L}\,dx} \quad \text{for } n = 1, 2, 3, \ldots \tag{14}$$

$$\mathbf{B_n = \frac{2}{cn\pi}\int_0^L g(x)\sin\frac{n\pi x}{L}\,dx} \tag{15}$$

Application: Figure 13.6 shows a stretched string of length 50 cm which is set oscillating by displacing its mid-point a distance of 2 cm from its rest position and releasing it with zero velocity. Solve the wave equation: $\frac{\partial^2 u}{\partial x^2} = \frac{1}{c^2}\frac{\partial^2 u}{\partial t^2}$ where $c^2 = 1$, to determine the resulting motion $u(x, t)$.

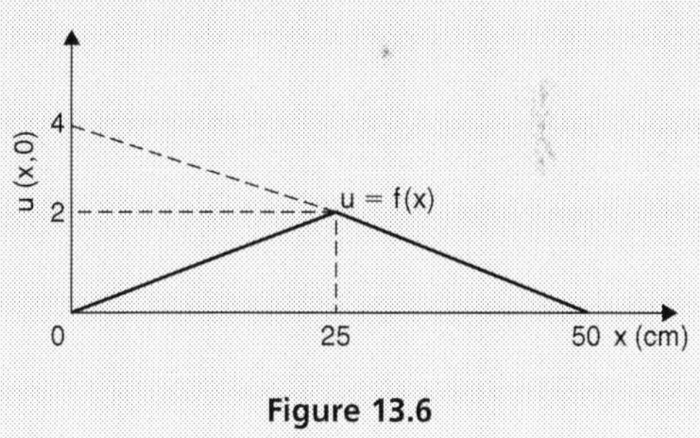

Figure 13.6

Following the above procedure:

1. The boundary and initial conditions given are:

$$\left.\begin{aligned} u(0, t) &= 0 \\ u(50, t) &= 0 \end{aligned}\right\} \quad \text{i.e. fixed end points}$$

$$u(x, 0) = \mathbf{f(x) = \frac{2}{25}x} \quad 0 \leq x \leq 25$$

$$= -\frac{2}{25}x + 4 = \mathbf{\frac{100 - 2x}{25}} \quad 25 \leq x \leq 50$$

(Note: $y = mx + c$ is a straight line graph, so the gradient, m, between 0 and 25 is 2/25 and the y-axis intercept is zero, thus $y = f(x) = \frac{2}{25}x + 0$; between 25 and 50, the gradient $= -2/25$ and the y-axis intercept is at 4, thus $f(x) = -\frac{2}{25}x + 4$).

$$\left[\frac{\partial u}{\partial t}\right]_{t=0} = 0 \text{ i.e. zero initial velocity}$$

2. Assuming a solution $u = XT$, where X is a function of x only, and T is a function of t only,

then $\frac{\partial u}{\partial x} = X'T$ and $\frac{\partial^2 u}{\partial x^2} = X''T$ and $\frac{\partial u}{\partial y} = XT'$ and

$$\frac{\partial^2 u}{\partial y^2} = XT''$$

Substituting into the partial differential equation,

$$\frac{\partial^2 u}{\partial x^2} = \frac{1}{c^2}\frac{\partial^2 u}{\partial t^2}$$

gives: $X''T = \frac{1}{c^2}XT''$ i.e. $X''T = XT''$ since $c^2 = 1$

3. Separating the variables gives: $\frac{X''}{X} = \frac{T''}{T}$

Let constant, $\mu = \frac{X''}{X} = \frac{T''}{T}$ then $\mu = \frac{X''}{X}$ and $\mu = \frac{T''}{T}$

from which, $X'' - \mu X = 0$ and $T'' - \mu T = 0$

4. Letting $\mu = -p^2$ to give an oscillatory solution gives

   $X'' + p^2X = 0$ and $T'' + p^2 T = 0$

   The auxiliary equation for each is: $m^2 + p^2 = 0$ from which,

   $m = \sqrt{-p^2} = \pm jp$

5. Solving each equation gives: $X = A\cos px + B\sin px$ and $T = C\cos pt + D\sin pt$

   Thus, $u(x, t) = \{A\cos px + B\sin px\}\{C\cos pt + D\sin pt\}$

6. Applying the boundary conditions to determine constants A and B gives:

   (i) $u(0, t) = 0$, hence $0 = A\{C\cos pt + D\sin pt\}$ from which we conclude that $A = 0$

   Therefore, $u(x, t) = B\sin px\,\{C\cos pt + D\sin pt\}$ (a)

   (ii) $u(50, t) = 0$, hence $0 = B\sin 50p\{C\cos pt + D\sin pt\}$

   $B \neq 0$ hence $\sin 50p = 0$ from which, $50p = n\pi$ and $p = \frac{n\pi}{50}$

7. Substituting in equation (a) gives:

$$u(x, t) = B\sin\frac{n\pi x}{50}\left\{C\cos\frac{n\pi t}{50} + D\sin\frac{n\pi t}{50}\right\}$$

   or, more generally,

$$u_n(x, t) = \sum_{n=1}^{\infty}\sin\frac{n\pi x}{50}\left\{A_n\cos\frac{n\pi t}{50} + B_n\sin\frac{n\pi t}{50}\right\} \quad \text{(b)}$$

   where $A_n = BC$ and $B_n = BD$

8. From equation (14),

$$A_n = \frac{2}{L}\int_0^L f(x)\sin\frac{n\pi x}{L}\,dx$$

$$= \frac{2}{50}\left[\int_0^{25}\left(\frac{2}{25}x\right)\sin\frac{n\pi x}{50}\,dx + \int_{25}^{50}\left(\frac{100-2x}{25}\right)\sin\frac{n\pi x}{50}\,dx\right]$$

Each integral is determined using integration by parts (see chapter 12, page 323) with the result:

$$\mathbf{A_n} = \frac{16}{n^2\pi^2}\sin\frac{n\pi}{2}$$

From equation (15), $B_n = \frac{2}{cn\pi}\int_0^L g(x)\sin\frac{n\pi x}{L}\,dx$

$$\left[\frac{\partial u}{\partial t}\right]_{t=0} = 0 = g(x) \text{ thus, } B_n = 0$$

Substituting into equation (b) gives:

$$u_n(x,t) = \sum_{n=1}^{\infty}\sin\frac{n\pi x}{50}\left\{A_n\cos\frac{n\pi t}{50} + B_n\sin\frac{n\pi t}{50}\right\}$$

$$= \sum_{n=1}^{\infty}\sin\frac{n\pi x}{50}\left\{\frac{16}{n^2\pi^2}\sin\frac{n\pi}{2}\cos\frac{n\pi t}{50} + (0)\sin\frac{n\pi t}{50}\right\}$$

Hence,

$$\mathbf{u(x, t) = \frac{16}{\pi^2}\sum_{n=1}^{\infty}\frac{1}{n^2}\sin\frac{n\pi x}{50}\sin\frac{n\pi}{2}\cos\frac{n\pi t}{50}}$$

For stretched string problems as above, the main parts of the procedure are:

1. Determine A_n from equation (14).

   Note that $\frac{2}{L}\int_0^L f(x)\sin\frac{n\pi x}{L}\,dx$

   is **always** equal to $\frac{8d}{n^2\pi^2}\sin\frac{n\pi}{2}$ (see Figure 13.7)

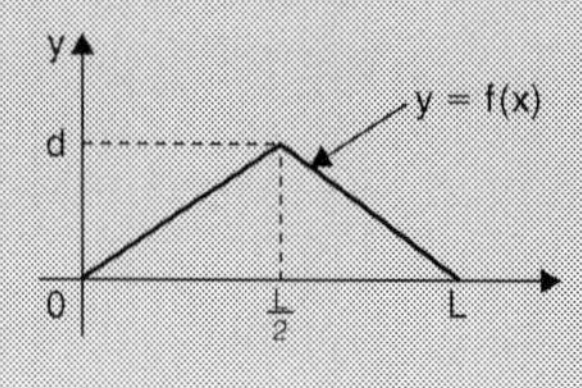

Figure 13.7

2. Determine B_n from equation (15)
3. Substitute in equation (b) to determine u(x, t)

The heat conduction equation

The **heat conduction equation** is of the form: $\frac{\partial^2 u}{\partial x^2} = \frac{1}{c^2}\frac{\partial u}{\partial t}$ where $c^2 = \frac{h}{\sigma\rho}$, with h being the thermal conductivity of the material, σ the specific heat of the material, and ρ the mass/unit length of material.

Application: A metal bar, insulated along its sides, is 1 m long. It is initially at room temperature of 15°C and at time t = 0, the ends are placed into ice at 0°C. Find an expression for the temperature at a point P at a distance x m from one end at any time t seconds after t = 0

The temperature u along the length of bar is shown in Figure 13.8

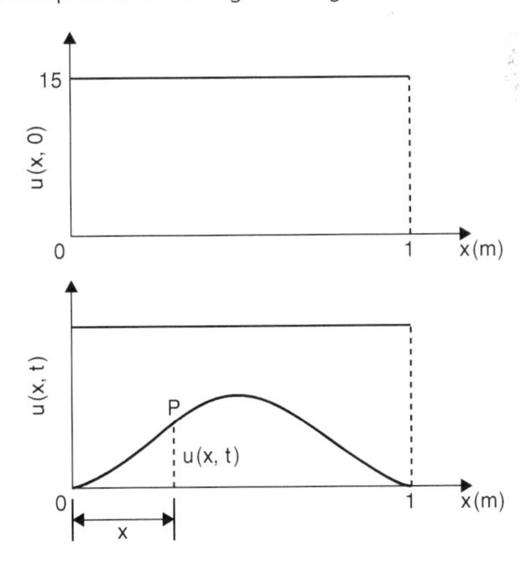

Figure 13.8

The heat conduction equation is $\frac{\partial^2 u}{\partial x^2} = \frac{1}{c^2}\frac{\partial u}{\partial t}$ and the given boundary conditions are:

$$u(0, t) = 0,\ u(1, t) = 0 \text{ and } u(x, 0) = 15$$

Assuming a solution of the form $u = XT$, then it may be shown that

$$X = A\cos px + B\sin px$$

and $\quad T = ke^{-p^2c^2t}$

Thus, the general solution is given by:

$$u(x, t) = \{P\cos px + Q\sin px\}\ e^{-p^2c^2t}$$

$u(0, t) = 0$ thus $0 = P\ e^{-p^2c^2t}$ from which, $P = 0$
and $\quad u(x, t) = \{Q\sin px\}\,e^{-p^2c^2t}$

Also, $u(1, t) = 0$ thus $0 = \{Q\sin p\}\ e^{-p^2c^2t}$

Since $Q \neq 0$, $\sin p = 0$ from which, $p = n\pi \quad$ where $n = 1, 2, 3, \ldots$

Hence, $u(x, t) = \sum_{n=1}^{\infty}\left\{Q_n e^{-p^2c^2t}\sin n\pi x\right\}$

The final initial condition given was that at $t = 0$, $u = 15$,
i.e. $u(x, 0) = f(x) = 15$

Hence, $15 = \sum_{n=1}^{\infty}\{Q_n\sin n\pi x\}$

where, from Fourier coefficients, $Q_n = 2 \times$ mean value of $15\sin n\pi x$ from $x = 0$ to $x = 1$,

$$\text{i.e.}\quad Q_n = \frac{2}{1}\int_0^1 15\sin n\pi x\,dx = 30\left[-\frac{\cos n\pi x}{n\pi}\right]_0^1$$

$$= -\frac{30}{n\pi}\left[\cos n\pi - \cos 0\right] = \frac{30}{n\pi}(1 - \cos n\pi)$$

$$= 0 \text{ (when n is even) and } \frac{60}{n\pi} \text{ (when n is odd)}$$

Hence, the required solution is:

$$\mathbf{u(x, t)} = \sum_{n=1}^{\infty}\left\{Q_n e^{-p^2c^2t} \sin n\pi x\right\}$$

$$= \frac{\mathbf{60}}{\pi} \sum_{n(odd)=1}^{\infty} \frac{\mathbf{1}}{\mathbf{n}} \mathbf{(\sin n\pi x) e^{-n^2\pi^2c^2t}}$$

Laplace's equation

Laplace's equation, used extensively with electrostatic fields, is of the form:

$$\frac{\partial^2 u}{\partial x^2} + \frac{\partial^2 u}{\partial y^2} + \frac{\partial^2 u}{\partial z^2} = 0$$

Application: A square plate is bounded by the lines $x = 0$, $y = 0$, $x = 1$ and $y = 1$. Apply the Laplace equation $\frac{\partial^2 u}{\partial x^2} + \frac{\partial^2 u}{\partial y^2} = 0$ to determine the potential distribution $u(x, y)$ over the plate, subject to the following boundary conditions:

$u = 0$ when $x = 0$ $0 \le y \le 1$,

$u = 0$ when $x = 1$ $0 \le y \le 1$,

$u = 0$ when $y = 0$ $0 \le x \le 1$,

$u = 4$ when $y = 1$ $0 \le x \le 1$

Initially a solution of the form $u(x, y) = X(x)Y(y)$ is assumed, where X is a function of x only, and Y is a function of y only. Simplifying to $u = XY$, determining partial derivatives, and substituting into $\frac{\partial^2 u}{\partial x^2} + \frac{\partial^2 u}{\partial y^2} = 0$ gives: $X''Y + XY'' = 0$

Separating the variables gives: $\frac{X''}{X} = -\frac{Y''}{Y}$

Letting each side equal a constant, $-p^2$, gives the two equations:

$$X'' + p^2X = 0 \quad \text{and} \quad Y'' - p^2Y = 0$$

from which, $X = A\cos px + B\sin px$

and $Y = Ce^{py} + De^{-py}$ or $Y = C\cosh py + D\sinh py$ or
$Y = E\sinh p(y + \phi)$

Hence $u(x, y) = XY = \{A\cos px + B\sin px\}\{E\sinh p(y + \phi)\}$

or $u(x, y) = \{P\cos px + Q\sin px\}\{\sinh p(y + \phi)\}$
where $P = AE$ and $Q = BE$

The first boundary condition is: $u(0, y) = 0$,

hence $0 = P\sinh p(y + \phi)$

from which, $P = 0$

Hence, $u(x, y) = Q\sin px \sinh p(y + \phi)$

The second boundary condition is: $u(1, y) = 0$,

hence $0 = Q\sin p(1) \sinh p(y + \phi)$

from which, $\sin p = 0$,

hence, $p = n\pi$ for $n = 1, 2, 3, \ldots$

The third boundary condition is: $u(x, 0) = 0$,

hence, $0 = Q\sin px \sinh p(\phi)$

from which, $\sinh p(\phi) = 0$ and $\phi = 0$

Hence, $u(x, y) = Q\sin px \sinh py$

Since there are many solutions for integer values of n,

$$u(x, y) = \sum_{n=1}^{\infty} Q_n \sin px \sinh py = \sum_{n=1}^{\infty} Q_n \sin n\pi x \sinh n\pi y \qquad \text{(a)}$$

The fourth boundary condition is: $u(x, 1) = 4 = f(x)$,

hence, $f(x) = \sum_{n=1}^{\infty} Q_n \sin n\pi x \sinh n\pi(1)$

From Fourier series coefficients,

$Q_n \sinh n\pi = 2 \times$ the mean value of $f(x) \sin n\pi x$ from $x=0$ to $x=1$

i.e. $$= \frac{2}{1}\int_0^1 4 \sin n\pi x \, dx = 8\left[-\frac{\cos n\pi x}{n\pi}\right]_0^1$$

$$= -\frac{8}{n\pi}(\cos n\pi - \cos 0) = \frac{8}{n\pi}(1 - \cos n\pi)$$

$$= 0 \text{ (for even values of n)}, = \frac{16}{n\pi} \text{ (for odd values of n)}$$

Hence, $$Q_n = \frac{16}{n\pi(\sinh n\pi)} = \frac{16}{n\pi} \operatorname{cosech} n\pi$$

Hence, from equation (a),

$$\mathbf{u(x, y)} = \sum_{n=1}^{\infty} Q_n \sin n\pi x \sinh n\pi y$$

$$= \boldsymbol{\frac{16}{\pi} \sum_{n(\text{odd})=1}^{\infty} \frac{1}{n} (\operatorname{cosech} n\pi \, \sin n\pi x \sinh n\pi y)}$$

---

14 Statistics and Probability

14.1 Presentation of ungrouped data

Ungrouped data can be presented diagrammatically by:

(a) **pictograms**, in which pictorial symbols are used to represent quantities,

(b) **horizontal bar charts**, having data represented by equally spaced horizontal rectangles,

(c) **vertical bar charts**, in which data are represented by equally spaced vertical rectangles,

(d) **percentage component bar chart**, where rectangles are subdivided into values corresponding to the percentage relative frequencies of the members, and

(e) **pie diagrams**, where the area of a circle represents the whole, and the areas of the sectors of the circle are made proportional to the parts that make up the whole.

Application: The number of television sets repaired in a workshop by a technician in six, one-month periods is as shown below.

Month	January	February	March	April	May	June
Number repaired	11	6	15	9	13	8

Present the data in a pictogram

This data is represented as a pictogram as shown in Figure 14.1 where each symbol represents two television sets repaired. Thus, in January, $5\frac{1}{2}$ symbols are used to represent the 11 sets repaired, in February, 3 symbols are used to represent the 6 sets repaired, and so on.

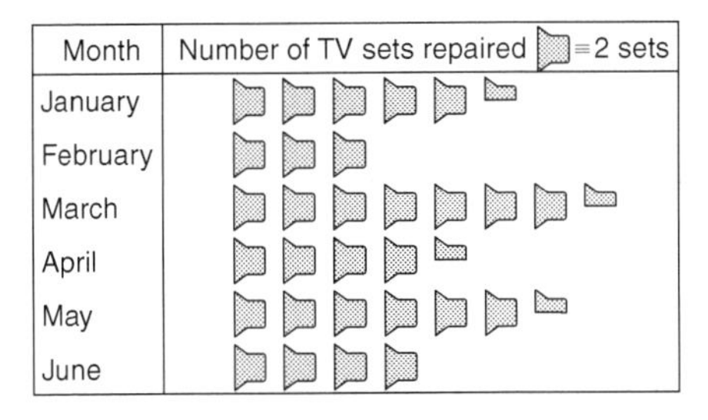

Figure 14.1

Application: The distance in miles travelled by four salesmen in a week are as shown below.

Salesmen	P	Q	R	S
Distance travelled (miles)	413	264	597	143

Represent the data by a horizontal bar chart

To represent these data diagrammatically by a horizontal bar chart, equally spaced horizontal rectangles of any width, but whose length is proportional to the distance travelled, are used. Thus, the length of the rectangle for salesman P is proportional to 413 miles, and so on. The horizontal bar chart depicting these data is shown in Figure 14.2.

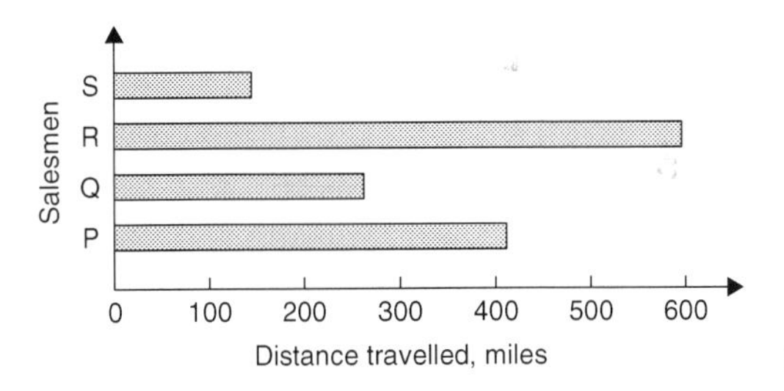

Figure 14.2

Application: The number of issues of tools or materials from a store in a factory is observed for seven, one-hour periods in a day, and the results of the survey are as follows:

Period	1	2	3	4	5	6	7
Number of issues	34	17	9	5	27	13	6

Represent the data by a vertical bar chart

In a vertical bar chart, equally spaced vertical rectangles of any width, but whose height is proportional to the quantity being represented, are used. Thus the height of the rectangle for period 1 is proportional to 34 units, and so on. The vertical bar chart depicting these data is shown in Figure 14.3.

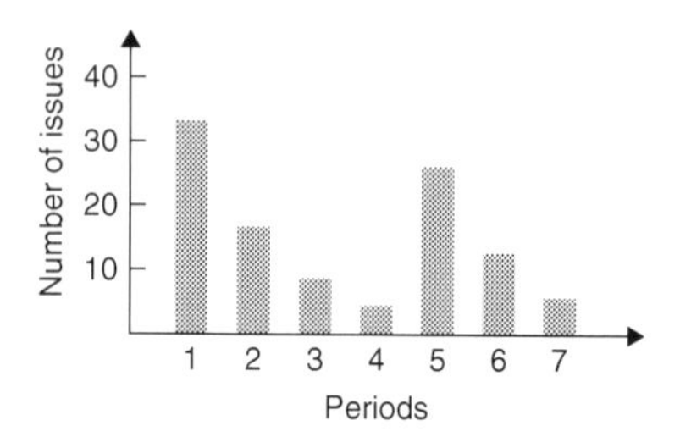

Figure 14.3

Application: The numbers of various types of dwellings sold by a company annually over a three-year period are as shown below.

	Year 1	Year 2	Year 3
4-roomed bungalows	24	17	7
5-roomed bungalows	38	71	118
4-roomed houses	44	50	53
5-roomed houses	64	82	147
6-roomed houses	30	30	25

Draw a percentage component bar chart to represent the above data

To draw a percentage component bar chart to present these data, a table of percentage relative frequency values, correct to the nearest 1%, is the first requirement. Since, percentage relative frequency = $\dfrac{\text{frequency of member} \times 100}{\text{total frequency}}$ then for 4-roomed bungalows in year 1:

$$\text{percentage relative frequency} = \frac{24 \times 100}{24 + 38 + 44 + 64 + 30} = 12\%$$

The percentage relative frequencies of the other types of dwellings for each of the three years are similarly calculated and the results are as shown in the table below.

	Year 1	Year 2	Year 3
4-roomed bungalows	12%	7%	2%
5-roomed bungalows	19%	28%	34%
4-roomed houses	22%	20%	15%
5-roomed houses	32%	33%	42%
6-roomed houses	15%	12%	7%

The percentage component bar chart is produced by constructing three equally spaced rectangles of any width, corresponding to the three years. The heights of the rectangles correspond to 100% relative frequency, and are subdivided into the values in the table of percentages shown above. A key is used (different types of shading or different colour schemes) to indicate corresponding percentage values in the rows of the table of percentages. The percentage component bar chart is shown in Figure 14.4.

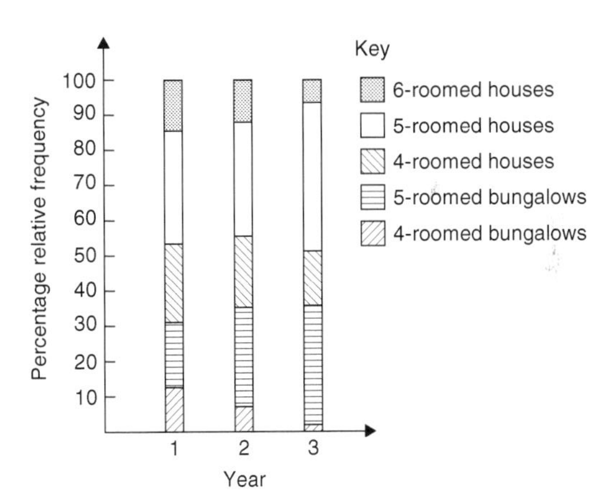

Figure 14.4

Application: The retail price of a product costing £2 is made up as follows: materials 10p, labour 20p, research and development 40p, overheads 70p, profit 60p.

Present this data on a pie diagram

To present these data on a pie diagram, a circle of any radius is drawn, and the area of the circle represents the whole, which in this case is £2. The circle is subdivided into sectors so that the areas of the sectors are proportional to the parts, i.e. the parts that make up the total retail price. For the area of a sector to be proportional to a part, the angle at the centre of the circle must be proportional to that part. The whole, £2 or 200p, corresponds to 360°. Therefore,

$$10\text{p corresponds to } 360 \times \frac{10}{200} \text{ degrees, i.e. } 18^\circ$$

$$20\text{p corresponds to } 360 \times \frac{20}{200} \text{ degrees, i.e. } 36^\circ$$

and so on, giving the angles at the centre of the circle for the parts of the retail price as: 18°, 36°, 72°, 126° and 108°, respectively.

The pie diagram is shown in Figure 14.5.

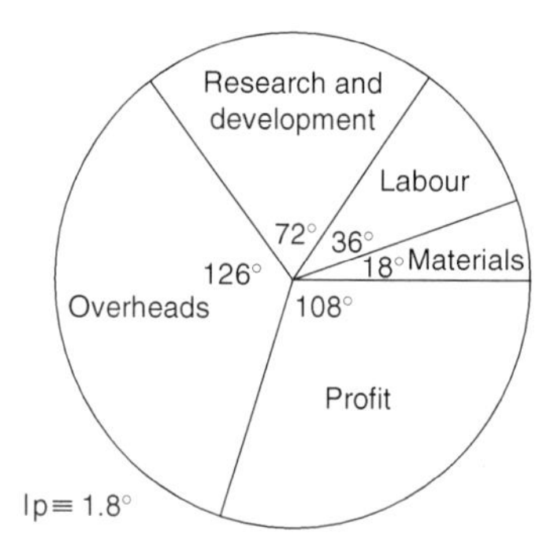

Figure 14.5

14.2 Presentation of grouped data

Grouped data can be presented diagrammatically by:

(a) a **histogram**, in which the **areas** of vertical, adjacent rectangles are made proportional to frequencies of the classes,
(b) a **frequency polygon**, which is the graph produced by plotting frequency against class mid-point values and joining the co-ordinates with straight lines,

(c) a **cumulative frequency distribution**, which is a table showing the cumulative frequency for each value of upper class boundary, and
(d) an **ogive** or a **cumulative frequency distribution curve**, which is a curve obtained by joining the co-ordinates of cumulative frequency (vertically) against upper class boundary (horizontally).

Application: The masses of 50 ingots, in kilograms, are measured correct to the nearest 0.1 kg and the results are as shown below.

8.0	8.6	8.2	7.5	8.0	9.1	8.5	7.6	8.2	7.8
8.3	7.1	8.1	8.3	8.7	7.8	8.7	8.5	8.4	8.5
7.7	8.4	7.9	8.8	7.2	8.1	7.8	8.2	7.7	7.5
8.1	7.4	8.8	8.0	8.4	8.5	8.1	7.3	9.0	8.6
7.4	8.2	8.4	7.7	8.3	8.2	7.9	8.5	7.9	8.0

Produce for this data (a) a frequency distribution for 7 classes, (b) a frequency polygon, (c) a histogram, (d) a cumulative frequency distribution, and (e) an ogive.

(a) The **range** of the data is the member having the largest value minus the member having the smallest value. Inspection of the set of data shows that: range $= 9.1 - 7.1 = 2.0$

The size of each class is given approximately by

$$\frac{\text{range}}{\text{number of classes}}$$

If about seven classes are required, the size of each class is 2.0/7, that is approximately 0.3, and thus the **class limits** are selected as 7.1 to 7.3, 7.4 to 7.6, 7.7 to 7.9, and so on.

The **class mid-point** for the 7.1 to 7.3 class is $\frac{7.35 + 7.05}{2}$, i.e. 7.2, for the 7.4 to 7.6 class is $\frac{7.65 + 7.35}{2}$ i.e. 7.5, and so on.

To assist with accurately determining the number in each class, a **tally diagram** is produced as shown in Table 14.1. This is obtained by listing the classes in the left-hand column and then inspecting each of the 50 members of the set of data in turn and allocating it to the appropriate class by putting a '1' in the appropriate row. Each fifth '1' allocated to a particular row is marked as an oblique line to help with final counting.

A **frequency distribution** for the data is shown in Table 14.2 and lists classes and their corresponding frequencies. Class mid-points are also shown in this table, since they are used when constructing the frequency polygon and histogram.

Table 14.1

Class	Tally
7.1 to 7.3	111
7.4 to 7.6	~~1111~~
7.7 to 7.9	~~1111~~ 1111
8.0 to 8.2	~~1111~~ ~~1111~~ 1111
8.3 to 8.5	~~1111~~ ~~1111~~ 1
8.6 to 8.8	~~1111~~ 1
8.9 to 9.1	11

Table 14.2

Class	Class mid-point	Frequency
7.1 to 7.3	7.2	3
7.4 to 7.6	7.5	5
7.7 to 7.9	7.8	9
8.0 to 8.2	8.1	14
8.3 to 8.5	8.4	11
8.6 to 8.8	8.7	6
8.9 to 9.1	9.0	2

(b) A **frequency polygon** is shown in Figure 14.6, the co-ordinates corresponding to the class mid-point/frequency values, given in Table 14.2. The co-ordinates are joined by straight lines and the polygon is 'anchored-down' at each end by joining to the next class mid-point value and zero frequency.

(c) A **histogram** is shown in Figure 14.7, the width of a rectangle corresponding to (upper class boundary value – lower class boundary value) and height corresponding to the class frequency. The easiest way to draw a histogram is to mark class mid-point values on the horizontal scale and to draw the rectangles symmetrically about the appropriate class mid-point values and touching one another. A histogram for the data given in Table 14.2 is shown in Figure 14.7.

(d) A **cumulative frequency distribution** is a table giving values of cumulative frequency for the values of upper class boundaries, and is shown in Table 14.3. Columns 1 and 2 show the classes and

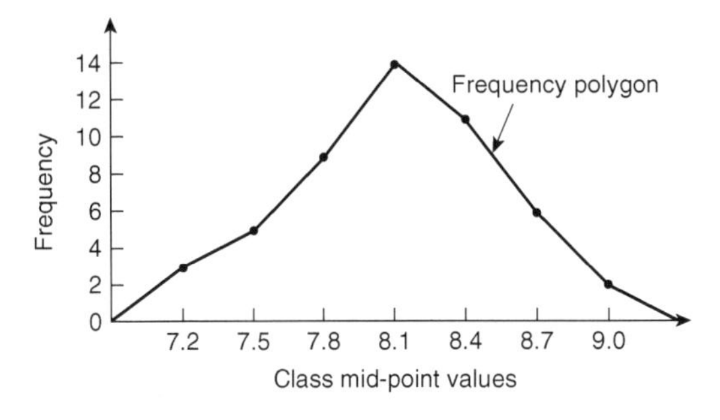

Figure 14.6

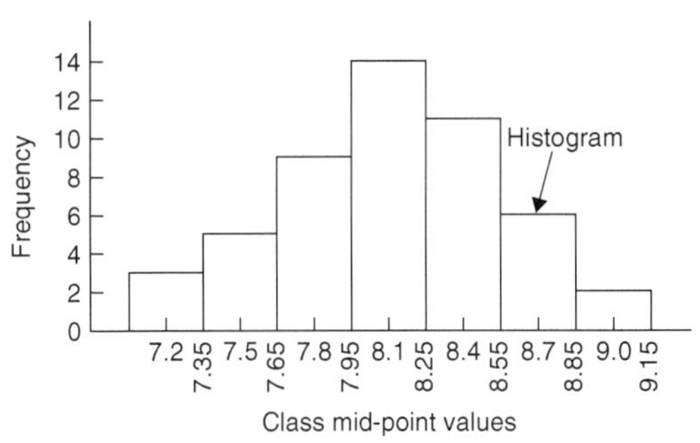

Figure 14.7

Table 14.3

1 **Class**	**2** **Frequency**	**3** **Upper class boundary**	**4** **Cumulative frequency**
		Less than	
7.1–7.3	3	7.35	3
7.4–7.6	5	7.65	8
7.7–7.9	9	7.95	17
8.0–8.2	14	8.25	31
8.3–8.5	11	8.55	42
8.6–8.8	6	8.85	48
8.9–9.1	2	9.15	50

their frequencies. Column 3 lists the upper class boundary values for the classes given in column 1. Column 4 gives the cumulative frequency values for all frequencies less than the upper class boundary values given in column 3. Thus, for example, for the 7.7 to 7.9 class shown in row 3, the cumulative frequency value is the sum of all frequencies having values of less than 7.95, i.e. $3 + 5 + 9 = 17$, and so on.

(e) The **ogive** for the cumulative frequency distribution given in Table 14.3 is shown in Figure 14.8. The co-ordinates corresponding to each upper class boundary/cumulative frequency value are plotted and the co-ordinates are joined by straight lines (– not the best curve drawn through the co-ordinates as in experimental work). The ogive is 'anchored' at its start by adding the co-ordinate (7.05, 0).

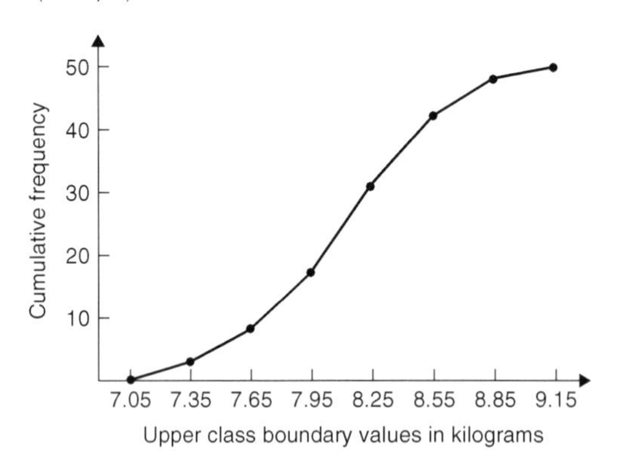

Figure 14.8

14.3 Measures of central tendency

(a) Discrete data

mean value, $\bar{x} = \dfrac{\sum x}{n}$

the **median** is the middle term of a ranked set of data,

the **mode** is the most commonly occurring value in a set of data, and

standard deviation, $\sigma = \sqrt{\left[\dfrac{\sum (x - \bar{x})^2}{n}\right]}$

Application: Find the median of the set {7, 5, 74, 10}

The set: {7, 5, 74, 10} is ranked as {5, 7, 10, 74}, and since it contains an even number of members (four in this case), the mean of 7 and 10 is taken, giving a median value of **8.5**

Application: Find the median of the set {3, 81, 15, 7, 14}

The set: {3, 81, 15, 7, 14} is ranked as {3, 7, 14, 15, 81} and the median value is the value of the middle member, i.e. **14**

Application: Find the modal value of the set {5, 6, 8, 2, 5, 4, 6, 5, 3}

The set: {5, 6, 8, 2, 5, 4, 6, 5, 3} has a modal value of **5**, since the member having a value of 5 occurs three times.

Application: Find the mean, median and modal values for the set {2, 3, 7, 5, 5, 13, 1, 7, 4, 8, 3, 4, 3}

For the set {2, 3, 7, 5, 5, 13, 1, 7, 4, 8, 3, 4, 3}

mean value,

$$\bar{x} = \frac{2+3+7+5+5+13+1+7+4+8+3+4+3}{13} = \frac{65}{13} = \mathbf{5}$$

To obtain the median value the set is ranked, that is, placed in ascending order of magnitude, and since the set contains an odd number of members the value of the middle member is the median value. Ranking the set gives: {1, 2, 3, 3, 3, 4, 4, 5, 5, 7, 7, 8, 13}

The middle term is the seventh member, i.e. 4, thus the **median value is 4**.

The **modal value** is the value of the most commonly occurring member and is **3**, which occurs three times, all other members only occurring once or twice.

Application: Determine the standard deviation from the mean of the set of numbers:

{5, 6, 8, 4, 10, 3}, correct to 4 significant figures

The arithmetic mean, $\bar{x} = \dfrac{\sum x}{n} = \dfrac{5+6+8+4+10+3}{6} = 6$

Standard deviation, $\sigma = \sqrt{\left\{\dfrac{\sum (x-\bar{x})^2}{n}\right\}}$

The $(x - \bar{x})^2$ values are: $(5-6)^2, (6-6)^2, (8-6)^2, (4-6)^2, (10-6)^2$ and $(3-6)^2$

The sum of the $(x - \bar{x})^2$ values,

i.e. $\sum (x-\bar{x})^2 = 1+0+4+4+16+9 = 34$

and $\dfrac{\sum (x-\bar{x})^2}{n} = \dfrac{34}{6} = 5.\dot{6}$ since there are 6 members in the set.

Hence, **standard deviation,**

$$\sigma = \sqrt{\left\{\frac{\sum (x-\bar{x})^2}{n}\right\}} = \sqrt{5.\dot{6}}$$

$= \mathbf{2.380}$, correct to 4 significant figures.

(b) Grouped data

$$\textbf{mean value, } \bar{x} = \frac{\sum (fx)}{\sum f}$$

$$\textbf{standard deviation, } \sigma = \sqrt{\left\{\frac{\sum \{f(x-\bar{x})^2\}}{\sum f}\right\}}$$

Application: Find (a) the mean value, and (b) the standard deviation for the following values of resistance, in ohms, of 48 resistors:

20.5–20.9	3,	21.0–21.4	10,	21.5–21.9	11,
22.0–22.4	13,	22.5–22.9	9,	23.0–23.4	2

(a) The class mid-point/frequency values are:

20.7 3, 21.2 10, 21.7 11, 22.2 13, 22.7 9 and 23.2 2

For grouped data, the mean value is given by: $\bar{x} = \dfrac{\sum (f\,x)}{\sum f}$

where f is the class frequency and x is the class mid-point value. Hence

$$\text{mean value, } \bar{x} = \frac{(3 \times 20.7) + (10 \times 21.2) + (11 \times 21.7) + (13 \times 22.2) + (9 \times 22.7) + (2 \times 23.2)}{48}$$

$$= \frac{1052.1}{48} = 21.919..$$

i.e. **the mean value is 21.9 ohms**, correct to 3 significant figures.

(b) From part (a), mean value, $\bar{x} = 21.92$, correct to 4 significant figures.

The 'x-values' are the class mid-point values, i.e. 20.7, 21.2, 21.7, ...

Thus the $(x - \bar{x})^2$ values are $(20.7 - 21.92)^2$, $(21.2 - 21.92)^2$, $(21.7 - 21.92)^2$, ...,

and the $f(x - \bar{x})^2$ values are $3(20.7 - 21.92)^2$, $10(21.2 - 21.92)^2$, $11(21.7 - 21.92)^2$, ...

The $\sum f(x - \bar{x})^2$ values are $4.4652 + 5.1840 + 0.5324 + 1.0192 + 5.4756 + 3.2768 = 19.9532$

$$\frac{\sum \{f(x - \bar{x})^2\}}{\sum f} = \frac{19.9532}{48} = 0.41569$$

and **standard deviation,**

$$\sigma = \sqrt{\left\{\frac{\sum \left\{f(x - \bar{x})^2\right\}}{\sum f}\right\}} = \sqrt{0.41569}$$

$= \mathbf{0.645}$, correct to 3 significant figures

Application: The time taken in minutes to assemble a device is measured 50 times and the results are as shown below:

14.5–15.5	5,	16.5–17.5	8,	18.5–19.5	16,
20.5–21.5	12,	22.5–23.5	6,	24.5–25.5	3

Determine the mean, median and modal values of the distribution by depicting the data on a histogram

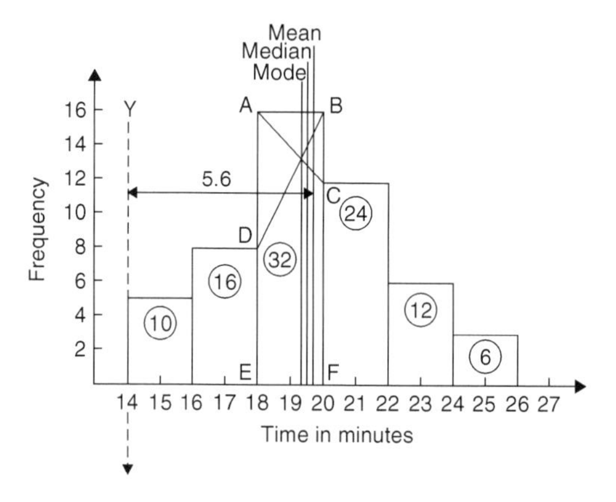

Figure 14.9

The histogram is shown in Figure 14.9. The mean value lies at the centroid of the histogram. With reference to any arbitrary axis, say YY shown at a time of 14 minutes, the position of the horizontal value of the centroid can be obtained from the relationship $AM = \sum(am)$, where A is the area of the histogram, M is the horizontal distance of the centroid from the axis YY, a is the area of a rectangle of the histogram and m is the distance of the centroid of the rectangle from YY. The areas of the individual rectangles are shown circled on the histogram giving a total area of 100 square units. The positions, m, of the centroids of the individual rectangles are 1, 3, 5, ... units from YY. Thus

$$100M = (10 \times 1) + (16 \times 3) + (32 \times 5) + (24 \times 7) + (12 \times 9) + (6 \times 11)$$

i.e. $$M = \frac{560}{100} = 5.6 \text{ units from YY}$$

Thus the position of the **mean** with reference to the time scale is $14 + 5.6$, i.e. **19.6 minutes**.

The median is the value of time corresponding to a vertical line dividing the total area of the histogram into two equal parts. The total area is 100 square units; hence the vertical line must be drawn to give 50 units of area on each side. To achieve this with reference to Figure 14.9, rectangle ABFE must be split so that 50 − (10 + 16) units of area lie on one side and 50 − (24 + 12 + 6) units of area lie on the other. This shows that the area of ABFE is split so that 24 units of area lie to the left of the line and 8 units of area lie to the right, i.e. the vertical line must pass through 19.5 minutes. Thus the **median value** of the distribution is **19.5 minutes**.

The mode is obtained by dividing the line AB, which is the height of the highest rectangle, proportionally to the heights of the adjacent rectangles. With reference to Figure 14.9, this is done by joining AC and BD and drawing a vertical line through the point of intersection of these two lines. This gives the **mode** of the distribution and is **19.3 minutes**.

14.4 Quartiles, deciles and percentiles

The **quartile values** of a set of discrete data are obtained by selecting the values of members which divide the set into four equal parts.

When a set contains a large number of members, the set can be split into ten parts, each containing an equal number of members; these ten parts are then called **deciles**.

For sets containing a very large number of members, the set may be split into one hundred parts, each containing an equal number of members; one of these parts is called a **percentile**.

Application: The frequency distribution given below refers to the overtime worked by a group of craftsmen during each of 48 working weeks in a year.

25–29 5, 30–34 4, 35–39 7, 40–44 11,
45–49 12, 50–54 8, 55–59 1

Draw an ogive for this data and hence determine the quartiles values

The cumulative frequency distribution (i.e. upper class boundary/cumulative frequency values) is:

29.5 5, 34.5 9, 39.5 16, 44.5 27,
49.5 39, 54.5 47, 59.5 48

The ogive is formed by plotting these values on a graph, as shown in Figure 14.10.

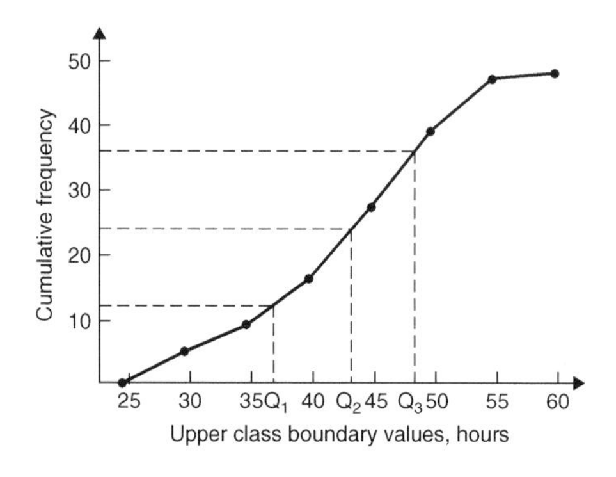

Figure 14.10

The total frequency is divided into four equal parts, each having a range of 48/4, i.e. 12. This gives cumulative frequency values of 0 to 12 corresponding to the first quartile, 12 to 24 corresponding to the second quartile, 24 to 36 corresponding to the third quartile and 36 to 48 corresponding to the fourth quartile of the distribution, i.e. the distribution is divided into four equal parts. The quartile values are those of the variable corresponding to cumulative frequency values of 12, 24 and 36, marked Q_1, Q_2 and Q_3 in Figure 14.10. These values, correct to the nearest hour, are **37 hours, 43 hours and 48 hours**, respectively. The Q_2 value is also equal to the median value of the distribution. One measure of the dispersion of a distribution is called the **semi-interquartile range** and is given by $(Q_2 - Q_1)/2$, and is $(48 - 37)/2$ in this case, i.e. **5 ½ hours.**

Application: Determine the numbers contained in the (a) 41st to 50th percentile group, and (b) 8th decile group of the set of numbers shown below:

14 22 17 21 30 28 37 7 23 32
24 17 20 22 27 19 26 21 15 29

The set is ranked, giving:

7	14	15	17	17	19	20	21	21	22
22	23	24	26	27	28	29	30	32	37

(a) There are 20 numbers in the set, hence the first 10% will be the two numbers 7 and 14, the second 10% will be 15 and 17, and so on

Thus the 41st to 50th percentile group will be the numbers **21 and 22**

(b) The first decile group is obtained by splitting the ranked set into 10 equal groups and selecting the first group, i.e. the numbers 7 and 14. The second decile group are the numbers 15 and 17, and so on.

Thus the 8th decile group contains the numbers **27 and 28**

14.5 Probability

The probability of events **A or B or C or .... . N** happening is given by

$$\mathbf{p_A + p_B + p_C + \cdots + p_N}$$

The probability of events **A and B and C and ... N** happening is given by

$$\mathbf{p_A \times p_B \times p_C \times \cdots \times p_N}$$

Application: Determine the probability of selecting at random the winning horse in a race in which 10 horses are running

Since only one of the ten horses can win, the probability of selecting at random the winning horse is $\frac{\text{number of winners}}{\text{number of horses}}$, i.e. $\mathbf{\frac{1}{10}}$ or **0.10**

Application: Determine the probability of selecting at random the winning horses in both the first and second races if there are 10 horses in each race

The probability of selecting the winning horse in the first race is $\frac{1}{10}$

The probability of selecting the winning horse in the second race is $\frac{1}{10}$

The probability of selecting the winning horses in the first **and** second race is given by the multiplication law of probability, i.e.

$$\textbf{probability} = \frac{1}{10} \times \frac{1}{10} = \frac{1}{100} \text{ or } \mathbf{0.01}$$

Application: The probability of a component failing in one year due to excessive temperature is $\frac{1}{20}$, due to excessive vibration is $\frac{1}{25}$ and due to excessive humidity is $\frac{1}{50}$. Determine the probabilities that during a one year period a component: (a) fails due to excessive temperature and excessive vibration, (b) fails due to excessive vibration or excessive humidity, and (c) will not fail due to excessive temperature and excessive humidity

Let p_A be the probability of failure due to excessive temperature, then

$$p_A = \frac{1}{20} \text{ and } \overline{p}_A = \frac{19}{20} \quad \text{(where } \overline{p}_A \text{ is the probability of not failing)}$$

Let p_B be the probability of failure due to excessive vibration, then

$$p_B = \frac{1}{25} \quad \text{and} \quad \overline{p}_B = \frac{24}{25}$$

Let p_C be the probability of failure due to excessive humidity, then

$$p_C = \frac{1}{50} \quad \text{and} \quad \overline{p}_C = \frac{49}{50}$$

(a) The probability of a component failing due to excessive temperature **and** excessive vibration is given by:

$$p_A \times p_B = \frac{1}{20} \times \frac{1}{25} = \mathbf{\frac{1}{500}} \quad \text{or} \quad \mathbf{0.002}$$

(b) The probability of a component failing due to excessive vibration **or** excessive humidity is:

$$p_B + p_C = \frac{1}{25} + \frac{1}{50} = \mathbf{\frac{3}{50}} \quad \text{or} \quad \mathbf{0.06}$$

(c) The probability that a component will not fail due to excessive temperature **and** will not fail due to excess humidity is:

$$\bar{p}_A \times \bar{p}_C = \frac{19}{20} \times \frac{49}{50} = \mathbf{\frac{931}{1000}} \quad \text{or} \quad \mathbf{0.931}$$

Application: A batch of 40 components contains 5 which are defective. If a component is drawn at random from the batch and tested and then a second component is drawn at random, calculate the probability of having one defective component, both with and without replacement.

The probability of having one defective component can be achieved in two ways. If p is the probability of drawing a defective component and q is the probability of drawing a satisfactory component, then the probability of having one defective component is given by drawing a satisfactory component and then a defective component **or** by drawing a defective component and then a satisfactory one, i.e. by $q \times p + p \times q$

With replacement:

$$p = \frac{5}{40} = \frac{1}{8} \text{ and } q = \frac{35}{40} = \frac{7}{8}$$

Hence, probbility of having one defective component is:

$$\frac{1}{8} \times \frac{7}{8} + \frac{7}{8} \times \frac{1}{8}, \text{ i.e. } \frac{7}{64} + \frac{7}{64} = \mathbf{\frac{7}{32}} \text{ or } \mathbf{0.2188}$$

Without replacement:

$p_1 = \frac{1}{8}$ and $q_1 = \frac{7}{8}$ on the first of the two draws. The batch number is now 39 for the second draw, thus, $p_2 = \frac{5}{39}$ and $q_2 = \frac{35}{39}$

$$p_1q_2 + q_1p_2 = \frac{1}{8} \times \frac{35}{39} + \frac{7}{8} \times \frac{5}{39} = \frac{35 + 35}{312} = \mathbf{\frac{70}{312}} \text{ or } \mathbf{0.2244}$$

14.6 The binomial distribution

If p is the probability that an event will happen and q is the probability that the event will not happen, then the probabilities that the event will happen 0, 1, 2, 3,..., n times in n trials are given by the successive terms of the expansion of $(q + p)^n$, taken from left to right, i.e.

$$q^n,\ nq^{n-1}p,\ \frac{n(n-1)}{2!}q^{n-2}p^2,\ \frac{n(n-1)(n-2)}{3!}q^{n-3}p^3,\ \ldots$$

Industrial inspection

The probabilities that 0, 1, 2, 3, ... , n components are defective in a sample of n components, drawn at random from a large batch of components, are given by the successive terms of the expansion of $(q + p)^n$, taken from left to right.

Application: A dice is rolled 9 times. Find the probabilities of having a 4 upwards (a) 3 times and (b) less than 4 times

Let p be the probability of having a 4 upwards. Then p = 1/6, since dice have six sides.

Let q be the probability of not having a 4 upwards. Then q = 5/6. The probabilities of having a 4 upwards 0, 1, 2.. n times are given by the successive terms of the expansion of $(q + p)^n$, taken from left to right.

From the binomial expansion:

$$(q + q)^9 = q^9 + 9q^8p + 36q^7p^2 + 84q^6p^3 + ..$$

The probability of having a 4 upwards no times is

$$q^9 = (5/6)^9 = \mathbf{0.1938}$$

The probability of having a 4 upwards once is $9q^8p = 9(5/6)^8\,(1/6)$

$$= \mathbf{0.3489}$$

The probability of having a 4 upwards twice is

$$36q^7p^2 = 36(5/6)^7(1/6)^2 = \mathbf{0.2791}$$

(a) The probability of having a 4 upwards 3 times is

$$84q^6p^3 = 84(5/6)^6\,(1/6)^3 = \mathbf{0.1302}$$

(b) The probability of having a 4 upwards less than 4 times is the sum of the probabilities of having a 4 upwards 0,1, 2, and 3 times, i.e.

$$0.1938 + 0.3489 + 0.2791 + 0.1302 = \mathbf{0.9520}$$

Application: A package contains 50 similar components and inspection shows that four have been damaged during transit. If six components are drawn at random from the contents of the package, determine the probabilities that in this sample (a) one and (b) less than three are damaged

The probability of a component being damaged, p, is 4 in 50, i.e. 0.08 per unit. Thus, the probability of a component not being damaged, q, is 1 − 0.08, i.e. 0.92

The probability of there being 0, 1, 2,..., 6 damaged components is given by the successive terms of $(q + p)^6$, taken from left to right.

$$(q + p)^6 = q^6 + 6q^5p + 15q^4p^2 + 20q^3p^3 + \cdots$$

(a) The probability of one damaged component is

$$6q^5p = 6 \times 0.92^5 \times 0.08 = \mathbf{0.3164}$$

(b) The probability of less than three damaged components is given by the sum of the probabilities of 0, 1 and 2 damaged components, i.e.

$$\begin{aligned} q^6 + 6q^5p + 15q^4p^2 &= 0.92^6 + 6 \times 0.92^5 \times 0.08 \\ &\quad + 15 \times 0.92^4 \times 0.08^2 \\ &= 0.6064 + 0.3164 + 0.0688 = \mathbf{0.9916} \end{aligned}$$

14.7 The Poisson distribution

If λ is the expectation of the occurrence of an event then the probability of 0, 1, 2, 3, ... occurrences is given by:

$$e^{-\lambda}, \lambda e^{-\lambda}, \lambda^2 \frac{e^{-\lambda}}{2!}, \lambda^3 \frac{e^{-\lambda}}{3!}, \ldots$$

Application: If 3% of the gearwheels produced by a company are defective, determine the probabilities that in a sample of 80 gearwheels (a) two and (b) more than two will be defective

The sample number, n, is large, the probability of a defective gearwheel, p, is small and the product np is 80×0.03, i.e. 2.4, which is less than 5. Hence a Poisson approximation to a binomial distribution may be used. The expectation of a defective gearwheel, $\lambda = np = 2.4$

The probabilities of 0, 1, 2,... defective gearwheels are given by the successive terms of the expression $e^{-\lambda}\left(1 + \lambda + \frac{\lambda^2}{2!} + \frac{\lambda^3}{3!} + \ldots\right)$ taken from left to right, i.e. by $e^{-\lambda}$, $\lambda e^{-\lambda}$, $\frac{\lambda^2 e^{-\lambda}}{2!}$, ..

The probability of no defective gearwheels is $e^{-\lambda} = e^{-2.4} = \mathbf{0.0907}$

The probability of 1 defective gearwheel is $\lambda e^{-\lambda} = 2.4e^{-2.4}$
$= \mathbf{0.2177}$

(a) the probability of 2 defective gearwheels is $\frac{\lambda^2 e^{-\lambda}}{2!} = \frac{2.4^2 e^{-2.4}}{2 \times 1}$
$= \mathbf{0.2613}$

(b) The probability of having more than 2 defective gearwheels is 1 – (the sum of the probabilities of having 0, 1, and 2 defective gearwheels), i.e.

$$1 - (0.0907 + 0.2177 + 0.2613), \text{ that is, } \mathbf{0.4303}$$

Application: A production department has 35 similar milling machines. The number of breakdowns on each machine averages 0.06 per week. Determine the probabilities of having (a) one, and (b) less than three machines breaking down in any week

Since the average occurrence of a breakdown is known but the number of times when a machine did not break down is unknown, a Poisson distribution must be used.

The expectation of a breakdown for 35 machines is 35×0.06, i.e. 2.1 breakdowns per week. The probabilities of a breakdown occurring 0, 1, 2, ... times are given by the successive terms of the expression $e^{-\lambda}\left(1+\lambda+\frac{\lambda^2}{2!}+\frac{\lambda^3}{3!}+\cdots\right)$, taken from left to right.

Hence the probability of no breakdowns $e^{-\lambda} = e^{-2.1} = \mathbf{0.1225}$

(a) The probability of 1 breakdown is $\lambda e^{-\lambda} = 2.1e^{-2.1} = \mathbf{0.2572}$

(b) The probability of 2 breakdowns is $\frac{\lambda^2 e^{-\lambda}}{2!} = \frac{2.1^2 e^{-2.1}}{2 \times 1} = \mathbf{0.2700}$

The probability of less than 3 breakdowns per week is the sum of the probabilities of 0, 1 and 2 breakdowns per week,

i.e. $0.1225 + 0.2572 + 0.2700 = \mathbf{0.6497}$

14.8 The normal distribution

A table of partial areas under the standardised normal curve is shown in Table 14.4.

Application: The mean height of 500 people is 170 cm and the standard deviation is 9 cm. Assuming the heights are normally distributed, determine (a) the number of people likely to have heights between 150 cm and 195 cm, (b) the number of people likely to have heights of less than 165 cm, and (c) the number of people likely to have heights of more than 194 cm

(a) The mean value, $\bar{x}$, is 170 cm and corresponds to a normal standard variate value, z, of zero on the standardised normal curve. A height of 150 cm has a z-value given by $z = \frac{x - \bar{x}}{\sigma}$ standard deviations, i.e. $\frac{150 - 170}{9}$ or -2.22 standard deviations.

Using a table of partial areas beneath the standardised normal curve (see Table 14.4), a z-value of -2.22 corresponds to an area of 0.4868 between the mean value and the ordinate $z = -2.22$.

Table 14.4 Partial areas under the standardised normal curve

$z = \frac{x - \bar{x}}{\sigma}$	0	1	2	3	4	5	6	7	8	9
0.0	0.0000	0.0040	0.0080	0.0120	0.0159	0.0199	0.0239	0.0279	0.0319	0.0359
0.1	0.0398	0.0438	0.0478	0.0517	0.0557	0.0596	0.0636	0.0678	0.0714	0.0753
0.2	0.0793	0.0832	0.0871	0.0910	0.0948	0.0987	0.1026	0.1064	0.1103	0.1141
0.3	0.1179	0.1217	0.1255	0.1293	0.1331	0.1388	0.1406	0.1443	0.1480	0.1517
0.4	0.1554	0.1591	0.1628	0.1664	0.1700	0.1736	0.1772	0.1808	0.1844	0.1879
0.5	0.1915	0.1950	0.1985	0.2019	0.2054	0.2086	0.2123	0.2157	0.2190	0.2224
0.6	0.2257	0.2291	0.2324	0.2357	0.2389	0.2422	0.2454	0.2486	0.2517	0.2549
0.7	0.2580	0.2611	0.2642	0.2673	0.2704	0.2734	0.2760	0.2794	0.2823	0.2852
0.8	0.2881	0.2910	0.2939	0.2967	0.2995	0.3023	0.3051	0.3078	0.3106	0.3133
0.9	0.3159	0.3186	0.3212	0.3238	0.3264	0.3289	0.3315	0.3340	0.3365	0.3389
1.0	0.3413	0.3438	0.3451	0.3485	0.3508	0.3531	0.3554	0.3577	0.3599	0.3621
1.1	0.3643	0.3665	0.3686	0.3708	0.3729	0.3749	0.3770	0.3790	0.3810	0.3830
1.2	0.3849	0.3869	0.3888	0.3907	0.3925	0.3944	0.3962	0.3980	0.3997	0.4015
1.3	0.4032	0.4049	0.4066	0.4082	0.4099	0.4115	0.4131	0.4147	0.4162	0.4177

1.4	0.4192	0.4207	0.4222	0.4236	0.4251	0.4265	0.4279	0.4292	0.4306	0.4319
1.5	0.4332	0.4345	0.4357	0.4370	0.4382	0.4394	0.4406	0.4418	0.4430	0.4441
1.6	0.4452	0.4463	0.4474	0.4484	0.4495	0.4505	0.4515	0.4525	0.4535	0.4545
1.7	0.4554	0.4564	0.4573	0.4582	0.4591	0.4599	0.4608	0.4616	0.4625	0.4633
1.8	0.4641	0.4649	0.4656	0.4664	0.4671	0.4678	0.4686	0.4693	0.4699	0.4706
1.9	0.4713	0.4719	0.4726	0.4732	0.4738	0.4744	0.4750	0.4756	0.4762	0.4767
2.0	0.4772	0.4778	0.4783	0.4785	0.4793	0.4798	0.4803	0.4808	0.4812	0.4817
2.1	0.4821	0.4826	0.4830	0.4834	0.4838	0.4842	0.4846	0.4850	0.4854	0.4857
2.2	0.4861	0.4864	0.4868	0.4871	0.4875	0.4878	0.4881	0.4884	0.4887	0.4890
2.3	0.4893	0.4896	0.4898	0.4901	0.4904	0.4906	0.4909	0.4911	0.4913	0.4916
2.4	0.4918	0.4920	0.4922	0.4925	0.4927	0.4929	0.4931	0.4932	0.4934	0.4936
2.5	0.4938	0.4940	0.4941	0.4943	0.4945	0.4946	0.4948	0.4949	0.4951	0.4952
2.6	0.4953	0.4955	0.4956	0.4957	0.4959	0.4960	0.4961	0.4962	0.4963	0.4964
2.7	0.4965	0.4966	0.4967	0.4968	0.4969	0.4970	0.4971	0.4972	0.4973	0.4974
2.8	0.4974	0.4975	0.4076	0.4977	0.4977	0.4978	0.4979	0.4980	0.4980	0.4981
2.9	0.4981	0.4982	0.4982	0.4983	0.4984	0.4984	0.4985	0.4985	0.4986	0.4986
3.0	0.4987	0.4987	0.4987	0.4988	0.4988	0.4989	0.4989	0.4989	0.4990	0.4990
3.1	0.4990	0.4991	0.4991	0.4991	0.4992	0.4992	0.4992	0.4992	0.4993	0.4993
3.2	0.4993	0.4993	0.4994	0.4994	0.4994	0.4994	0.4994	0.4995	0.4995	0.4995
3.3	0.4995	0.4995	0.4995	0.4996	0.4996	0.4996	0.4996	0.4996	0.4996	0.4997
3.4	0.4997	0.4997	0.4997	0.4997	0.4997	0.4997	0.4997	0.4997	0.4997	0.4998
3.5	0.4998	0.4998	0.4998	0.4998	0.4998	0.4998	0.4998	0.4998	0.4998	0.4998
3.6	0.4998	0.4998	0.4999	0.4999	0.4999	0.4999	0.4999	0.4999	0.4999	0.4999
3.7	0.4999	0.4999	0.4999	0.4999	0.4999	0.4999	0.4999	0.4999	0.4999	0.4999
3.8	0.4999	0.4999	0.4999	0.4999	0.4999	0.4999	0.4999	0.4999	0.4999	0.4999
3.9	0.5000	0.5000	0.5000	0.5000	0.5000	0.5000	0.5000	0.5000	0.5000	0.5000

The negative z-value shows that it lies to the left of the $z = 0$ ordinate.

This area is shown shaded in Figure 14.11(a). Similarly, 195 cm has a z-value of $\frac{195 - 170}{9}$ that is 2.78 standard deviations. From Table 14.4, this value of z corresponds to an area of 0.4973, the positive value of z showing that it lies to the right of the $z = 0$ ordinate. This area is shown shaded in Figure 14.11(b). The total area shaded in Figures 14.11(a) and (b) is shown in Figure 14.11(c) and is $0.4868 + 0.4973$, i.e. 0.9841 of the total area beneath the curve.

However, the area is directly proportional to probability. Thus, the probability that a person will have a height of between 150 and 195 cm is 0.9841. For a group of 500 people, 500×0.9841, i.e. **492 people are likely to have heights in this range**.

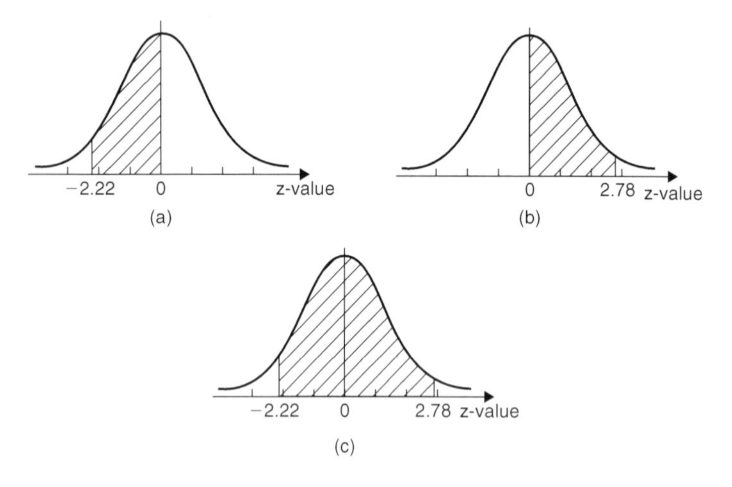

Figure 14.11

(b) A height of 165 cm corresponds to $\frac{165 - 170}{9}$, i.e. -0.56 standard deviations.

The area between $z = 0$ and $z = -0.56$ (from Table 14.4) is 0.2123, shown shaded in Figure 14.12(a). The total area under the standardised normal curve is unity and since the curve is symmetrical, it follows that the total area to the left of the $z = 0$ ordinate is 0.5000. Thus the area to the left of the $z = -0.56$

ordinate ('left' means 'less than', 'right' means 'more than') is 0.5000 − 0.2123, i.e. 0.2877 of the total area, which is shown shaded in Figure 14.12(b). The area is directly proportional to probability and since the total area beneath the standardised normal curve is unity, the probability of a person's height being less than 165 cm is 0.2877. For a group of 500 people, 500 × 0.2877, i.e. **144 people are likely to have heights of less than 165 cm.**

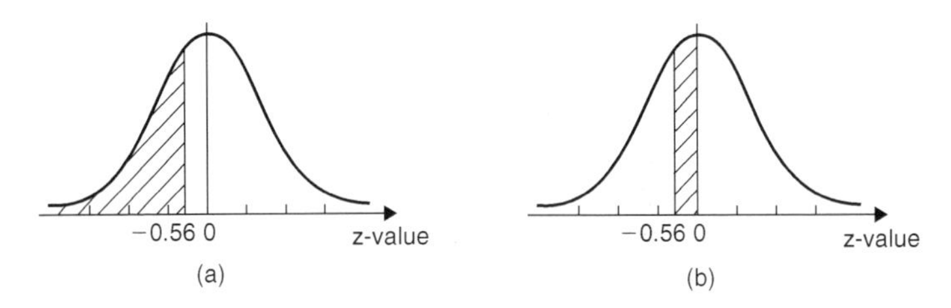

Figure 14.12

(c) 194 cm corresponds to a z-value of $\frac{194 - 170}{9}$ that is, 2.67 standard deviations. From Table 14.4, the area between $z = 0$, $z = 2.67$ and the standardised normal curve is 0.4962, shown shaded in Figure 14.13(a). Since the standardised normal curve is symmetrical, the total area to the right of the $z = 0$ ordinate is 0.5000, hence the shaded area shown in Figure 14.13(b) is 0.5000 − 0.4962, i.e. 0.0038. This area represents the probability of a person having a height of more than 194 cm, and for 500 people, the number of people likely to have a height of more than 194 cm is 0.0038 × 500, i.e. **2 people**.

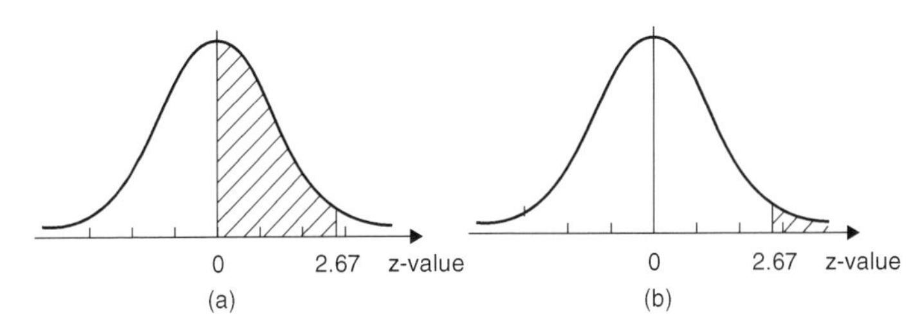

Figure 14.13

Testing for a normal distribution

Application: Use normal probability paper to determine whether the data given below, which refers to the masses of 50 copper ingots, is approximately normally distributed. If the data is normally distributed, determine the mean and standard deviation of the data from the graph drawn.

Class mid-point value (kg)	29.5	30.5	31.5	32.5	33.5	34.5	35.5	36.5	37.5	38.5
Frequency	2	4	6	8	9	8	6	4	2	1

To test the normality of a distribution, the upper class boundary/percentage cumulative frequency values are plotted on normal probability paper. The upper class boundary values are: 30, 31, 32,..., 38, 39. The corresponding cumulative frequency values (for 'less than' the upper class boundary values) are: 2, $(4 + 2) = 6$, $(6 + 4+2) = 12$, 20, 29, 37, 43, 47, 49 and 50. The corresponding percentage cumulative frequency values are $\frac{2}{50} \times 100 = 4$, $\frac{6}{50} \times 100 = 12$, 24, 40, 58, 74, 86, 94, 98 and 100%

The co-ordinates of upper class boundary/percentage cumulative frequency values are plotted as shown in Figure 14.14. When plotting these values, it will always be found that the co-ordinate for the 100% cumulative frequency value cannot be plotted, since the maximum value on the probability scale is 99.99. **Since the points plotted in Figure 14.14 lie very nearly in a straight line, the data is approximately normally distributed.**

The mean value and standard deviation can be determined from Figure 14.14. Since a normal curve is symmetrical, the mean value is the value of the variable corresponding to a 50% cumulative frequency value, shown as point P on the graph. This shows that **the mean value is 33.6 kg**. The standard deviation is determined using the 84% and 16% cumulative frequency values, shown as Q and R in Figure 14.14. The variable values for Q and R are 35.7 and 31.4 respectively; thus two standard deviations correspond to $35.7 - 31.4$, i.e. 4.3, showing that the standard deviation of the distribution is approximately $\frac{4.3}{2}$ i.e. **2.15 standard deviations**.

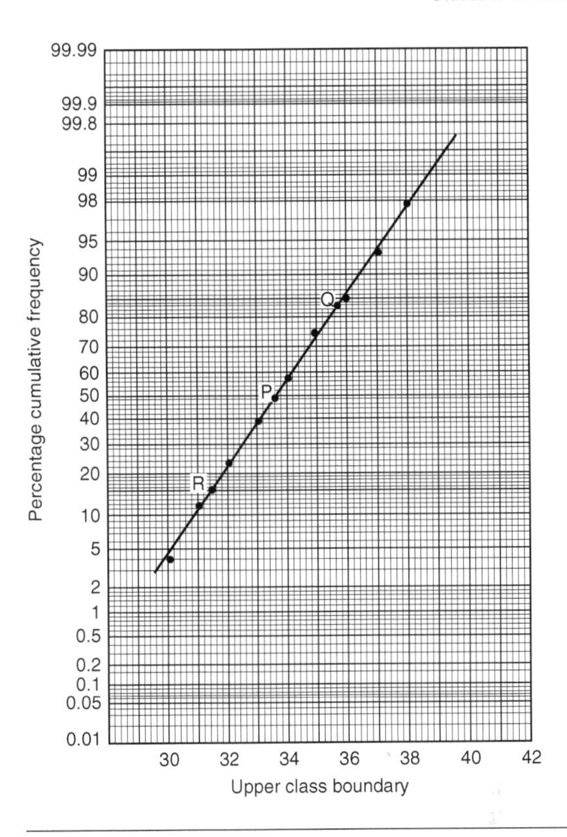

Figure 14.14

14.9 Linear correlation

The product-moment formula for determining the linear correlation coefficient, coefficient of correlation,

$$r = \frac{\sum xy}{\sqrt{\left\{\left(\sum x^2\right)\left(\sum y^2\right)\right\}}}$$

where $x = (X - \overline{X})$ and $y = (Y - \overline{Y})$

Application: In an experiment to determine the relationship between force on a wire and the resulting extension, the following data is obtained:

Force (N)	10	20	30	40	50	60	70
Extension (mm)	0.22	0.40	0.61	0.85	1.20	1.45	1.70

Determine the linear coefficient of correlation for this data

Let X be the variable force values and Y be the dependent variable extension values, respectively. Using a tabular method to determine the quantities of this formula gives:

X	Y	$x = (X - \bar{X})$	$y = (Y - \bar{Y})$	xy	x^2	y^2
10	0.22	−30	−0.699	20.97	900	0.489
20	0.40	−20	−0.519	10.38	400	0.269
30	0.61	−10	−0.309	3.09	100	0.095
40	0.85	0	−0.069	0	0	0.005
50	1.20	10	0.281	2.81	100	0.079
60	1.45	20	0.531	10.62	400	0.282
70	1.70	30	0.781	23.43	900	0.610
$\sum X = 280$ $\bar{X} = \frac{280}{7} = 40$	$\sum Y = 6.43$ $\bar{Y} = \frac{6.43}{7} = 0.919$			$\sum xy = 71.30$	$\sum x^2 = 2800$	$\sum y^2 = 1.829$

Thus, **coefficient of correlation,**

$$\mathbf{r} = \frac{\sum xy}{\sqrt{\left\{\left(\sum x^2\right)\left(\sum y^2\right)\right\}}} = \frac{71.3}{\sqrt{[2800 \times 1.829]}} = \mathbf{0.996}$$

This shows that a **very good direct correlation exists** between the values of force and extension.

14.10 Linear regression

The least-squares regression lines

If the equation of the least-squares regression line is of the form: $\mathbf{Y = a_0 + a_1X}$ the values of regression coefficient a_0 and a_1 are obtained from the equations:

$$\sum Y = a_0 N + a_1 \sum X \qquad (1)$$

$$\sum (XY) = a_0 \sum X + a_1 \sum X^2 \qquad (2)$$

If the equation of the regression line is of the form: $\mathbf{X = b_0 + b_1Y}$ the values of regression coefficient b_0 and b_1 are obtained from the equations:

$$\sum X = b_0 N + b_1 \sum Y \qquad (3)$$

$$\sum (XY) = b_0 \sum Y + b_1 \sum Y^2 \qquad (4)$$

Application: The experimental values relating centripetal force and radius, for a mass travelling at constant velocity in a circle, are as shown:

Force (N)	5	10	15	20	25	30	35	40
Radius (cm)	55	30	16	12	11	9	7	5

Determine the equations of (a) the regression line of force on radius and (b) the regression line of radius on force. Hence, calculate the force at a radius of 40 cm and the radius corresponding to a force of 32 N

(a) Let the radius be the independent variable X, and the force be the dependent variable Y.

The equation of the regression line of force on radius is of the form $Y = a_0 + a_1X$

Using a tabular approach to determine the values of the summations gives:

Radius, X	Force, Y	X^2	XY	Y^2
55	5	3025	275	25
30	10	900	300	100
16	15	256	240	225
12	20	144	240	400
11	25	121	275	625
9	30	81	270	900
7	35	49	245	1225
5	40	25	200	1600
$\sum X = 145$	$\sum Y = 180$	$\sum X^2 = 4601$	$\sum XY = 2045$	$\sum Y^2 = 5100$

Thus, from equations (1) and (2), $180 = 8a_0 + 145a_1$ and $2045 = 145a_0 + 4601a_1$

Solving these simultaneous equations gives $a_0 = 33.7$ and $a_1 = -0.617$, correct to 3 significant figures. Thus the equation of the regression line of force on radius is:

$$\mathbf{Y = 33.7 - 0.617\,X}$$

Thus the force, Y, at a radius of 40 cm, is:
$Y = 33.7 - 0.617(40) = 9.02$

i.e. **the force at a radius of 40 cm is 9.02 N**

(b) The equation of the regression line of radius on force is of the form $X = b_0 + b_1Y$

From equations (3) and (4), $145 = 8b_0 + 180b_1$ and
$2045 = 180b_0 + 5100b_1$

Solving these simultaneous equations gives $b_0 = 44.2$ and $b_1 = -1.16$, correct to 3 significant figures. Thus the equation of the regression line of radius on force is:

$$\mathbf{X = 44.2 - 1.16Y}$$

Thus, the radius, X, when the force is 32 N is:
$X = 44.2 - 1.16(32) = 7.08$,

i.e. **the radius when the force is 32 N is 7.08 cm**

14.11 Sampling and estimation theories

Theorem 1

If all possible samples of size N are drawn from a finite population, N_p, without replacement, and the standard deviation of the mean values of the sampling distribution of means is determined, then:

$$\textbf{standard error of the means, } \sigma_{\bar{x}} = \frac{\sigma}{\sqrt{N}}\sqrt{\left(\frac{N_p - N}{N_p - 1}\right)} \quad (5)$$

where $\sigma_{\bar{x}}$ is the standard deviation of the sampling distribution of means and σ is the standard deviation of the population

For an infinite population and/or for sampling with replacement:

$$\sigma_{\bar{x}} = \frac{\sigma}{\sqrt{N}} \quad (6)$$

Theorem 2

If all possible samples of size N are drawn from a population of size N_p and the mean value of the sampling distribution of means $\mu_{\bar{x}}$ is determined then

$$\mu_{\bar{x}} = \mu \quad (7)$$

where μ is the mean value of the population

Application: The heights of 3000 people are normally distributed with a mean of 175 cm and a standard deviation of 8 cm. If random samples are taken of 40 people, predict the standard deviation and the mean of the sampling distribution of means if sampling is done (a) with replacement, and (b) without replacement

For the population: number of members, $N_p = 3000$;

standard deviation, $\sigma = 8$ cm; mean, $\mu = 175$ cm

For the samples: number in each sample, $N = 40$

(a) When sampling is done **with replacement**, the total number of possible samples (two or more can be the same) is infinite.

Hence, from equation (6) the **standard error of the mean (i.e. the standard deviation of the sampling distribution of means)**

$$\sigma_{\bar{x}} = \frac{\sigma}{\sqrt{N}} = \frac{8}{\sqrt{40}} = \mathbf{1.265\,cm}$$

From equation (7), **the mean of the sampling distribution**

$$\mu_{\bar{x}} = \mu = \mathbf{175\ cm}$$

(b) When sampling is done **without replacement**, the total number of possible samples is finite and hence equation (5) applies. Thus **the standard error of the means,**

$$\sigma_{\bar{x}} = \frac{\sigma}{\sqrt{N}}\sqrt{\left(\frac{N_p - N}{N_p - 1}\right)} = \frac{8}{\sqrt{40}}\sqrt{\left(\frac{3000 - 40}{3000 - 1}\right)}$$

$$= (1.265)(0.9935) = \mathbf{1.257\,cm}$$

Provided the sample size is large, the mean of the sampling distribution of means is the same for both finite and infinite populations. Hence, from equation (3), $\mu_{\bar{x}} = \mathbf{175\,cm}$

The estimation of population parameters based on a large sample size

Table 14.5 Confidence levels

Confidence level, %	99	98	96	95	90	80	50
Confidence coefficient, z_C	2.58	2.33	2.05	1.96	1.645	1.28	0.6745

Application: Determine the confidence coefficient corresponding to a confidence level of 98.5%

98.5% is equivalent to a per unit value of 0.9850. This indicates that the area under the standardised normal curve between $-z_C$ and $+z_C$, i.e. corresponding to $2z_C$, is 0.9850 of the total area. Hence the area between the mean value and z_C is 0.9850/2 i.e. 0.4925 of the total area. The z-value corresponding to a partial area of 0.4925 is 2.43 standard deviations from Table 14.4. Thus, **the confidence coefficient corresponding to a confidence limit of 98.5% is 2.43**

Estimating the mean of a population when the standard deviation of the population is known

The confidence limits of the mean of a population are:

$$\bar{x} \pm \frac{z_C \sigma}{\sqrt{N}} \sqrt{\left(\frac{N_p - N}{N_p - 1}\right)} \qquad (8)$$

for a finite population of size N_p

The **confidence limits for the mean of the population are:**

$$\bar{x} \pm \frac{z_C \sigma}{\sqrt{N}} \qquad (9)$$

for an infinite population.

Application: It is found that the standard deviation of the diameters of rivets produces by a certain machine over a long period of time is 0.018 cm. The diameters of a random sample of 100 rivets produced by this machine in a day have a mean value of 0.476 cm. If the machine produces 2500 rivets a day, determine (a) the 90% confidence limits, and (b) the 97% confidence limits for an estimate of the mean diameter of all the rivets produced by the machine in a day

For the population: standard deviation, $\sigma = 0.018$ cm

number in the population, $N_p = 2500$

For the sample: number in the sample, $N = 100$

mean, $\bar{x} = 0.476$ cm

There is a finite population and the standard deviation of the population is known, hence expression (8) is used.

(a) For a 90% confidence level, the value of z_C, the confidence coefficient, is 1.645 from Table 14.5. Hence, the estimate of the confidence limits of the population mean, μ, is:

$$0.476 \pm \left(\frac{(1.645)(0.018)}{\sqrt{100}}\right)\sqrt{\left(\frac{2500-100}{2500-1}\right)}$$

i.e. $0.476 \pm (0.00296)(0.9800) = 0.476 \pm 0.0029\,\text{cm}$

Thus, **the 90% confidence limits are 0.473 cm and 0.479 cm**

This indicates that if the mean diameter of a sample of 100 rivets is 0.476 cm, then it is predicted that the mean diameter of all the rivets will be between 0.473 cm and 0.479 cm and this prediction is made with confidence that it will be correct nine times out of ten.

(b) For a 97% confidence level, the value of z_C has to be determined from a table of partial areas under the standardised normal curve given in Table 14.4, as it is not one of the values given in Table 14.5. The total area between ordinates drawn at $-z_C$ and $+z_C$ has to be 0.9700. Because the standardised normal curve is symmetrical, the area between $z_C = 0$ and z_C is 0.9700/2, i.e. 0.4850. From Table 14.4 an area of 0.4850 corresponds to a z_C value of 2.17.

Hence, the estimated value of the confidence limits of the population mean is between

$$\bar{x} \pm \frac{z_C\,\sigma}{\sqrt{N}}\sqrt{\left(\frac{N_p - N}{N_p - 1}\right)} = 0.476 \pm \left(\frac{(2.17)(0.018)}{\sqrt{100}}\right)\sqrt{\left(\frac{2500-100}{2500-1}\right)}$$
$$= 0.476 \pm (0.0039)(0.9800)$$
$$= 0.476 \pm 0.0038$$

Thus, **the 97% confidence limits are 0.472 cm and 0.480 cm**

It can be seen that the higher value of confidence level required in part (b) results in a larger confidence interval.

Estimating the mean and standard deviation of a population from sample data

The confidence limits of the mean value of the population, μ, are given by:

$$\mu_{\bar{x}} \pm z_C\,\sigma_{\bar{x}} \qquad (10)$$

If s is the standard deviation of a sample, then the confidence limits of the standard deviation of the population are given by:

$$s \pm z_C \, \sigma_s \quad (11)$$

Application: Several samples of 50 fuses selected at random from a large batch are tested when operating at a 10% overload current and the mean time of the sampling distribution before the fuses failed is 16.50 minutes. The standard error of the means is 1.4 minutes. Determine the estimated mean time to failure of the batch of fuses for a confidence level of 90%

For the sampling distribution: the mean, $\mu_{\bar{x}} = 16.50$,

the standard error of the means, $\sigma_{\bar{x}} = 1.4$

The estimated mean of the population is based on sampling distribution data only and so expression (10) is used.

For an 90% confidence level, $z_C = 1.645$ (from Table 14.5),

thus $\mu_{\bar{x}} \pm z_C \, \sigma_{\bar{x}} = 16.50 \pm (1.645)(1.4) = 16.50 \pm 2.30$ minutes.

Thus, **the 90% confidence level of the mean time to failure is from 14.20 minutes to 18.80 minutes.**

Estimating the mean of a population based on a small sample size

Table 14.6 Percentile values (t_p) for Student's t distribution with ν degrees of freedom (shaded area = p)

t_p

ν	$t_{0.995}$	$t_{0.99}$	$t_{0.975}$	$t_{0.95}$	$t_{0.90}$	$t_{0.80}$	$t_{0.75}$	$t_{0.70}$	$t_{0.60}$	$t_{0.55}$
1	63.66	31.82	12.71	6.31	3.08	1.376	1.000	0.727	0.325	0.158
2	9.92	6.96	4.30	2.92	1.89	1.061	0.816	0.617	0.289	0.142

Table 14.6 Continued

ν	$t_{0.995}$	$t_{0.99}$	$t_{0.975}$	$t_{0.95}$	$t_{0.90}$	$t_{0.80}$	$t_{0.75}$	$t_{0.70}$	$t_{0.60}$	$t_{0.55}$
3	5.84	4.54	3.18	2.35	1.64	0.978	0.765	0.584	0.277	0.137
4	4.60	3.75	2.78	2.13	1.53	0.941	0.741	0.569	0.271	0.134
5	4.03	3.36	2.57	2.02	1.48	0.920	0.727	0.559	0.267	0.132
6	3.71	3.14	2.45	1.94	1.44	0.906	0.718	0.553	0.265	0.131
7	3.50	3.00	2.36	1.90	1.42	0.896	0.711	0.549	0.263	0.130
8	3.36	2.90	2.31	1.86	1.40	0.889	0.706	0.546	0.262	0.130
9	3.25	2.82	2.26	1.83	1.38	0.883	0.703	0.543	0.261	0.129
10	3.17	2.76	2.23	1.81	1.37	0.879	0.700	0.542	0.260	0.129
11	3.11	2.72	2.20	1.80	1.36	0.876	0.697	0.540	0.260	0.129
12	3.06	2.68	2.18	1.78	1.36	0.873	0.695	0.539	0.259	0.128
13	3.01	2.65	2.16	1.77	1.35	0.870	0.694	0.538	0.259	0.128
14	2.98	2.62	2.14	1.76	1.34	0.868	0.692	0.537	0.258	0.128
15	2.95	2.60	2.13	1.75	1.34	0.866	0.691	0.536	0.258	0.128
16	2.92	2.58	2.12	1.75	1.34	0.865	0.690	0.535	0.258	0.128
17	2.90	2.57	2.11	1.74	1.33	0.863	0.689	0.534	0.257	0.128
18	2.88	2.55	2.10	1.73	1.33	0.862	0.688	0.534	0.257	0.127
19	2.86	2.54	2.09	1.73	1.33	0.861	0.688	0.533	0.257	0.127
20	2.84	2.53	2.09	1.72	1.32	0.860	0.687	0.533	0.257	0.127
21	2.83	2.52	2.08	1.72	1.32	0.859	0.686	0.532	0.257	0.127
22	2.82	2.51	2.07	1.72	1.32	0.858	0.686	0.532	0.256	0.127
23	2.81	2.50	2.07	1.71	1.32	0.858	0.685	0.532	0.256	0.127
24	2.80	2.49	2.06	1.71	1.32	0.857	0.685	0.531	0.256	0.127
25	2.79	2.48	2.06	1.71	1.32	0.856	0.684	0.531	0.256	0.127
26	2.78	2.48	2.06	1.71	1.32	0.856	0.684	0.531	0.256	0.127
27	2.77	2.47	2.05	1.70	1.31	0.855	0.684	0.531	0.256	0.127
28	2.76	2.47	2.05	1.70	1.31	0.855	0.683	0.530	0.256	0.127
29	2.76	2.46	2.04	1.70	1.31	0.854	0.683	0.530	0.256	0.127
30	2.75	2.46	2.04	1.70	1.31	0.854	0.683	0.530	0.256	0.127
40	2.70	2.42	2.02	1.68	1.30	0.851	0.681	0.529	0.255	0.126
60	2.66	2.39	2.00	1.67	1.30	0.848	0.679	0.527	0.254	0.126
120	2.62	2.36	1.98	1.66	1.29	0.845	0.677	0.526	0.254	0.126
∞	2.58	2.33	1.96	1.645	1.28	0.842	0.674	0.524	0.253	0.126

The confidence limits of the mean value of a population based on a small sample drawn at random from the population are given by

$$\bar{x} \pm \frac{t_C s}{\sqrt{(N-1)}} \qquad (12)$$

Application: A sample of 12 measurements of the diameter of a bar are made and the mean of the sample is 1.850 cm. The standard deviation of the samples is 0.16 mm. Determine (a) the 90% confidence limits and (b) the 70% confidence limits for an estimate of the actual diameter of the bar

For the sample: the sample size, N = 12; mean, $\bar{x} = 1.850$ cm; standard deviation, s = 0.16 mm = 0.016 cm

Since the sample number is less than 30, the small sample estimate as given in expression (12) must be used. The number of degrees of freedom, i.e. sample size minus the number of estimations of population parameters to be made, is 12 − 1, i.e. 11

(a) The percentile value corresponding to a confidence coefficient value of $t_{0.90}$ and a degree of freedom value of $\nu = 11$ can be found by using Table 14.6, and is 1.36, i.e. $t_C = 1.36$. The estimated value of the mean of the population is given by:

$$\bar{x} \pm \frac{t_C\, s}{\sqrt{(N-1)}} = 1.850 \pm \frac{(1.36)(0.016)}{\sqrt{11}}$$

$$= 1.850 \pm 0.0066 \text{ cm}$$

Thus, **the 90% confidence limits are 1.843 cm and 1.857 cm**

This indicates that the actual diameter is likely to lie between 1.843 cm and 1.857 cm and that this prediction stands a 90% chance of being correct.

(b) The percentile value corresponding to $t_{0.70}$ and to $\nu = 11$ is obtained from Table 14.6, and is 0.540, i.e. $t_C = 0.540$.

The estimated value of the 70% confidence limits is given by:

$$\bar{x} \pm \frac{t_C\, s}{\sqrt{(N-1)}} = 1.850 \pm \frac{(0.540)(0.016)}{\sqrt{11}}$$

$$= 1.850 \pm 0.0026 \text{ cm}$$

Thus, **the 70% confidence limits are 1.847 cm and 1.853 cm**, i.e. the actual diameter of the bar is between 1.847 cm and 1.853 cm and this result has a 70% probability of being correct.

14.12 Chi-square values

Table 14.7 Chi-square distribution

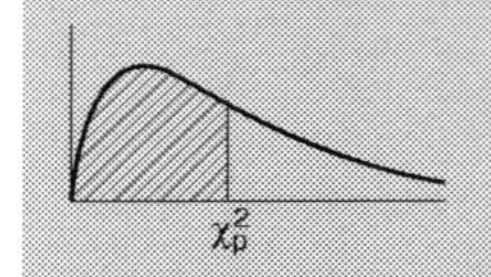

Percentile values (χ_p^2) for the Chi-square distribution with ν degrees of freedom

ν	$\chi^2_{0.995}$	$\chi^2_{0.99}$	$\chi^2_{0.975}$	$\chi^2_{0.95}$	$\chi^2_{0.90}$	$\chi^2_{0.75}$	$\chi^2_{0.50}$	$\chi^2_{0.25}$	$\chi^2_{0.10}$	$\chi^2_{0.05}$	$\chi^2_{0.025}$	$\chi^2_{0.001}$	$\chi^2_{0.005}$
1	7.88	6.63	5.02	3.84	2.71	1.32	0.455	0.102	0.0158	0.0039	0.0010	0.0002	0.0000
2	10.6	9.21	7.38	5.99	4.61	2.77	1.39	0.575	0.211	0.103	0.0506	0.0201	0.0100
3	12.8	11.3	9.35	7.81	6.25	4.11	2.37	1.21	0.584	0.352	0.216	0.115	0.072
4	14.9	13.3	11.1	9.49	7.78	5.39	3.36	1.92	1.06	0.711	0.484	0.297	0.207
5	16.7	15.1	12.8	11.1	9.24	6.63	4.35	2.67	1.61	1.15	0.831	0.554	0.412
6	18.5	16.8	14.4	12.6	10.6	7.84	5.35	3.45	2.20	1.64	1.24	0.872	0.676
7	20.3	18.5	16.0	14.1	12.0	9.04	6.35	4.25	2.83	2.17	1.69	1.24	0.989
8	22.0	20.1	17.5	15.5	13.4	10.2	7.34	5.07	3.49	2.73	2.18	1.65	1.34
9	23.6	21.7	19.0	16.9	14.7	11.4	8.34	5.90	4.17	3.33	2.70	2.09	1.73
10	25.2	23.2	20.5	18.3	16.0	12.5	9.34	6.74	4.87	3.94	3.25	2.56	2.16
11	26.8	24.7	21.9	19.7	17.3	13.7	10.3	7.58	5.58	4.57	3.82	3.05	2.60
12	28.3	26.2	23.3	21.0	18.5	14.8	11.3	8.44	6.30	5.23	4.40	3.57	3.07
13	29.8	27.7	24.7	22.4	19.8	16.0	12.3	9.30	7.04	5.89	5.01	4.11	3.57

14	31.3	29.1	26.1	23.7	21.1	17.1	13.3	10.2	7.79	6.57	5.63	4.66	4.07
15	32.8	30.6	27.5	25.0	22.3	18.2	14.3	11.0	8.55	7.26	6.26	5.23	4.60
16	34.3	32.0	28.8	26.3	23.5	19.4	15.3	11.9	9.31	7.96	6.91	5.81	5.14
17	35.7	33.4	30.2	27.6	24.8	20.5	16.3	12.8	10.1	8.67	7.56	6.41	5.70
18	37.2	34.8	31.5	28.9	26.0	21.6	17.3	13.7	10.9	9.39	8.23	7.01	6.26
19	38.6	36.2	32.9	30.1	27.2	22.7	18.3	14.6	11.7	10.1	8.91	7.63	6.84
20	40.0	37.6	34.4	31.4	28.4	23.8	19.3	15.5	12.4	10.9	9.59	8.26	7.43
21	41.4	38.9	35.5	32.7	29.6	24.9	20.3	16.3	13.2	11.6	10.3	8.90	8.03
22	42.8	40.3	36.8	33.9	30.8	26.0	21.3	17.2	14.0	12.3	11.0	9.54	8.64
23	44.2	41.6	38.1	35.2	32.0	27.1	22.3	18.1	14.8	13.1	11.7	10.2	9.26
24	45.6	43.0	39.4	36.4	33.2	28.2	23.3	19.0	15.7	13.8	12.4	10.9	9.89
25	46.9	44.3	40.6	37.7	34.4	29.3	24.3	19.9	16.5	14.6	13.1	11.5	10.5
26	48.3	45.9	41.9	38.9	35.6	30.4	25.3	20.8	17.3	15.4	13.8	12.2	11.2
27	49.6	47.0	43.2	40.1	36.7	31.5	26.3	21.7	18.1	16.2	14.6	12.9	11.8
28	51.0	48.3	44.5	41.3	37.9	32.6	27.3	22.7	18.9	16.9	15.3	13.6	12.5
29	52.3	49.6	45.7	42.6	39.1	33.7	28.3	23.6	19.8	17.7	16.0	14.3	13.1
30	53.7	50.9	47.7	43.8	40.3	34.8	29.3	24.5	20.6	18.5	16.8	15.0	13.8
40	66.8	63.7	59.3	55.8	51.8	45.6	39.3	33.7	29.1	26.5	24.4	22.2	20.7
50	79.5	76.2	71.4	67.5	63.2	56.3	49.3	42.9	37.7	34.8	32.4	29.7	28.0
60	92.0	88.4	83.3	79.1	74.4	67.0	59.3	52.3	46.5	43.2	40.5	37.5	35.5
70	104.2	100.4	95.0	90.5	85.5	77.6	69.3	61.7	55.3	51.7	48.8	45.4	43.3
80	116.3	112.3	106.6	101.9	96.6	88.1	79.3	71.1	64.3	60.4	57.2	53.5	51.2
90	128.3	124.1	118.1	113.1	107.6	98.6	89.3	80.6	73.3	69.1	65.6	61.8	59.2
100	140.2	135.8	129.6	124.3	118.5	109.1	99.3	90.1	82.4	77.9	74.2	70.1	67.3

Application: As a result of a survey carried out of 200 families, each with five children, the distribution shown below was produced. Test the null hypothesis that the observed frequencies are consistent with male and female births being equally probable, assuming a binomial distribution, a level of significance of 0.05 and a 'too good to be true' fit at a confidence level of 95%

Number of boys (B) and girls (G)	5B,0G	4B,1G	3B,2G	2B,3G	1B,4G,	0B,5G
Number of families	11	35	69	55	25	5

To determine the expected frequencies

Using the usual binomial distribution symbols, let p be the probability of a male birth and $q = 1 - p$ be the probability of a female birth. The probabilities of having 5 boys, 4 boys,.., 0 boys are given by the successive terms of the expansion of $(q + p)^n$. Since there are 5 children in each family, $n = 5$, and $(q + p)^5 = q^5 + 5q^4 p + 10q^3p^2 + 10q^2p^3 + 5qp^4 + p^5$

When $q = p = 0.5$, the probabilities of 5 boys, 4 boys,..., 0 boys are 0.03125, 0.15625, 0.3125, 0.3125, 0.15625 and 0.03125

For 200 families, the expected frequencies, rounded off to the nearest whole number are: 6, 31, 63, 63, 31 and 6 respectively.

To determine the χ^2-value

Using a tabular approach, the χ^2-value is calculated using

$$\chi^2 = \sum\left\{\frac{(o - e)^2}{e}\right\}$$

Number of boys(B) and girls(G)	Observed frequency, o	Expected frequency, e	o − e	$(o - e)^2$	$\frac{(o-e)^2}{e}$
5B, 0G	11	6	5	25	4.167
4B, 1G	35	31	4	16	0.516
3B, 2G	69	63	6	36	0.571

2B, 3G	55	63	−8	64	1.016
1B, 4G	25	31	−6	36	1.161
0B, 5G	5	6	−1	1	0.167

$$\chi^2 = \sum\left\{\frac{(o-e)^2}{e}\right\} = \mathbf{7.598}$$

To test the significance of the χ^2-value

The number of degrees of freedom is given by $\nu = N - 1$ where N is the number of rows in the table above, thus $\nu = 6 - 1 = 5$. For a level of significance of 0.05, the confidence level is 95%, i.e. 0.95 per unit. From Table 14.7, for the $\chi^2_{0.95}$, $\nu = 5$ value, the percentile value χ^2_p is 11.1. Since the calculated value of χ^2 is less than χ^2_p **the null hypothesis that the observed frequencies are consistent with male and female births being equally probable is accepted**.

For a confidence level of 95%, the $\chi^2_{0.05}$, $\nu = 5$ value from Table 14.7 is 1.15 and because the calculated value of χ^2 (i.e. 7.598) is greater than this value, **the fit is not so good as to be unbelievable**.

14.13 The sign test

Table 14.8 Critical values for the sign test

	$\alpha_1 = 5\%$	$2\frac{1}{2}\%$	1%	$\frac{1}{2}\%$
n	$\alpha_2 = 10\%$	5%	2%	1%
1	—	—	—	—
2	—	—	—	—
3	—	—	—	—
4	—	—	—	—
5	0	—	—	—
6	0	0	—	—
7	0	0	0	—
8	1	0	0	0
9	1	1	0	0
10	1	1	0	0
11	2	1	1	0
12	2	2	1	1

Table 14.8 Continued

n	$\alpha_1 = 5\%$ $\alpha_2 = 10\%$	$2\frac{1}{2}\%$ 5%	1% 2%	$\frac{1}{2}\%$ 1%
12	2	2	1	1
13	3	2	1	1
14	3	2	2	1
15	3	3	2	2
16	4	3	2	2
17	4	4	3	2
18	5	4	3	3
19	5	4	4	3
20	5	5	4	3
21	6	5	4	4
22	6	5	5	4
23	7	6	5	4
24	7	6	5	5
25	7	7	6	5
26	8	7	6	6
27	8	7	7	6
28	9	8	7	6
29	9	8	7	7
30	10	9	8	7
31	10	9	8	7
32	10	9	8	8
33	11	10	9	8
34	11	10	9	9
35	12	11	10	9
36	12	11	10	9
37	13	12	10	10
38	13	12	11	10
39	13	12	11	11
40	14	13	12	11
41	14	13	12	11
42	15	14	13	12
43	15	14	13	12
44	16	15	13	13
45	16	15	14	13
46	16	15	14	13
47	17	16	15	14
48	17	16	15	14
49	18	17	15	15
50	18	17	16	15

Procedure for sign test

1. State for the data the null and alternative hypotheses, H_0 and H_1
2. Know whether the stated significance level, α , is for a one-tailed or a two-tailed test. Let, for example, H_0: $x = \phi$, then if H_1: $x \neq \phi$ then a two-tailed test is suggested because x could be less than or more than ϕ (thus use α_2 in Table 14.8), but if say H_1: $x < \phi$ or H_1: $x > \phi$ then a one-tailed test is suggested (thus use α_1 in Table 14.8)
3. Assign plus or minus signs to each piece of data – compared with ϕ or assign plus and minus signs to the difference for paired observations
4. Sum either the number of plus signs or the number of minus signs. For the two-tailed test, whichever is the smallest is taken; for a one-tailed test, the one which would be expected to have the smaller value when H_1 is true is used. The sum decided upon is denoted by S
5. Use Table 14.8 for given values of n, and α_1 or α_2 to read the critical region of S. For example, if, say, $n = 16$ and $\alpha_1 = 5\%$, then from Table 14.8, $S \leq 4$. Thus if S in part (iv) is greater than 4 we accept the null hypothesis H_0 and if S is less than or equal to 4 we accept the alternative hypothesis H_1

Application: A manager of a manufacturer is concerned about suspected slow progress in dealing with orders. He wants at least half of the orders received to be processed within a working day (i.e. 7 hours). A little later he decides to time 17 orders selected at random, to check if his request had been met.

The times spent by the 17 orders being processed were as follows:

$$4\frac{3}{4}\text{h} \quad 9\frac{3}{4}\text{h} \quad 15\frac{1}{2}\text{h} \quad 11\text{h} \quad 8\frac{1}{4}\text{h} \quad 6\frac{1}{2}\text{h} \quad 9\text{h} \quad 8\frac{3}{4}\text{h} \quad 10\frac{3}{4}\text{h}$$

$$3\frac{1}{2}\text{h} \quad 8\frac{1}{2}\text{h} \quad 9\frac{1}{2}\text{h} \quad 15\frac{1}{4}\text{h} \quad 13\text{h} \quad 8\text{h} \quad 7\frac{3}{4}\text{h} \quad 6\frac{3}{4}\text{h}$$

Use the sign test at a significance level of 5% to check if the managers request for quicker processing is being met

Using the above procedure:

1. The hypotheses are $\mathbf{H_0: t = 7\,h}$ and $\mathbf{H_1: t > 7\,h}$, where t is time.
2. Since H_1 is $t > 7\,h$, a one-tail test is assumed, i.e. $\alpha_1 = \mathbf{5\%}$
3. In the sign test each value of data is assigned a + or − sign. For the above data let us assign a + for times greater than 7 hours and a – for less than 7 hours. This gives the following pattern:

   − + + + + − + + +
   − + + + + + + −

4. The test statistic, S, in this case is the number of minus signs (−if H_0 were true there would be an equal number of + and − signs). Table 14.8 gives critical values for the sign test and is given in terms of small values; hence in this case S is the number of − signs, i.e. **S = 4**
5. From Table 14.8, with a sample size n = 17, for a significance level of α_1 = 5%, $\mathbf{S \leq 4}$. Since S = 4 in our data, the result **is significant** at α_1 = 5%, i.e. **the alternative hypothesis is accepted – it appears that the managers request for quicker processing of orders is not being met.**

14.14 Wilcoxon signed-rank test

Table 14.9 Critical values for the Wilcoxon signed-rank test

	$\alpha_1 = 5\%$	$2\frac{1}{2}\%$	1%	$\frac{1}{2}\%$
n	$\alpha_2 = 10\%$	5%	2%	1%
1	—	—	—	—
2	—	—	—	—
3	—	—	—	—
4	—	—	—	—
5	0	—	—	—
6	2	0	—	—
7	3	2	0	—
8	5	3	1	0

Table 14.9 Continued

n	$\alpha_1 = 5\%$ $\alpha_2 = 10\%$	$2\frac{1}{2}\%$ 5%	1% 2%	$\frac{1}{2}\%$ 1%
9	8	5	3	1
10	10	8	5	3
11	13	10	7	5
12	17	13	9	7
13	21	17	12	9
14	25	21	15	12
15	30	25	19	15
16	35	29	23	19
17	41	34	27	23
18	47	40	32	27
19	53	46	37	32
20	60	52	43	37
21	67	58	49	42
22	75	65	55	48
23	83	73	62	54
24	91	81	69	61
25	100	89	76	68
26	110	98	84	75
27	119	107	92	83
28	130	116	101	91
29	140	126	110	100
30	151	137	120	109
31	163	147	130	118
32	175	159	140	128
33	187	170	151	138
34	200	182	162	148
35	213	195	173	159
36	227	208	185	171
37	241	221	198	182
38	256	235	211	194
39	271	249	224	207
40	286	264	238	220
41	302	279	252	233
42	319	294	266	247
43	336	310	281	261

Table 14.9 Continued

n	$\alpha_1 = 5\%$ $\alpha_2 = 10\%$	$2\frac{1}{2}\%$ 5%	1% 2%	$\frac{1}{2}\%$ 1%
44	353	327	296	276
45	371	343	312	291
46	389	361	328	307
47	407	378	345	322
48	426	396	362	339
49	446	415	379	355
50	466	434	397	373

Procedure for the Wilcoxon signed-rank test

1. State for the data the null and alternative hypotheses, H_0 and H_1
2. Know whether the stated significance level, α, is for a one-tailed or a two-tailed test (see 2. in the procedure for the sign test on page 459)
3. Find the difference of each piece of data compared with the null hypothesis or assign plus and minus signs to the difference for paired observations
4. Rank the differences, ignoring whether they are positive or negative
5. The Wilcoxon signed-rank statistic T is calculated as the sum of the ranks of either the positive differences or the negative differences – whichever is the smaller for a two-tailed test, and the one which would be expected to have the smaller value when H_1 is true for a one-tailed test
6. Use Table 14.9 for given values of n, and α_1 or α_2 to read the critical region of T. For example, if, say, $n = 16$ and $\alpha_1 = 5\%$, then from Table 14.9, $t \leq 35$. Thus if T in part 5 is greater than 35 we accept the null hypothesis H_0 and if T is less than or equal to 35 we accept the alternative hypothesis H_1

Application: The following data represents the number of hours that a portable car vacuum cleaner operates before recharging is required.

Operating time (h)	1.4	2.3	0.8	1.4	1.8	1.5	1.9	1.4	2.1	1.1	1.6

Use the Wilcoxon signed-rank test to test the hypothesis, at a 5% level of significance, that this particular vacuum cleaner operates, on average, 1.7 hours before needing a recharge

Using the above procedure:

1. $\mathbf{H_0}$: **t = 1.7 h** and $\mathbf{H_1}$: **t ≠ 1.7 h**
2. Significance level, α_2 **= 5%** (since this is a two-tailed test)
3. Taking the difference between each operating time and 1.7 h gives:

   −0.3 h   +0.6 h   −0.9 h   −0.3 h   +0.1 h   −0.2 h
   +0.2 h   −0.3 h   +0.4 h   −0.6 h   −0.1 h

4. These differences may now be ranked from 1 to 11 (ignoring whether they are positive or negative).

   Some of the differences are equal to each other. For example, there are two 0.1's (ignoring signs) that would occupy positions 1 and 2 when ordered. We average these as far as rankings are concerned i.e. each is assigned a ranking of $\frac{1+2}{2}$ i.e. 1.5. Similarly the two 0.2 values in positions 3 and 4 when ordered are each assigned rankings of $\frac{3+4}{2}$ i.e. 3.5, and the three 0.3 values in positions 5, 6, and 7 are each assigned a ranking of $\frac{5+6+7}{3}$ i.e. 6, and so on. The rankings are therefore:

Rank	1.5	1.5	3.5	3.5	6	6
Difference	+0.1	−0.1	−0.2	+0.2	−0.3	−0.3

Rank	6	8	9.5	9.5	11
Difference	−0.3	+0.4	+0.6	−0.6	−0.9

5. There are 4 positive terms and 7 negative terms. Taking the smaller number, the four positive terms have rankings of 1.5, 3.5, 8 and 9.5.
   Summing the positive ranks gives: **T** = 1.5 + 3.5 + 8+9.5 = **22.5**
6. From Table 14.9, when n = 11 and α_2 = 5%, $\mathbf{T \le 10}$

Since T = 22.5 falls in the acceptance region (i.e. in this case is greater than 10), **the null hypothesis is accepted, i.e. the average operating time is not significantly different from 1.7 h**

[Note that if, say, a piece of the given data was 1.7 h, such that the difference was zero, that data is ignored and n would be 10 instead of 11 in this case.]

14.15 The Mann-Whitney test

Table 14.10 Critical values for the Mann-Whitney test

		$\alpha_1 = 5\%$	$2\frac{1}{2}\%$	1%	$\frac{1}{2}\%$
n_1	n_2	$\alpha_2 = 10\%$	5%	2%	1%
2	2	—	—	—	—
2	3	—	—	—	—
2	4	—	—	—	—
2	5	0	—	—	—
2	6	0	—	—	—
2	7	0	—	—	—
2	8	1	0	—	—
2	9	1	0	—	—
2	10	1	0	—	—
2	11	1	0	—	—
2	12	2	1	—	—
2	13	2	1	0	—
2	14	3	1	0	—
2	15	3	1	0	—
2	16	3	1	0	—
2	17	3	2	0	—
2	18	4	2	0	—
2	19	4	2	1	0
2	20	4	2	1	0
3	3	0	—	—	—
3	4	0	—	—	—
3	5	1	0	—	—
3	6	2	1	—	—
3	7	2	1	0	—
3	8	3	2	0	—
3	9	4	2	1	0

Table 14.10 Continued

n_1	n_2	$\alpha_1 = 5\%$ $\alpha_2 = 10\%$	$2\frac{1}{2}\%$ 5%	1% 2%	$\frac{1}{2}\%$ 1%
3	10	4	3	1	0
3	11	5	3	1	0
3	12	5	4	2	1
3	13	6	4	2	1
3	14	7	5	2	1
3	15	7	5	3	2
3	16	8	6	3	2
3	17	9	6	4	2
3	18	9	7	4	2
3	19	10	7	4	3
3	20	11	8	5	3
4	4	1	0	—	—
4	5	2	1	0	—
4	6	3	2	1	0
4	7	4	3	1	0
4	8	5	4	2	1
4	9	6	4	3	1
4	10	7	5	3	2
4	11	8	6	4	2
4	12	9	7	5	3
4	13	10	8	5	3
4	14	11	9	6	4
4	15	12	10	7	5
4	16	14	11	7	5
4	17	15	11	8	6
4	18	16	12	9	6
4	19	17	13	9	7
4	20	18	14	10	8
5	5	4	2	1	0
5	6	5	3	2	1
5	7	6	5	3	1
5	8	8	6	4	2
5	9	9	7	5	3
5	10	11	8	6	4
5	11	12	9	7	5
5	12	13	11	8	6
5	13	15	12	9	7

Table 14.10 Continued

n_1	n_2	$\alpha_1 = 5\%$ $\alpha_2 = 10\%$	$2\frac{1}{2}\%$ 5%	1% 2%	$\frac{1}{2}\%$ 1%
5	14	16	13	10	7
5	15	18	14	11	8
5	16	19	15	12	9
5	17	20	17	13	10
5	18	22	18	14	11
5	19	23	19	15	12
5	20	25	20	16	13
6	6	7	5	3	2
6	7	8	6	4	3
6	8	10	8	6	4
6	9	12	10	7	5
6	10	14	11	8	6
6	11	16	13	9	7
6	12	17	14	11	9
6	13	19	16	12	10
6	14	21	17	13	11
6	15	23	19	15	12
6	16	25	21	16	13
6	17	26	22	18	15
6	18	28	24	19	16
6	19	30	25	20	17
6	20	32	27	22	18
7	7	11	8	6	4
7	8	13	10	7	6
7	9	15	12	9	7
7	10	17	14	11	9
7	11	19	16	12	10
7	12	21	18	14	12
7	13	24	20	16	13
7	14	26	22	17	15
7	15	28	24	19	16
7	16	30	26	21	18
7	17	33	28	23	19
7	18	35	30	24	21
7	19	37	32	26	22
7	20	39	34	28	24

Table 14.10 Continued

n_1	n_2	$\alpha_1 = 5\%$ $\alpha_2 = 10\%$	$2\frac{1}{2}\%$ 5%	1% 2%	$\frac{1}{2}\%$ 1%
8	8	15	13	9	7
8	9	18	15	11	9
8	10	20	17	13	11
8	11	23	19	15	13
8	12	26	22	17	15
8	13	28	24	20	17
8	14	31	26	22	18
8	15	33	29	24	20
8	16	36	31	26	22
8	17	39	34	28	24
8	18	41	36	30	26
8	19	44	38	32	28
8	20	47	41	34	30
9	9	21	17	14	11
9	10	24	20	16	13
9	11	27	23	18	16
9	12	30	26	21	18
9	13	33	28	23	20
9	14	36	31	26	22
9	15	39	34	28	24
9	16	42	37	31	27
9	17	45	39	33	29
9	18	48	42	36	31
9	19	51	45	38	33
9	20	54	48	40	36
10	10	27	23	19	16
10	11	31	26	22	18
10	12	34	29	24	21
10	13	37	33	27	24
10	14	41	36	30	26
10	15	44	39	33	29
10	16	48	42	36	31
10	17	51	45	38	34
10	18	55	48	41	37
10	19	58	52	44	39
10	20	62	55	47	42

Table 14.10 Continued

n_1	n_2	$\alpha_1 = 5\%$ $\alpha_2 = 10\%$	$2\frac{1}{2}\%$ 5%	1% 2%	$\frac{1}{2}\%$ 1%
11	11	34	30	25	21
11	12	38	33	28	24
11	13	42	37	31	27
11	14	46	40	34	30
11	15	50	44	37	33
11	16	54	47	41	36
11	17	57	51	44	39
11	18	61	55	47	42
11	19	65	58	50	45
11	20	69	62	53	48
12	12	42	37	31	27
12	13	47	41	35	31
12	14	51	45	38	34
12	15	55	49	42	37
12	16	60	53	46	41
12	17	64	57	49	44
12	18	68	61	53	47
12	19	72	65	56	51
12	20	77	69	60	54
13	13	51	45	39	34
13	14	56	50	43	38
13	15	61	54	47	42
13	16	65	59	51	45
13	17	70	63	55	49
13	18	75	67	59	53
13	19	80	72	63	57
13	20	84	76	67	60
14	14	61	55	47	42
14	15	66	59	51	46
14	16	71	64	56	50
14	17	77	69	60	54
14	18	82	74	65	58
14	19	87	78	69	63
14	20	92	83	73	67
15	15	72	64	56	51
15	16	77	70	61	55

Table 14.10 Continued

n_1	n_2	$\alpha_1 = 5\%$ $\alpha_2 = 10\%$	$2\frac{1}{2}\%$ 5%	1% 2%	$\frac{1}{2}\%$ 1%
15	17	83	75	66	60
15	18	88	80	70	64
15	19	94	85	75	69
15	20	100	90	80	73
16	16	83	75	66	60
16	17	89	81	71	65
16	18	95	86	76	70
16	19	101	92	82	74
16	20	107	98	87	79
17	17	96	87	77	70
17	18	102	92	82	75
17	19	109	99	88	81
17	20	115	105	93	86
18	18	109	99	88	81
18	19	116	106	94	87
18	20	123	112	100	92
19	19	123	112	101	93
19	20	130	119	107	99
20	20	138	127	114	105

Procedure for the Mann-Whitney test

1. State for the data the null and alternative hypotheses, H_0 and H_1
2. Know whether the stated significance level, α, is for a one-tailed or a two-tailed test (see 2. in the procedure for the sign test on page 459)
3. Arrange all the data in ascending order whilst retaining their separate identities
4. If the data is now a mixture of, say, A's and B's, write under each letter A the number of B's that precede it in the sequence (or vice-versa)

5. Add together the numbers obtained from 4 and denote total by U.
   U is defined as whichever type of count would be expected to be smallest when H_1 is true
6. Use Table 14.10 for given values of n_1 and n_2, and α_1 or α_2 to read the critical region of U. For example, if, say, $n_1 = 10$ and $n_2 = 16$ and $\alpha_2 = 5\%$, then from Table 14.10, $U \leq 42$. If U in part 5 is greater than 42 we accept the null hypothesis H_0, and if U is equal or less than 42, we accept the alternative hypothesis H_1

Application: 10 British cars and 8 non-British cars are compared for faults during their first 10000 miles of use. The percentage of cars of each type developing faults were as follows:

Non-British cars, P	5	8	14	10	15	7	12	4		
British cars, Q	18	9	25	6	21	20	28	11	16	34

Use the Mann-Whitney test, at a level of significance of 1%, to test whether non-British cars have better average reliability than British models

Using the above procedure:

1. The hypotheses are:
   H_0: Equal proportions of British and non-British cars have breakdowns
   H_1: A higher proportion of British cars have breakdowns
2. Level of significance $\alpha_1 = 1\%$
3. Let the sizes of the samples be n_P and n_Q, where $n_P = 8$ and $n_Q = 10$
   The Mann-Whitney test compares every item in sample P in turn with every item in sample Q, a record being kept of the number of times, say, that the item from P is greater than Q, or vice-versa. In this case there are $n_P n_Q$, i.e. (8)(10) = 80 comparisons to be made. All the data is arranged into ascending order whilst retaining their separate identities – an easy way is to arrange a linear scale as shown in Figure 14.15.

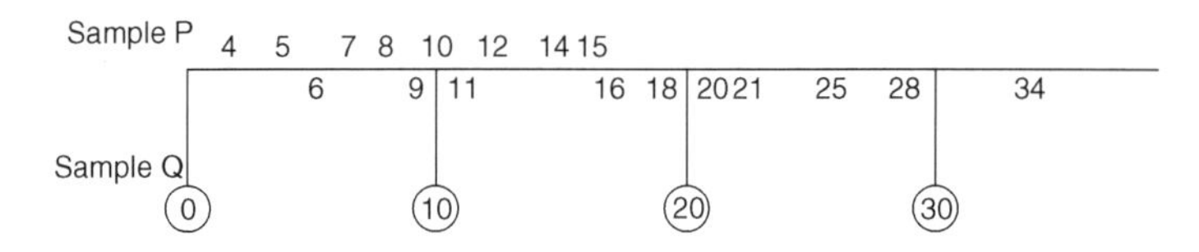

Figure 14.15

From Figure 14.15, a list of P's and Q's can be ranked giving:

P P Q P P Q P Q P P P Q Q Q Q Q Q Q

4. Write under each letter P the number of Q's that precede it in the sequence, giving:

   P P Q P P Q P Q P P P Q Q Q Q Q Q Q
   0 0   1 1   2   3 3 3

5. Add together these 8 numbers, denoting the sum by U, i.e.

   $$\mathbf{U} = 0 + 0 + 1 + 1 + 2 + 3 + 3 + 3 = \mathbf{13}$$

6. The critical regions are of the form $U \leq$ critical region
   From Table 14.10, for a sample size 8 and 10 at significance level $\alpha_1 = 1\%$ the critical regions is $\mathbf{U \leq 13}$
   The value of U in our case, from 5, is 13 which is significant at 1% significance level.

The Mann-Whitney test has therefore confirmed that **there is evidence that the non-British cars have better reliability than the British cars in the first 10,000 miles, i.e. the alternative hypothesis applies.**

15 Laplace Transforms

15.1 Standard Laplace transforms

Table 15.1

	Time function f(t)	**Laplace transform** $\mathscr{L}\{f(t)\} = \int_0^\infty e^{-st} f(t)\,dt$
1.	δ (unit impulse)	1
2.	1 (unit step function)	$\frac{1}{s}$
3.	k (step function)	$\frac{k}{s}$
4.	e^{at} (exponential function)	$\frac{1}{s-a}$
5.	unit step delayed by T	$\frac{e^{-sT}}{s}$
6.	$\sin \omega t$ (sine wave)	$\frac{\omega}{s^2+\omega^2}$
7.	$\cos \omega t$ (cosine wave)	$\frac{s}{s^2+\omega^2}$
8.	t (unit ramp function)	$\frac{1}{s^2}$
9.	t^2	$\frac{2!}{s^3}$

Table 15.1

Time function f(t)	Laplace transform $\mathcal{L}\{f(t)\} = \int_0^\infty e^{-st} f(t)\,dt$
10. t^n (n = positive integer)	$\frac{n!}{s^{n+1}}$
11. cosh ωt	$\frac{s}{s^2 - \omega^2}$
12. sinh ωt	$\frac{\omega}{s^2 - \omega^2}$
13. $e^{at} t^n$	$\frac{n!}{(s-a)^{n+1}}$
14. $e^{-at} \sin \omega t$ (damped sine wave)	$\frac{\omega}{(s+a)^2 + \omega^2}$
15. $e^{-at} \cos \omega t$ (damped cosine wave)	$\frac{s+a}{(s+a)^2 + \omega^2}$
16. $e^{-at} \sinh \omega t$	$\frac{\omega}{(s+a)^2 - \omega^2}$
17. $e^{-at} \cosh \omega t$	$\frac{s+a}{(s+a)^2 - \omega^2}$

Common notations used for the Laplace transform

There are various commonly used notations for the Laplace transform of f(t) and these include:

(i) $\mathcal{L}\{f(t)\}$ or $L\{f(t)\}$

(ii) $\mathcal{L}(f)$ or Lf

(iii) $\bar{f}(s)$ or f(s)

Also, the letter p is sometimes used instead of s as the parameter.

Application: Determine $\mathcal{L}\left\{1 + 2t - \frac{1}{3}t^4\right\}$

$$\mathcal{L}\{1+2t-\frac{1}{3}t^4\} = \mathcal{L}\{1\} + 2\mathcal{L}\{t\} - \frac{1}{3}\mathcal{L}\{t^4\}$$

$$= \frac{1}{s} + 2\left(\frac{1}{s^2}\right) - \frac{1}{3}\left(\frac{4!}{s^{4+1}}\right) \text{ from 2, 8 and 10 of Table 15.1}$$

$$= \frac{1}{s} + \frac{2}{s^2} - \frac{1}{3}\left(\frac{4.3.2.1}{s^5}\right) = \mathbf{\frac{1}{s} + \frac{2}{s^2} - \frac{8}{s^5}}$$

Application: Determine $\mathcal{L}\{5e^{2t} - 3e^{-t}\}$

$$\mathcal{L}\{5e^{2t} - 3e^{-t}\} = 5\mathcal{L}(e^{2t}) - 3\mathcal{L}\{e^{-t}\}$$

$$= 5\left(\frac{1}{s-2}\right) - 3\left(\frac{1}{s--1}\right) \text{ from 4 of Table 15.1}$$

$$= \frac{5}{s-2} - \frac{3}{s+1} = \frac{5(s+1) - 3(s-2)}{(s-2)(s+1)} = \mathbf{\frac{2s+11}{s^2-s-2}}$$

Application: Determine $\mathcal{L}\{6\sin 3t - 4\cos 5t\}$

$$\mathcal{L}\{6\sin 3t - 4\cos 5t\} = 6\mathcal{L}\{\sin 3t\} - 4\mathcal{L}\{\cos 5t\}$$

$$= 6\left(\frac{3}{s^2+3^2}\right) - 4\left(\frac{s}{s^2+5^2}\right) \text{ from 6 and 7 of Table 15.1}$$

$$= \mathbf{\frac{18}{s^2+9} - \frac{4s}{s^2+25}}$$

Application: Determine $\mathcal{L}\{2\cosh 2\theta - \sinh 3\theta\}$

$$\mathcal{L}\{2\cosh 2\theta - \sinh 3\theta\} = 2\mathcal{L}\{\cosh 2\theta\} - \mathcal{L}\{\sinh 3\theta\}$$

$$= 2\left(\frac{s}{s^2-2^2}\right) - \left(\frac{3}{s^2-3^2}\right) \text{ from 11 and 12 of Table 15.1}$$

$$= \mathbf{\frac{2s}{s^2-4} - \frac{3}{s^2-9}}$$

Application: Determine $\mathscr{L}\{\sin^2 t\}$

$$\mathscr{L}\{\sin^2 t\} = \mathscr{L}\left\{\frac{1}{2}(1 - \cos 2t\right\} \text{ since } \cos 2t = 1 - 2\sin^2 t \text{ and}$$

$$\sin^2 t = \frac{1}{2}(1 - \cos 2t)$$

$$= \frac{1}{2}\mathscr{L}\{1\} - \frac{1}{2}\mathscr{L}\{\cos 2t\}$$

$$= \frac{1}{2}\left(\frac{1}{s}\right) - \frac{1}{2}\left(\frac{s}{s^2 + 2^2}\right) \text{ from 2 and 7 of Table 15.1}$$

$$= \frac{(s^2 + 4) - s^2}{2s(s^2 + 4)} = \frac{4}{2s(s^2 + 4)} = \mathbf{\frac{2}{s(s^2 + 4)}}$$

Application: Determine $\mathscr{L}\{2t^4 e^{3t}\}$

$$\mathscr{L}\{2t^4 e^{3t}\} = 2\mathscr{L}\{t^4 e^{3t}\} = 2\left(\frac{4!}{(s - 3)^{4+1}}\right) \text{ from 13 of Table 15.1}$$

$$= \frac{2(4)(3)(2)}{(s - 3)^5} = \mathbf{\frac{48}{(s - 3)^5}}$$

Application: Determine $\mathscr{L}\{4e^{3t} \cos 5t\}$

$$\mathscr{L}\{4e^{3t} \cos 5t\} = 4\mathscr{L}\{e^{3t} \cos 5t\}$$

$$= 4\left(\frac{s - 3}{(s - 3)^2 + 5^2}\right) \text{ from 15 of Table 15.1}$$

$$= \frac{4(s - 3)}{s^2 - 6s + 9 + 25} = \mathbf{\frac{4(s - 3)}{s^2 - 6s + 34}}$$

Application: Determine $\mathscr{L}\{5e^{-3t} \sinh 2t\}$

$$\mathscr{L}\{5e^{-3t} \sinh 2t\} = 5\mathscr{L}\{e^{-3t} \sinh 2t\}$$

$$= 5\left(\frac{2}{(s - -3)^2 - 2^2}\right) \text{ from 16 of Table 15.1}$$

$$= \frac{10}{(s + 3)^2 - 2^2} = \frac{10}{s^2 + 6s + 9 - 4} = \mathbf{\frac{10}{s^2 + 6s + 5}}$$

Application: Determine the Laplace transform of a step function of 10 volts which is delayed by $t = 5\,s$, and sketch the function

The Laplace transform of a step function of 10 volts, shown in Figure 15.1(a), is given by:

$$\mathscr{L}\{10\} = \frac{\mathbf{10}}{\mathbf{s}} \qquad \text{from 3 of Table 15.1}$$

The Laplace transform of a step function of 10 volts which is delayed by $t = 5\,s$ is given by:

$$10\left(\frac{e^{-sT}}{s}\right) = 10\left(\frac{e^{-5s}}{s}\right) = \frac{\mathbf{10}}{\mathbf{s}}\,\mathbf{e^{-5s}} \qquad \text{from 5 of Table 15.1}$$

The function is shown sketched in Figure 15.1(b).

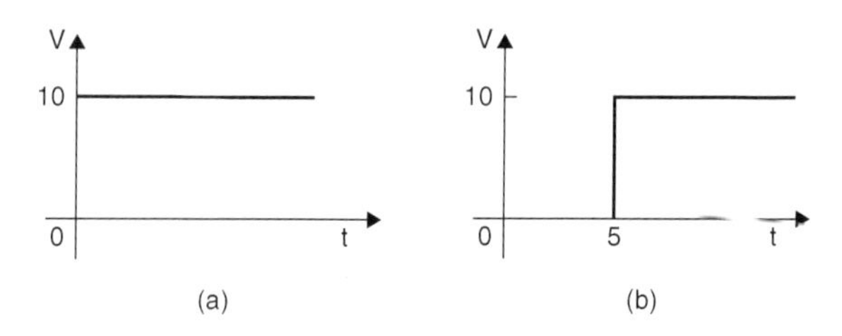

Figure 15.1

Application: Determine the Laplace transform of a ramp function which is delayed by 1 s and increases at 4 V/s. Sketch the function.

The Laplace transform of a ramp function which starts at zero and increases at 4 V/s, shown in Figure 15.2(a), is given by:

$$4\mathscr{L}\{t\} = \frac{\mathbf{4}}{\mathbf{s^2}} \qquad \text{from 8 of Table 15.1}$$

The Laplace transform of a ramp function which is delayed by 1 s and increases at 4 V/s is given by:

$$\left(\frac{\mathbf{4}}{\mathbf{s^2}}\right)\mathbf{e^{-s}} \qquad \text{from 5 of Table 15.1}$$

A sketch of the ramp function is shown in Figure 15.2(b).

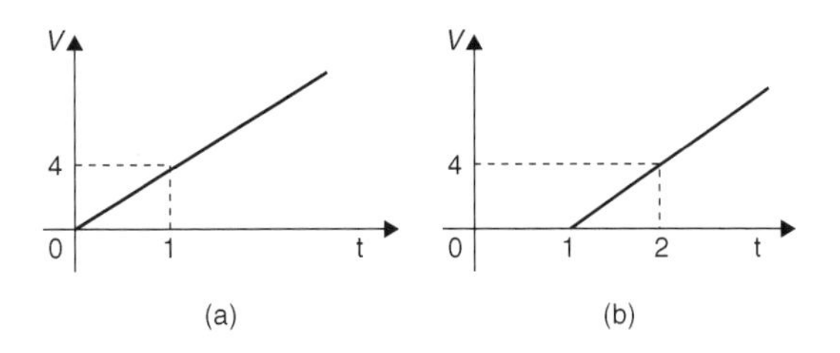

Figure 15.2

Application: Determine the Laplace transform of an impulse voltage of 8 volts which is delayed by 2 s. Sketch the function

The Laplace transform of an impulse voltage of 8 V which starts at time t = 0, shown in Figure 15.3(a), is given by:

$$8\mathscr{L}\{\delta\} = \mathbf{8} \qquad \text{from 1 of Table 15.1}$$

The Laplace transform of an impulse voltage of 8 volts which is delayed by 2 s is given by:

$$\mathbf{8\,e^{-2s}} \qquad \text{from 5 of Table 15.1}$$

A sketch of the delayed impulse function is shown in Figure 15.3(b).

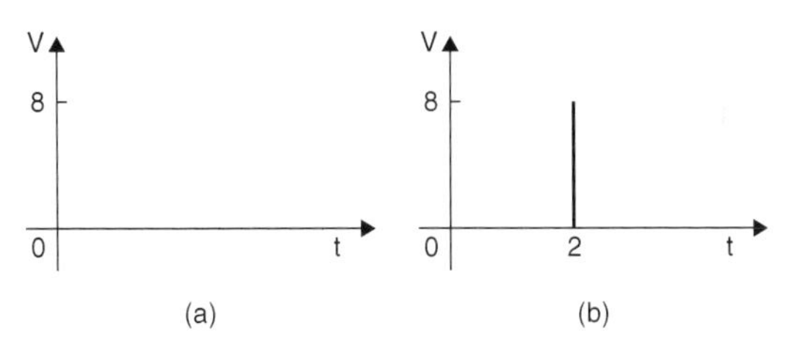

Figure 15.3

15.2 Initial and final value theorems

The initial value theorem

$$\lim_{t\to 0} \text{it}\,[f(t)] = \lim_{s\to\infty} \text{it}\,[s\mathscr{L}\{f(t)\}]$$

The final value theorem

$$\lim_{t \to \infty} [f(t)] = \lim_{s \to 0} [s\mathcal{L}\{f(t)\}]$$

The initial and final value theorems are used in pulse circuit applications where the response of the circuit for small periods of time, or the behaviour immediately after the switch is closed, are of interest. The final value theorem is particularly useful in investigating the stability of systems (such as in automatic aircraft-landing systems) and is concerned with the steady state response for large values of time t, i.e. after all transient effects have died away.

Application: Verify the initial value theorem when $f(t) = 3e^{4t}$

If $f(t) = 3e^{4t}$ then $\mathcal{L}\{3e^{4t}\} = \dfrac{3}{s-4}$ from 4 of Table 15.1

By the initial value theorem, $\displaystyle\lim_{t \to 0} [3e^{4t}] = \lim_{s \to \infty} \left[s\left(\frac{3}{s-4}\right) \right]$

i.e. $\quad 3e^0 = \infty\left(\dfrac{3}{\infty - 4}\right)$

i.e. $\quad$ **3 = 3**, which illustrates the theorem.

Application: Verify the initial value theorem for the voltage function $(5 + 2\cos 3t)$ volts:

Let $f(t) = 5 + 2\cos 3t$

$\mathcal{L}\{f(t)\} = \mathcal{L}\{5 + 2\cos 3t\} = \dfrac{5}{s} + \dfrac{2s}{s^2 + 9}$ from 3 and 7 of Table 15.1

By the initial value theorem, $\displaystyle\lim_{t \to 0} [f(t)] = \lim_{s \to \infty} [s\mathcal{L}\{f(t)\}]$

i.e. $\displaystyle\lim_{t \to 0} [5 + 2\cos 3t] = \lim_{s \to \infty} \left[s\left(\frac{5}{s} + \frac{2s}{s^2 + 9}\right) \right] = \lim_{s \to \infty} \left[5 + \frac{2s^2}{s^2 + 9} \right]$

i.e. $5 + 2(1) = 5 + \dfrac{2\infty^2}{\infty^2 + 9} = 5 + 2$

i.e. **7 = 7**, which verifies the theorem in this case.

The initial value of the voltage is thus **7 V**

Application: Verify the final value theorem when $f(t) = 3e^{-4t}$

$$\lim_{t\to\infty}[3e^{-4t}] = \lim_{s\to 0}\left[s\left(\frac{3}{s+4}\right)\right]$$

i.e. $$3e^{-\infty} = (0)\left(\frac{3}{0+4}\right)$$

i.e. **0 = 0**, which illustrates the theorem.

Application: Verify the final value theorem for the function $(2 + 3e^{-2t}\sin 4t)$ cm, which represents the displacement of a particle

Let $f(t) = 2 + 3e^{-2t}\sin 4t$

$$\mathscr{L}\{f(t)\} = \mathscr{L}\{2 + 3e^{-2t}\sin 4t\} = \frac{2}{s} + 3\left(\frac{4}{(s - -2)^2 + 4^2}\right)$$

$$= \frac{2}{s} + \frac{12}{(s+2)^2 + 16} \quad \text{from 3 and 14 of Table 15.1}$$

By the final value theorem, $\lim_{t\to\infty}[f(t)] = \lim_{s\to 0}[s\mathscr{L}\{f(t)\}]$

i.e. $$\lim_{t\to\infty}[2 + 3e^{-2t}\sin 4t] = \lim_{s\to 0}\left[s\left(\frac{2}{s} + \frac{12}{(s+2)^2 + 16}\right)\right]$$

$$= \lim_{s\to 0}\left[2 + \frac{12s}{(s+2)^2 + 16}\right]$$

i.e. $2 + 0 = 2 + 0$

i.e. **2 = 2**, which verifies the theorem in this case.

The final value of the displacement is thus 2 cm.

15.3 Inverse Laplace transforms

If the Laplace transform of a function f(t) is F(s), i.e. $\mathcal{L}\{f(t)\} = F(s)$, then f(t) is called the **inverse Laplace transform** of F(s) and is written as

$$\mathbf{f(t) = \mathcal{L}^{-1}\{F(s)\}}$$

Table 15.1 is used to determine inverse Laplace transforms.

Application: Determine $\mathcal{L}^{-1}\left\{\frac{1}{s^2+9}\right\}$

$$\mathcal{L}^{-1}\left\{\frac{1}{s^2+9}\right\} = \mathcal{L}^{-1}\left\{\frac{1}{s^2+3^2}\right\}$$

$$= \frac{1}{3}\mathcal{L}^{-1}\left\{\frac{3}{s^2+3^2}\right\} = \mathbf{\frac{1}{3}\sin 3t} \text{ from 6 of Table 15.1}$$

Application: Determine $\mathcal{L}^{-1}\left\{\frac{5}{3s-1}\right\}$

$$\mathcal{L}^{-1}\left\{\frac{5}{3s-1}\right\} = \mathcal{L}^{-1}\left\{\frac{5}{3\left(s-\frac{1}{3}\right)}\right\}$$

$$= \frac{5}{3}\mathcal{L}^{-1}\left\{\frac{1}{\left(s-\frac{1}{3}\right)}\right\} = \mathbf{\frac{5}{3}e^{\frac{1}{3}t}} \quad \text{from 4 of Table 15.1}$$

Application: Determine $\mathcal{L}^{-1}\left\{\frac{3}{s^4}\right\}$

$$\mathcal{L}^{-1}\left\{\frac{3}{s^4}\right\} = \frac{3}{3!}\mathcal{L}^{-1}\left\{\frac{3!}{s^{3+1}}\right\} = \mathbf{\frac{1}{2}t^3} \quad \text{from 10 of Table 15.1}$$

Application: Determine $\mathcal{L}^{-1}\left\{\frac{7s}{s^2+4}\right\}$

$$\mathcal{L}^{-1}\left\{\frac{7s}{s^2+4}\right\} = 7\mathcal{L}^{-1}\left\{\frac{s}{s^2+2^2}\right\} = \mathbf{7\cos 2t} \text{ from 7 of Table 15.1}$$

Application: Determine $\mathcal{L}^{-1}\left\{\frac{3}{s^2-7}\right\}$

$$\mathcal{L}^{-1}\left\{\frac{3}{s^2-7}\right\} = 3\mathcal{L}^{-1}\left\{\frac{1}{s^2-(\sqrt{7})^2}\right\}$$

$$= \frac{3}{\sqrt{7}}\mathcal{L}^{-1}\left\{\frac{\sqrt{7}}{s^2-(\sqrt{7})^2}\right\}$$

$$= \mathbf{\frac{3}{\sqrt{7}}\sinh\sqrt{7}t} \quad \text{from 12 of Table 15.1}$$

Application: Determine $\mathcal{L}^{-1}\left\{\frac{2}{(s-3)^5}\right\}$

$$\mathcal{L}^{-1}\left\{\frac{2}{(s-3)^5}\right\} = \frac{2}{4!}\mathcal{L}^{-1}\left\{\frac{4!}{(s-3)^{4+1}}\right\}$$

$$= \mathbf{\frac{1}{12}e^{3t}t^4} \quad \text{from 13 of Table 15.1}$$

Application: Determine $\mathcal{L}^{-1}\left\{\frac{3}{s^2-4s+13}\right\}$

$$\mathcal{L}^{-1}\left\{\frac{3}{s^2-4s+13}\right\} = \mathcal{L}^{-1}\left\{\frac{3}{(s-2)^2+3^2}\right\}$$

$$= \mathbf{e^{2t}\sin 3t} \quad \text{from 14 of Table 15.1}$$

Application: Determine $\mathcal{L}^{-1}\left\{\dfrac{4s-3}{s^2-4s-5}\right\}$

$$\mathcal{L}^{-1}\left\{\frac{4s-3}{s^2-4s-5}\right\} = \mathcal{L}^{-1}\left\{\frac{4s-3}{(s-2)^2-3^2}\right\} = \mathcal{L}^{-1}\left\{\frac{4(s-2)+5}{(s-2)^2-3^2}\right\}$$

$$= \mathcal{L}^{-1}\left\{\frac{4(s-2)}{(s-2)^2-3^2}\right\} + \mathcal{L}^{-1}\left\{\frac{5}{(s-2)^2-3^2}\right\}$$

$$= 4e^{2t}\cosh 3t + \mathcal{L}^{-1}\left\{\frac{\frac{5}{3}(3)}{(s-2)^2-3^2}\right\}$$

from 17 of Table 15.1

$$= \mathbf{4e^{2t}\cosh 3t + \frac{5}{3}e^{2t}\sinh 3t}$$

from 16 of Table 15.1

Inverse Laplace transforms using partial fractions

Application: Determine $\mathcal{L}^{-1}\left\{\dfrac{4s-5}{s^2-s-2}\right\}$

Let

$$\frac{4s-5}{s^2-s-2} \equiv \frac{4s-5}{(s-2)(s+1)} \equiv \frac{A}{(s-2)} + \frac{B}{(s+1)} \equiv \frac{A(s+1)+B(s-2)}{(s-2)(s+1)}$$

Hence, $\quad 4s-5 \equiv A(s+1) + B(s-2)$

When $s = 2$, $\quad 3 = 3A$, from which, $A = 1$

When $s = -1$, $\quad -9 = -3B$, from which, $B = 3$

Hence

$$\mathcal{L}^{-1}\left\{\frac{4s-5}{s^2-s-2}\right\} \equiv \mathcal{L}^{-1}\left\{\frac{1}{s-2} + \frac{3}{s+1}\right\} = \mathcal{L}^{-1}\left\{\frac{1}{s-2}\right\} + \mathcal{L}^{-1}\left\{\frac{3}{s+1}\right\}$$

$$= \mathbf{e^{2t} + 3e^{-t}}$$

from 4 of Table 15.1

Application: Determine $\mathscr{L}^{-1}\left\{\frac{5s^2+8s-1}{(s+3)(s^2+1)}\right\}$

Let

$$\frac{5s^2+8s-1}{(s+3)(s^2+1)} \equiv \frac{A}{s+3}+\frac{Bs+C}{(s^2+1)} \equiv \frac{A(s^2+1)+(Bs+C)(s+3)}{(s+3)(s^2+1)}$$

Hence, $5s^2+8s-1 \equiv A(s^2+1)+(Bs+C)(s+3)$

When $s=-3$, $20=10A$, from which, $A=2$

Equating s^2 terms gives: $5=A+B$, from which, $B=3$, since $A=2$

Equating s terms gives: $8=3B+C$, from which, $C=-1$, since $B=3$

Hence

$$\mathscr{L}^{-1}\left\{\frac{5s^2+8s-1}{(s+3)(s^2+1)}\right\} \equiv \mathscr{L}^{-1}\left\{\frac{2}{s+3}+\frac{3s-1}{s^2+1}\right\}$$

$$\equiv \mathscr{L}^{-1}\left\{\frac{2}{s+3}\right\}+\mathscr{L}^{-1}\left\{\frac{3s}{s^2+1}\right\}-\mathscr{L}^{-1}\left\{\frac{1}{s^2+1}\right\}$$

$$= \mathbf{2e^{-3t}+3\cos t-\sin t}$$

from 4, 7 and 6 of Table 15.1

15.4 Solving differential equations using Laplace transforms

The Laplace transforms of derivatives

First derivatives: $\mathscr{L}\{f'(t)\} = s\mathscr{L}\{f(t)\} - f(0)$

or $$\mathscr{L}\left\{\frac{dy}{dx}\right\} = s\mathscr{L}\{y\} - y(0) \qquad (1)$$

where y(0) is the value of y at $x = 0$

Second derivative: $\mathscr{L}\{f''(t)\} = s^2\mathscr{L}\{f(t)\} - sf(0) - f'(0)$

or $$\mathscr{L}\left\{\frac{d^2y}{dx^2}\right\} = s^2\mathscr{L}\{y\} - s\,y(0) - y'(0) \qquad (2)$$

where y′(0) is the value of $\frac{dy}{dx}$ at $x = 0$

Higher derivatives:

$$\mathscr{L}\{f^n(t)\} = s^n\mathscr{L}\{f(t)\} - s^{n-1}f(0) - s^{n-2}f'(0) \ldots - f^{n-1}(0)$$

$$\text{or } \mathscr{L}\left\{\frac{d^ny}{dx^n}\right\} = s^n\mathscr{L}\{y\} - s^{n-1}y(0) - s^{n-2}y'(0) \ldots - y^{n-1}(0)$$

Procedure to solve differential equations by using Laplace transforms

1. Take the Laplace transform of both sides of the differential equation by applying the formulae for the Laplace transforms of derivatives (i.e. equations (1) and (2)) and, where necessary, using a list of standard Laplace transforms, as in Tables 15.1
2. Put in the given initial conditions, i.e. $y(0)$ and $y'(0)$
3. Rearrange the equation to make $\mathscr{L}\{y\}$ the subject.
4. Determine y by using, where necessary, partial fractions, and taking the inverse of each term.

Application: Solve the differential equation $2\frac{d^2y}{dx^2} + 5\frac{dy}{dx} - 3y = 0$, given that when $x = 0$, $y = 4$ and $\frac{dy}{dx} = 9$

Using the above procedure:

1. $$2\mathscr{L}\left\{\frac{d^2y}{dx^2}\right\} + 5\mathscr{L}\left\{\frac{dy}{dx}\right\} - 3\mathscr{L}\{y\} = \mathscr{L}\{0\}$$

   $$2[s^2\mathscr{L}\{y\} - sy(0) - y'(0)] + 5[s\mathscr{L}\{y\} - y(0)] - 3\mathscr{L}\{y\} = 0,$$
   from equations (1) and (2)

2. $y(0) = 4$ and $y'(0) = 9$

   Thus $2[s^2\mathscr{L}\{y\} - 4s - 9] + 5[s\mathscr{L}\{y\} - 4] - 3\mathscr{L}\{y\} = 0$

   i.e. $2s^2\mathscr{L}\{y\} - 8s - 18 + 5s\mathscr{L}\{y\} - 20 - 3\mathscr{L}\{y\} = 0$

3. Rearranging gives: $(2s^2 + 5s - 3)\mathscr{L}\{y\} = 8s + 38$

   i.e. $$\mathscr{L}\{y\} = \frac{8s + 38}{2s^2 + 5s - 3}$$

4. $y = \mathscr{L}^{-1}\left\{\dfrac{8s + 38}{2s^2 + 5s - 3}\right\}$

Let $$\frac{8s + 38}{2s^2 + 5s - 3} \equiv \frac{8s + 38}{(2s - 1)(s + 3)}$$
$$\equiv \frac{A}{2s - 1} + \frac{B}{s + 3} \equiv \frac{A(s + 3) + B(2s - 1)}{(2s - 1)(s + 3)}$$

Hence, $8s + 38 = A(s + 3) + B(2s - 1)$

When $s = 0.5$, $42 = 3.5A$, from which, $A = 12$

When $s = -3$, $14 = -7B$, from which, $B = -2$

Hence, $$y = \mathscr{L}^{-1}\left\{\frac{8s + 38}{2s^2 + 5s - 3}\right\} = \mathscr{L}^{-1}\left\{\frac{12}{2s - 1} - \frac{2}{s + 3}\right\}$$
$$= \mathscr{L}^{-1}\left\{\frac{12}{2\left(s - \frac{1}{2}\right)}\right\} - \mathscr{L}^{-1}\left\{\frac{2}{s + 3}\right\}$$

Hence, $\mathbf{y = 6e^{\frac{1}{2}x} - 2e^{-3x}}$ from 4 of Table 15.1

Application: Solve $\dfrac{d^2y}{dx^2} - 3\dfrac{dy}{dx} = 9$, given that when $x = 0$, $y = 0$ and $\dfrac{dy}{dx} = 0$

1. $\mathscr{L}\left\{\dfrac{d^2y}{dx^2}\right\} - 3\mathscr{L}\left\{\dfrac{dy}{dx}\right\} = \mathscr{L}\{9\}$

   Hence, $[s^2\mathscr{L}\{y\} - sy(0) - y'(0)] - 3[s\mathscr{L}\{y\} - y(0)] = \dfrac{9}{s}$

2. $y(0) = 0$ and $y'(0) = 0$

   Hence, $s^2\mathscr{L}\{y\} - 3s\mathscr{L}\{y\} = \dfrac{9}{s}$

3. Rearranging gives: $(s^2 - 3s)\mathscr{L}\{y\} = \dfrac{9}{s}$

   i.e. $\mathscr{L}\{y\} = \dfrac{9}{s(s^2 - 3s)} = \dfrac{9}{s^2(s - 3)}$

4. $y = \mathscr{L}^{-1}\left\{\frac{9}{s^2(s-3)}\right\}$

Let $\frac{9}{s^2(s-3)} \equiv \frac{A}{s} + \frac{B}{s^2} + \frac{C}{s-3} \equiv \frac{A(s)(s-3) + B(s-3) + Cs^2}{s^2(s-3)}$

Hence, $9 \equiv A(s)(s-3) + B(s-3) + Cs^2$

When $s = 0$, $9 = -3B$, from which, $B = -3$

When $s = 3$, $9 = 9C$, from which, $C = 1$

Equating s^2 terms gives: $0 = A + C$, from which, $A = -1$, since $C = 1$

Hence, $\mathscr{L}^{-1}\left\{\frac{9}{s^2(s-3)}\right\} = \mathscr{L}^{-1}\left\{-\frac{1}{s} - \frac{3}{s^2} + \frac{1}{s-3}\right\}$

$= -1 - 3x + e^{3x}$

from 2, 8 and 4 of Table 15.1

i.e. $\mathbf{y = e^{3x} - 3x - 1}$

Application: Solve $\frac{d^2y}{dx^2} - 7\frac{dy}{dx} + 10y = e^{2x} + 20$, given that when $x = 0$, $y = 0$ and $\frac{dy}{dx} = -\frac{1}{3}$

1. $\mathscr{L}\left\{\frac{d^2y}{dx^2}\right\} - 7\mathscr{L}\left\{\frac{dy}{dx}\right\} + 10\mathscr{L}\{y\} = \mathscr{L}\{e^{2x} + 20\}$

Hence, $[s^2\mathscr{L}\{y\} - sy(0) - y'(0)] - 7[s\mathscr{L}\{y\} - y(0)] + 10\mathscr{L}\{y\}$

$= \frac{1}{s-2} + \frac{20}{s}$

2. $y(0) = 0$ and $y'(0) = -\frac{1}{3}$

Hence, $s^2\mathscr{L}\{y\} - 0 - \left(-\frac{1}{3}\right) - 7s\mathscr{L}\{y\} + 0 + 10\mathscr{L}\{y\}$

$= \frac{s + 20(s-2)}{s(s-2)} = \frac{21s - 40}{s(s-2)}$

3. $(s^2 - 7s + 10)\mathscr{L}\{y\}$

$= \frac{21s - 40}{s(s-2)} - \frac{1}{3} = \frac{3(21s - 40) - s(s-2)}{3s(s-2)} = \frac{-s^2 + 65s - 120}{3s(s-2)}$

$$\text{Hence, } \mathcal{L}\{y\} = \frac{-s^2 + 65s - 120}{3s(s-2)(s^2 - 7s + 10)} = \frac{1}{3}\left[\frac{-s^2 + 65s - 120}{s(s-2)(s-2)(s-5)}\right]$$

$$= \frac{1}{3}\left[\frac{-s^2 + 65s - 120}{s(s-5)(s-2)^2}\right]$$

4. $y = \frac{1}{3}\mathcal{L}^{-1}\left\{\frac{-s^2 + 65s - 120}{s(s-5)(s-2)^2}\right\}$

$$\text{Let } \frac{-s^2 + 65s - 120}{s(s-5)(s-2)^2} \equiv \frac{A}{s} + \frac{B}{s-5} + \frac{C}{s-2} + \frac{D}{(s-2)^2}$$

$$\equiv \frac{A(s-5)(s-2)^2 + B(s)(s-2)^2 + C(s)(s-5)(s-2) + D(s)(s-5)}{s(s-5)(s-2)^2}$$

Hence, $-s^2 + 65s - 120 \equiv A(s-5)(s-2)^2 + B(s)(s-2)^2 + C(s)(s-5)(s-2) + D(s)(s-5)$

When $s = 0$, $-120 = -20A$, from which, $A = 6$

When $s = 5$, $180 = 45B$, from which, $B = 4$

When $s = 2$, $6 = -6D$, from which, $D = -1$

Equating s^3 terms gives: $0 = A + B + C$, from which, $C = -10$

Hence,

$$\frac{1}{3}\mathcal{L}^{-1}\left\{\frac{-s^2 + 65s - 120}{s(s-5)(s-2)^2}\right\} = \frac{1}{3}\mathcal{L}^{-1}\left\{\frac{6}{s} + \frac{4}{s-5} - \frac{10}{s-2} - \frac{1}{(s-2)^2}\right\}$$

$$= \frac{1}{3}[6 + 4e^{5x} - 10e^{2x} - xe^{2x}]$$

Thus, $\mathbf{y = 2 + \frac{4}{3}e^{5x} - \frac{10}{3}e^{2x} - \frac{x}{3}e^{2x}}$

15.5 Solving simultaneous differential equations using Laplace transforms

Procedure to solve simultaneous differential equations using Laplace transforms

1. Take the Laplace transform of both sides of each simultaneous equation by applying the formulae for the Laplace transforms

of derivatives (i.e. equations (1) and (2), page 483) and using a list of standard Laplace transforms, as in Table 15.1

2. Put in the initial conditions, i.e. $x(0)$, $y(0)$, $x'(0)$, $y'(0)$
3. Solve the simultaneous equations for $\mathscr{L}\{y\}$ and $\mathscr{L}\{x\}$ by the normal algebraic method.
4. Determine y and x by using, where necessary, partial fractions, and taking the inverse of each term.

Application: Solve the following pair of simultaneous differential equations

$$\frac{dy}{dt} + x = 1$$

$$\frac{dx}{dt} - y + 4e^t = 0$$

given that at $t = 0$, $x = 0$ and $y = 0$

Using the above procedure:

1. $$\mathscr{L}\left\{\frac{dy}{dt}\right\} + \mathscr{L}\{x\} = \mathscr{L}\{1\} \qquad (3)$$

$$\mathscr{L}\left\{\frac{dx}{dt}\right\} - \mathscr{L}\{y\} + 4\mathscr{L}\{e^t\} = \mathscr{L}\{0\} \qquad (4)$$

Equation (3) becomes:

$$[s\mathscr{L}\{y\} - y(0)] + \mathscr{L}\{x\} = \frac{1}{s} \qquad (3')$$

from equation (1), page 483 and Table 15.1

Equation (4) becomes:

$$[s\mathscr{L}\{x\} - x(0)] - \mathscr{L}\{y\} = -\frac{4}{s-1} \qquad (4')$$

2. $x(0) = 0$ and $y(0) = 0$ hence

Equation (3′) becomes:

$$s\mathscr{L}\{y\} + \mathscr{L}\{x\} = \frac{1}{s} \qquad (3'')$$

and equation (4′) becomes: $s\mathcal{L}\{x\} - \mathcal{L}\{y\} = -\frac{4}{s-1}$

or $$-\mathcal{L}\{y\} + s\mathcal{L}\{x\} = -\frac{4}{s-1} \qquad (4'')$$

3. 1 × equation (3″) and s × equation (4″) gives:

$$s\mathcal{L}\{y\} + \mathcal{L}\{x\} = \frac{1}{s} \qquad (5)$$

$$-s\mathcal{L}\{y\} + s^2\mathcal{L}\{x\} = -\frac{4s}{s-1} \qquad (6)$$

Adding equations (5) and (6) gives:

$$(s^2+1)\mathcal{L}\{x\} = \frac{1}{s} - \frac{4s}{s-1} = \frac{(s-1)-s(4s)}{s(s-1)} = \frac{-4s^2+s-1}{s(s-1)}$$

from which, $\mathcal{L}\{x\} = \dfrac{-4s^2+s-1}{s(s-1)(s^2+1)}$

Using partial fractions

$$\frac{-4s^2+s-1}{s(s-1)(s^2+1)} \equiv \frac{A}{s} + \frac{B}{(s-1)} + \frac{Cs+D}{(s^2+1)}$$

$$= \frac{A(s-1)(s^2+1) + B\,s\,(s^2+1) + (Cs+D)\,s\,(s-1)}{s(s-1)(s^2+1)}$$

Hence,

$-4s^2 + s - 1 = A(s-1)(s^2+1) + Bs(s^2+1) + (Cs+D)s(s-1)$

When $s = 0$, $-1 = -A$ hence, $\mathbf{A = 1}$

When $s = 1$, $-4 = 2B$ hence, $\mathbf{B = -2}$

Equating s^3 coefficients:

$0 = A + B + C$ hence, $\mathbf{C = 1}$ (since $A = 1$ and $B = -2$)

Equating s^2 coefficients:

$-4 = -A + D - C$ hence $\mathbf{D = -2}$ (since $A = 1$ and $C = 1$)

Thus, $$\mathcal{L}\{x\} = \frac{-4s^2+s-1}{s(s-1)(s^2+1)} = \frac{1}{s} - \frac{2}{(s-1)} + \frac{s-2}{(s^2+1)}$$

4. Hence, $x = \mathcal{L}^{-1}\left\{\frac{1}{s} - \frac{2}{(s-1)} + \frac{s-2}{(s^2+1)}\right\}$

$$= \mathcal{L}^{-1}\left\{\frac{1}{s} - \frac{2}{(s-1)} + \frac{s}{(s^2+1)} - \frac{2}{(s^2+1)}\right\}$$

i.e. $\mathbf{x = 1 - 2e^t + \cos t - 2\sin t}$ from Table 15.1, page 472

The second equation given originally is $\frac{dx}{dt} - y + 4e^t = 0$

from which, $y = \frac{dx}{dt} + 4e^t = \frac{d}{dt}(1 - 2e^t + \cos t - 2\sin t) + 4e^t$

$$= -2e^t - \sin t - 2\cos t + 4e^t$$

i.e. $\mathbf{y = 2e^t - \sin t - 2\cos t}$

[Alternatively, to determine y, return to equations (3″) and (4″)]

Application: Solve the following pair of simultaneous differential equations

$$\frac{d^2x}{dt^2} - x = y$$

$$\frac{d^2y}{dt^2} + y = -x$$

given that at $t = 0$, $x = 2$, $y = -1$, $\frac{dx}{dt} = 0$ and $\frac{dy}{dt} = 0$

1. $[s^2\mathcal{L}\{x\} - s\,x(0) - x'(0)] - \mathcal{L}\{x\} = \mathcal{L}\{y\}$ (7)

   $[s^2\mathcal{L}\{y\} - s\,y(0) - y'(0)] + \mathcal{L}\{y\} = -\mathcal{L}\{x\}$ (8)

2. $x(0) = 2$, $y(0) = -1$, $x'(0) = 0$ and $y'(0) = 0$

   hence $s^2\mathcal{L}\{x\} - 2s - \mathcal{L}\{x\} = \mathcal{L}\{y\}$ (7′)

   $s^2\mathcal{L}\{y\} + s + \mathcal{L}\{y\} = -\mathcal{L}\{x\}$ (8′)

3. Rearranging gives:

   $(s^2 - 1)\mathcal{L}\{x\} - \mathcal{L}\{y\} = 2s$ (9)

   $\mathcal{L}\{x\} + (s^2 + 1)\mathcal{L}\{y\} = -s$ (10)

Equation (9) $\times$ (s^2+1) and equation (10) $\times$ 1 gives:

$$(s^2+1)(s^2-1)\mathscr{L}\{x\} - (s^2+1)\mathscr{L}\{y\} = (s^2+1)2s \quad (11)$$

$$\mathscr{L}\{x\} + (s^2+1)\mathscr{L}\{y\} = -s \quad (12)$$

Adding equations (11) and (12) gives:

$$[(s^2+1)(s^2-1)+1]\mathscr{L}\{x\} = (s^2+1)2s - s$$

i.e. $s^4\mathscr{L}\{x\} = 2s^3 + s = s(2s^2+1)$

from which, $\mathscr{L}\{x\} = \dfrac{s(2s^2+1)}{s^4} = \dfrac{2s^2+1}{s^3} = \dfrac{2s^2}{s^3} + \dfrac{1}{s^3} = \dfrac{2}{s} + \dfrac{1}{s^3}$

4. Hence $x = \mathscr{L}^{-1}\left\{\dfrac{2}{s} + \dfrac{1}{s^3}\right\}$

i.e. $\mathbf{x = 2 + \frac{1}{2}t^2}$

Returning to equations (9) and (10) to determine y:

1 $\times$ equation (9) and (s^2-1) $\times$ equation (10) gives:

$$(s^2-1)\mathscr{L}\{x\} - \mathscr{L}\{y\} = 2s \quad (13)$$

$$(s^2-1)\mathscr{L}\{x\} + (s^2-1)(s^2+1)\mathscr{L}\{y\} = -s(s^2-1) \quad (14)$$

Equation (13) $-$ equation (14) gives:

$$[-1-(s^2-1)(s^2+1)]\mathscr{L}\{y\} = 2s + s(s^2-1)$$

i.e. $-s^4\mathscr{L}\{y\} = s^3 + s$

and $\mathscr{L}\{y\} = \dfrac{s^3+s}{-s^4} = -\dfrac{1}{s} - \dfrac{1}{s^3}$

from which, $y = \mathscr{L}^{-1}\left\{-\dfrac{1}{s} - \dfrac{1}{s^3}\right\}$

i.e. $\mathbf{y = -1 - \frac{1}{2}t^2}$

16 Fourier Series

16.1 Fourier series for periodic functions of period 2π

The basis of a **Fourier series** is that all functions of practical significance which are defined in the interval $-\pi \leq x \leq \pi$ can be expressed in terms of a convergent trigonometric series of the form:

$$f(x) = a_0 + a_1 \cos x + a_2 \cos 2x + a_3 \cos 3x + \cdots + b_1 \sin x + b_2 \sin 2x + b_3 \sin 3x + \cdots$$

when $a_0, a_1, a_2, \ldots b_1, b_2, \ldots$ are real constants, i.e.

$$\mathbf{f(x) = a_0 + \sum_{n=1}^{\infty} (a_n \cos nx + b_n \sin nx)} \qquad (1)$$

where for the range $-\pi$ to π:

$$a_0 = \frac{1}{2\pi}\int_{-\pi}^{\pi} f(x)dx$$

$$a_n = \frac{1}{\pi}\int_{-\pi}^{\pi} f(x)\cos nx\, dx \quad (n = 1, 2, 3, \ldots)$$

and $$b_n = \frac{1}{\pi}\int_{-\pi}^{\pi} f(x)\sin nx\, dx \quad (n = 1, 2, 3, \ldots)$$

Fourier series provides a method of analysing periodic functions into their constituent components. Alternating currents and voltages, displacement, velocity and acceleration of slider-crank mechanisms and acoustic waves are typical practical examples in engineering and science where periodic functions are involved and often requiring analysis.

For an exact representation of a complex wave, an infinite number of terms are, in general, required. In many practical cases, however, it is sufficient to take the first few terms only.

Application: Obtain a Fourier series for the periodic function f(x) defined as:

$$f(x) = \begin{cases} -k, \text{ when } -\pi \langle x \langle 0 \\ +k, \text{ when } 0 \langle x \langle \pi \end{cases}$$

(The function is periodic outside of this range with period 2π)

The square wave function defined is shown in Figure 16.1. Since f(x) is given by two different expressions in the two halves of the range the integration is performed in two parts, one from $-\pi$ to 0 and the other from 0 to π.

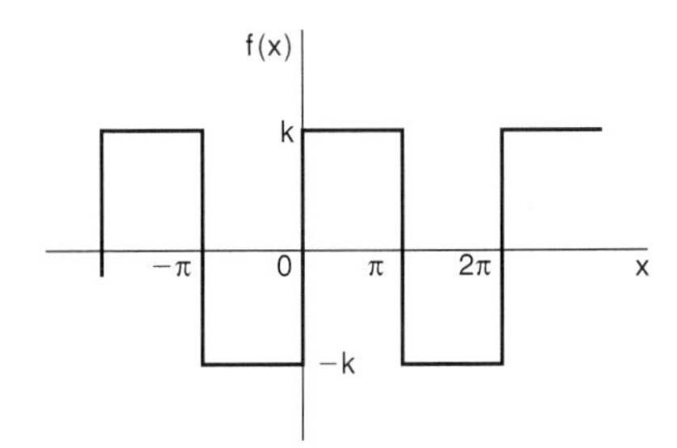

Figure 16.1

From above: $$a_0 = \frac{1}{2\pi}\int_{-\pi}^{\pi} f(x)dx = \frac{1}{2\pi}\left[\int_{-\pi}^{0} -k\,dx + \int_{0}^{\pi} k\,dx\right]$$

$$= \frac{1}{2\pi}\left\{[-kx]_{-\pi}^{0} + [kx]_{0}^{\pi}\right\}$$

$$= 0$$

[a_0 is in fact the **mean value** of the waveform over a complete period of 2π and this could have been deduced on sight from Figure 16.1]

$$a_n = \frac{1}{\pi}\int_{-\pi}^{\pi} f(x)\cos nx\,dx = \frac{1}{\pi}\left\{\int_{-\pi}^{0} -k\cos nx\,dx + \int_{0}^{\pi} k\cos nx\,dx\right\}$$

$$= \frac{1}{\pi}\left\{\left[\frac{-k\sin nx}{n}\right]_{-\pi}^{0} + \left[\frac{k\sin nx}{n}\right]_{0}^{\pi}\right\}$$

$$= 0$$

Hence $a_1, a_2, a_3, \ldots$ are all zero (since $\sin 0 = \sin(-n\pi) = \sin n\pi = 0$), and therefore no cosine terms will appear in the Fourier series.

$$b_n = \frac{1}{\pi}\int_{-\pi}^{\pi} f(x)\sin nx\,dx = \frac{1}{\pi}\left\{\int_{-\pi}^{0} -k\sin nx\,dx + \int_{0}^{\pi} k\sin nx\,dx\right\}$$

$$= \frac{1}{\pi}\left\{\left[\frac{k\cos nx}{n}\right]_{-\pi}^{0} + \left[\frac{-k\cos nx}{n}\right]_{0}^{\pi}\right\}$$

When n is odd:

$$b_n = \frac{k}{\pi}\left\{\left[\left(\frac{1}{n}\right) - \left(-\frac{1}{n}\right)\right] + \left[-\left(-\frac{1}{n}\right) - \left(-\frac{1}{n}\right)\right]\right\} = \frac{k}{\pi}\left\{\frac{2}{n} + \frac{2}{n}\right\} = \frac{4k}{n\pi}$$

Hence, $b_1 = \dfrac{4k}{\pi}$, $b_3 = \dfrac{4k}{3\pi}$, $b_5 = \dfrac{4k}{5\pi}$, and so on

When n is even: $b_n = \dfrac{k}{\pi}\left\{\left[\dfrac{1}{n} - \dfrac{1}{n}\right] + \left[-\dfrac{1}{n} - \left(-\dfrac{1}{n}\right)\right]\right\} = 0$

Hence, from equation (1), the Fourier series for the function shown in Figure 16.1 is given by:

$$f(x) = a_0 + \sum_{n=1}^{\infty}(a_n\cos nx + b_n\sin nx) = 0 + \sum_{n=1}^{\infty}(0 + b_n\sin nx)$$

i.e. $$f(x) = \frac{4k}{\pi}\sin x + \frac{4k}{3\pi}\sin 3x + \frac{4k}{5\pi}\sin 5x + \cdots$$

i.e. $$\mathbf{f(x) = \frac{4k}{\pi}\left(\sin x + \frac{1}{3}\sin 3x + \frac{1}{5}\sin 5x + \cdots\right)}$$

If $k = \pi$ in the above Fourier series then:

$$f(x) = 4\left(\sin x + \frac{1}{3}\sin 3x + \frac{1}{5}\sin 5x + \cdots\right)$$

$4\sin x$ is termed the first partial sum of the Fourier series of f(x),

$\left(4\sin x + \dfrac{4}{3}\sin 3x\right)$ is termed the second partial sum of the Fourier series, and

$\left(4\sin x + \frac{4}{3}\sin 3x + \frac{4}{5}\sin 5x\right)$ is termed the third partial sum, and so on.

Let $P_1 = 4\sin x$, $P_2 = \left(4\sin x + \frac{4}{3}\sin 3x\right)$ and

$$P_3 = \left(4\sin x + \frac{4}{3}\sin 3x + \frac{4}{5}\sin 5x\right).$$

Graphs of P_1, P_2 and P_3, obtained by drawing up tables of values, and adding waveforms, are shown in Figures 16.2(a) to (c) and they show that the series is convergent, i.e. continually approximating towards a definite limit as more and more partial sums are taken, and in the limit will have the sum $f(x) = \pi$.

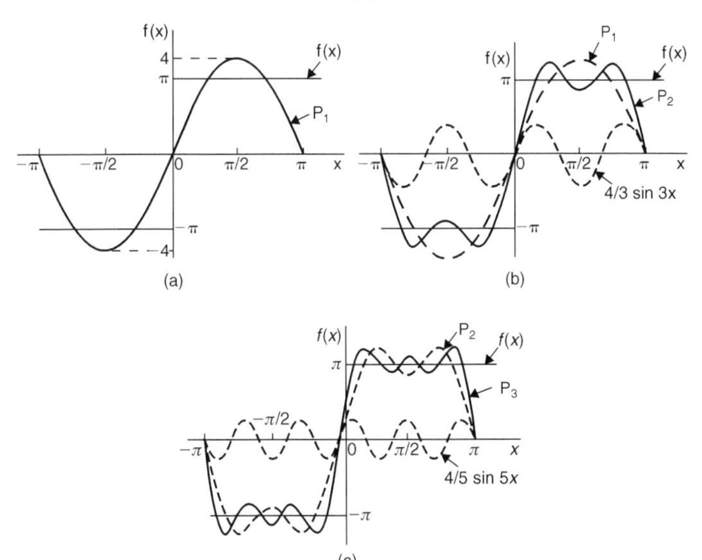

Figure 16.2

Even with just three partial sums, the waveform is starting to approach the **rectangular wave** the Fourier series is representing. Thus, a rectangular wave is comprised of a fundamental and an infinite number of odd harmonics.

16.2 Fourier series for a non-periodic function over range 2π

If a function f(x) is not periodic then it cannot be expanded in a Fourier series for **all** values of x.

However, it is possible to determine a Fourier series to represent the function over any range of width 2π.

For determining a Fourier series of a non-periodic function over a range 2π, exactly the same formulae for the Fourier coefficients are used as in equation (1), page 492.

Application: Determine the Fourier series to represent the function $f(x) = 2x$ in the range $-\pi$ to $+\pi$

The function $f(x) = 2x$ is not periodic. The function is shown in the range $-\pi$ to π in Figure 16.3 and is then constructed outside of that range so that it is periodic of period 2π (see broken lines) with the resulting saw-tooth waveform.

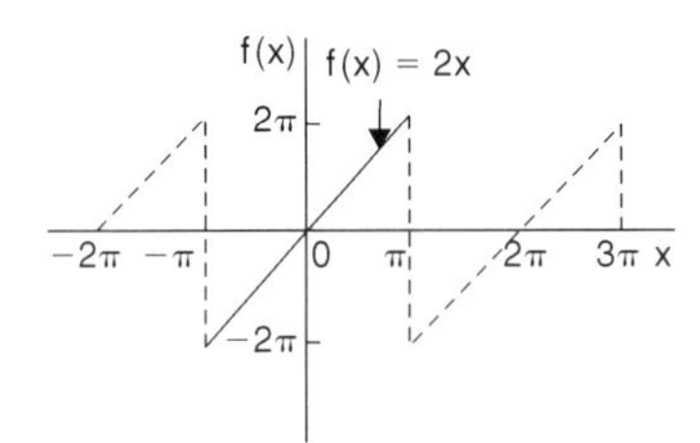

Figure 16.3

For a Fourier series: $f(x) = a_0 + \sum_{n=1}^{\infty}(a_n \cos nx + b_n \sin nx)$

$$a_0 = \frac{1}{2\pi}\int_{-\pi}^{\pi} f(x)dx = \frac{1}{2\pi}\int_{-\pi}^{\pi} 2x\,dx = \frac{1}{2\pi}\left[x^2\right]_{-\pi}^{\pi} = 0$$

$$a_n = \frac{1}{\pi}\int_{-\pi}^{\pi} f(x)\cos nx\,dx = \frac{1}{\pi}\int_{-\pi}^{\pi} 2x\cos nx\,dx$$

$$= \frac{2}{\pi}\left[\frac{x\sin nx}{n} - \int\frac{\sin nx}{n}dx\right]_{-\pi}^{\pi} \quad \text{by parts (see Chapter 12)}$$

$$= \frac{2}{\pi}\left[\frac{x \sin nx}{n} + \frac{\cos nx}{n^2}\right]_{-\pi}^{\pi} = \frac{2}{\pi}\left[\left(0 + \frac{\cos n\pi}{n^2}\right) - \left(0 + \frac{\cos n(-\pi)}{n^2}\right)\right] = 0$$

$$b_n = \frac{1}{\pi}\int_{-\pi}^{\pi} f(x) \sin nx \, dx = \frac{1}{\pi}\int_{-\pi}^{\pi} 2x \sin nx \, dx$$

$$= \frac{2}{\pi}\left[\frac{-x \text{ cox } nx}{n} - \int\left(\frac{-\cos nx}{n}\right)dx\right]_{-\pi}^{\pi} \qquad \text{by parts}$$

$$= \frac{2}{\pi}\left[\frac{-x \cos nx}{n} + \frac{\sin nx}{n^2}\right]_{-\pi}^{\pi}$$

$$= \frac{2}{\pi}\left[\left(\frac{-\pi \cos n\pi}{n} + \frac{\sin n\pi}{n^2}\right) - \left(\frac{-(-\pi)\cos n(-\pi)}{n} + \frac{\sin n(-\pi)}{n^2}\right)\right]$$

$$= \frac{2}{\pi}\left[\frac{-\pi \cos n\pi}{n} - \frac{\pi \cos(-n\pi)}{n}\right] = \frac{-4}{n}\cos n\pi$$

since $\cos n\pi = \cos(-n\pi)$

When n is odd, $b_n = \frac{4}{n}$. Thus $b_1 = 4$, $b_3 = \frac{4}{3}$, $b_5 = \frac{4}{5}$, and so on.

When n is even, $b_n = -\frac{4}{n}$. Thus $b_2 = -\frac{4}{2}$, $b_4 = -\frac{4}{4}$, $b_6 = -\frac{4}{6}$, and so on.

Thus, $$f(x) = 2x = 4 \sin x - \frac{4}{2}\sin 2x + \frac{4}{3}\sin 3x - \frac{4}{4}\sin 4x + \frac{4}{5}\sin 5x - \frac{4}{6}\sin 6x + \cdots$$

i.e. $$\mathbf{2x = 4\left(\sin x - \frac{1}{2}\sin 2x + \frac{1}{3}\sin 3x - \frac{1}{4}\sin 4x + \frac{1}{5}\sin 5x - \frac{1}{6}\sin 6x + \cdots\right)}$$

for values of f(x) between $-\pi$ and π.

16.3 Even and odd functions

A function $y = f(x)$ is said to be **even** if $f(-x) = f(x)$ for all values of x. Graphs of even functions are always **symmetrical about the y-axis** (i.e. a mirror image). Two examples of even functions are $y = x^2$ and $y = \cos x$ as shown in Figure 6.38, page 181.

A function $y = f(x)$ is said to be **odd** if $f(-x) = -f(x)$ for all values of x. Graphs of odd functions are always **symmetrical about the origin.** Two examples of odd functions are $y = x^3$ and $y = \sin x$ as shown in Figure 6.39, page 181.

Many functions are neither even nor odd, two such examples being $y = \ln x$ and $y = e^x$.

Fourier cosine series

The Fourier series of an **even** periodic function f(x) having period 2π contains **cosine terms only** (i.e. contains no sine terms) and may contain a constant term.

Hence $$\mathbf{f(x) = a_0 + \sum_{n=1}^{\infty} a_n \cos nx} \qquad (2)$$

where $$a_0 = \frac{1}{2\pi}\int_{-\pi}^{\pi} f(x)\,dx = \frac{1}{\pi}\int_0^{\pi} f(x)\,dx \quad \text{(due to symmetry)}$$

and $$a_n = \frac{1}{\pi}\int_{-\pi}^{\pi} f(x)\cos nx\,dx = \frac{2}{\pi}\int_0^{\pi} f(x)\cos nx\,dx$$

Fourier sine series

The Fourier series of an **odd** periodic function f(x) having period 2π contains **sine terms only** (i.e. contains no constant term and no cosine terms).

Hence $$\mathbf{f(x) = \sum_{n=1}^{\infty} b_n \sin nx} \qquad (3)$$

where $$b_n = \frac{1}{\pi}\int_{-\pi}^{\pi} f(x)\sin nx\,dx = \frac{2}{\pi}\int_0^{\pi} f(x)\sin nx\,dx$$

Application: Determine the Fourier series for the periodic function defined by:

$$f(x) = \begin{cases} -2, \text{ when } -\pi < x < -\dfrac{\pi}{2} \\ 2, \text{ when } -\dfrac{\pi}{2} < x < \dfrac{\pi}{2} \\ -2, \text{ when } \dfrac{\pi}{2} < x < \pi \end{cases} \quad \text{and has a period of } 2\pi$$

The square wave shown in Figure 16.4 is an **even function** since it is symmetrical about the f(x) axis.

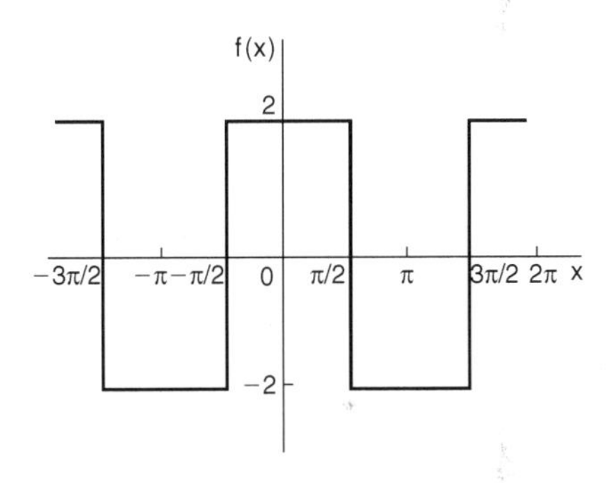

Figure 16.4

Hence from equation (2), the Fourier series is given by:

$$f(x) = a_0 + \sum_{n=1}^{\infty} a_n \cos nx$$ (i.e. the series contains no sine terms).

$$\begin{aligned} a_0 &= \frac{1}{\pi}\int_0^{\pi} f(x)\,dx = \frac{1}{\pi}\left\{\int_0^{\pi/2} 2\,dx + \int_{\pi/2}^{\pi} -2\,dx\right\} \\ &= \frac{1}{\pi}\left\{[2x]_0^{\pi/2} + [-2x]_{\pi/2}^{\pi}\right\} \\ &= \frac{1}{\pi}\left[(\pi) + [(-2\pi) - (-\pi)]\right] \\ &= 0 \end{aligned}$$

$$a_n = \frac{2}{\pi}\int_0^{\pi} f(x)\cos nx\,dx = \frac{2}{\pi}\left\{\int_0^{\pi/2} 2\cos nx\,dx + \int_{\pi/2}^{\pi} -2\cos nx\,dx\right\}$$

$$= \frac{4}{\pi}\left\{\left[\frac{\sin nx}{n}\right]_0^{\pi/2} + \left[\frac{-\sin nx}{n}\right]_{\pi/2}^{\pi}\right\} = \frac{4}{\pi}\left\{\left(\frac{\sin(\pi/2)n}{n} - 0\right) + \left(0 - \frac{-\sin(\pi/2)n}{n}\right)\right\}$$

$$= \frac{4}{\pi}\left(\frac{2\sin(\pi/2)n}{n}\right) = \frac{8}{\pi n}\left(\sin\frac{n\pi}{2}\right)$$

When n is even, $a_n = 0$

When n is odd, $a_n = \dfrac{8}{\pi n}$ for $n = 1, 5, 9, \ldots$

and $\quad a_n = \dfrac{-8}{\pi n}$ for $n = 3, 7, 11, \ldots$

Hence, $a_1 = \dfrac{8}{\pi}, a_3 = \dfrac{-8}{3\pi}, a_5 = \dfrac{8}{5\pi}$, and so on

Hence the Fourier series for the waveform of Figure 16.4 is given by:

$$\mathbf{f(x) = \frac{8}{\pi}\left(\cos x - \frac{1}{3}\cos 3x + \frac{1}{5}\cos 5x - \frac{1}{7}\cos 7x + \cdots\right)}$$

Application: Obtain the Fourier series for the square wave shown in Figure 16.5.

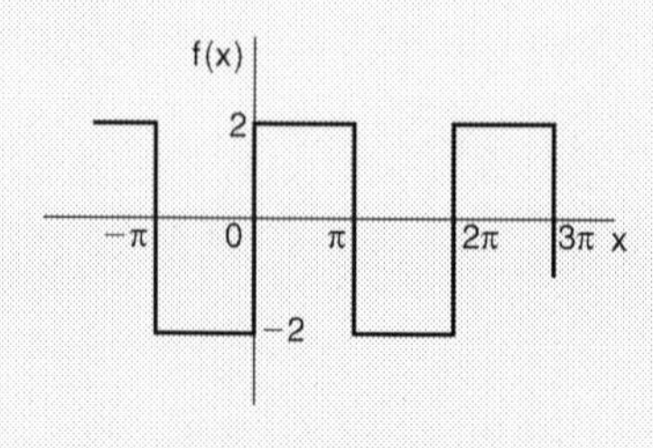

Figure 16.5

The square wave is an **odd function** since it is symmetrical about the origin.

Hence, from equation (3), the Fourier series is given by:

$$f(x) = \sum_{n=1}^{\infty} b_n \sin nx$$

The function is defined by: $f(x) = \begin{cases} -2, \text{ when } -\pi < x < 0 \\ 2, \text{ when } \quad 0 < x < \pi \end{cases}$

$$b_n = \frac{2}{\pi}\int_0^{\pi} f(x)\sin nx\,dx = \frac{2}{\pi}\int_0^{\pi} 2\sin nx\,dx = \frac{4}{\pi}\left[\frac{-\cos nx}{n}\right]_0^{\pi}$$

$$= \frac{4}{\pi}\left[\left(\frac{-\cos n\pi}{n}\right) - \left(-\frac{1}{n}\right)\right] = \frac{4}{\pi n}(1 - \cos n\pi)$$

When n is even, $b_n = 0$. When n is odd, $b_n = \frac{4}{\pi n}[1 - (-1)] = \frac{8}{\pi n}$

Hence, $b_1 = \frac{8}{\pi}, b_3 = \frac{8}{3\pi}, b_5 = \frac{8}{5\pi}$, and so on

Hence the Fourier series is:

$$\mathbf{f(x) = \frac{8}{\pi}\left(\sin x + \frac{1}{3}\sin 3x + \frac{1}{5}\sin 5x + \frac{1}{7}\sin 7x + \cdots\right)}$$

16.4 Half range Fourier series

When a function is defined over the range say 0 to π instead of from 0 to 2π it may be expanded in a series of sine terms only or of cosine terms only. The series produced is called a **half-range Fourier series**.

When a **half range cosine series** is required then:

$$\mathbf{f(x) = a_0 + \sum_{n=1}^{\infty} a_n \cos nx} \qquad (4)$$

where $\mathbf{a_0 = \frac{1}{\pi}\int_0^{\pi} f(x)\,dx}$ and $\mathbf{a_n = \frac{2}{\pi}\int_0^{\pi} f(x)\cos nx\,dx}$

If a **half-range cosine series** is required for the function $f(x) = x$ in the range 0 to π then an **even** periodic function is required. In Figure 16.6, $f(x) = x$ is shown plotted from $x = 0$ to $x = \pi$. Since an even function is symmetrical about the f(x) axis the line AB is constructed as shown. If the triangular waveform produced is assumed to be periodic of period 2π outside of this range then the waveform is as shown in Figure 16.6.

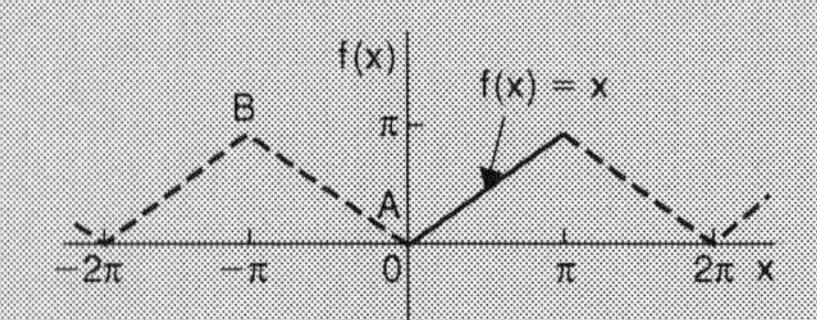

Figure 16.6

When a **half-range sine series** is required then the Fourier coefficient b_n is calculated as earlier, i.e.

$$\mathbf{f(x) = \sum_{n=1}^{\infty} b_n \sin nx} \tag{5}$$

where $$\mathbf{b_n = \frac{2}{\pi}\int_0^{\pi} f(x)\sin nx\,dx}$$

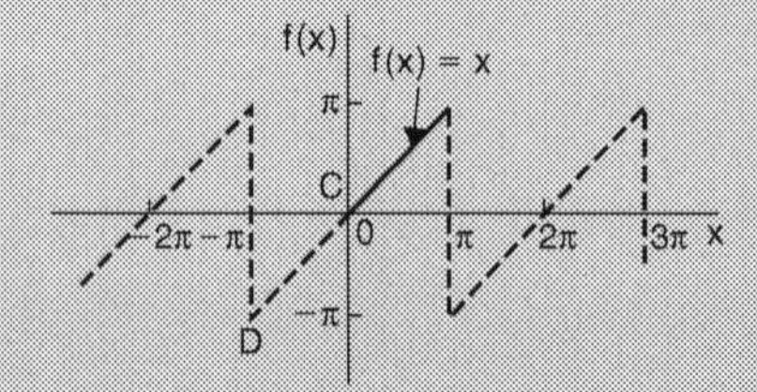

Figure 16.7

If a **half-range sine series** is required for the function $f(x) = x$ in the range 0 to π then an odd periodic function is required. In Figure 16.7, $f(x) = x$ is shown plotted from $x = 0$ to $x = \pi$. Since an odd function is symmetrical about the origin the line CD is constructed as shown. If the sawtooth waveform produced is assumed to be periodic of period 2π outside of this range, then the waveform is as shown in Figure 16.7

Application: Determine the half-range Fourier cosine series to represent the function $f(x) = x$ in the range $0 \le x \le \pi$

The function is shown in Figure 16.6.

When $f(x) = x$, $a_0 = \frac{1}{\pi}\int_0^{\pi} f(x)dx = \frac{1}{\pi}\int_0^{\pi} x\,dx = \frac{1}{\pi}\left[\frac{x^2}{2}\right]_0^{\pi} = \frac{\pi}{2}$

$$a_n = \frac{2}{\pi}\int_0^{\pi} f(x)\cos nx\,dx = \frac{2}{\pi}\int_0^{\pi} x\cos nx\,dx$$

$$= \frac{2}{\pi}\left[\frac{x\sin nx}{n} + \frac{\cos nx}{n^2}\right]_0^{\pi} \qquad \text{by parts}$$

$$= \frac{2}{\pi}\left[\left(\frac{\pi\sin n\pi}{n} + \frac{\cos n\pi}{n^2}\right) - \left(0 + \frac{\cos 0}{n^2}\right)\right]$$

$$= \frac{2}{\pi}\left(0 + \frac{\cos n\pi}{n^2} - \frac{\cos 0}{n^2}\right) = \frac{2}{\pi n^2}(\cos n\pi - 1)$$

When n is even, $a_n = 0$

When n is odd, $a_n = \frac{2}{\pi n^2}(-1-1) = \frac{-4}{\pi n^2}$

Hence, $a_1 = \frac{-4}{\pi}$, $a_3 = \frac{-4}{\pi 3^2}$, $a_5 = \frac{-4}{\pi 5^2}$, and so on

Hence, the half-range Fourier cosine series is given by:

$$\mathbf{f(x) = x = \frac{\pi}{2} - \frac{4}{\pi}\left(\cos x + \frac{1}{3^2}\cos 3x + \frac{1}{5^2}\cos 5x + \cdots\right)}$$

Application: Determine the half-range Fourier sine series to represent the function $f(x) = x$ in the range $0 \le x \le \pi$

The function is shown in Figure 16.7.

When f(x) = x,

$$b_n = \frac{2}{\pi}\int_0^{\pi} f(x)\sin nx\,dx = \frac{2}{\pi}\int_0^{\pi} x\sin nx\,dx$$

$$= \frac{2}{\pi}\left[\frac{-x\cos nx}{n} + \frac{\sin nx}{n^2}\right]_0^{\pi} \quad \text{by parts}$$

$$= \frac{2}{\pi}\left[\left(\frac{-\pi\cos n\pi}{n} + \frac{\sin n\pi}{n^2}\right) - (0+0)\right] = -\frac{2}{n}\cos n\pi$$

When n is odd, $b_n = \frac{2}{n}$. Hence, $b_1 = \frac{2}{1}$, $b_3 = \frac{2}{3}$, $b_5 = \frac{2}{5}$ and so on.

When n is even, $b_n = -\frac{2}{n}$. Hence $b_2 = -\frac{2}{2}$, $b_4 = -\frac{2}{4}$, $b_6 = -\frac{2}{6}$ and so on

Hence the half-range Fourier sine series is given by:

$$\mathbf{f(x) = x}$$
$$\mathbf{= 2\left(\sin x - \frac{1}{2}\sin 2x + \frac{1}{3}\sin 3x - \frac{1}{4}\sin 4x + \frac{1}{5}\sin 5x - \ldots\right)}$$

16.5 Expansion of a periodic function of period L

If f(x) is a function of period L, then its Fourier series is given by:

$$\mathbf{f(x) = a_0 + \sum_{n=1}^{\infty}\left[a_n\cos\left(\frac{2\pi nx}{L}\right) + b_n\sin\left(\frac{2\pi nx}{L}\right)\right]} \quad (6)$$

where, in the range $-\frac{L}{2}$ to $+\frac{L}{2}$:

$$\mathbf{a_0 = \frac{1}{L}\int_{-L/2}^{L/2} f(x)dx, \quad a_n = \frac{2}{L}\int_{-L/2}^{L/2} f(x)\cos\left(\frac{2\pi nx}{L}\right)dx}$$

and $\mathbf{b_n = \frac{2}{L}\int_{-L/2}^{L/2} f(x)\sin\left(\frac{2\pi nx}{L}\right) dx}$

(The limits of integration may be replaced by any interval of length L, such as from 0 to L)

Application: The voltage from a square wave generator is of the form:

$$v(t) = \begin{cases} 0, -4 \langle t \langle 0 \\ 10, \ 0 \langle t \langle 4 \end{cases} \quad \text{and has a period of 8 ms.}$$

Find the Fourier series for this periodic function

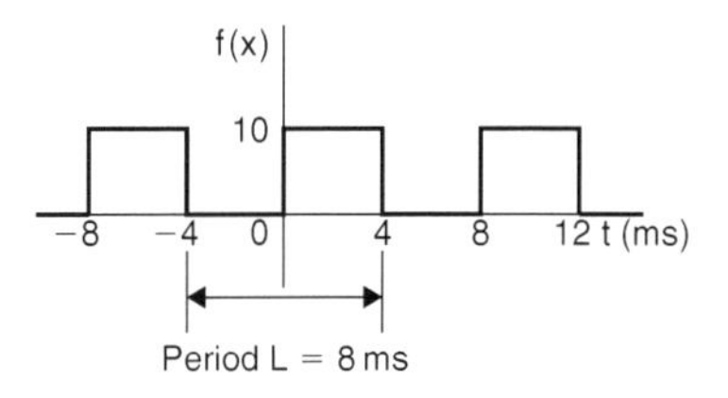

Figure 16.8

The square wave is shown in Figure 16.8. From above, the Fourier series is of the form:

$$v(t) = a_0 + \sum_{n=1}^{\infty}\left[a_n \cos\left(\frac{2\pi nt}{L}\right) + b_n \sin\left(\frac{2\pi nt}{L}\right)\right]$$

$$a_0 = \frac{1}{L}\int_{-L/2}^{L/2} v(t)dt = \frac{1}{8}\int_{-4}^{4} v(t)dt = \frac{1}{8}\left\{\int_{-4}^{0} 0\,dt + \int_{0}^{4} 10\,dt\right\}$$

$$= \frac{1}{8}\left[10\,t\right]_0^4 = 5$$

$$a_n = \frac{2}{L}\int_{-L/2}^{L/2} v(t)\cos\left(\frac{2\pi nt}{L}\right)dt = \frac{2}{8}\int_{-4}^{4} v(t)\cos\left(\frac{2\pi nt}{8}\right)dt$$

$$= \frac{1}{4}\left[\int_{-4}^{0} 0\cos\left(\frac{\pi nt}{4}\right)dt + \int_{0}^{4} 10\cos\left(\frac{\pi nt}{4}\right)dt\right]$$

$$= \frac{1}{4}\left[\frac{10\sin\left(\frac{\pi n t}{4}\right)}{\left(\frac{\pi n}{4}\right)}\right]_0^4 = \frac{10}{\pi n}[\sin \pi n - \sin 0] = 0 \quad \text{for } n = 1, 2, 3, \ldots$$

$$b_n = \frac{2}{L}\int_{-L/2}^{L/2} v(t)\sin\left(\frac{2\pi n t}{L}\right)dt$$

$$= \frac{2}{8}\int_{-4}^{4} v(t)\sin\left(\frac{2\pi n t}{8}\right)dt$$

$$= \frac{1}{4}\left\{\int_{-4}^{0} 0\sin\left(\frac{\pi n t}{4}\right)dt + \int_{0}^{4} 10\sin\left(\frac{\pi n t}{4}\right)dt\right\}$$

$$= \frac{1}{4}\left[\frac{-10\cos\left(\frac{\pi n t}{4}\right)}{\left(\frac{\pi n}{4}\right)}\right]_0^4 = \frac{-10}{\pi n}[\cos \pi n - \cos 0]$$

When n is even, $b_n = 0$

When n is odd, $b_1 = \frac{-10}{\pi}(-1-1) = \frac{20}{\pi}$,

$$b_3 = \frac{-10}{3\pi}(-1-1) = \frac{20}{3\pi},\; b_5 = \frac{20}{5\pi}, \text{ and so on}$$

Thus the Fourier series for the function v(t) is given by:

$$\mathbf{v(t) = 5 + \frac{20}{\pi}\left[\sin\left(\frac{\pi t}{4}\right) + \frac{1}{3}\sin\left(\frac{3\pi t}{4}\right) + \frac{1}{5}\sin\left(\frac{5\pi t}{4}\right) + \cdots\right]}$$

Application: Obtain the Fourier series for the function defined by:

$$f(x) = \begin{cases} 0, & \text{when } -2 \langle x \langle -1 \\ 5, & \text{when } -1 \langle x \langle 1 \\ 0, & \text{when } \quad 1 \langle x \langle 2 \end{cases}$$

The function is periodic outside of this range of period 4

The function f(x) is shown in Figure 16.9 where period, L = 4. Since the function is symmetrical about the f(x) axis it is an even function and the Fourier series contains no sine terms (i.e. $b_n = 0$)

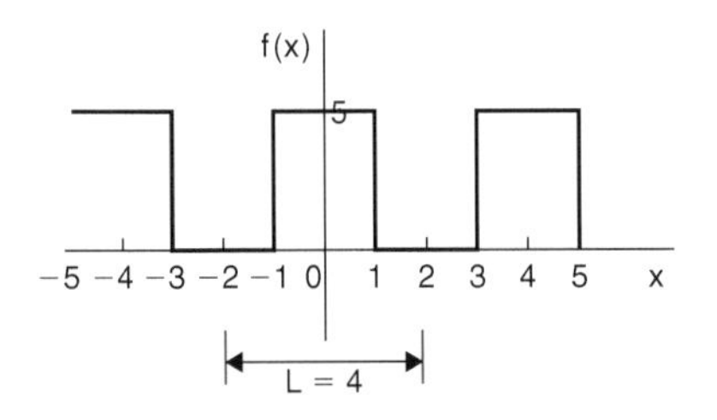

Figure 16.9

Thus, from equation (6), $f(x) = a_0 + \sum_{n=1}^{\infty} a_n \cos\left(\frac{2\pi nx}{L}\right)$

$$a_0 = \frac{1}{L}\int_{-L/2}^{L/2} f(x)dx = \frac{1}{4}\int_{-2}^{2} f(x)dx$$

$$= \frac{1}{4}\left\{\int_{-2}^{-1} 0\,dx + \int_{-1}^{1} 5\,dx + \int_{1}^{2} 0\,dx\right\}$$

$$= \frac{1}{4}[5x]_{-1}^{1} = \frac{1}{4}[(5) - (-5)] = \frac{10}{4} = \frac{5}{2}$$

$$a_n = \frac{2}{L}\int_{-L/2}^{L/2} f(x)\cos\left(\frac{2\pi nx}{L}\right) dx$$

$$= \frac{2}{4}\int_{-2}^{2} f(x)\cos\left(\frac{2\pi nx}{4}\right) dx$$

$$= \frac{1}{2}\left\{\int_{-2}^{-1} 0\cos\left(\frac{\pi nx}{2}\right) dx + \int_{-1}^{1} 5\cos\left(\frac{\pi nx}{2}\right) dx + \int_{1}^{2} 0\cos\left(\frac{\pi nx}{2}\right) dx\right\}$$

$$= \frac{5}{2}\left[\frac{\sin\frac{\pi nx}{2}}{\frac{\pi n}{2}}\right]_{-1}^{1} = \frac{5}{\pi n}\left[\sin\left(\frac{\pi n}{2}\right) - \sin\left(\frac{-\pi n}{2}\right)\right]$$

When n is even, $a_n = 0$

When n is odd, $a_1 = \frac{5}{\pi}(1 - -1) = \frac{10}{\pi}$, $a_3 = \frac{5}{3\pi}(-1 - 1) = \frac{-10}{3\pi}$,

$a_5 = \frac{5}{5\pi}(1 - -1) = \frac{10}{5\pi}$, and so on

Hence the Fourier series for the function f(x) is given by:

$$\mathbf{f(x) = \frac{5}{2} + \frac{10}{\pi}\left[\cos\left(\frac{\pi x}{2}\right) - \frac{1}{3}\cos\left(\frac{3\pi x}{2}\right) + \frac{1}{5}\cos\left(\frac{5\pi x}{2}\right) - \frac{1}{7}\cos\left(\frac{7\pi x}{2}\right) + \ldots\right]}$$

16.6 Half-range Fourier series for functions defined over range L

A **half-range cosine series** in the range 0 to L can be expanded as:

$$f(x) = a_0 + \sum_{n=1}^{\infty} a_n \cos\left(\frac{n\pi x}{L}\right) \qquad (7)$$

where

$$a_0 = \frac{1}{L}\int_0^L f(x)\,dx \quad \text{and} \quad a_n = \frac{2}{L}\int_0^L f(x)\cos\left(\frac{n\pi x}{L}\right)dx$$

A **half-range sine series** in the range 0 to L can be expanded as:

$$f(x) = \sum_{n=1}^{\infty} b_n \sin\left(\frac{n\pi x}{L}\right) \qquad (8)$$

where $$b_n = \frac{2}{L}\int_0^L f(x)\sin\left(\frac{n\pi x}{L}\right)dx$$

Application: Determine the half-range Fourier cosine series for the function $f(x) = x$ in the range $0 \le x \le 2$

A half-range Fourier cosine series indicates an even function. Thus the graph of $f(x) = x$ in the range 0 to 2 is shown in Figure 16.10 and is extended outside of this range so as to be symmetrical about the $f(x)$ axis as shown by the broken lines.

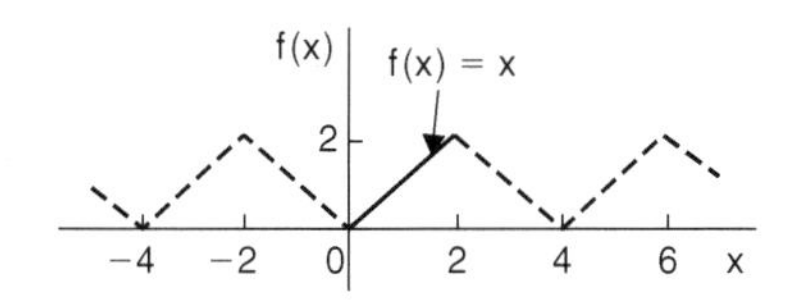

Figure 16.10

For a half-range cosine series: $f(x) = a_0 + \sum_{n=1}^{\infty} a_n \cos\left(\frac{n\pi x}{L}\right)$ from equation (7)

$$a_0 = \frac{1}{L}\int_0^L f(x)dx = \frac{1}{2}\int_0^2 x\,dx = \frac{1}{2}\left[\frac{x^2}{2}\right]_0^2 = 1$$

$$a_n = \frac{2}{L}\int_0^L f(x)\cos\left(\frac{n\pi x}{L}\right)dx$$

$$= \frac{2}{2}\int_0^2 x\cos\left(\frac{n\pi x}{2}\right)dx = \left[\frac{x\sin\left(\frac{n\pi x}{2}\right)}{\left(\frac{n\pi}{2}\right)} + \frac{\cos\left(\frac{n\pi x}{2}\right)}{\left(\frac{n\pi}{2}\right)^2}\right]_0^2$$

$$= \left[\left(\frac{2\sin n\pi}{\left(\frac{n\pi}{2}\right)} + \frac{\cos n\pi}{\left(\frac{n\pi}{2}\right)^2}\right) - \left(0 + \frac{\cos 0}{\left(\frac{n\pi}{2}\right)^2}\right)\right] = \left[\frac{\cos n\pi}{\left(\frac{n\pi}{2}\right)^2} - \frac{1}{\left(\frac{n\pi}{2}\right)^2}\right]$$

$$= \left(\frac{2}{\pi n}\right)^2(\cos n\pi - 1)$$

When n is even, $a_n = 0$, $a_1 = \frac{-8}{\pi^2}$, $a_3 = \frac{-8}{\pi^2 3^2}$, $a_5 = \frac{-8}{\pi^2 5^2}$, and so on.

Hence the half-range Fourier cosine series for f(x) in the range 0 to 2 is given by:

$$\mathbf{f(x) = 1 - \frac{8}{\pi^2}\left[\cos\left(\frac{\pi x}{2}\right) + \frac{1}{3^2}\cos\left(\frac{3\pi x}{2}\right) + \frac{1}{5^2}\cos\left(\frac{5\pi x}{2}\right) + \cdots\right]}$$

Application: Determine the half-range Fourier sine series for the function $f(x) = x$ in the range $0 \le x \le 2$

A half-range Fourier sine series indicates an odd function. Thus the graph of $f(x) = x$ in the range 0 to 2 is shown in Figure 16.11 and is extended outside of this range so as to be symmetrical about the origin, as shown by the broken lines.

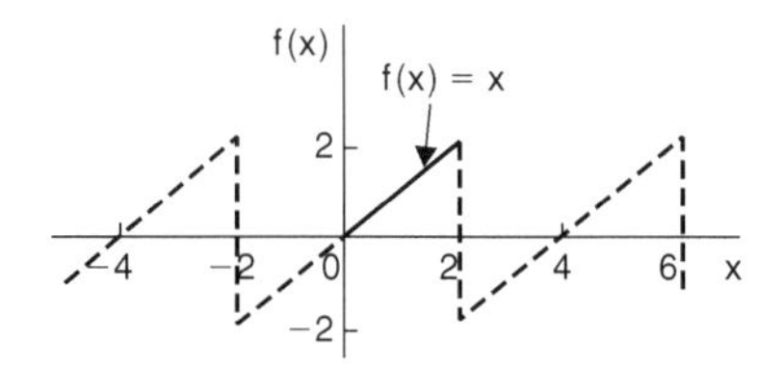

Figure 16.11

For a half-range sine series: $f(x) = \sum_{n=1}^{\infty} b_n \sin\left(\frac{n\pi x}{L}\right)$ from equation (8)

$$b_n = \frac{2}{L}\int_0^L f(x)\sin\left(\frac{n\pi x}{L}\right)dx = \frac{2}{2}\int_0^2 x\sin\left(\frac{n\pi x}{L}\right)dx$$

$$= \left[\frac{-x\cos\left(\frac{n\pi x}{2}\right)}{\left(\frac{n\pi}{2}\right)} + \frac{\sin\left(\frac{n\pi x}{2}\right)}{\left(\frac{n\pi}{2}\right)^2}\right]_0^2$$

$$= \left[\left(\frac{-2\cos n\pi}{\left(\frac{n\pi}{2}\right)} + \frac{\sin n\pi}{\left(\frac{n\pi}{2}\right)^2}\right) - \left(0 + \frac{\sin 0}{\left(\frac{n\pi}{2}\right)^2}\right)\right] = \frac{-2\cos n\pi}{\frac{n\pi}{2}} = \frac{-4}{n\pi}\cos n\pi$$

Hence, $b_1 = \frac{-4}{\pi}(-1) = \frac{4}{\pi}, b_2 = \frac{-4}{2\pi}(1) = \frac{-4}{2\pi}, b_3 = \frac{-4}{3\pi}(-1) = \frac{4}{3\pi},$

and so on

Thus the half-range Fourier sine series in the range 0 to 2 is given by:

$$\mathbf{f(x) = \frac{4}{\pi}\left[\sin\left(\frac{\pi x}{2}\right) - \frac{1}{2}\sin\left(\frac{2\pi x}{2}\right) + \frac{1}{3}\sin\left(\frac{3\pi x}{2}\right) - \frac{1}{4}\sin\left(\frac{4\pi x}{2}\right) + \cdots\right]}$$

16.7 The complex or exponential form of a Fourier series

The form used for the Fourier series considered previously consisted of cosine and sine terms. However, there is another form that is commonly used – one that directly gives the amplitude terms in the frequency spectrum and relates to phasor notation. This form involves the use of complex numbers (see Chapter 8). It is called the **exponential** or **complex form** of a Fourier series.

$$e^{j\theta} = \cos\theta + j\sin\theta \quad \text{and} \quad e^{-j\theta} = \cos\theta - j\sin\theta$$

$$e^{j\theta} + e^{-j\theta} = 2\cos\theta \quad \text{from which,} \quad \mathbf{\cos\theta = \frac{e^{j\theta} + e^{-j\theta}}{2}} \qquad (9)$$

$$e^{j\theta} - e^{-j\theta} = 2j\sin\theta \quad \text{from which,} \quad \mathbf{\sin\theta = \frac{e^{j\theta} - e^{-j\theta}}{2j}} \qquad (10)$$

The **complex** or **exponential form** of the Fourier series.

$$\mathbf{f(x) = \sum_{n=-\infty}^{\infty} c_n e^{j\frac{2\pi nx}{L}}} \qquad (11)$$

where $$c_n = \frac{1}{L}\int_{-\frac{L}{2}}^{\frac{L}{2}} f(x)\, e^{-j\frac{2\pi nx}{L}}\, dx \qquad (12)$$

Care needs to be taken when determining c_0. If n appears in the denominator of an expression the expansion can be invalid when n = 0. In such circumstances it is usually simpler to evaluate c_0 by using the relationship:

$$c_0 = a_0 = \frac{1}{L}\int_{-\frac{L}{2}}^{\frac{L}{2}} f(x)\, dx \qquad (13)$$

Application: Determine the complex Fourier series for the function defined by:

$$f(x) = \begin{cases} 0, \text{ when } -2 \leq x \leq -1 \\ 5, \text{ when } -1 \leq x \leq 1 \\ 0, \text{ when } \quad 1 \leq x \leq 2 \end{cases}$$

The function is periodic outside this range of period 4.

This is the same Application Problem as on page 506 and we can use this to demonstrate that the two forms of Fourier series are equivalent.

The function f(x) was shown in Figure 16.9, where the period, L = 4.

From equation (11), the complex Fourier series is given by:

$$f(x) = \sum_{n=-\infty}^{\infty} c_n\, e^{j\frac{2\pi nx}{L}}$$

where c_n is given by: $c_n = \frac{1}{L}\int_{-\frac{L}{2}}^{\frac{L}{2}} f(x)\, e^{-j\frac{2\pi nx}{L}}\, dx$ (from equation (12))

With reference to Figure 16.9, when L = 4,

$$c_n = \frac{1}{4}\left\{\int_{-2}^{-1} 0\,dx + \int_{-1}^{1} 5\,e^{-j\frac{2\pi nx}{4}}dx + \int_{1}^{2} 0\,dx\right\}$$

$$= \frac{1}{4}\int_{-1}^{1} 5\,e^{-\frac{j\pi nx}{2}}dx = \frac{5}{4}\left[\frac{e^{-\frac{j\pi nx}{2}}}{-\frac{j\pi n}{2}}\right]_{-1}^{1} = \frac{-5}{j2\pi n}\left[e^{-\frac{j\pi nx}{2}}\right]_{-1}^{1}$$

$$= \frac{-5}{j2\pi n}\left(e^{-\frac{j\pi n}{2}} - e^{\frac{j\pi n}{2}}\right) = \frac{5}{\pi n}\left(\frac{e^{j\frac{\pi n}{2}} - e^{-j\frac{\pi n}{2}}}{2j}\right)$$

$$= \frac{5}{\pi n}\sin\frac{\pi n}{2} \qquad \text{(from equation (10))}$$

Hence, from equation (11), **the complex form of the Fourier series** is given by:

$$\mathbf{f(x)} = \sum_{n=-\infty}^{\infty} c_n\,e^{j\frac{2\pi nx}{L}} = \sum_{\mathbf{n=-\infty}}^{\infty} \frac{\mathbf{5}}{\boldsymbol{\pi}\mathbf{n}}\,\mathbf{sin}\,\frac{\boldsymbol{\pi}\mathbf{n}}{\mathbf{2}}\,\mathbf{e}^{\mathbf{j}\frac{\boldsymbol{\pi}\mathbf{nx}}{\mathbf{2}}} \qquad (14)$$

Let us show how this result is equivalent to the result involving sine and cosine terms determined on page 508.

From equation (13),

$$\mathbf{c_0} = a_0 = \frac{1}{L}\int_{-\frac{L}{2}}^{\frac{L}{2}} f(x)dx = \frac{1}{4}\int_{-1}^{1} 5\,dx = \frac{5}{4}[x]_{-1}^{1} = \frac{5}{4}[1 - -1] = \frac{\mathbf{5}}{\mathbf{2}}$$

Since $c_n = \frac{5}{\pi n}\sin\frac{\pi n}{2}$, then $\mathbf{c_1} = \frac{5}{\pi}\sin\frac{\pi}{2} = \frac{\mathbf{5}}{\boldsymbol{\pi}}$

$\mathbf{c_2} = \frac{5}{2\pi}\sin\pi = \mathbf{0}$ (in fact, **all even terms will be zero** since $\sin n\pi = 0$)

$$\mathbf{c_3} = \frac{5}{\pi n}\sin\frac{\pi n}{2} = \frac{5}{3\pi}\sin\frac{3\pi}{2} = -\frac{\mathbf{5}}{\mathbf{3}\boldsymbol{\pi}}$$

By similar substitution, $\mathbf{c_5} = \frac{\mathbf{5}}{\mathbf{5}\boldsymbol{\pi}}$, $\mathbf{c_7} = -\frac{\mathbf{5}}{\mathbf{7}\boldsymbol{\pi}}$, and so on.

Similarly, $\mathbf{c_{-1}} = \dfrac{5}{-\pi}\sin\dfrac{-\pi}{2} = \mathbf{\dfrac{5}{\pi}}$

$$\mathbf{c_{-2}} = -\frac{5}{2\pi}\sin\frac{-2\pi}{2} = \mathbf{0} = \mathbf{c_{-4}} = \mathbf{c_{-6}}, \text{ and so on}$$

$$\mathbf{c_{-3}} = -\frac{5}{3\pi}\sin\frac{-3\pi}{2} = \mathbf{-\frac{5}{3\pi}}$$

$$\mathbf{c_{-5}} = -\frac{5}{5\pi}\sin\frac{-5\pi}{2} = \mathbf{\frac{5}{5\pi}}, \text{ and so on.}$$

Hence, the extended complex form of the Fourier series shown in equation (14) becomes:

$$f(x) = \frac{5}{2} + \frac{5}{\pi}e^{j\frac{\pi x}{2}} - \frac{5}{3\pi}e^{j\frac{3\pi x}{2}} + \frac{5}{5\pi}e^{j\frac{5\pi x}{2}} - \frac{5}{7\pi}e^{j\frac{7\pi x}{2}} + \cdots$$

$$+ \frac{5}{\pi}e^{-j\frac{\pi x}{2}} - \frac{5}{3\pi}e^{-j\frac{3\pi x}{2}} + \frac{5}{5\pi}e^{-j\frac{5\pi x}{2}} - \frac{5}{7\pi}e^{-j\frac{7\pi x}{2}} + \cdots$$

$$= \frac{5}{2} + \frac{5}{\pi}\left(e^{j\frac{\pi x}{2}} + e^{-j\frac{\pi x}{2}}\right) - \frac{5}{3\pi}\left(e^{j\frac{3\pi x}{2}} + e^{-j\frac{3\pi x}{2}}\right)$$

$$+ \frac{5}{5\pi}\left(e^{j\frac{5\pi x}{2}} + e^{-j\frac{5\pi x}{2}}\right) - \cdots$$

$$= \frac{5}{2} + \frac{5}{\pi}(2)\left(\frac{e^{j\frac{\pi x}{2}} + e^{-j\frac{\pi x}{2}}}{2}\right) - \frac{5}{3\pi}(2)\left(\frac{e^{j\frac{3\pi x}{2}} + e^{-j\frac{3\pi x}{2}}}{2}\right)$$

$$+ \frac{5}{5\pi}(2)\left(\frac{e^{j\frac{5\pi x}{2}} + e^{-j\frac{5\pi x}{2}}}{2}\right) - \cdots$$

$$= \frac{5}{2} + \frac{10}{\pi}\cos\left(\frac{\pi x}{2}\right) - \frac{10}{3\pi}\cos\left(\frac{3\pi x}{2}\right) + \frac{10}{5\pi}\cos\left(\frac{5\pi x}{2}\right) - \cdots$$

(from equation 9)

i.e. $$\mathbf{f(x) = \frac{5}{2} + \frac{10}{\pi}\left[\cos\left(\frac{\pi x}{2}\right) - \frac{1}{3}\cos\left(\frac{3\pi x}{2}\right) + \frac{1}{5}\cos\left(\frac{5\pi x}{2}\right) - \cdots\right]}$$

which is the same as obtained on page 508.

Hence, $\sum_{n=-\infty}^{\infty} \frac{5}{\pi n} \sin \frac{n\pi}{2} e^{j\frac{\pi n x}{2}}$ is equivalent to:

$$\frac{5}{2} + \frac{10}{\pi}\left[\cos\left(\frac{\pi x}{2}\right) - \frac{1}{3}\cos\left(\frac{3\pi x}{2}\right) + \frac{1}{5}\cos\left(\frac{5\pi x}{2}\right) - \ldots\right]$$

Symmetry relationships

If even or odd symmetry is noted in a function, then time can be saved in determining coefficients.

The Fourier coefficients present in the complex Fourier series form are affected by symmetry.

For **even symmetry**:

$$c_n = \frac{a_n}{2} = \frac{2}{L}\int_0^{\frac{L}{2}} f(x)\cos\left(\frac{2\pi n x}{L}\right) dx \tag{15}$$

For **odd symmetry**:

$$c_n = \frac{-jb_n}{2} = -j\frac{2}{L}\int_0^{\frac{L}{2}} f(x)\sin\left(\frac{2\pi n x}{L}\right) dx \tag{16}$$

For example, in the Application Problem on page 512, the function f(x) is even, since the waveform is symmetrical about the f(x) axis. Thus equation (15) could have been used, giving:

$$c_n = \frac{2}{L}\int_0^{\frac{L}{2}} f(x)\cos\left(\frac{2\pi n x}{L}\right) dx$$

$$= \frac{2}{4}\int_0^{2} f(x)\cos\left(\frac{2\pi n x}{4}\right) dx = \frac{1}{2}\left\{\int_0^{1} 5\cos\left(\frac{\pi n x}{2}\right) dx + \int_1^{2} 0\, dx\right\}$$

$$= \frac{5}{2}\left[\frac{\sin\left(\frac{\pi n x}{2}\right)}{\frac{\pi n}{2}}\right]_0^1 = \frac{5}{2}\left(\frac{2}{\pi n}\right)\left(\sin\frac{n\pi}{2} - 0\right) = \frac{5}{\pi n}\sin\frac{n\pi}{2}$$

which is the same answer as on page 513; however, a knowledge of even functions has produced the coefficient more quickly.

Application: Obtain the Fourier series, in complex form, for the square wave shown in Figure 16.12

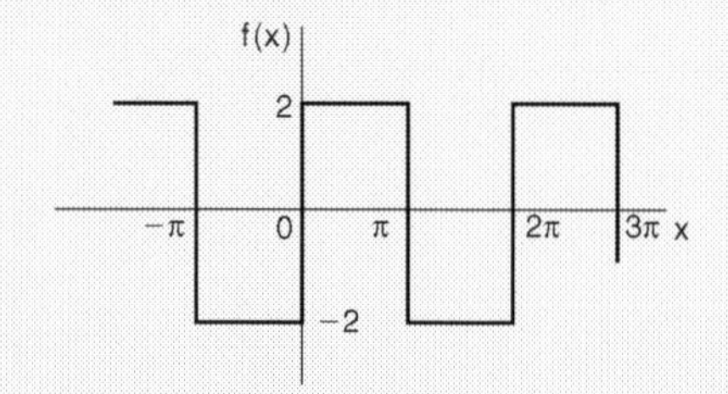

Figure 16.12

The square wave shown in Figure 16.12 is an **odd function** since it is symmetrical about the origin.

The period of the waveform, $L = 2\pi$.

Thus, using equation (16): $c_n = -j\frac{2}{L}\int_0^{\frac{L}{2}} f(x)\sin\left(\frac{2\pi nx}{L}\right)dx$

$$= -j\frac{2}{2\pi}\int_0^{\pi} 2\sin\left(\frac{2\pi nx}{2\pi}\right)dx$$

$$= -j\frac{2}{\pi}\int_0^{\pi} \sin nx\,dx$$

$$= -j\frac{2}{\pi}\left[\frac{-\cos nx}{n}\right]_0^{\pi}$$

$$= -j\frac{2}{\pi n}\left((-\cos\pi n) - (-\cos 0)\right)$$

i.e.
$$\mathbf{c_n = -j\frac{2}{\pi n}[1 - \cos\pi n]} \qquad (17)$$

From equation (11), the complex Fourier series is given by:

$$\mathbf{f(x)} = \sum_{n=-\infty}^{\infty} c_n e^{j\frac{2\pi nx}{L}} = \sum_{n=-\infty}^{\infty} -j\frac{2}{n\pi}(1-\cos n\pi)e^{jnx} \qquad (18)$$

This is the same as that obtained on page 501, i.e.

$$f(x) = \frac{8}{\pi}\left(\sin x + \frac{1}{3}\sin 3x + \frac{1}{5}\sin 5x + \frac{1}{7}\sin 7x + \cdots\right)$$

which is demonstrated below.

From equation (17), $c_n = -j\dfrac{2}{n\pi}(1 - \cos n\pi)$

When n = 1, $c_1 = -j\dfrac{2}{(1)\pi}(1 - \cos \pi) = -j\dfrac{2}{\pi}(1 - -1) = -\dfrac{j4}{\pi}$

When n = 2, $c_2 = -j\dfrac{2}{2\pi}(1 - \cos 2\pi) = 0$; in fact, all even values of c_n will be zero.

When n = 3, $c_3 = -j\dfrac{2}{3\pi}(1 - \cos 3\pi) = -j\dfrac{2}{3\pi}(1 - -1) = -\dfrac{j4}{3\pi}$

By similar reasoning, $c_5 = -\dfrac{j4}{5\pi}, c_7 = -\dfrac{j4}{7\pi}$, and so on.

When n = −1, $c_{-1} = -j\dfrac{2}{(-1)\pi}\left(1 - \cos(-\pi)\right) = +j\dfrac{2}{\pi}(1 - -1) = +\dfrac{j4}{\pi}$

When n = −3, $c_{-3} = -j\dfrac{2}{(-3)\pi}(1 - \cos(-3\pi)) = +j\dfrac{2}{3\pi}(1 - -1) = +\dfrac{j4}{3\pi}$

By similar reasoning, $c_{-5} = +\dfrac{j4}{5\pi}, c_{-7} = +\dfrac{j4}{7\pi}$, and so on.

Since the waveform is odd, $c_0 = a_0 = 0$

From equation (18), $f(x) = \sum_{n=-\infty}^{\infty} -j\dfrac{2}{n\pi}(1 - \cos n\pi)\, e^{jnx}$

Hence,

$$f(x) = -\frac{j4}{\pi}e^{jx} - \frac{j4}{3\pi}e^{j3x} - \frac{j4}{5\pi}e^{j5x} - \frac{j4}{7\pi}e^{j7x} - \ldots$$

$$+ \frac{j4}{\pi}e^{-jx} + \frac{j4}{3\pi}e^{-j3x} + \frac{j4}{5\pi}e^{-j5x} + \frac{j4}{7\pi}e^{-j7x} + \ldots$$

$$= \left(-\frac{j4}{\pi}e^{jx} + \frac{j4}{\pi}e^{-jx}\right) + \left(-\frac{j4}{3\pi}e^{3x} + \frac{j4}{3\pi}e^{-3x}\right)$$

$$+ \left(-\frac{j4}{5\pi}e^{5x} + \frac{j4}{5\pi}e^{-5x}\right) + \cdots$$

$$= -\frac{j4}{\pi}\left(e^{jx} - e^{-jx}\right) - \frac{j4}{3\pi}\left(e^{3x} - e^{-3x}\right) - \frac{j4}{5\pi}\left(e^{5x} - e^{-5x}\right) + \cdots$$

$$= \frac{4}{j\pi}(e^{jx} - e^{-jx}) + \frac{4}{j3\pi}(e^{3x} - e^{-3x}) + \frac{4}{j5\pi}(e^{5x} - e^{-5x}) + \cdots$$

by multiplying top and bottom by j

$$= \frac{8}{\pi}\left(\frac{e^{jx} - e^{-jx}}{2j}\right) + \frac{8}{3\pi}\left(\frac{e^{j3x} - e^{-j3}}{2j}\right) + \frac{8}{5\pi}\left(\frac{e^{j5x} - e^{-j5x}}{2j}\right) + \cdots \quad \text{by rearranging}$$

$$= \frac{8}{\pi}\sin x + \frac{8}{3\pi}\sin 3x + \frac{8}{3x}\sin 5x + \cdots$$

from equation 10, page 511

i.e. $$\mathbf{f(x) = \frac{8}{\pi}\left(\sin x + \frac{1}{3}\sin 3x + \frac{1}{5}\sin 5x + \frac{1}{7}\sin 7x + \cdots\right)}$$

Hence, $$\mathbf{f(x) = \sum_{n=-\infty}^{\infty} -j\frac{2}{n\pi}(1 - \cos n\pi)\, e^{jnx}}$$

$$\mathbf{\equiv \frac{8}{\pi}\left(\sin x + \frac{1}{3}\sin 3x + \frac{1}{5}\sin 5x + \frac{1}{7}\sin 7x + \cdots\right)}$$

16.8 A numerical method of harmonic analysis

Many practical waveforms can be represented by simple mathematical expressions, and, by using Fourier series, the magnitude of their harmonic components determined, as above. For waveforms not in this category, analysis may be achieved by numerical methods.

Harmonic analysis is the process of resolving a periodic, non-sinusoidal quantity into a series of sinusoidal components of ascending order of frequency.

The **trapezoidal rule** can be used to evaluate the Fourier coefficients, which are given by:

$$\mathbf{a_0 \approx \frac{1}{p}\sum_{k=1}^{p} y_k} \qquad (19)$$

$$\mathbf{a_n \approx \frac{2}{P}\sum_{k=1}^{p} y_k \cos nx_k} \qquad (20)$$

$$\mathbf{b_n \approx \frac{2}{P}\sum_{k=1}^{p} y_k \sin nx_k} \qquad (21)$$

Application: A graph of voltage V against angle θ is shown in Figure 16.13. Determine a Fourier series to represent the graph.

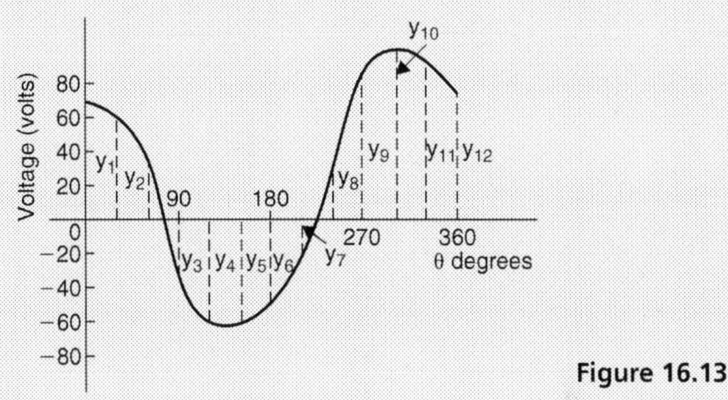

Figure 16.13

The values of the ordinates y_1, y_2, y_3, ... are 62, 35, −38, −64, −63, −52, −28, 24, 80, 96, 90 and 70, the 12 equal intervals each being of width 30°. (If a larger number of intervals are used, results having a greater accuracy are achieved).

The voltage may be analysed into its first three constituent components as follows:

The data is tabulated in the proforma shown in Table 16.1.

From equation (19), $a_0 \approx \frac{1}{p}\sum_{k=1}^{p} y_k = \frac{1}{12}(212) = 17.67$ (since $p = 12$)

From equation (20), $a_n \approx \frac{2}{p}\sum_{k=1}^{p} y_k \cos nx_k$

hence $a_1 \approx \frac{2}{12}(417.94) = 69.66$,

$$a_2 \approx \frac{2}{12}(-39) = -6.50 \quad \text{and} \quad a_3 \approx \frac{2}{12}(-49) = -8.17$$

From equation (21), $b_n \approx \frac{2}{p}\sum_{k=1}^{p} y_k \sin nx_k$

hence $b_1 \approx \frac{2}{12}(-278.53) = -46.42$,

$$b_2 \approx \frac{2}{12}(29.43) = 4.91 \quad \text{and} \quad b_3 \approx \frac{2}{12}(55) = 9.17$$

Table 16.1

Ordinates	θ	V	cos θ	V cos θ	sin θ	V sin θ	cos 2θ	V cos 2θ	sin 2θ	V sin 2θ	cos 3θ	V cos 3θ	sin 3θ	V sin 3θ
Y_1	30	62	0.866	53.69	0.5	31	0.5	31	0.866	53.69	0	0	1	62
Y_2	60	35	0.5	17.5	0.866	30.31	−0.5	−17.5	0.866	30.31	−1	−35	0	0
Y_3	90	−38	0	0	1	−38	−1	38	0	0	0	0	−1	38
Y_4	120	−64	−0.5	32	0.866	−55.42	−0.5	32	−0.866	55.42	1	−64	0	0
Y_5	150	−63	−0.866	54.56	0.5	−31.5	0.5	−31.5	−0.866	54.56	0	0	1	−63
Y_6	180	−52	−1	52	0	0	1	−52	0	0	−1	52	0	0
Y_7	210	−28	−0.866	24.25	−0.5	14	0.5	−14	0.866	−24.25	0	0	−1	28
Y_8	240	24	−0.5	−12	−0.866	−20.78	−0.5	−12	0.866	20.78	1	24	0	0
Y_9	270	80	0	0	−1	−80	−1	−80	0	0	0	0	1	80
Y_{10}	300	96	0.5	48	−0.866	−83.14	−0.5	−48	−0.866	−83.14	−1	−96	0	0
Y_{11}	330	90	0.866	77.94	−0.5	−45	0.5	45	−0.866	−77.94	0	0	−1	−90
y_{12}	360	70	1	70	0	0	1	70	0	0	1	70	0	0
$\sum_{k=1}^{12} y_k = 212$			$\sum_{k=1}^{12} y_k \cos \theta_k = 417.94$		$\sum_{k=1}^{12} y_k \sin \theta_k = -278.53$		$\sum_{k=1}^{12} y_k \cos 2\theta_k = -39$		$\sum_{k=1}^{12} y_k \sin 2\theta_k = 29.43$		$\sum_{k=1}^{12} y_k \cos 3\theta_k = -49$		$\sum_{k=1}^{12} y_k \sin 3\theta_k = 55$	

Substituting these values into the Fourier series:

$$f(x) = a_0 + \sum_{n=1}^{\infty} (a_n \cos nx + b_n \sin nx)$$

gives: $\mathbf{v = 17.67 + 69.66 \cos\theta - 6.50 \cos 2\theta - 8.17 \cos 3\theta + \ldots}$

$$\mathbf{-46.42 \sin\theta + 4.91 \sin 2\theta + 9.17 \sin 3\theta + \cdots} \qquad (22)$$

Note that in equation (22), $(-46.42 \sin\theta + 69.66 \cos\theta)$ comprises the fundamental, $(4.91 \sin 2\theta - 6.50 \cos 2\theta)$ comprises the second harmonic and $(9.17 \sin 3\theta - 8.17 \cos 3\theta)$ comprises the third harmonic.

It is shown in Chapter 5 that: $a \sin\omega t + b \cos\omega t = R \sin(\omega t + \alpha)$

where $a = R\cos\alpha$, $b = R\sin\alpha$, $R = \sqrt{a^2 + b^2}$ and $\alpha = \tan^{-1}\frac{b}{a}$

For the fundamental, $R = \sqrt{(-46.42)^2 + (69.66)^2} = 83.71$

If $a = R\cos\alpha$, then $\cos\alpha = \frac{a}{R} = \frac{-46.42}{83.71}$ which is negative,

and if $b = R\sin\alpha$, then $\sin\alpha = \frac{b}{R} = \frac{69.66}{83.71}$ which is positive.

The only quadrant where $\cos\alpha$ is negative and $\sin\alpha$ is positive is the second quadrant.

Hence, $\alpha = \tan^{-1}\frac{b}{a} = \tan^{-1}\frac{69.66}{-46.42} = 123.68^\circ$ or 2.16 rad

Thus, $(-46.42 \sin\theta + 69.66 \cos\theta) = 83.71 \sin(\theta + 2.16)$

By a similar method it may be shown that the second harmonic

$(4.91 \sin 2\theta - 6.50 \cos 2\theta) = 8.15 \sin(2\theta - 0.92)$ and the third harmonic

$(9.17 \sin 3\theta - 8.17 \cos 3\theta) = 12.28 \sin(3\theta - 0.73)$

Hence equation (22) may be re-written as:

$$\mathbf{v = 17.67 + 83.71 \sin(\theta + 2.16) + 8.15 \sin(2\theta - 0.92) + 12.28 \sin(3\theta - 0.73)} \textbf{ volts}$$

which is the form normally used with complex waveforms.

16.9 Complex waveform considerations

It is sometimes possible to predict the harmonic content of a waveform on inspection of particular waveform characteristics.

1. If a periodic waveform is such that the area above the horizontal axis is equal to the area below then the mean value is zero. Hence $a_0 = 0$ (see Figure 16.14(a)).
2. An **even function** is symmetrical about the vertical axis and contains **no sine terms** (see Figure 16.14(b)).
3. An **odd function** is symmetrical about the origin and contains **no cosine terms** (see Figure 16.14(c)).
4. $f(x) = f(x + \pi)$ represents a waveform which repeats after half a cycle and **only even harmonics** are present (see Figure 16.14(d)).
5. $f(x) = -f(x + \pi)$ represents a waveform for which the positive and negative cycles are identical in shape and **only odd harmonics** are present (see Figure 16.14(e)).

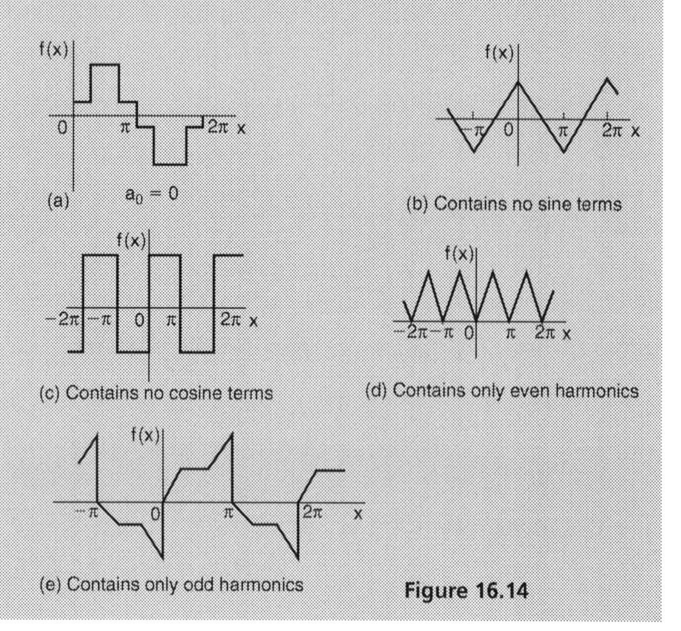

Figure 16.14

Application: An alternating current i amperes is shown in Figure 16.15. Analyse the waveform into its constituent harmonics as far as and including the fifth harmonic, taking 30° intervals.

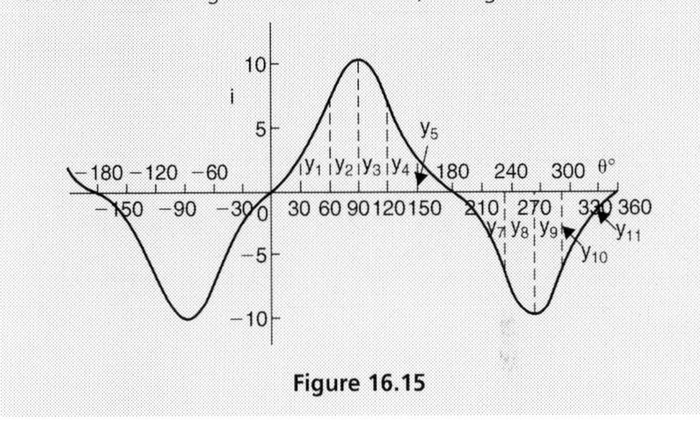

Figure 16.15

With reference to Figure 16.15, the following characteristics are noted:

(i) The mean value is zero since the area above the θ axis is equal to the area below it. Thus the constant term, or d.c. component, $a_0 = 0$

(ii) Since the waveform is symmetrical about the origin the function i is odd, which means that there are no cosine terms present in the Fourier series.

(iii) The waveform is of the form $f(\theta) = -f(\theta + \pi)$ which means that only odd harmonics are present.

Investigating waveform characteristics has thus saved unnecessary calculations and in this case the Fourier series has only odd sine terms present, i.e.

$$i = b_1 \sin \theta + b_3 \sin 3\theta + b_5 \sin 5\theta + \cdots$$

A proforma, similar to Table 16.1, but without the 'cosine terms' columns and without the 'even sine terms' columns is shown in Table 16.2 up to, and including, the fifth harmonic, from which the Fourier coefficients b_1, b_3 and b_5 can be determined. Twelve co-ordinates are chosen and labelled $y_1, y_2, y_3, \ldots y_{12}$ as shown in Figure 16.15.

Table 16.2

Ordinate	θ	i	sin θ	i sin θ	sin 3θ	i sin 3θ	sin 5θ	i sin 5θ
Y_1	30	2	0.5	1	1	2	0.5	1
Y_2	60	7	0.866	6.06	0	0	−0.866	−6.06
Y_3	90	10	1	10	−1	−10	1	10
Y_4	120	7	0.866	6.06	0	0	−0.866	−6.06
Y_5	150	2	0.5	1	1	2	0.5	1
Y_6	180	0	0	0	0	0	0	0
Y_7	210	−2	−0.5	1	−1	2	−0.5	1
Y_8	240	−7	−0.866	6.06	0	0	0.866	−6.06
Y_9	270	−10	−1	10	1	−10	−1	10
Y_{10}	300	−7	−0.866	6.06	0	0	0.866	−6.06
Y_{11}	330	−2	−0.5	1	−1	2	−0.5	1
Y_{12}	360	0	0	0	0	0	0	0
			$\sum_{k=1}^{12} y_k \sin \theta_k$	$= 48.24$	$\sum_{k=1}^{12} y_k \sin 3\theta_k$	$= -12$	$\sum_{k=1}^{12} y_k \sin 5\theta_k$	$= -0.24$

From equation (21), $b_n = \dfrac{2}{p}\sum_{k=1}^{p} i_k \sin n\theta_k$ where $p = 12$

Hence, $b_1 \approx \dfrac{2}{12}(48.24) = 8.04$, $b_3 \approx \dfrac{2}{12}(-12) = -2.00$ and

$b_5 \approx \dfrac{2}{12}(-0.24) = -0.04$

Thus the Fourier series for current i is given by:

$$\mathbf{i = 8.04 \sin \theta - 2.00 \sin 3\theta - 0.04 \sin 5\theta}$$

Index